Physical and Physiological Forest Ecology

Pertti Hari • Kari Heliövaara • Liisa Kulmala
Editors

Physical and Physiological Forest Ecology

Springer

Editors
Pertti Hari
Department of Forest Sciences
University of Helsinki
Helsinki
Finland

Kari Heliövaara
Department of Forest Sciences
University of Helsinki
Helsinki
Finland

Liisa Kulmala
Department of Forest Sciences
University of Helsinki
Helsinki
Finland

ISBN 978-94-007-5602-1 ISBN 978-94-007-5603-8 (eBook)
DOI 10.1007/978-94-007-5603-8
Springer Dordrecht Heidelberg New York London

Library of Congress Control Number: 2012951122

Printed on acid-free paper

Springer is part of Springer Science+Business Media (www.springer.com)

Preface

Photosynthesis is the source of energy for nearly all life on Earth and the source of the carbon in all the organic compounds within the bodies of organisms. Photosynthesis is the biochemical reaction through which all green plants, including trees, live and grow. We started field measurements of photosynthesis in Hyytiälä, the Forestry Field Station at the University of Helsinki, in the year 1972, and this was the beginning of a still continuing work in the forests around the station. We had the vision that physical and physiological knowledge in addition to the quantitative methods, especially field measurements and dynamic modelling, should have an important role in forest ecology. We introduced into our vision the important role of mass and energy fluxes when we studied the Chernobyl fallout in the summer of 1986. The construction of the SMEAR II measuring station clarified our vision, and we included the fluxes in the soil and between the soil and trees into our approach.

Some members in our team have had good knowledge in tree physiology, and they introduced the physiological aspect into the research. We have been lucky to work with physicists who are deeply interested in the role of physical phenomena in forests. Dr. Taisto Raunemaa (1939–2006) started common projects with us, and the physicists become a natural component in our research team. The co-operation between the Academy of Finland and Soviet Academy of Sciences enabled common research in Finland, Soviet Karelia and Estonia. Physicists, Academician Juhan Ross (1925–2002), Dr. Leo Kaipiainen (1932–2004) and Dr. Tiit Nilson have been the key persons in the team.

The co-operation with physicists was intensified when Dr. Markku Kulmala began to lead the team of young aerosol physics at the Department of Physics, Helsinki University. He was the other principal planner of the measuring systems SMEAR I and II. In addition, he is interested in the role of boreal forest in the formation and growth of aerosols. Dr. Timo Vesala introduced the important role of measurements of ecosystem fluxes with the Eddy Covariance method. He is also interested in the role of nitrogen in forest ecosystems.

The feedback from our measurements has been clear and positive. The developed model structures have been able to explain well the behaviour of trees and ground vegetation in field conditions. We have published our results in international literature, although some of the most relevant papers have remained unpublished in scientific journals. During the last 20 years, we have actively participated into European research co-operation.

The Academy of Finland and Helsinki University have been the main financial supporters during the whole long period.

Our first reader, Dr. John Grace from Edinburgh, has made valuable comments on our manuscript and we express our gratitude to him. We acknowledge Dr. Kourosh Kabiri Koupaei, for his contribution to the editing of the book.

Our book is the summary of our work during 40 years at the present Department of Forest Sciences, previously Department of Forest Ecology and Department of Silviculture at Helsinki University. Our vision stressing the importance of physics and physiological knowledge, quantitative methods, material and energy fluxes and field working in the forests has had the central role in the planning of the research.

Helsinki
6th December 2012

Pertti (Pepe) Hari, D. Sc. (Agr. & For.), Prof.
Kari Heliövaara, D. Sc. (Agr. & For.), Prof.
Liisa Kulmala, D. Sc. (Agr. & For.)

Contents

Contributors

Tuomas Aakala Department of Forest Sciences, University of Helsinki, Helsinki, Finland

Juho Aalto Hyytiälä Forestry Field Station, University of Helsinki, Helsinki, Finland

Jaana Bäck Department of Forest Sciences, University of Helsinki, Helsinki, Finland

John Grace Institute for Atmospheric and Environmental Science, University of Edinburgh, Edinburgh, UK

Beñat Olascoaga Gracia Department of Forest Sciences, University of Helsinki, Helsinki, Finland

Risto Häkkinen Finnish Forest Research Institute, Vantaa, Finland

Heikki Hänninen Department of Biosciences, University of Helsinki, Helsinki, Finland

Pertti Hari Department of Forest Sciences, University of Helsinki, Helsinki, Finland

Mikko Havimo Department of Forest Sciences, University of Helsinki, Helsinki, Finland

Jussi Heinonsalo Department of Forest Sciences, Department of Physics and Department of Food and Environmental Sciences, University of Helsinki, Helsinki, Finland

Kari Heliövaara Department of Forest Sciences, University of Helsinki, Helsinki, Finland

Heljä-Sisko Helmisaari Department of Forest Sciences, University of Helsinki, Helsinki, Finland

Teemu Hölttä Department of Forest Sciences, University of Helsinki, Helsinki, Finland

Kalev Jõgiste Institute of Forestry and Rural Engineering, Estonian University of Life Sciences, Tartu, Estonia

Johanna Joensuu Department of Forest Sciences, University of Helsinki, Helsinki, Finland

Eija Juurola Department of Forest Sciences and Department of Physics, University of Helsinki, Helsinki, Finland

Mika Kähkönen Department of Food and Environmental Sciences, University of Helsinki, Helsinki, Finland

Tuomo Kalliokoski Finnish Forest Research Institute, Vantaa, Finland

Ahto Kangur Institute of Forestry and Rural Engineering, Estonian University of Life Sciences, Tartu, Estonia

Antti-Jussi Kieloaho Department of Physics, University of Helsinki, Helsinki, Finland

Pasi Kolari Department of Forest Sciences and Department of Physics, University of Helsinki, Helsinki, Finland

Janne F.J. Korhonen Department of Physics, University of Helsinki, Helsinki, Finland

Kourosh Kabiri Koupaei Department of Forest Sciences, University of Helsinki, Helsinki, Finland

Liisa Kulmala Department of Forest Sciences, University of Helsinki, Helsinki, Finland

Markku Kulmala Department of Physics, University of Helsinki, Helsinki, Finland

Jukka Kurola Department of Ecological and Environmental Sciences and Department of Applied Chemistry and Microbiology, University of Helsinki, Helsinki, Finland

Samuli Launiainen Finnish Forest Research Institute, Joensuu, Finland

Tapio Linkosalo Finnish Forest Research Institute, Vantaa, Finland

Anna Lintunen Department of Forest Sciences, University of Helsinki, Helsinki, Finland

Jari Liski Finnish Environment Institute, Helsinki, Finland

Federico Magnani Dipartimento Colture Arboree, University of Bologna, Bologna, Italy

Miikka dal Maso Department of Physics, University of Helsinki, Helsinki, Finland

Eero Nikinmaa Department of Forest Sciences, University of Helsinki, Helsinki, Finland

Twan van Noije Royal Netherlands Meteorological Institute, De Bilt, The Netherlands

Tuukka Petäjä Department of Physics, University of Helsinki, Helsinki, Finland

Mari Pihlatie Department of Physics, University of Helsinki, Helsinki, Finland

Toivo Pohja Hyytiälä Forestry Field Station, University of Helsinki, Helsinki, Finland

Albert Porcar-Castell Department of Forest Sciences, University of Helsinki, Helsinki, Finland

Minna Pulkkinen Department of Forest Sciences, University of Helsinki, Helsinki, Finland

Jukka Pumpanen Department of Forest Sciences, University of Helsinki, Helsinki, Finland

Jouni Räisänen Department of Physics, University of Helsinki, Helsinki, Finland

Maarit Raivonen Department of Physics, University of Helsinki, Helsinki, Finland

Üllar Rannik Department of Physics, University of Helsinki, Helsinki, Finland

Ilona Riipinen Department of Applied Environmental Science & Bert Bolin Center for Climate Research, Stockholm University, Stockholm, Sweden

Mirja Salkinoja-Salonen Department of Food and Environmental Sciences, University of Helsinki, Helsinki, Finland

Pauliina Schiestl-Aalto Hyytiälä Forestry Field Station, University of Helsinki, Helsinki, Finland

Erkki Siivola Department of Physics, University of Helsinki, Helsinki, Finland

Asko Simojoki Department of Food and Environmental Sciences, University of Helsinki, Helsinki, Finland

Sampo Smolander Department of Physics, University of Helsinki, Helsinki, Finland

David Stevenson Institute for Atmospheric and Environmental Science, University of Edinburgh, Edinburgh, UK

Inge Stupak Forest & Landscape Denmark/Applied Ecology, University of Copenhagen, Copenhagen, Denmark

Sari Timonen Department of Food and Environmental Sciences, University of Helsinki, Helsinki, Finland

Heikki Tuomenvirta Finnish Meteorological Institute, Helsinki, Finland

Timo Vesala Department of Physics, University of Helsinki, Helsinki, Finland

Chapter 1
Introduction to Physical, Physiological and Causal Forest Ecology

Pertti Hari, Liisa Kulmala, and Mikko Havimo

Contents

Abstract H. T. Odum proposed that energy flows are the key factors in understanding the interactions between ecosystems and their environment. He used electric analogue models and analogue computers in his approach. The development of quantitative methods, measurements, dynamic modelling and digital computing, has opened novel possibilities to analyse the energy and material flows between living organisms and their environment. In addition, the applications of ecological knowledge to the responses of forests to present climate change and forestry should be based on causal effects in trees, atmosphere and soil. The conservation principle of mass and energy, a cornerstone of Newtonian physics, results in dynamic models that can utilise physiological and physical background knowledge to provide causal explanations for the theory and models. The aim of our book is to develop and test a quantitative and causal theory of forest ecology that utilises physics and physiology as background knowledge.

Keywords Mass and energy fluxes • Physical knowledge • Physiological knowledge • Conservation principle

---

P. Hari (✉)
Department of Forest Sciences, University of Helsinki, P.O. Box 27, 00014 University of Helsinki, Helsinki, Finland
e-mail: pertti.hari@helsinki.fi

P. Hari et al. (eds.), *Physical and Physiological Forest Ecology*,
DOI 10.1007/978-94-007-5603-8_1,

1.1 Climate Change: A Challenge for Forest Ecology

The atmospheric CO_2 concentration has increased from 280 ppm (parts per million) to nearly 400 ppm during the last 200 years due to use of fossil fuels and deforestation. The increase continues at a rate of nearly 2 ppm annually. The extra CO_2 in the atmosphere reduces the thermal radiation escaping from the planet, and this inevitably causes an increase in temperature: the observed increase in the global mean temperature is nearly 1°C over the last century, very similar to the increase expected from climate modelling. As concentrations of CO_2 in the atmosphere continue to increase, the rise in temperature is expected to continue (IPCC 2007).

The forests of the world respond to the changes in the atmospheric CO_2 concentration and temperature: photosynthesis, for example, is enhanced and decomposition of organic matter in the soil is accelerated. Thus, trees have more raw materials for growth and the carbon pool in the soil is reduced. These changes generate a complicated web of responses to climate change in forests.

The forests have several important effects on the atmosphere by, for example, storing carbon, transpiring water and absorbing solar radiation. All these responses affect the climate change generating feedbacks from forests to climate change. The role of the feedbacks is under debate.

Ecology may be defined in several different ways, with different emphasis on the biological versus the physical and chemical aspects. The idea is, however, rather consistent in the different versions. Ecology studies the interactions between living organisms and their environment and the distribution and abundance of living organisms.

Knowledge on the interactions between forests and their environment is needed for the proper analysis of climate change, for example, because the responses of forests and other carbon-rich ecosystems play an important role in the global climate models (GCM). However, the forests are treated very roughly in the GCMs, and they introduce additional uncertainty in the simulations of the expected climate change.

1.2 Quantitative Knowledge in Ecology

For ages, human activities such as agriculture, fishing and hunting have been built on practical experience of the natural world. Ecological knowledge as a science grew slowly within biology during the early twentieth century gaining momentum in the later part, to some extent due to environmental movement. The present climate change and its effects on life on the Earth have further increased the need of causal understanding of the interactions between living organisms and their environment.

A qualitative approach dominated ecology in the early phase of the discipline. Typical studies of this era were the exploration and classification of plant communities, which resulted in definitions of vegetation zones and forest site types.

The still ongoing classification studies have successfully systematised the versatile life formations, but on the other hand, they have rarely given causal explanations on the interaction between organisms and their environment.

The quantitative ecological tradition developed parallel with the qualitative one. Alfred J. Lotka (1880–1949) pioneered in the field of quantitative ecology already in the early twentieth century. His famous predator–prey model is still in common use. System analysis and dynamic modelling were emerging methods in studying complex systems in 1950s. At the time, H. T. Odum (1924–2002) greatly influenced the development of ecology by applying physical principles and studying energy flows in ecosystems. He stressed the importance of emergent properties in large and complex systems. He also realised that it is not possible to derive the behaviour of ecosystems only from detailed knowledge concerning small objects in the system. The classical physics, especially electronics, played an important role in his thinking.

H. T. Odum based his models on the flows of energy using electric analogue, especially Ohm's law that describes electric current with resistors and voltages. These models were rather easy to construct, but the analogue of ecosystems with electric systems was not explicit and required rather strange concepts. The causal interpretation of the models was, in most cases, impossible, and the applications of the models were problematic. In addition, the analogue computers of his day were inadequate for further development of this technique. Moreover, the present numerical methods were not available 50 years ago. Therefore, the proper analysis of the energy flows in ecosystems was impossible with the methods available in 1950s and 1960s.

H. T. Odum widely utilised physical knowledge and methodology and applied them in an ecological context, but he knew rather little about the metabolism of living objects. It was not until the 1970s that the new research disciplines of ecophysiology and physiological ecology were introduced. Sporadic field measurements and experiments were used to study the relationships between metabolic activity and environment. Photosynthesis, for example, received much attention following the development of the infrared gas analyser. The results were used in an attempt to understand ecosystems, and they were applied even at the global scale, but the problems with transitions between scales and emergent properties were omitted rather commonly.

The lack of causal and emergent features is the major problem when extrapolating sporadic experiments and measurements on ecosystem and global scale. Applications such as studying the responses of forests to climate change and silvicultural actions require causal interpretation of the available knowledge. In addition, the emergent properties should be properly understood so that they can be introduced, too. Theoretical knowledge is needed to bridge fieldwork with the applications on very aggregated levels. However, theoretical work has received rather limited attention in forest ecology during the last decades.

1.3 Energy and Material Fluxes Through Ecosystems

Ecology includes research of a very versatile set of phenomena from predator–prey interactions to photosynthesis. Earlier, the focus was often on the interactions between animals whilst the vegetation played a minor role. However, the role of mammals and birds in forest ecosystems, for example, is generally small as their masses per hectare are negligible when compared with that of vegetation. The material and energy flows through mammals and birds are very small as well, although of course these organisms may exert a disproportionate effect on ecosystems as, for example, through dispersal of seeds and their role as ecological engineers.

Metabolism of living organisms synthesises numerous and often complicated molecules from simple substances such as carbon dioxide (CO_2) and water, utilising the energy captured from the solar radiation. The synthesised molecules are used by organisms to construct new leaves, roots, etc. or store energy for metabolic needs.

Metabolism and physical phenomena generate concentration, temperature and pressure differences that call for mass and energy fluxes between vegetation and environment at several spatial scales. These fluxes convey the interactions between living organisms and their environment. Water uptake by roots and loss in transpiration are the biggest mass fluxes, whilst carbon dioxide intake by leaves is the second biggest one. The nutrient flux from soil to vegetation is rather small in terms of mass, but very crucial in the functioning of plants. All the fluxes originate from the energy input by solar radiation. The heat exchange between vegetation and the atmosphere is also large (Fig. 1.1).

Long distances between the locations of material and energy inflows and their usage call for a need of transport within living organisms. The water uptake takes place in roots whilst transpiration occurs in leaves. The distance between these places is often over 10 m in trees and up to 100 m in the tallest trees; thus, an efficient water transport system is needed. Sugars, formed in photosynthesis, are the main building block in the synthesis of large molecules, and sugars are also used to store and release energy for metabolism. Thus, the distance between the formation and usage of sugars is large, and so an efficient transport system for sugars is necessary as well.

The forest stand components – trees, ground vegetation and soil – are interacting with each other and the atmosphere. A stand interacts very little with the neighbouring forests since the connecting material and energy fluxes are usually small. The material and energy fluxes between large forested areas and the atmosphere can be obtained as the sum of the fluxes from the separate ecosystems in the area. Thus, a forest ecosystem should be in the focus of forest ecological theory.

In several milliard years of evolution, organisms have developed the means to control the flows of material and energy between the organisms and their surroundings as well as within their own structures. For example, there are small pores in leaves that serve to reduce the evaporative loss of water in transpiration

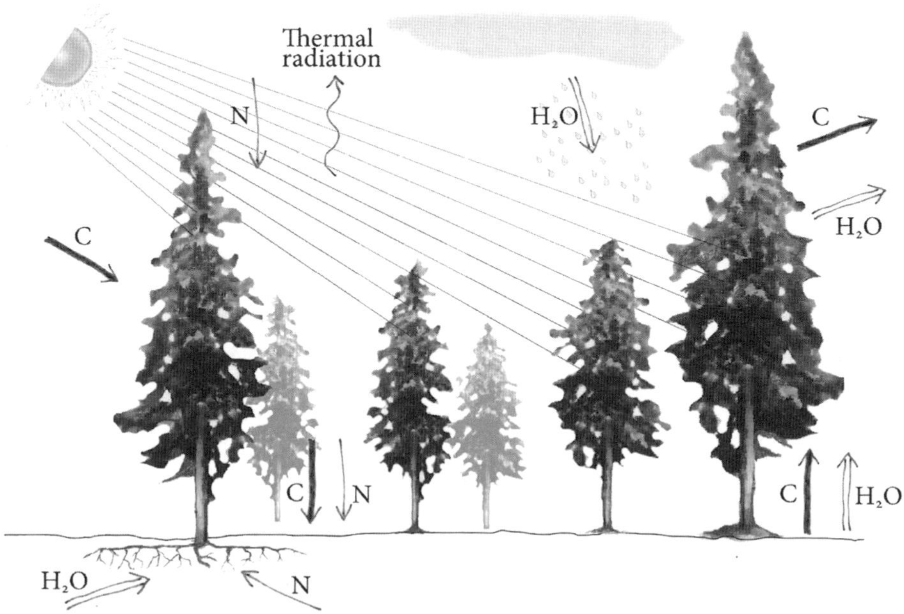

Fig. 1.1 Interactions in a forest ecosystem. Solar energy, carbon (C), nitrogen (N), water (H_2O) and thermal fluxes connect the forest with the atmosphere

and simultaneously allow the intake of carbon dioxide. The exchanges of water and carbon between leaves and atmosphere are thus intimately linked. The surfaces of roots have special structures for capturing water and nutrient, whilst in the leaves pigments are present to capture the radiation energy. Complex biophysical and biochemical systems convert the captured energy into a chemical form that can be used by the plant to build its structure.

Novelty in metabolism and transport arises during evolution. These novelties are tested against the prevailing ones. If the novelty is more efficient than the present one, it has a tendency to become more common and later it may replace the older one. This development of novelties and their continual testing against the prevailing ones lead to slow improvement, and therefore, efficient mechanisms for metabolism and transport have developed in evolution.

Conservations of mass and energy are cornerstones of Newtonian physics. The theory of relativity replaces these conservation principles with more general ideas, but Newtonian physics and the theory of relativity do not differ from each other under normal macroscopic conditions on the Earth. Evolution has a similar fundamental role in ecology. Biological entities are results of evolution during the past milliard years, and their properties should be considered in the evolutionary framework. If any consideration in ecology is in conflict with the conservations of mass or energy or with evolution, it should be rejected as fundamentally erroneous.

1.4 Towards Physical and Physiological Theory of Forest Ecology

The methodological and technological progress during the last few decades has created novel possibilities for ecological research. System analysis and dynamic modelling enables better processing of theoretical ideas. The efficiency of digital computers is usually sufficient for calculations. Numeric methods allow solution of equations, and automatic monitoring systems can be used to measure material and energy fluxes in ecosystems and between ecosystems and the atmosphere. These novel possibilities should be fully utilised in ecological research.

We are facing unprecedented climate change: the forests will respond to the changes in the composition of the atmosphere, in temperature and in rainfall. The responses of forests generate feedbacks to the climate system. To quantify these feedbacks, we need understanding of the metabolic and physical responses to environmental change as well as interpretation of the changes on the ecosystem and global levels. Providing the urgently needed knowledge, with a sound physiological and physical basis, is the challenge for ecological research in the near future.

We wanted to explore the possibilities that the development of research methodology and technology together with physical and physiological knowledge has opened for forest ecology. *The aim of our book is to develop and test a quantitative and causal theory of forest ecology that utilises physics and physiology as background knowledge*. The flows of material and energy combined with the conservation principle play a key role in the analysis of different phenomena such as photosynthesis, respiration and transport of water or sugars. Efficient metabolism and transport structures have arisen in the evolution.

Reference

IPCC (2007) Contribution of working group I to the fourth assessment report of the intergovernmental panel on climate change. In: Solomon S, Qin D, Manning M, Chen Z, Marquis M, Averyt K, Tignor MMB, Miller HL (eds) Climate change 2007: the physical basis. Cambridge University Press, Cambridge

Chapter 2
The Approach to Construct and Test the Theory of Forest Ecology

Pertti Hari

Contents

Abstract This chapter gives the main lines of biological and methodological thinking applied in the whole book. The vision that mass and energy fluxes convey the interactions between forest ecosystem and its surroundings at different levels is the starting point of the theory formation. We formulate a common framework, called cover theory, to deal with different phenomena in the ecosystem. We analyse the fluxes generated by metabolism, and we use conservation of mass and energy for derivation of differential equations to describe each phenomenon under study. We test the dynamic models with field data to obtain feedback from nature. Finally, we combine the theories to form the physical and physiological theory of forest ecology.

P. Hari (✉)
Department of Forest Sciences, University of Helsinki, P.O. Box 27, 00014 University of Helsinki, Helsinki, Finland
e-mail: pertti.hari@helsinki.fi

P. Hari et al. (eds.), *Physical and Physiological Forest Ecology*,
DOI 10.1007/978-94-007-5603-8_2,

Keywords Vision • Cover theory • Background knowledge • Basic concepts and ideas • Dynamic models • Field test • Corroboration

2.1 Outline

Forests are complex ecosystems including very different groups of organisms ranging from trees to soil microbes. Thus simultaneous, comprehensive and coherent treatment of the versatile phenomena in ecosystems is needed, and it can be obtained with a hierarchy of theories: *The cover theory* outlines the great lines of research, and *specific theories* deal with different phenomena within the cover theory. From the coherent framework of theories, it follows that the methods, i.e. modelling and measurements, share a common basis, and the tests of specific theories with field measurements are based on the same approach. The construction of specific theories under the guidance of the cover theory and common methodology allows proper flow of knowledge in all research concerning these versatile and complex forest ecosystems and the formation of the general theory of forest ecology.

The construction of the cover theory begins with the formulation of the vision that outlines roughly the most essential features in the interactions between forest ecosystems and their environment. It introduces in general terms the framework for the theory. Theories of physics, physiology, chemistry, anatomy and evolution provide essential background knowledge. The definition of *basic concepts* introduces tools for clarifying the vision. Basic concepts allow the formulation of the *basic ideas* to express the most essential relationships in a more exact form.

The cover theory is too general for research into all the different phenomena in trees, in ground vegetation and in soil. Therefore, specific theories are needed to be formulated and specified under the guidance of the general ideas presented in the cover theory. The specific basic concepts and ideas outline the thinking in the studies of each phenomenon, and they are intended to capture the most essential features of the study object. The construction of theoretical models is based on the analysis of material and energy fluxes and conservation of mass and energy resulting in differential or difference equations that can be tested with field measurements. The relationship between specific theories and the cover theory is visualised in Fig. 2.1.

The formation of specific theories proceeds via basic concepts and ideas to the formation of theoretical models. We need feedback from forests to evaluate and develop the theory formation. The severe testing of theoretical models puts strict requirements for the data to be used. The arrangement of measurements is a highly technical but necessary sidetrack in the theory formation. Tests of theoretical model with high-quality data are the last phase in clarifying the theory. The feedback from test to the previous phases in the theory formation is essential. The different steps in the theory formation are visualised in Fig. 2.2. The feedback from the test will result in improvements of the theory. Our methodological approach follows the ideas presented by Tuomivaara et al. (1994), Bunge (1996) and Hari (2008).

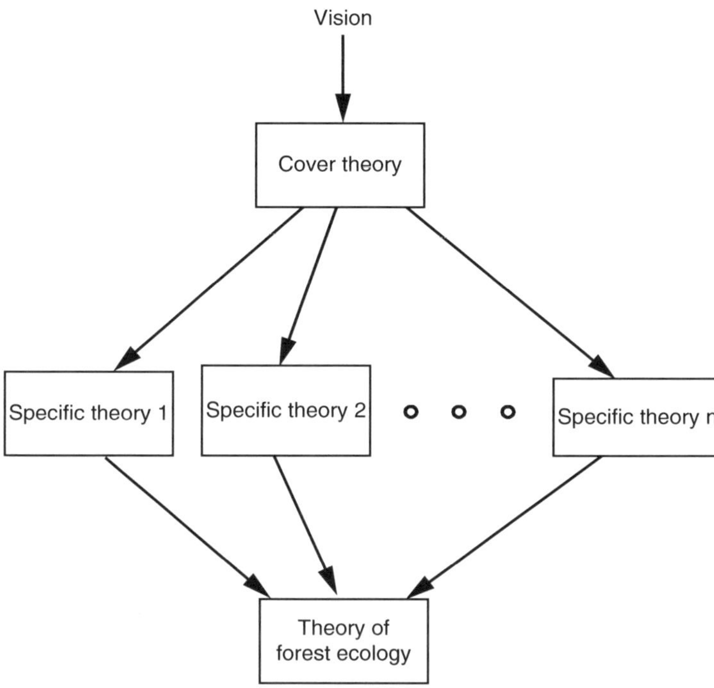

Fig. 2.1 The relationship between the cover theory and specific theories. We construct specific theories in the common framework provided by the cover theory. In this way, we can combine a coherent set of specific theories that enable knowledge flow between specific theories and construction of the theory of forest ecology. *Boxes* describe theories and *arrows* the flow of knowledge

2.2 Cover Theory of Forest Ecosystems

2.2.1 Vision

Solar radiation is the source of energy in forests. Vegetation has specialised structures that are able to convert energy in light quanta into a chemical form that is stored as carbohydrates such as sugars and starch. The chemical energy in sugars is further utilised in the synthesis of large carbon-, hydrogen- and oxygen-rich molecules. In addition, small amounts of nitrogen, phosphorus, potassium and other elements are needed for biosynthesis and the formation of special structures in vegetation. The metabolism of vegetation and microbes consumes O_2 and produces CO_2, H_2O, NH_4 and other simple compounds or ions generating material fluxes between forests and their environment. Already over 50 years ago, H. T. Odum stressed the important role of material and energy fluxes between ecosystem and their environment.

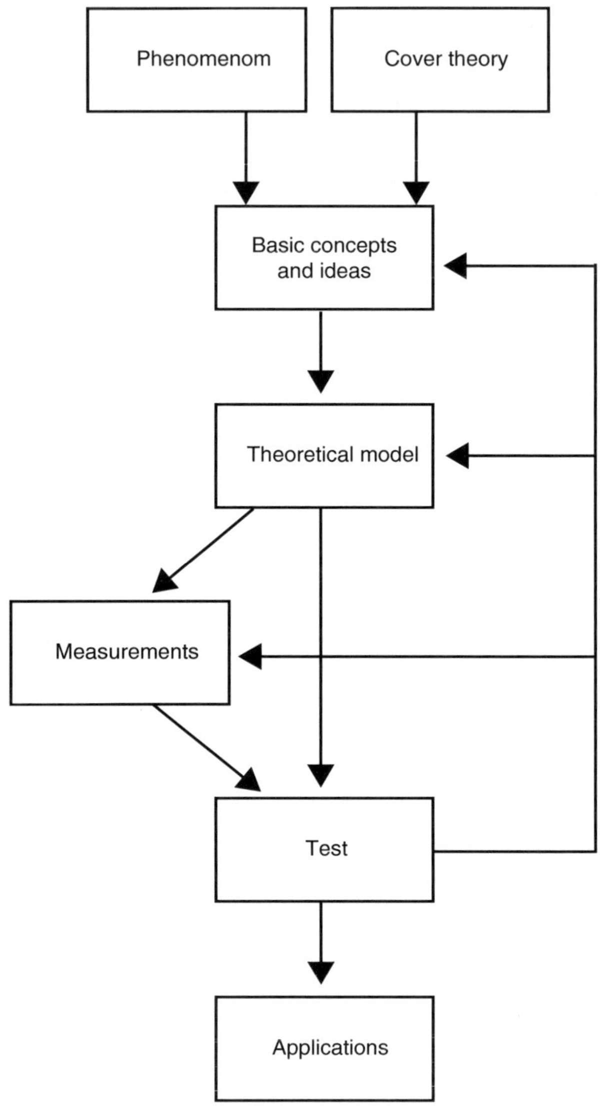

Fig. 2.2 The phases of formation of the specific theories utilising the cover theory as backbone. The feedback from the test to theory development and measurements is essential for the progress of the research. *Boxes* describe phases in the research and *arrows* the flow of knowledge

The physical distance for transport between sites of material intake and usage is in tall trees often over 20 m. Trees have effective transport structures connecting the roots and leaves with each other.

The strong annual cycle caused by the orbiting of the globe in a tilted position around the Sun is a stable characteristic for the Earth's environment. Forests have to tolerate the regular alternation of favourable and critical conditions during each year, and living organisms have means to cope with the cyclic behaviour in the environment.

Vegetation and microbes in soil have developed by evolution since the beginning of life on the Earth. The slow development during milliards of years has generated strong regularities in the structure and metabolism of living organisms. Evolution provides important insights into the present metabolism and structure of forests.

Conservation of energy and material, the cornerstone of Newtonian physics, is also a valid principle in the metabolism of vegetation and microbes in forests. Metabolism and physical phenomena convert material and energy in other forms, but the atoms and energy do not disappear. Both the environment and the state of the vegetation have effect on the conversion of material and energy, and by that they provide the causal explanations for the phenomena in forests.

The variation in the involved temporal and spatial scales is huge: from picoseconds in photosynthetic light reactions to century in the growth of a forest stand and from micrometres in the cell to hundreds of kilometres in a vegetation zone. Individuals are the basic units of the forests; they have complicated fine structure for metabolic tasks.

The vision outlines the basic features in forest ecosystems in very general form. These general ideas guide the formation of the theory of physical and physiological forest ecology. We proceed stepwise in the theory formation from background information, via definition of concepts, basic ideas and formulation of dynamic models, to their testing to give more concrete form to the vision.

2.2.2 Background

The physical and physiological theory of forest ecology utilises knowledge from very different disciplines as background. However, physics and physiology play dominating role in the argumentation.

Living organisms in forests have special structures to convert energy from light to chemical form and to utilise that chemical form in their metabolism. The intake and release of material and transport within trees take place as well via specialised structures. The discipline of anatomy provides an important insight into these rather complicated features of living organisms.

Several steps are involved in the conversion of solar energy into chemical energy as sugar. Vegetation synthesises a large number of macromolecules from sugars and ions. Each step in photosynthesis and in the synthesis of macromolecules is based on specialised enzymes, membrane pumps or pigments. The chains of steps are well balanced in such way that the sequence runs smoothly from the first step to the final product.

The enzymes, membrane pumps and pigments needed for the synthesis of new compounds are unstable proteins. They are unstable in the sense that they are fugitive over time. Their amounts and activation states fluctuate; thus new ones must be synthesised to replace the damaged ones and maintain the proper amount or activation state. This is a very demanding task for the metabolism of cells.

The annual cycle in light, temperature and other properties in the environment is an additional challenge for the metabolism of vegetation and microbes. Living processes and life in general must utilise effectively the favourable periods and tolerate the hard conditions during winter or drought. This is an additional complication in the control of the concentrations and activities of enzymes, membrane pumps and pigments. Physiology is the discipline that provides insight into the metabolism of vegetation and microbes.

Metabolism and physical phenomena generate concentration, temperature and pressure differences that cause diffusion or convection flows of mass and energy within individuals and between forest ecosystems and the environment.

The fundamental principle, conservation of mass and energy, allows the combination of metabolic and physical phenomena in a chain of conversion of energy or material. In this way, the causal relationships are introduced into the theory of forest ecology.

Effective structures, metabolism and control of enzymes, membrane pumps and pigments have developed in evolution.

Instrumentation has developed very rapidly during the last decades, and several phenomena can rather easily be measured in forests. Thus we can obtain important feedbacks from material and energy fluxes to theoretical thinking. This novel instrumentation is based on the discipline of electronics. Physical knowledge is essential for the proper understanding of fluxes, causality and electronics.

2.2.3 Basic Concepts

The construction of the cover theory begins with the definition of basic concepts. They are needed to cover the most important aspects in the interactions between forests and their environment. We should be able to introduce such concepts that characterise the most important ecological phenomena and that enable effective utilisation of the background knowledge.

The definition of basic concepts begins with the introduction of process. Large carbon molecules are characteristic for trees and other living organisms in forests. They are formed in the metabolism in long chains of biochemical reactions. The concept process is defined to characterise the conversion of material and energy in the metabolism and physical phenomena. Thus a *process*, by definition, converts material and/or energy into a new form or moves material through a membrane. For example, sugars are formed in photosynthesis from carbon dioxide and water using solar energy. Thus photosynthesis is a process according to the definition in the cover theory. Similarly, formation of new tissues and nitrogen uptake by roots are processes. They are characterised by the amounts of material converted to new forms or penetrated through a biological membrane. Processes are quantified by the fluxes of material or energy generated by the process under consideration.

The processes respond to the environment and also to the state of the living organisms. Thus concepts are needed to characterise the environment and the state of the living individual. *Environmental factors*, by definition, are those features in the environment that have effect on the processes. For example, light intensity, temperature and carbon dioxide concentration have substantial effects on photosynthesis. Thus at least these three are environmental factors.

Stable carbon compounds form the *structure* of living organisms. Cells are basic structural units of living organisms. Similar cells form tissues: They build up organs and finally organisms to form ecosystems. The structure involves several levels that have their characteristic phenomena. Geometrical features, chemical composition and mass characterise the structure.

Metabolic processes are non-spontaneous and special biological structures make them possible. *Functional substances* enable the metabolic processes in living organisms. Most of the structure of plant cells is formed by cell walls and lipid bilayers, which are passive skeletons for the metabolism. All biochemical processes have specific catalysing enzymes, membrane pumps actively transport material through membranes, and pigments capture light quanta. Thus enzymes, membrane pumps and pigments form the functional substances, and they play a key role in the utilisation of physiological knowledge in the formation of our forest ecological theory.

Functional substances are often non-stable large molecules; thus their lifetime is shorter than that of cell walls. In addition, the functional substances have often active and passive forms. Thus synthesis, decomposition, activation and deactivation of functional substances must take place in all living cells. Several processes and subprocesses are interrelated, and the functional substances of different processes and subprocesses have to match with each other to produce balanced metabolism, and the functional substances form large chains and webs to fulfil the metabolic needs. Living organisms have proteins, under the control of genetic system, that synthesise, decompose, activate and deactivate the functional substances. We call, by definition, these proteins and their genetic background as a *biochemical regulation system* and their action as *regulation*.

The metabolism of living organisms and physical phenomena convert material and energy to other forms. Thus metabolism and physical phenomena require new material or energy often from distant places. *Transport* moves material or energy in space. The long distances to be traversed by materials within trees have intrigued scientists for a long time, and transport phenomena are still being actively researched. *Fluxes* are material or energy flows per a unit of area during a unit time. Transport plays an important role when physical knowledge is utilised as background knowledge in ecology.

Forest ecosystems are very versatile and size differences between acting individuals are huge, ranging from microbes to large trees. In addition, the timescale runs from picoseconds to 100 years. Thus we need means to handle all these different space and timescales. An ordered structure is called a *hierarchy*. Proper knowledge flow between the different hierarchical levels of our analysis allows utilisation of knowledge gained on detailed level in the analysis at more aggregated one.

Forest ecosystems are complex hierarchical systems, and several simultaneous phenomena occur in the hierarchy. The interactions at lower levels generate new phenomena and properties at higher levels. These novelties are *emergent properties* in the system. There are several definitions of the emergence, and our definition is often called as weak emergence. The characteristic emergent properties at each level in the hierarchy should be identified and introduced into the treatment of phenomena.

The globe has orbited the Sun during the whole period of development of life on the Earth. This is why biochemical regulation systems have developed in evolution to cope with the very regular and strong annual cycle of light and temperature that is a characteristic feature of the environmental factors. The *annual cycle of metabolism* covers the changes in the biochemical regulation systems, in functional substances and in structures matching vegetation with the annual cycle of environment. The processes in living organisms have their own annual cycle since biochemical regulations system changes the concentrations and activities of the functional substances.

Besides the annual cycle, there are slow (i.e. during the lifetime of individuals) and irregular changes in the environment such as extreme weather events. *Acclimations of processes* are changes in functional substances and in the biochemical regulation system matching processes in vegetation with the slow and irregular changes in environment. *Acclimations of structure* are slow changes in structure matching it with irregular changes in the environment such as changes in light environment after collapse of a neighbour tree. In contrast to adaptations, which occur over evolutionary timescales, acclimations occur over days, months and years, and they do not involve alterations in the genetic composition.

Vegetation synthesises annually large amounts of energy-rich macromolecules, especially into cell walls. The senescent tissues are the source of energy and raw material for life in the soil. In *decomposition* microbes transform the large organic molecules into smaller molecules that are utilised in the microbial metabolism. The nutrients in the senescent tissues are released into the soil, and they are available for reuse by the vegetation.

Individuals are the fundamental functional units in a forest ecosystem, they are first born, and they metabolise and finally die. The ecosystem is not only the sum of the individuals but novel features emerge. The effects between individuals in an ecosystem are called *interactions*. The behaviour of the individuals determines the interactions, but they are characteristic for the ecosystem. The reduction of photosynthesis in the canopy caused by shading is the dominating interaction in forest ecosystems.

The regularities in phenomena in forests can now be analysed with the basic concepts: process, transport, environmental factors, hierarchy, regulation, annual cycle, interaction, emergence, acclimation and decomposition. The basic concepts are used to characterise the essential features of the study object. Visual symbols for most of the basic concepts are introduced (Fig. 2.3) to clarify the ideas dealing with the behaviour of the object. Material/energy flows, process, biochemical regulation system, transport, structure and functional substances have their own symbols in the visual language; thus flows can be visualised utilising physiological, physical and anatomical knowledge as background in each phenomenon under study.

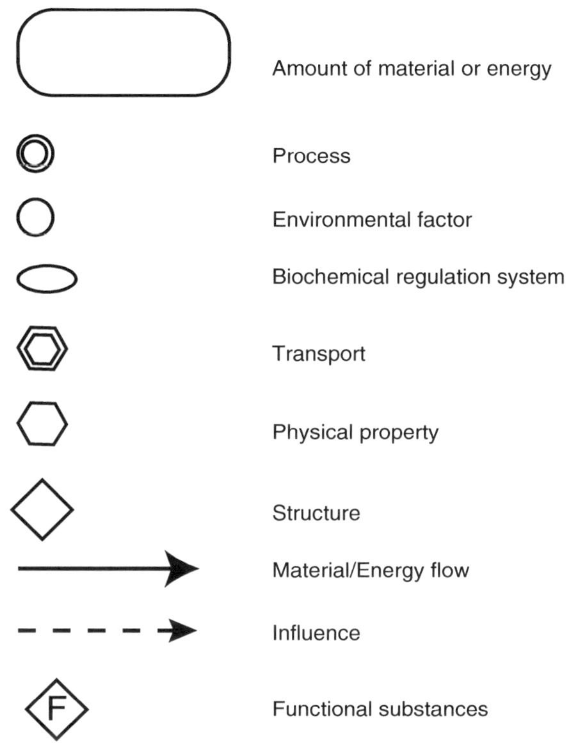

Fig. 2.3 Symbols describing the basic concepts to be used in the visualisation of theoretical ideas

2.2.4 *Basic Ideas*

The basic ideas link the basic concepts with each other and they enable the clarification of the vision. The basic ideas in the cover theory are:

1. The environmental factors vary greatly both in time and space. The orbiting of the globe around the Sun generates strong annual cycle, whereas the spinning of the globe creates strong daily cycle in environmental factors such as light and temperature. Spatial processes give rise to spatial and temporal variation in environmental factors. For example, photosynthesis decreases and respiration increases the surrounding CO_2 concentration near leaf surface.
2. The determination of environmental factors is problematic due to their great variability in time and space. Environmental factors can be well defined only in sufficiently small space and time elements. The scales of the environmental factors determine the most detailed level in the hierarchy. The great variability is characteristic for light, but also, e.g. concentrations and temperatures vary strongly in space and time.
3. The properties of the environment and of living organisms are reflected in the metabolism of living organisms. Environmental factors and the functional substances determine the conversion of material and energy in processes.

The relationships between processes and environmental factors can be determined properly only at the level of space and time elements due to the great variability of environment. The space and time element has to be so small that the variation in environmental factors is negligible within the element.

4. Cells (space element), individuals and ecosystems are natural spatial units. The metabolism responds very rapidly to environmental factors, i.e. during a short time element. The other evident timescales are year and rotation period. Thus the spatial levels of hierarchy are space elements of tissue, individual and ecosystem. The temporal levels of hierarchy are time element, year and rotation period. The flow of knowledge between the hierarchical levels, especially from detailed to more aggregated ones, is important in the analysis of the interactions between ecosystems and their environment.
5. The short lifetime of functional substances calls for a need to maintain the concentrations of the enzymes, membrane pumps and pigments in a long chain of steps generating the metabolic process. The biochemical regulation system synthesises, decomposes, activates and deactivates the functional substances. The actions of biochemical regulation systems keep the concentrations of the functional substances stable or change them, and the resulting regularities in the involved functional substances are emergent properties. The changes in the concentration or activity of functional substances are reflected in the effects of environmental factors on processes.
6. Material and energy fluxes mediate the interactions between ecosystems and their environment. In addition, the material fluxes within individuals are important due to the long distances between the locations of production and consumption. This is why transport plays a key role in forest ecology. Physical properties such as pressure, temperature or concentration difference generate material and energy fluxes, and in this way, physical knowledge is necessary in forest ecology.
7. Living organisms are the consequence of evolution over milliards of years. The emergence of novel features and their test against the existing ones result in effective metabolic, structural and regulation solutions. Thus efficient processes, transport structures and biochemical regulation systems have developed during evolution. Efficient solutions can, at least in some cases, be found as mathematical solutions of optimisation problems.
8. The strong and regular annual cycle in environmental factors has been a characteristic feature in the environment during the entire period of evolution, and a clear annual cycle in metabolism has developed to utilise the regular sequence of favourable and critical periods. In addition, the annual cycle of metabolism has been well tested during evolution, and it enables individuals to utilise the very regular alternation of warm and freezing or rain and dry periods in the climate of forests.
9. Acclimations are common in forests and they benefit the individuals since most of them have been tested in evolution. The acclimations of processes, structures and biochemical regulations have been tested during evolution only if the environmental factors generating the acclimation have varied during

the lifetime of individuals. Responses to increasing shading and drought are examples of well-tested acclimations. The atmospheric CO_2 concentration has varied considerably during the last million years, but during the lifetime of individuals, the concentration has been stable due to slow concentration changes. The present increase, about 100 ppm/100 years, is outside the range tested during evolution. Thus acclimation to CO_2 can benefit or harm the individual, and the predictions of the acclimations of forests to increasing CO_2 are problematic.

10. Important metabolic processes and transport phenomena have their own specialised structures. Actions of the biochemical regulation systems, together with the genetic code, determine the properties of the structure. The resulting regularities in the tree structure are emergent properties. The genetic code determines the great features, and regulation tunes it to the prevailing situation resulting in acclimations of the structure.
11. The interactions between individuals in a forest ecosystem are numerous and versatile. However, some interactions are more important and dominate the development of the forest ecosystem. Environmental factors convey the interactions in a plant community. Extinction of light and reduced photosynthesis in lower parts of a canopy is the most important interaction. The spatial variation in other environmental factors, especially between individuals, is small, and the different plants experience about the same environment. Thus spatially homogeneous environment is unable to convey interactions between individuals.
12. Conservation principles are cornerstones of Newtonian physics, and huge evidence has piled to support this theoretical idea. The living objects cannot violate the fundamental physical laws. The form of chemical compounds and energy may change, but the overall number of atoms and the total amount of energy do not change during processes or in transport.
13. The flow of information from the space and time element to the ecosystem level plays an important role in the analysis of the fluxes between forest ecosystems and their environment. However, it is very problematic to utilise detailed knowledge at the aggregated level, because the analyses become heavy and the small details easily hinder the researcher's view of important aspects. There are emergent properties at the various levels in ecosystems, and these new features allow the utilisation of the detailed knowledge at lower level in the hierarchy in condensed forms. The emergent features should be carefully analysed at each level, and the possibilities to utilise knowledge on more detailed level as emergent properties should be carefully analysed.
14. The carbon flow through ecosystem and circulation of nitrogen within the ecosystem, via photosynthesis, synthesis of macromolecules and finally decomposition of macromolecules in soil by soil invertebrates and microbes, are characteristic features for forests. In the soil, microbes decompose the macromolecules through the action of extracellular enzymes and utilise the resulting small energy-rich molecules in their metabolism. Without microbial activity, the organic matter would accumulate in the soil and nutrient cycling would be impossible.

A general description of vegetation and soil structures is also included in the cover theory to give the necessary background for theories of specific phenomena.

The cover theory outlines the research dealing with different phenomena in forest ecosystems. However, it is too general to be applied, and it must be developed to meet the properties of each single phenomenon under study within specific theories.

2.3 Constructing a Specific Theory

2.3.1 Phenomenon

There are several processes in forest ecosystems that contribute to the interactions between the ecosystem and its environment. In addition, the emergent properties, generated by the action of the biochemical regulation systems in the functional substances, in the tree structure and in the interactions between individuals, play an important role in the interactions. These phenomena serve closer attention and formation of specific theories within the cover theory, i.e. the rather general concepts and ideas in the cover theory are specified to meet the properties of each phenomenon under study.

The construction of the specific theory begins with the general description of the phenomenon under study, especially the connections with the material and energy fluxes in the ecosystems. The history of the research and the literature in the field is shortly described.

2.3.2 Structural, Metabolic and Physical Background

Metabolic processes in the different structures in ecosystems consume, convert and produce material and energy, whereas material and energy fluxes are physical phenomena. This structural, metabolic and physical background of the phenomenon is analysed.

There are highly specialised structures for each important phenomenon in forest ecosystems at each hierarchical level: The cells and the functional substances at tissue level, transporting structures at tree level and the organisms form the structural background at ecosystem level. Anatomy can provide useful knowledge for description of the structures involved.

The consumed/produced amounts of material and energy are in the focus of the treatment of metabolism. The action of the regulation systems changes the functional substances. The relationship between environment factors and metabolism respond to the change in the enzymes, membrane pumps and pigments. Physiological knowledge provides necessary understanding of the regularities in the metabolism.

The metabolism generates concentration, pressure and temperature differences, which give rise to material and energy fluxes at cellular, individual and ecosystem level. Physical knowledge contributes to the understanding of the fluxes and their connection to the metabolism.

The analysis of the structures, metabolism and transport phenomena is needed as background for proper understanding of the fluxes generating the phenomenon under study.

2.3.3 Specification of Basic Concepts and Ideas

The basic concepts in the cover theory deal with all phenomena in the ecosystem, and most of them can be omitted when dealing with some single phenomenon in the ecosystem. The necessary concepts are identified and clarified to correspond with the special features of the phenomenon under study in the framework of the general basic concepts in the cover theory.

The detailed analysis of the structures, metabolism and transport and specific concepts enables rather precise formulation of the basic ideas dealing with the phenomenon under study. These obtained specifications of the basic ideas in the cover theory are utilised in the formation of the theoretical model.

2.3.4 Theoretical Model

The basic ideas outline the fluxes and the factors affecting on the fluxes in a forest ecosystem. They must be clarified to be able to predict phenomena, especially fluxes. Then quantitative measures are introduced, and the phenomenon is studied in quantitative terms, because it allows more exact formulation of the ideas, it is more effective, and it enables utilisation of measurements.

The phenomenon under study includes often a space element having inflow, i, and outflow, o, of material or energy (Fig. 2.4). Metabolic processes and transport phenomena generate usually the inflows and outflows; thus physics and physiology provide causal explanations for the flows. Let X denote the amount of material or energy in the space element. Let Δt be such a short time interval that the inflow and outflows remain constant during Δt. Then the content of the space element at moment $t + \Delta t$, $X(t + \Delta t)$, can be obtained with the content at t, $X(t)$, and the in- and outflows utilising conservation of mass (Basic idea 10):

$$X(t + \Delta t) = X(t) + (i - o)\Delta t. \tag{2.1}$$

When we move $X(t)$ to the left-hand side of the equation and divide it with Δt, we get

$$\frac{X\,(t + \Delta t) - X(t)}{\Delta t} = i - o. \tag{2.2}$$

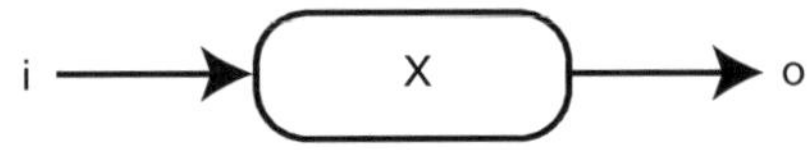

Fig. 2.4 Inflow and outflow in a container

When Δt approaches zero, the Eq. (2.2) results in differential equation

$$\frac{dX}{dt} = i - o. \tag{2.3}$$

Metabolic and physical processes and transport determine the in- and outflows. Thus physiological and physical knowledge can provide causal explanations for the material and energy fluxes between living organisms and their environment. In addition, physiological and physical knowledge determines the model structures linking the in- and outflows with environmental factors and functional substances (basic idea 3) and with concentration, pressure and temperature differences (basic idea 6). In this way, we can derive causal theoretical models, based on physical and physiological knowledge, to describe the fluxes between living organisms and their environment. The derivation of the models for in- and outflows requires additional simplifying assumptions and idealisations that can introduce shortcomings into model structures.

We meet quite often the situation where one single volume element is not sufficient, but a chain of interconnected volume elements are needed. The difference models, analogous to the model (Eq. 2.3), can often be constructed utilising physiological and physical knowledge. This approach leads to partial differential equations when the volume of the volume elements approaches zero. The chains of difference equations or partial differential equations result in the dependences of concentration or of temperature on space coordinates.

Often the fluxes between living organisms and their environment and the spatial patterns of concentration or temperature can be measured in the field. This provides the means to test the theoretical models, and in this way, we can get feedback from nature to the theory development.

2.3.5 *Measurements*

The aim of the measurements is to produce data to test specific theories and to provide descriptions of ecosystems. The theories deal with the fluxes between living organisms and their environment, and therefore, they should be measured in the field in addition to the environmental factors that influence them. After we have measured values of the response and its explaining factors, we can compare the model predictions with measured values enabling tests of the theory underlying the model.

Measurements of the response and explaining factors introduce an additional complication since measurements always include systematic and random errors. Let ε be the random, s the systematic measuring error of a theoretical quantity X and x the measured values of X. If we assume that the errors are additive with the signal to be measured, then

$$x = X + s + \varepsilon. \tag{2.4}$$

The noise and the systematic error in measurements reduce the quality of data, and they should be carefully analysed and reduced with technical arrangements.

We can never avoid systematic measuring errors, and their treatment is highly problematic. If we know the systematic error, it is easy to remove and correct the measurements, but in most cases, the systematic error is unknown and we have no means to correct it. However, we can often get an estimate of the magnitude of the systematic measuring error, and in this way, we can evaluate the possible bias in the results.

We can determine the statistical properties of the measuring noise with proper additional measurements. In addition, we can convert measuring noise in explaining factors with simulations utilising artificial data to random disturbance in the model behaviour. If the measuring noise of response or the random component in the model behaviour generated by noise in explaining factors is large when compared with the signal to be detected in the model behaviour, then the measuring noise smears the value of the data as test material.

2.3.6 *Test with Field Measurements*

We cannot prove a theory as a truthful description of the reality; we can only test the theory. Severe testing can detect shortcoming in the theory, it may provide explaining power of the theory, and it enables comparison with other theories in the field.

Determination of the values of the parameters. The test of the theory with field measurements begins with the determination of the values of unknown parameters in the theoretical model. The model fitting results in biased agreement between the modelled and measured values in the used data set. This is why we divide the data into two sets, estimation and testing data. We use the estimation data in the estimation of the parameter values, and we test the theory, whenever it is possible, with data that are not utilised in the model fitting to avoid the bias in conclusions caused by estimation.

The estimation is performed with normal statistical methods, e.g. minimising the residual sum of squares. This method is non-problematic if the explaining factors are independent from each other and if the parameters in the model do not compensate each other. In ecology, the explaining factors are often strongly interconnected and the parameters are interrelated. These features often make the estimation of the parameter values problematic, and the model structures must be simplified by reducing the number of parameters to be estimated; otherwise, the solution of the minimisation problem is not stable.

Prediction of characteristic features. Each phenomenon under study has its characteristic features that can be seen in the measurements. For example, the observed responses may have clear daily and annual patterns or rapid variation generated by changes in environmental factors. The first step in the test of the theory and the theoretical model is to answer the question: Is the theoretical model able to predict the characteristic features in the phenomenon under study in the test data set?

The focused study on characteristic features in the measurements enables qualitative evaluation of the explaining ability of the theory during informative episodes of the measurements. The analysis of the characteristic features focuses the test to critical part of the data, and in this way, shortcomings in the theory can be effectively seen.

Adequacy of model structure. The qualitative study of the prediction and explaining ability of the characteristic features in the behaviour of the study object is able to utilise only short episodes in the measured responses. Quantitative study of the agreement between observed and predicted behaviour of the research object enables treatment of large data sets, and it may detect rather small discrepancies between the theory and measurements, and it provides a measure of the agreement between prediction and observations.

Models are never truthful descriptions of the reality, and they always include shortcomings generated by the theory itself or the simplifying assumptions or idealisations in the derivation of the models for the fluxes involved. Measured field data enable the evaluation of the role of the shortcomings in the model structures within the accuracy and precision of the measurements. Special attention should be paid to the simplifying assumptions and to the idealisations done in the derivation of the models for the in- and outflows. The systematic measuring errors may confound the evaluation of the model behaviour.

The residuals, i.e. the differences between observed and modelled responses, are the basis of a tool to study the adequacy of the model structure. If there is some systematic behaviour in the residuals as a function of an explaining factor of the theoretical model or of any other factor, this is an indication of a shortcoming in the model structure or a sign of a systematic measuring error. The magnitudes of the systematic features in the residuals and their role in the reduction of the explaining power of the theory should be analysed. The reason of the systematic behaviour of the residuals should be carefully studied.

Explaining power. The theoretical explanation is the most important virtue of theories. The physical and physiological theory of forest ecology explains the observed behaviour of the system with the properties of the functional substances, with the action of the regulation system and with regularities in structures using evolution theory, physics, physiology, anatomy and chemistry as background knowledge. The crucial question is whether these concepts and background knowledge are sufficient to explain the observed behaviour in forest ecosystems.

Proportion of explained variance (PEV) is the quantitative measure of the power of the theoretical model to explain the observed variation in the behaviour of the research object. The remaining residual variation is caused by the shortcomings of the theoretical model, systematic measuring errors and random variation in the measurements. The PEV is calculated as one minus the ratio of variances of the residuals and of the measured responses. If measuring noises of response and explaining variables are small when compared with the residual variation, then small shortcomings in the theoretical model can be detected. Thus reduction of the measuring noises makes the test of the theoretical model more effective. The understanding of the reasons of the residual variation is important; thus we should analyse the roles of measuring noises, parameter estimation and shortcomings in the applied model structure in the observed residual variation.

Comparison with other theories. The last step in the theory testing is the comparison with other theories in the field. Then the basic concepts, ideas and theoretical models are compared. The field data are used to evaluate the relevance of the differences in field conditions. The performance of the models is based on their explaining and prediction power in the test data set. The differences in measurements, especially in measuring noise, and different environments are additional complications in the comparison.

Conclusion

The testing of the theory includes four aspects, i.e. (1) ability to predict and explain the characteristic features in the behaviour of the study object, (2) evaluation of the adequacy of the model structures, (3) determination of the explaining power and (4) comparison with competing theories. If the theory fails in some of the above four aspects, then it should be improved or rejected. If the theory is successful in all four aspects, then it gains corroboration and it is the best knowledge in its domain until a better theory is developed and tested. Development, testing and improvement of the forest ecological theories (Fig. 2.2) result in slow development and accumulation of knowledge dealing with the interactions between forests and their environment.

The four steps to test theories will play a major role in the Chaps. 4, 5, 6, 7, 9.

2.4 Formation of the Theory of Forest Ecology

Forest ecosystems are versatile and complex systems, and therefore, also the theoretical treatment of forest ecology is a demanding task. Knowledge can be gained about several phenomena at different levels in the hierarchy of space and time. Any forest

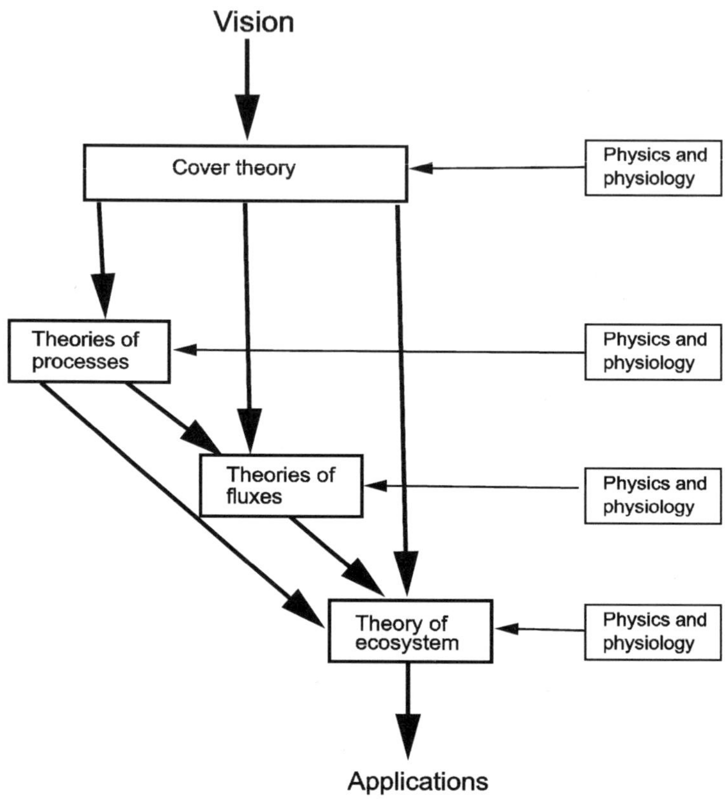

Fig. 2.5 Flow of knowledge in the formation of the physical and physiological theory of forest ecology. *Boxes* describe theory and *arrows* flow of knowledge

ecological theory should be able to combine the knowledge to a coherent system in which the understanding of different aspects is utilised at the ecosystem level.

Our solution to the problem of complexity is a web of specific theories under the guidance of the cover theory. The vision determines the main approach. Moreover, it outlines the objects, their characteristic properties and also the background knowledge. The definition of basic concepts allows the expression of the ideas in a more concrete form. The vision together with the basic concepts and ideas forms the cover theory, and they are applied as backbone in the formation of specific theories to deal with studied phenomena in forest ecosystems.

However, each phenomenon in the forest ecosystem requires its own specific theory including the specification of the basic concepts and ideas to meet the properties of the phenomenon. We can often derive a theoretical model using physiological and physical knowledge and the conservation of mass and energy. The theoretical model is interpreted as causal when it describes responses generated by metabolic and physical phenomenon. Field measurements allow testing of the theoretical model. If it predicts the flows well enough, then the model gains corroboration, and if it fails, then the approach should be improved.

The last step in the formation of the physical and physiological theory of forest ecology is to combine the specific theories to a coherent set of theories to cover the material and energy flows within an ecosystem and between the ecosystem and its surroundings (Fig. 2.5). This is possible since the specific theories 'speak the same language' due to the cover theory.

References

Bunge M (1996) Finding philosophy in social science. Yale University Press, New Haven, pp 6–7, 30–32, 49, 68–72, 79, 124–126, 179–180

Hari P (2008) Introduction. In: Hari P, Kulmala L (eds) Boreal forest and climate change: advances in global change research. Springer, Dordrecht/London

Tuomivaara T, Hari P, Rita H, Häkkinen R (1994) The guide-dog approach: a methodology for ecology. University of Helsinki, Department of Forest Ecology, Publ 11, Helsinki. http://hdl.handle.net/10138/26453

Chapter 3
Environmental Factors

Üllar Rannik, Samuli Launiainen, Jukka Pumpanen, Liisa Kulmala, Pasi Kolari, Timo Vesala, Janne F.J. Korhonen, and Pertti Hari

Contents

Abstract Solar radiation energy input is crucial for the environmental factors on the globe. The circulation of the globe around the Sun generates strong annual cycle, and the spinning of the globe around its own axis generates diurnal cycle in environmental factors. The annual and diurnal patterns are strong in irradiance, temperature and water vapour concentration but, on the other hand, the annual and daily cycles are relatively weak in CO_2 concentration. The water pool in the soil has also annual cycle reflecting the annual patterns in transpiration and in rainfall. The reduction of irradiance in the canopy is the dominating spatial feature in the environmental factors.

Keywords Annual and diurnal cycles • Solar radiation • Temperature • Carbon dioxide concentration • Water content • Soil properties

---

Ü. Rannik (✉)
Department of Physics, University of Helsinki, P.O. Box 38, 00014 University of Helsinki, Helsinki, Finland
e-mail: ullar.rannik@heuristica.ee

P. Hari et al. (eds.), *Physical and Physiological Forest Ecology*,
DOI 10.1007/978-94-007-5603-8_3,

3.1 Properties of Environment

Solar energy is the main driving force of the climate system and thus the strongest influence on other environmental factors.[1] A large fraction of the incoming radiation energy is converted to heat at the vegetation surfaces and in the atmosphere, resulting in spatial and temporal temperature differences that generate wind and turbulent mixing of the air. The resulting atmospheric motions transport the energy and atmospheric constituents and therefore reduce the local variations in temperature and concentrations of gases.

The properties of the environment affect the interactions between forest ecosystems and the atmosphere. Processes are the engines of phenomena in forest ecosystems, and we define environmental factors in relation to the processes. In Chap. 2, we introduced the idea of environmental factors which are, by definition, those features in the environment that have an effect on the processes of the ecosystem. There are several processes in the ecosystem that respond to different environmental properties, thus the number of environmental factors is rather large.

Photosynthesis is the most important process in ecosystems since it provides raw material and energy for metabolism and growth. Solar radiation is the source of energy in photosynthesis; thus, photosynthetically active radiation (see below for definition) is a key environmental factor. Atmospheric carbon dioxide is the source of carbon, and its concentration is also an environmental factor. Transpiration is an inevitable consequence of photosynthesis and also drives internal transport in trees and ground vegetation. The atmospheric water vapour concentration affects the rate of transpiration. Plant and microbial metabolism depends strongly on temperature. Vegetation takes nutrients and water from soil and the uptake depends on their concentrations and availability. There are several compounds that are toxic for the metabolism of vegetation and microbes. Therefore, the most important environmental factors for forest ecosystems are photosynthetically active radiation, temperature in air and soil, atmospheric CO_2 and water vapour concentration, water and nutrient concentrations in the soil and concentrations of toxic compounds in the air and in the soil. In this chapter, we describe their diurnal and annual pattern together with vertical profiles inside a canopy.

---

[1]The environmental factors are also affected by the climate system. The influence ranges from global to regional scales. On a global scale, transport of heat and humidity by ocean and atmosphere forms an important mechanism of redistribution of heat reaching unevenly the Earth's surface. Particularly for the Scandinavian climate, the Gulf Stream provides an important energy source. According to climate model simulations, the air temperatures over lands are few degrees higher due to oceanic heat transport. The influence to air temperature over ocean is even larger (Seager et al. 2002).

3.2 Diurnal Cycle of Environmental Factors

The rotation of the Earth on its axis that passes through the North and South Poles of the planet changes the solar elevation every day. This generates strong variation into the *solar energy input* per unit area. It is small at dawn and reaches maximum at midday and declines thereafter to zero at sunset. Solar energy is electromagnetic radiation emanating from the Sun, most of the energy being in the waveband 0.1–5 μm, which includes the ultraviolet (0.1–0.4 μm), the visible (0.4–0.7 μm) and the near-infrared (0.7–5 μm) radiation. Solar energy enters the canopy in the form of direct and diffuse radiation, summed together as *global radiation.* The amount of global radiation available at the top of the canopy depends strongly on (1) elevation of the Sun, (2) cloudiness and (3) aerosol particle concentration and size distribution. The aerosols in the air and clouds scatter light, especially at low solar altitude as the solar radiation beam has to travel further in the atmosphere leading to more absorption and scattering (including backscattering into space) of the photons and reduced global radiation at the Earth surface. Through scattering, the same factors determine also partitioning of global radiation into direct and diffuse components. 'Direct' refers to the strongly directional photon flux coming from the solar disc; 'diffuse' is the weaker intensity flux that comes from the entire sky as a result of scattering.

Flux density – amount of radiation energy through unit area per unit time ($W\ m^{-2}$)

Irradiance – radiation flux density incident on the surface ($W\ m^{-2}$)

Spectral irradiance – radiation flux density incident on the surface per unit radiation wavelength (or frequency) interval ($W\ m^{-2}\ nm^{-1}$)

Insolation – irradiance due to solar radiation ($W\ m^{-2}$)

Solar constant – the flux of solar radiation through a tangential plane above the atmosphere at the average distance between the Earth and the Sun, nearly constant at 1,370 $W\ m^{-2}$

Global radiation – the total of direct solar radiation and diffuse sky radiation received by a unit horizontal surface ($W\ m^{-2}$)

Direct radiation – solar radiation that passes through the Earth's atmosphere directly to the Earth's surface without scattering ($W\ m^{-2}$)

Diffuse radiation – the radiation that has been scattered out of the direct beam ($W\ m^{-2}$)

Photosynthetically active radiation (PAR) – the visible radiation in the waveband from 400 to 700 nm that is important from biological point of view (μmol photons $m^{-2}\ s^{-1}$)

From a biological point of view, the important part of the solar spectrum is the visible radiation in the waveband from 400 to 700 nm. This range of the solar energy spectrum coincides with the absorption spectrum of the photosynthetic pigments, and it is termed photosynthetically active radiation (PAR). It contains approximately 40% of total energy of the spectrum emitted by the Sun. Above the canopy, global PAR consists of two components, direct PAR similarly to global radiation the solar disc and diffuse PAR from all directions.

Solar Spectrum

Short-wave radiation (in contrast to terrestrial or long-wave radiation) contains most of the energy at the waveband from 0.1 to 5.0 μm, which includes ultraviolet (0.1–0.4 μm), visible (0.4–0.7 μm) and near-infrared (0.7–5 μm) spectral regions. The maximum of solar energy is located in the visible band at wavelength of about 500 nm. Altogether, about half of the radiation emitted by the Sun is in the short-wave part of the spectrum. As an example, the spectral irradiance at the Earth surface is presented for wavebands between 300 and 575 nm (Fig. 3.1). The atmosphere modifies the solar spectrum, and the spectral distribution reaching the Earth surface is different from that at the top of the atmosphere. The main absorbers of solar radiation are so-called greenhouse gases including water vapour, carbon dioxide, ozone and others, which each absorbs solar energy at distinct wavelengths.

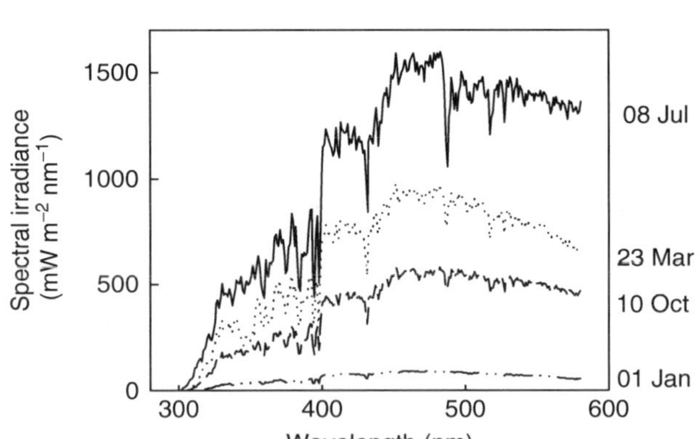

Fig. 3.1 Spectral irradiance on a horizontal surface at noon on four selected days for the SMEAR II in Hyytiälä, southern Finland, as an example for incoming solar radiation from clear sky in different seasons

There are three different atmospheric patterns during the day that cause characteristic features in photosynthetic active radiation at the Earth surface. During cloudless days, that are rather rare in boreal conditions, the solar elevation determines the

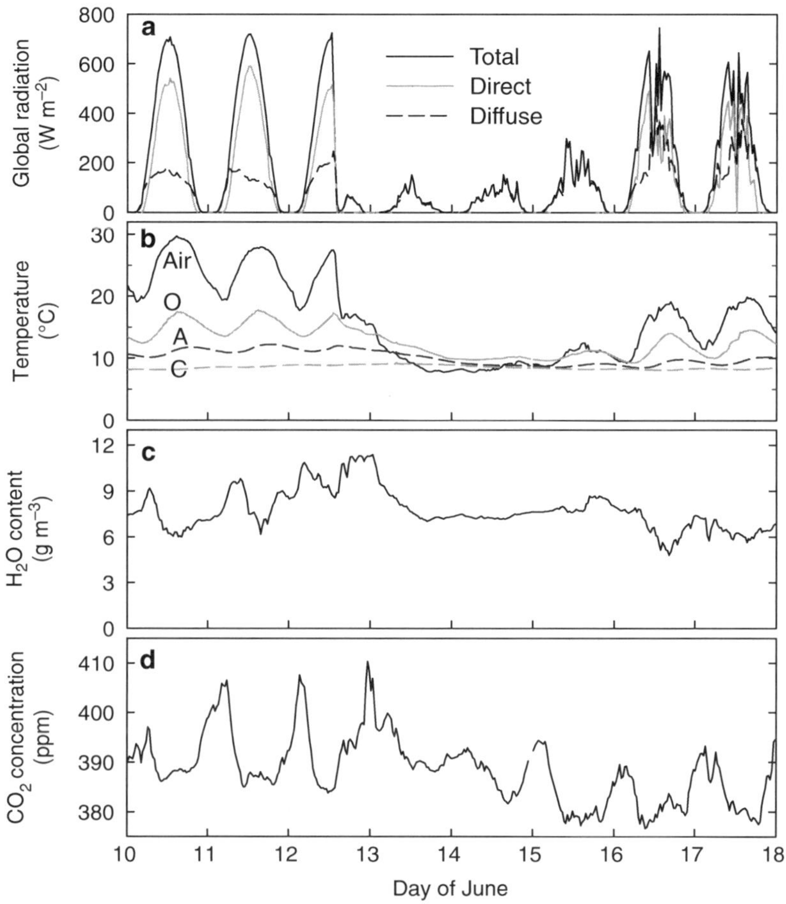

Fig. 3.2 Hourly average of components of global radiation (**a**), temperature in air and in soil horizons O, A and C (**b**), air water vapour content (**c**) and CO_2 concentration (**d**) recorded at SMEAR II, southern Finland, on 10–18 June 2011. The measurements in the panels (**a**, **c** and **d**) are above canopy. The soil horizons are introduced on p. 184

radiation and the daily pattern of photosynthetic active radiation is very regular (Fig. 3.2a). During normal summer days, the clouds travel across the sky. If a cloud covers the Sun, then the radiation near the surface is strongly reduced and the movement of clouds generates strong variation in radiation (Fig. 3.2a). Sometimes, the reflection from cloud sides enhances the radiation and then irradiance can be even larger than in clear sky conditions. Totally cloudy days (overcast) are also rather rare. Then the solar elevation and the thickness of cloud cover determine the photosynthetically active radiation. The direct solar radiation dominates the radiation in clear sky conditions at higher solar elevations (Fig. 3.2a), but the diffuse component increases when solar elevation is low and solar beam passes longer distance in the atmosphere. The share of direct radiation during cloudless days is circa 85% at solar elevations higher than 30°.

The diurnal cycle in solar radiation is mirrored by a cycle of *temperature* during each day. The absorption of solar radiation warms the absorbing media, and thus, the diurnal cycle of solar radiation input is the underlying reason for the temperature increase in the morning whilst thermal radiation streams into space and results the cooling in the evening when solar heating is weak or missing (see Chap. 5). A rapid and rather large temperature increase is characteristic for sunny days in the summer as well as for the days of intermittent cloudiness (Fig. 3.2b). Clouds reduce global radiation during cloudy days, and, consequently, also the temperature increase in the morning is smaller and the cooling in the evening is also smaller because water vapour in the clouds absorbs part of the thermal radiation that streams from the Earth into space and decreases radiative cooling (Fig. 3.2b).

Weather generates short-term variation in the soil temperature at a synoptic timescale, i.e. at the typical timescale of the variation of weather systems in northern latitudes (Fig. 3.2b). On sunny summer days, the diurnal amplitude of temperature in the humus (O layer) is 5°C whereas the temperature does not fluctuate as much during cloudy and rainy days. Deeper in the soil, the daily temperature fluctuation becomes smaller and variation is delayed relative to the topsoil temperature. The shading of the canopy has also an important effect on the diurnal course of soil temperature, reducing the daily variation.

Soil material and soil water content have an influence on the heat conductivity and *specific heat capacity*[2] of the soil. The higher the soil moisture content the greater is also the thermal conductivity because the conductivity of water is higher than that of air. However, high soil moisture content decreases the diurnal amplitude of soil temperature, because the specific heat capacity increases along with the increasing soil water content and elevated conductivity enhances heat transport between soil layers.

The diurnal cycle in the *water vapour concentration* of the atmosphere a few metres above the vegetation is less regular than that of temperature and global radiation (Fig. 3.2c). The turbulent transport efficiently mixes water vapour to higher atmospheric layers during periods of high transpiration and high solar radiation input. During nights, the transpiration is minimal due to small water vapour pressure deficit.[3]

The forest ecosystem is a CO_2 sink during summer days as a result of photosynthesis being much larger than ecosystem respiration, and source during nights, when photosynthesis is zero. The *CO_2 concentration* minimum occurs when ecosystems

---

[2]The amount of energy required to change the temperature of unit mass or volume of a material by given amount, usually 1 K.

[3]Vapour pressure deficit (VPD) can be defined as the difference (or deficit) between the pressure or concentration exerted by water vapour that could be held in saturated air (100% relative humidity) and the pressure or concentration exerted by the water vapour that is actually held in the air being measured.

act as a sink of CO_2 during strong solar energy input (Fig. 3.2d). Most of the nights are rather calm and then vertical concentration differences arise resulting in high CO_2 concentrations especially near soil surface (see later Fig. 3.12). If the night is windy, then turbulent mixing reduces these concentration differences between levels.

3.3 Annual Cycle of Environmental Factors

Our globe circles the Sun, and one orbit takes a year. In addition, the Earth spins on its own axis once in 24 h and the rotation axis tilts 23.5° from the perpendicular to the Earth–Sun plane (Fig. 3.3). This geometry generates a strong annual cycle of incoming solar radiation – especially at high and low latitudes. The tilted positioning of the Earth's rotation axis is the primary cause for seasonality of radiation input and other environmental variables whilst the varying distance between the globe and the Sun is of the secondary importance. We treat the dependence of solar radiation flux at the top of the atmosphere on time and location in more detail in Appendix 1.

The annual changes are particularly strong in the boreal zone. In summer, the day is extended and the Sun does not set beyond the Arctic Circle. In winter, the days are short, and beyond the Arctic Circle, the Sun does not rise daily during polar night.

The solar energy input at the top of the atmosphere depends on the variation of radiation emitted by the Sun and the position of the Earth in the solar system, as well as the angle of the Earth's axis with respect to the ecliptic plane. The position and orientation of the Earth are affected by the gravitational influence of the Sun and other planets of the solar system over very long periods of time (Appendix 1).

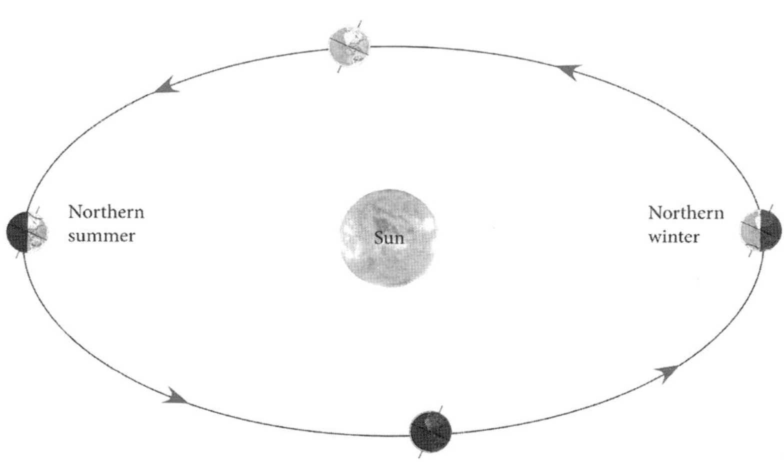

Fig. 3.3 The Earth revolves around the Sun and spins on its axis. Earth's axis is tilted 23.5° from the perpendicular to the Earth–Sun plane (ecliptic plane)

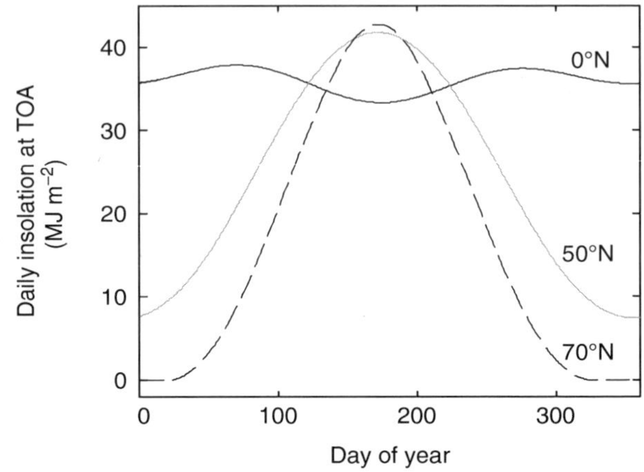

Fig. 3.4 Daily incoming solar radiation energy at the top of the atmosphere (daily insolation) as a function of time of the year for 0, 50 and 70° North

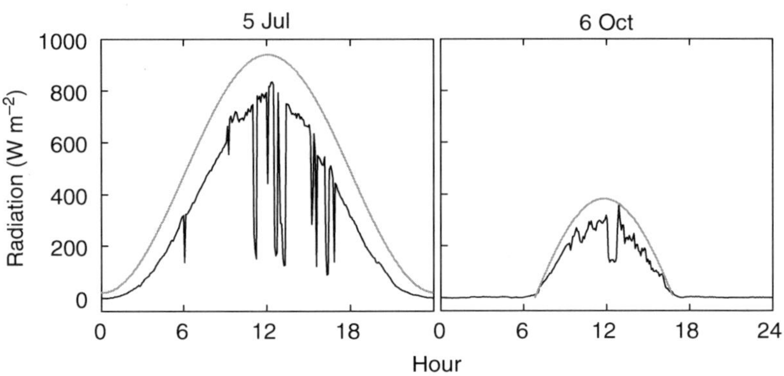

Fig. 3.5 Comparison of the insolation at the top of the atmosphere according to Eq. A1.6 in Appendix 1 (*grey line*) and measured values at SMEAR I (*black line*) on the dates 5 July 2007 and 6 October 2007

Latitudinal variation of solar insolation at the top of the atmosphere is the primary cause for seasonal variation of environmental factors, which is strongest at high latitudes (Fig. 3.4). The energy input from moon to the Earth is negligible. At the North Pole, the solar insolation equals zero between autumn and spring solstice (northern hemisphere). In the polar region in the northern hemisphere (from the Arctic Circle 66.56° northward), the polar night gets longer when moving northwards. In contrast, the northern areas get plenty of light in summer. It is noteworthy that the daily solar energy input for 70°N exceeds the values for 50°N and also for the equator close to and around midsummer.

At boreal latitudes, the global radiation values at the Earth surface in clear sky conditions at noon are usually 700–800 W m^{-2} in summer (Fig. 3.5). For the given clear sky example in Fig. 3.5 on 5 July 2007, the solar radiation perpendicular to the Earth surface at the top of the atmosphere is 940 W m^{-2} at noon, implying

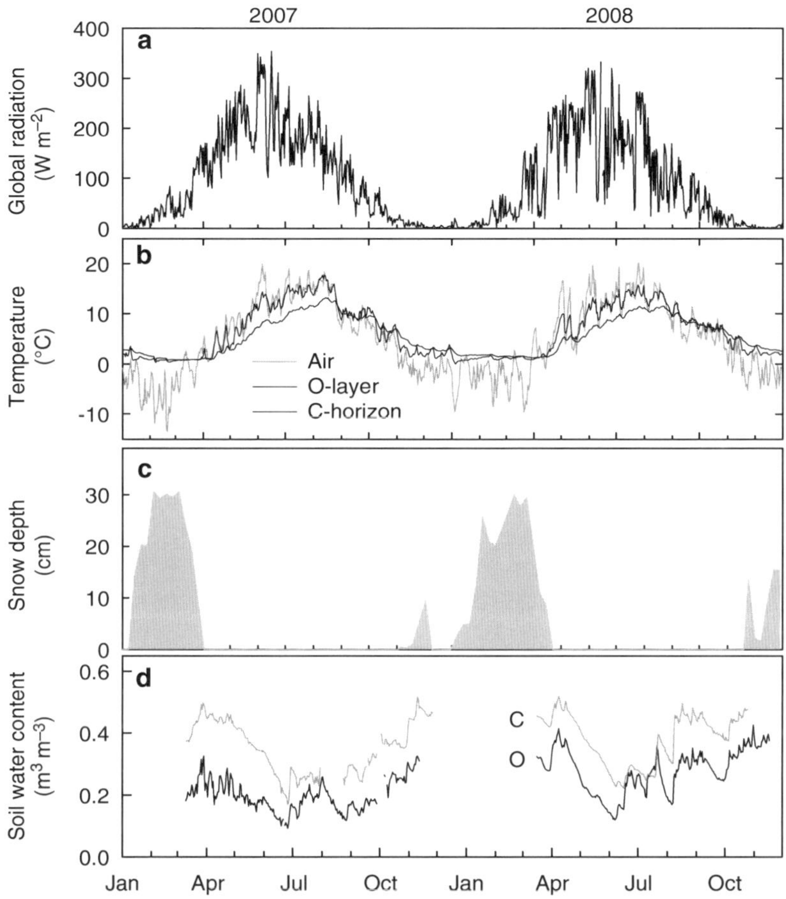

Fig. 3.6 Annual cycle of mean daily radiation (**a**), temperature in air, humus (O) and in soil horizon C (**b**), snow depth (**c**) and soil moisture in humus (O) and in soil horizon C (**d**) measured at SMEAR II in southern Finland during 2007–2008

approximately a 15% reduction as a result of passage through the atmosphere. Note that the global radiation is defined as the measure of solar energy through the plane parallel to the Earth's surface (the 'horizontal plane'), which is strongly affected by the elevation of the Sun. Below cloud cover, the values can be only fractions of the clear sky conditions. The difference between radiation energy above the atmosphere and at the surface level is caused by reflection and absorption in the atmosphere.

The annual cycle in the radiation energy input generates a corresponding annual cycle in temperature. However, there is delay and temperature reaches maxima and minima about 1 month later than radiation due to heat capacity of the Earth–atmosphere system (Fig. 3.6a, b).

In forest ecosystems, soil properties, such as soil fertility, are important but stable environmental factors. However, the annual cycle in atmospheric factors generates also a clear annual cycle in soil. On an annual scale, soil temperature is the most important environmental factor for biological processes occurring below-ground, such as root growth and microbial activity, because the biologically active metabolic processes are highly temperature dependent. In the boreal region, soil temperature has a strong seasonal pattern that follows the air temperature (Fig. 3.6b).

The topsoil in boreal forests usually reaches a daily maximum temperature in July–August. Soil temperature decreases towards deeper soil layers, which reach the maximum a couple of weeks later. In winter, the temperature gradient becomes inverse. The surface soil cools down faster than deeper soil layers along with decreasing air temperature in the autumn. When permanent snow cover has been established, the diurnal variation in soil temperature is very small. The snow cover insulates the soil effectively, and that prevents the soil from freezing in deep soil layers. Frost formation depends on the insulation of snow, soil heat storage and thermal conductivity of the soil.

The depth of the insulating snow cover depends on the canopy structure. In forests with a dense canopy, a substantial part of the snow is retained in the tree canopy and never reaches the soil. Therefore, the frost penetration may be much deeper below a dense canopy compared to open areas. Also the timing of snow cover formation affects soil temperatures during winter. If permanent snow cover is not established before very low air temperatures, the frost penetrates much deeper than during average years. In spring, the rate of snow melt and thawing of the soil in a forest is slower compared to open areas because the canopy reduces the radiative and turbulent energy exchange.

Soil water content has a typical seasonal pattern in boreal forests (Fig. 3.6d). It is highest during and after the snowmelt and decreases towards the end of the summer because evapotranspiration exceeds rainfall during this period. The maximum soil water content in soils depends on total porosity of the soil, but usually it is between 40 and 50% of the total soil volume. Soil water content may decrease during the course of the summer by about 50%, but this is variable depending on the amount of precipitation and evapotranspiration. Soil water content increases to near field capacity (the amount of soil water content held in soil after excess water has drained away) quite quickly during a normal, rainy autumn in September and October.

Soil water conditions are often characterised with water tension (see p. 178) that is generated by the curvature of water film on soil particles. The bigger the soil water tension is, the more tightly water is attached on the soil particles, and therefore, the harder it is to get water for the plant roots. During average years, the reduced soil water does not affect trees, but the decline of volumetric water content down to approximately 10% has been shown to limit transpiration and photosynthesis of the trees (Duursma et al. 2008).

The annual pattern in the CO_2 concentration is clear as can be seen in the Fig. 3.7c: the minimum occurs in late summer and long-lasting maximum in winter and early spring. During summer, the photosynthesis decreases the regional

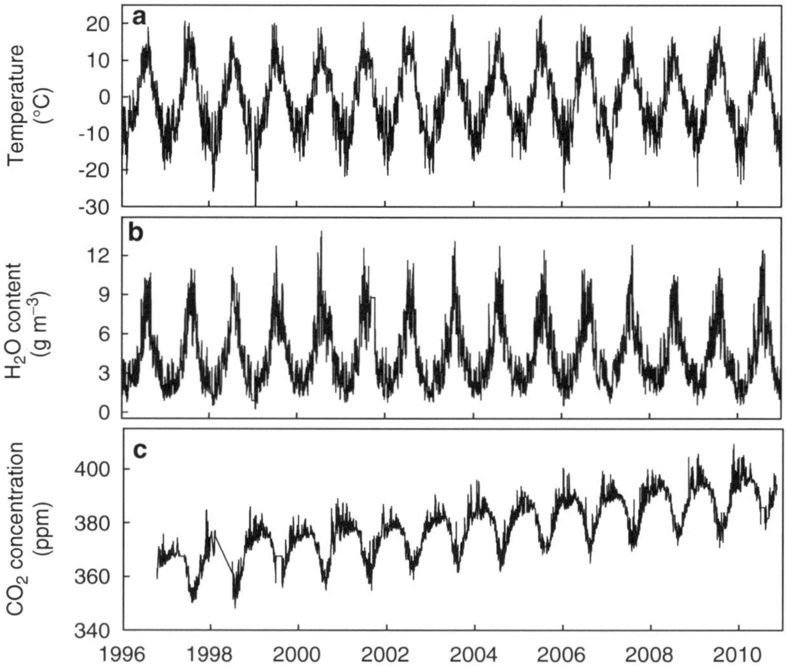

Fig. 3.7 Daily average temperature (**a**), water vapour content (**b**) and CO_2 concentration (**c**) records from northern Finland during 1996–2010. Measurements were recorded in Sammaltunturi (67°58′N, 24°07′E, elevation 565 m above sea level, see Aalto et al. 2002)

atmospheric CO_2 concentration until the respiration processes releasing CO_2 gain dominance in the autumn and CO_2 concentration starts to rise. Note the increasing trend in the atmospheric carbon dioxide concentration over the years.

The annual pattern in the atmospheric water vapour concentration is also very clear, showing minimum in the winter and maximum in the late summer (Fig. 3.7b). This annual pattern is generated by the strong temperature dependency of saturated water vapour pressure/concentration and the annual cycle of global radiation which drives the evapotranspiration. The warm air during summer is able to hold much more water vapour than cold winter air.

3.4 Environmental Factors Inside a Canopy

3.4.1 The Attenuation of Solar Radiation

The absorption of solar radiation by phytoelements (leaves, needles, stems, fruits, etc.) generates a strong spatial variation in the light inside a plant canopy. A single point is in full shade (umbra) if the solar disc is totally covered by canopy elements

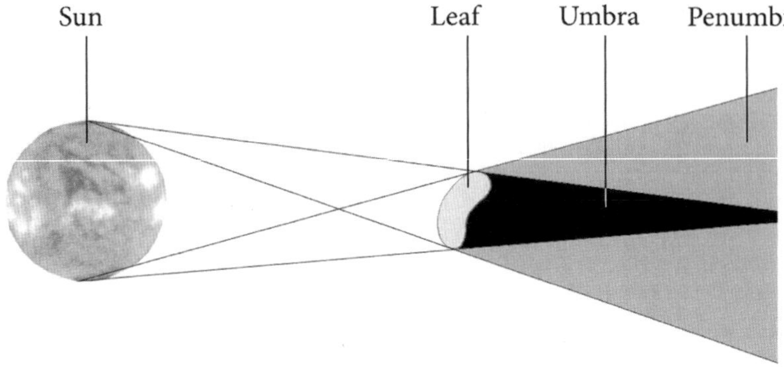

Fig. 3.8 Umbra and penumbra areas generated by a needle

and direct PAR does not illuminate the point. If the Sun is only partially covered, we call it a penumbra area. The situation where the Sun's disc is not shaded and direct sunlight is able to reach the point is called a sunfleck (Fig. 3.8). The distribution of PAR intensity inside a canopy depends on the height inside the canopy: the probability of sunflecks decreases and the probability of umbra increases at lower levels.

In a forest canopy, the Sun and shade patches alternate and light regime varies constantly below the fluttering leaves. Spatial distribution or arrangement of biomass affects the spatial distribution of environmental factors.

The global PAR inside a canopy consists of the following components:

- Direct PAR that penetrates through the canopy without scattering by phytoelements such as leaves and branches
- Diffuse PAR penetrating the canopy without scattering
- Diffuse PAR due to the scattering of direct PAR
- Diffuse PAR due to the scattering of diffuse radiation

Direct PAR has the highest and diffuse PAR the lowest irradiances. The share of direct PAR is highest in the top parts of the canopy whereas diffuse PAR is the most common light component in lower parts of the canopy (Fig. 3.9). The PAR value inside canopy is the sum of these components and varies strongly in time and space. The possibilities to capture this spatial and temporal variation depend on spatial averaging by a set of radiation sensors.

Inside the canopy at certain moment, a single value can characterise irradiance only at infinitesimally small point. At a larger scale, radiation intensities form a diverse distribution (Fig. 3.9). Below the forest canopy, oscillation of branches and movements of leaves cause rapid variation in the light during windy and sunny weather and the light component in a single point changes from sunfleck to penumbra or umbra. The leaves (especially the needles in the case of conifers) in canopies are rather small objects and a small movement is enough to change from

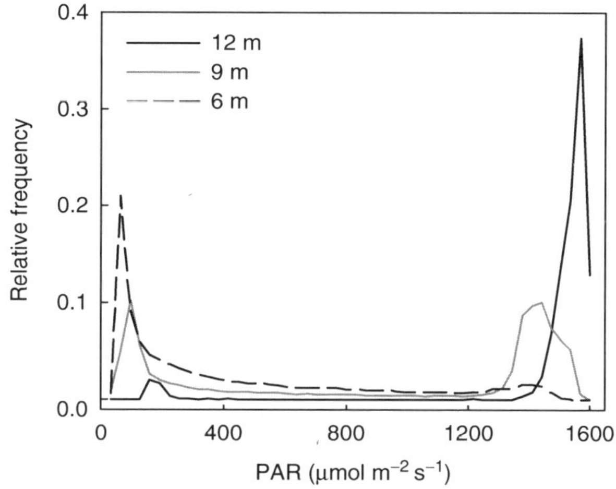

Fig. 3.9 Distribution of solar radiation values at different heights inside a Scots pine (*Pinus sylvestris*) canopy at SMEAR II on a summer day. The height of the canopy was 14 m. The peaks in low PAR intensities represent PAR intensities of full shade (diffuse radiation), whereas the peaks in high intensities are sunflecks (direct radiation). Penumbra is situated in between

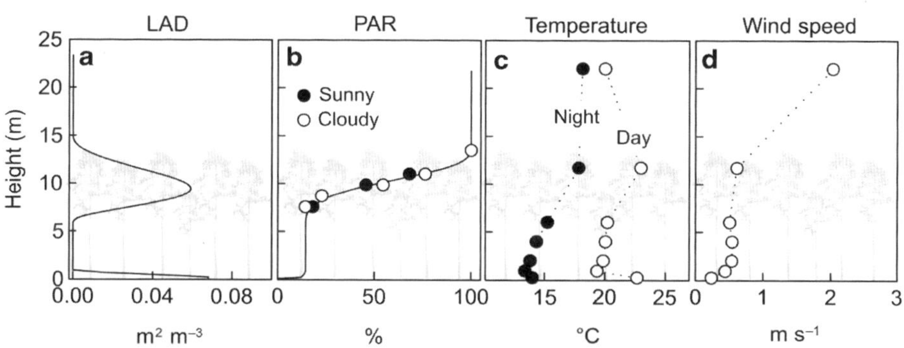

Fig. 3.10 Typical vertical variation of environmental factors in a pine forest. Vertical profiles of leaf area density (LAD), average photosynthetically active radiation (PAR, Vesala et al. 2000), temperature and wind speed (Launiainen et al. 2007) within a 15-m-high Scots pine forest at SMEAR II, southern Finland, during a summer day. The total one-sided leaf area index was 3.5

umbra to sunfleck. Thus, the spatial scale of variation in light intensity is often less than 1 mm. The timescale of the variation is as short as 0.1–1 s. In deciduous stands, the variation is slower due to larger shading elements, but flattering of leaves tends to reduce the variation timescale.

Within a forest, the light regime differs drastically from the conditions above the stand. Solar radiation interferes with canopy elements and is partly reflected and partly absorbed, the latter being a process of radiation transformation into heat and into chemical energy in photosynthesis. These processes affect solar radiation throughout the canopy down to the forest floor, where the global radiation is much reduced compared to above-canopy values (Fig. 3.10b).

3.4.2 Air Temperature and Wind Speed

The absorption of solar radiation, net emission of thermal radiation and evapotranspiration or condensation of water inside the canopy leads to heating or cooling of the canopy elements, giving rise to vertical as well as horizontal temperature differences (Chap. 5). The absorption and heating occurs differentially with height inside the canopy, most of the short-wave radiation being absorbed at the upper part of the canopy. At upper canopy, the needle surface temperature typically exceeds the air temperature by 0.6–1.5° but can become higher as much as 3.5° in calm wind and high radiation. Deeper in the forest, the temperature differences between the canopy elements and air are smaller. For broad-leaved species, the differences between leaf and air may be larger.

The heat exchange between canopy surfaces and the air along with the turbulent mixing of the air tends to reduce the spatial differences in temperature. Figure 3.10c shows temperature profile during a summer day. When temperature at the upper canopy is higher than the air above the canopy, then the air is unstably stratified, and that enhances the turbulent mixing. In night-time, the situation is opposite, stable stratification prevails (potential temperature increases with height) and turbulent mixing is weaker.

The drag forces of the foliage work against the airflow and decrease the wind speed inside the canopy. As a result, the vertical gradient in wind speed is strongest immediately above the canopy and in the upper layer of foliage. In the trunk space, where the foliage density is minimal, only the tree trunks create friction against the flow and the wind speed is rather constant with height. Closest to the ground, the friction performs again and the wind speed decreases rapidly above the understorey layer (Fig. 3.10d). Thus, the vertical gradients of wind speed depend mainly on the canopy structure and the magnitude is driven by the wind speed above the canopy. The relatively calm flow inside a forest is frequently broken by turbulent gusts which efficiently ventilate the canopy airspace and smooth the vertical and horizontal temperature, moisture and CO_2 gradients developed during calm periods.

The motion of air can be separated into average and randomly varying components represented by average horizontal wind and turbulent fluctuations of wind components. The motion of air inside and above canopy is generally turbulent. Heating by solar radiation and unstable thermal stratification generate more turbulence during day, and stable stratification acts to reduce mechanically generated turbulence at night.

The temperature at a single point in air, as measured with a very small sensor (small thermal mass), during the daytime varies a couple of Celsius degrees on the timescale of minutes (Fig. 3.11). This variation is mainly caused by turbulent mixing of air layers with different temperatures. Turbulent mixing occurs at timescales from hundredths of seconds to tens of minutes. The dominant timescale of turbulent temperature fluctuations is less than a minute. Much of the diurnal variation in temperature and concentrations is related to the cycle of solar heating (Fig. 3.5) and the development of the atmospheric boundary layer.

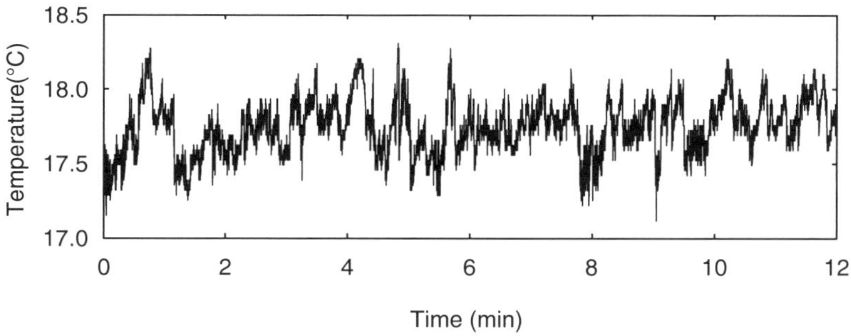

Fig. 3.11 Temperature measured with a fast-response ultrasonic anemometer (data recording rate 10 Hz) above a pine forest at SMEAR II on 4 August 2003, starting at 13:00

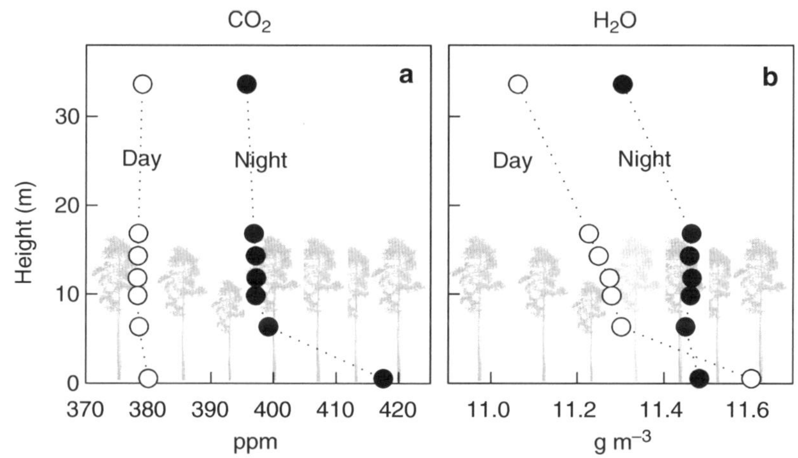

Fig. 3.12 Average CO_2 (**a**) and H_2O (**b**) profiles inside and above the forest at daytime and night-time. Data are from dry-canopy conditions in July 2010 (Launiainen et al. 2012)

3.4.3 *Carbon Dioxide and Water Vapour*

The vertical gradients of CO_2, H_2O (Fig. 3.12) and temperature (Fig. 3.10c) within a forest are strongly affected by biological and physical processes taking place within the canopy. The stems and branches respire CO_2, but the foliage itself is a major sink of CO_2 during the photosynthetically active period in summer daytime. This creates a local minimum in CO_2 concentration within the crown. Close to the ground, the soil efflux exceeds the photosynthetic sink of the ground vegetation leading to slightly higher concentrations. In well-ventilated open canopies such as most boreal pine forests, the turbulent mixing is strong and the forest floor is well coupled to the atmosphere in daytime conditions and strong vertical gradients do not exist within the forest. The concentration gradient of CO_2 is usually much stronger in the night-time when carbon dioxide, mainly released by microbes in the soil, accumulates in the trunk space above the forest floor (Fig. 3.12a).

At daytime, evapotranspiration of water occurs throughout the canopy and H_2O concentration increases monotonically towards the ground. This is because transpiration from the understorey is an important H_2O source and turbulent mixing is weaker than at higher layers of the canopy. At night, the transpiration through the stomata is small because of high air humidity and partial or full stomatal closure that limit transpiration. In night-time, condensation of water can occur resulting in dew formation and a small downward water flux. The water vapour concentration profile inside the forest is relatively uniform at night.

References

Aalto T, Hatakka J, Paatero I, Tuovinen J-P, Aurela M, Laurila T, Holmén K, Trivett N, Viisanen Y (2002) Tropospheric carbon dioxide concentrations at a northern boreal site in Finland: basic variations and source areas. Tellus B54:10–126

Duursma R, Kolari P, Perämäki M, Nikinmaa E, Hari P, Delzon S, Loustau D, Ilvesniemi H, Pumpanen J, Mäkelä A (2008) Predicting the decline in daily maximum transpiration rate of two pine stands during drought based on constant minimum leaf water potential and plant hydraulic conductance. Tree Physiol 28:265–276

Launiainen S, Vesala T, Mölder M, Mammarella I, Smolander S, Rannik Ü, Kolari P, Hari P, Lindroth A, Katul GG (2007) Vertical variability and effect of stability on turbulence characteristics down to the floor of a pine forest. Tellus B59:919–936

Launiainen S, Grönholm T, Vesala T, Katul GG (2012) Partitioning ozone fluxes between canopy and forest floor by measurements and a multi-layer model – importance of stomatal and non-stomatal pathways (manuscript)

Seager R, Battisti DS, Yin J, Gordon N, Naik N, Clement AC, Cane MA (2002) Is the Gulf Stream responsible for Europe's mild winters? Q J Roy Meteor Soc 128:2563–2586

Vesala T, Markkanen T, Palva L, Siivola E, Palmroth S, Hari P (2000) Effect of variations of PAR on CO_2 exchange estimation for Scots pine. Agric For Meteorol 100:337–347

Chapter 4
Processes in Living Structures

Jaana Bäck, Eero Nikinmaa, Liisa Kulmala, Asko Simojoki, Tuomo Kalliokoski, Pertti Hari, Risto Häkkinen, Tapio Linkosalo, Heikki Hänninen, Eija Juurola, Pasi Kolari, Kourosh Kabiri Koupaei, Albert Porcar-Castell, Beñat Olascoaga Gracia, Jussi Heinonsalo, Sari Timonen, Kari Heliövaara, Maarit Raivonen, Johanna Joensuu, Mari Pihlatie, Jukka Pumpanen, Jukka Kurola, Mirja Salkinoja-Salonen, and Mika Kähkönen

Contents

J. Bäck (✉)
Department of Forest Sciences, University of Helsinki, P.O. Box 27, 00014 University of Helsinki, Helsinki, Finland
e-mail: jaana.back@helsinki.fi

P. Hari et al. (eds.), *Physical and Physiological Forest Ecology*,
DOI 10.1007/978-94-007-5603-8_4,

Abstract Cells are the basic functional units in forest ecosystems. Plants have strong cell wall, formed by cellulose and lignin. Cell membrane isolates the cell from its surroundings, starch acts as storage and enzymes enable synthesis of new compounds. Membrane pumps allow penetration of cell membrane and pigments capture of light energy. We call enzymes, membrane pumps and pigments as functional substances. The biochemical regulation system changes the concentrations and activities of the functional substances: In summer, metabolism is very active, but in winter, vegetation is dormant and tolerates low temperatures. The action of the biochemical regulation system generates emergent regularities in the functional substances, called the state of the functional substances. The effect of environmental factors on metabolism is built in the complex chain of enzymes, membrane pumps and pigments, acting in each metabolic task. The process-specific state of functional substances and the environmental factors determine the rate of each metabolic process. Microbes have dominating role in the soil. Together with soil fauna, microbes break down macromolecules with extracellular enzymes to small molecules that can penetrate the microbial cell membrane through membrane pumps. The microbial metabolism utilises the small carbon-rich molecules for the energy needs, growth and synthesis of the extracellular enzymes.

Keywords Annual cycle • Biochemical regulation system • State of functional substances • Photosynthesis • Phenological events • Decomposition of litter • Extracellular enzymes

Introduction

Processes are the engines of the ecosystems. They produce material and energy for the living components to produce rather stable structures at the ecosystem level. In this chapter, we describe most important plant and soil structures and processes within the cover theory of forest ecosystems. In the construction of specific theories, this chapter elaborates the basic ideas (see Chap. 2): The tilted position and orbiting of the Earth generates annual cycle in environmental factors that vary in space and time and determines the conversion of material and energy to biomass through metabolic processes. However, vegetation controls the responses, and therefore, the reaction to environmental factors varies.

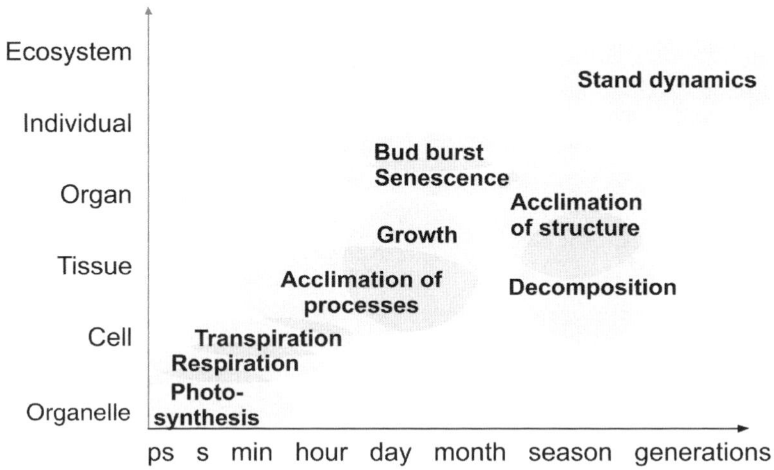

Fig. 4.1 Phenomena in forest ecosystems placed in spatiotemporal hierarchy

The vision, basic concepts and ideas of each specific theory stress the structural hierarchy of ecosystems (Fig. 4.1). The most detailed reasonable level of the hierarchy and theory formation is (1) plant tissue, (2) cell or (3) cell organelle because the anatomy of these entities is highly specialised to perform a specific process or transport. Ecosystems are at the most aggregated level of hierarchy. An ecosystem includes all living organisms and also the nonliving, physical environment such as atmosphere and soil. The different components are connected to each other through the flow of material and energy. The soil has a complicated structure for several living and nonliving processes to occur that are crucial to the functioning of a forested stand.

Biological systems are hierarchical also in time (Fig. 4.1). The temporal level begins from roughly a picosecond that is the timescale for photosynthetic light reactions to occur. At the other extreme, stand development and genetic adaptations occur on a timescale of hundreds and thousands of years in the case of trees. For the theory formation of metabolic processes, the time element has to be small enough for the assumption of stable environmental factors to be feasible.

This chapter first introduces the hierarchy of plant structure, and thereafter, the main plant processes that generate water and carbon flows such as photosynthesis, respiration and bud burst are described. Their seasonal cycles are driven by the biochemical regulation system that reacts also to several irregular changes in the environment resulting in acclimations of the processes where an organism adjusts its morphological and metabolic features in response to the change in its environment. In this chapter, we introduce process acclimation related to a decrease in soil moisture, an increase in CO_2 and a change in light intensity.

4.1 Structure of Vegetation

Jaana Bäck, Eero Nikinmaa, Liisa Kulmala, Asko Simojoki, Tuomo Kalliokoski, and Pertti Hari

4.1.1 *Cellular Level*

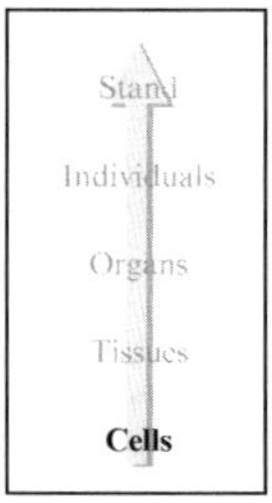

Cells are the basic structural units of all living organisms. They often have tissue-specific features, but within the plant kingdom, their basic structures are rather similar. Cells have their own metabolism, but are not independent in the sense that they need to take raw materials and information from other cells or surrounding environment. In response, they also may produce material to be used in other cells. The inner, subcellular structure of the cell reflects the specific metabolic task of the tissue formed by similar cells.

The general structure of plant cells and tissues in this book has been compiled from Buchanan et al. (2000), Evert (2006) and Taiz and Zeiger (2006). The following structural characterisations apply mostly for boreal coniferous tree species (and mainly the family Pinaceae), although in some cases, specific structural differences compared with deciduous species are also indicated. Specific references are added in places where detailed and/or updated information is given.

All plant cells are surrounded by a more or less rigid, compact *skeleton*, consisting of cell wall and plasma membrane. They allow separation of the cell interior from surrounding intercellular space but also participate in cellular metabolism by selectively allowing passage of molecules outwards and inwards.

Cell Wall

The cell wall (Fig. 4.2) is the strong, outermost layer of each cell in plant tissues, governing cell shape and size but also affecting cell expansion during growth (for a review, see Fry 2001). Walls allow free passage of most small, water-soluble molecules and ions such as oxygen, CO_2, nitrate, phosphate, sugars and amino acids. There are plasmodesmata, that is, small openings in the cell wall and plasma membrane, which form a continuum (symplastic space) between the cells and allow movement of water and solutes, ions and also many large molecules between the

J. Bäck (✉)
Department of Forest Sciences, University of Helsinki, P.O. Box 27, 00014 University of Helsinki, Helsinki, Finland
e-mail: jaana.back@helsinki.fi

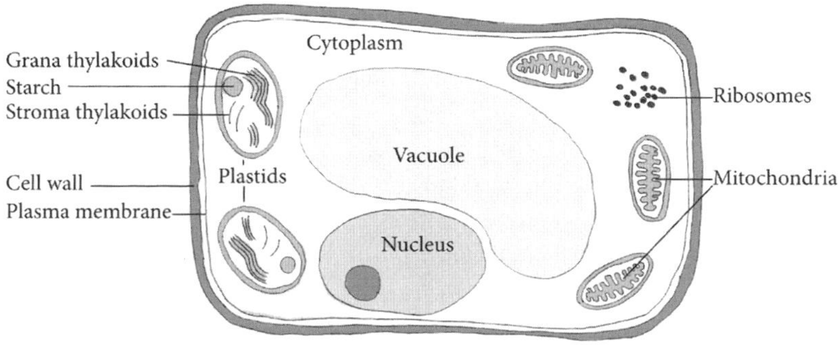

Fig. 4.2 A schematic presentation of the structural components typical of a plant cell

cells. Cell walls differ between tissue types and ontogenetic phase in their chemical composition, thickness, embedded substances and fine structure. Generally, an elastic primary wall is formed during cell division and elongation. Primary walls of plant cells contain typically 70% water, and 90% of the dry matter is polysaccharides such as cellulose, lignin and pectin (e.g. Sjöström 1993; Morrell and Gartner 1998). Secondary cell walls are much thicker and account for most of the carbohydrates in plant biomass. The secondary cell wall is usually formed after the cells have reached their final size and consist of three layers where cellulose microfibrils, lignin and pectin criss-cross and form a basketlike structure with high tensile strength in multiple dimensions.

Cell wall is composed of cellulose, pectins, hemicelluloses, lignin and, in the case of fungi and many insect and crustacean species, chitin. *Cellulose* molecules consist of numerous (from 2,000 to 20,000) D-glucose units arranged parallel to one another. The chains of glucose molecules bond to each other non-covalently with H–OH bonds, which make cellulose microfibrils insoluble in water or organic solvents, and provide the molecule's high tensile strength. Cellulose microfibrils vary in length and width and generally compose more than 30% of the dry weight of the wall. In Scots pine (*Pinus sylvestris*) heartwood, the cellulose concentration varies between 130 and 150 mg cm^{-3} (Harju et al. 2003).

Pectins are cell wall matrix polysaccharides comprising about 30% of the cell wall dry weight and participating in, for example, modulating wall pH and porosity. Pectins in the middle lamella bind the cells strongly together by branching, aggregating with calcium ions and cross-linkages between other cell wall components.

Hemicelluloses are a heterogeneous group of tightly bound polysaccharides, which do not assemble into microfibrils but are covalently linked to lignin molecules. The most abundant hemicelluloses are xyloglucans which can form about 20% of dry weight of the wall. In Scots pine heartwood, the hemicellulose content is ca. 60 mg cm^{-3} (Harju et al. 2003).

Lignin is an organic polymer, most common in the cell walls of supporting and conducting cells of vascular plants. Lignin is an important cell wall component due to its mechanical rigidity and hydrophobicity. Lignin molecules are formed of a complex mixture of several different phenylpropanoid alcohols, and their molecular weight can be as much as 15,000 Da.[1] Lignin fills the spaces between cellulose, hemicellulose and pectin components in cell wall matrix. Deposition of lignins into cell wall implies specific irreversible changes in the cells that ultimately lead to cell death and result in the formation of wood secondary xylem tissues. Lignin partially replaces the water in the wall structure, thus increasing the hydrophobicity of the wall structures. Lignin deposits may account for about 20–35% of the dry weight of wood, that is, 80–100 mg cm^{-3} (Harju et al. 2003).

Membranes

Cells as well as most cellular organelles (e.g. nucleus, chloroplast, mitochondria and plastids) are separated from their environment by a membrane (Fig. 4.2). There are special structures in the membrane that allow penetration of water, ions, sugars, amino acids, etc. The membranes facilitate and actively control the continuous molecular traffic as the cell/organelle takes up nutrients and metabolites, exports wastes and controls its turgor pressure.

Membranes are mainly bilayers, composed of lipids and proteins. They are very dynamic structures, responding to changes in environment with physical and chemical alterations. In the membrane double bilayers, the lipid molecules are organised so that the polar heads in the outward-facing region are hydrophilic, whereas the inner region (towards the centre of the membrane) fatty acids are hydrophobic. These properties are extremely important for the membrane functions, as they prevent random diffusion of solutes between cell compartments.

Molecules penetrate membranes with the help of a variety of intrinsic membrane proteins, which function either as active pumps for hydrophilic substances or as passive channels or carriers. Hydrophobic molecules can passively move through membranes proportional to their lipid solubility.

Membranes are composed mainly of proteins and lipids. Most lipids contain long-chain (16–18 carbons) saturated or unsaturated fatty acids which are esterified to glycerol, a 3-carbon alcohol. The composition of lipids varies between cellular organelles. Plasma membrane lipids are phospholipids, whereas in plastids, the membrane lipids are almost entirely glycosylglycerides. Many pigments, such as chlorophylls, plastoquinone, carotenoids and tocopherols, are actually also lipid-related substances, and they account for about 30% of the lipids in plant leaves. The epidermal cells are covered with an epicuticular layer (composed of a network of cutin and embedded polysaccharides and overlaid by wax crystals) to prevent uncontrolled flux of water from the leaves. Root endodermal cells also have a hydrophobic layer composed of suberin.

---

[1] Da = Dalton, the unified atomic mass unit based on the mass of a ^{12}C atom. 1 kDa = $1.66 \cdot 10^{-24}$ g. Also known as amu.

Membrane pumps are proteins which are essential in the exchange of molecules and ions through the semipermeable membranes separating the compartments from each other and in controlling cell and organelle pH and osmolality. They occur, for example, in mitochondrial membranes, plastid envelopes and plasma membranes.

Water transport through the lipid bilayer membranes occurs via aquaporins. They form a selective channel for water molecules across the plasma membrane or tonoplast, and their expression and functions are controlled in response to plant water availability (Maurel 1997). They have been assumed to form narrow pores of 0.3–0.4 nm in diameter, excluding all larger molecules than water, which would flow molecule by molecule through the pore.

Other Important Cell Organelles

Genetic codes determine all cellular functions and their responses to environmental factors. The *nucleus* (Fig. 4.2) is the organelle housing most of the plant genome, although plastids and mitochondria also contain a small genome of their own. The nucleic acids, DNA (deoxyribonucleic acid) and RNA (ribonucleic acid), are large polymers that store and transmit genetic information in cells. The nucleus is surrounded by a porous nuclear envelope, which is permeable to some macromolecules through the nuclear pore complexes housing an active membrane transporter.

Plastids are cylindrical, lens-shaped organelles which are one of the main characteristic features of plant cells (Fig. 4.2). Their most important function is to house the photosynthetic machinery (chloroplasts). Each mesophyll cell in a mature leaf or needle contains from several tens to several hundreds of chloroplasts. Chloroplasts are responsible for primary carbon assimilation, that is, photosynthesis, and also for many fundamental intermediary metabolic reactions involved, for example, in nitrogen metabolism.

Plastids are separated from their cellular environment by a double layer membrane, a plastid envelope. The outer and inner membranes in the chloroplast envelope differ in their composition and structure and thus also in their functions. Starch grains can periodically make up the majority of the chloroplast volume.

Photosynthetic light capture occurs in chloroplast inner thylakoids, which form a continuous network of membrane sacks, partially stacked in piles on top of each other (lamellae). *Thylakoid membranes* contain the photosynthetic pigments, mainly chlorophylls *a* and *b* and carotenoids. The integral membrane proteins include the photosynthetic reaction centres (PSI and PSII), ATPases and electron transport enzymes, positioned in highly specialised arrangements and unique orientation within the membrane (e.g. Allen and Forsberg 2001; Chitnis 2001).

Mitochondria are abundant in all living cells. They house the essential respiratory machinery which generates ATP by the citric acid cycle and electron transfer chain (see Sect. 4.2.3). Mitochondria are spherical or rod-like organelles which are surrounded by a double membrane and range from 0.5 to 1.0 μm in diameter and up to 3 μm in length. The smooth outer membrane completely surrounds the strongly invaginated inner membrane and the aqueous matrix located inside the double membrane. The membraneous formations of the inner membrane (cristae) contain high amounts of proteins.

Chemical Composition of a Cell

Plant cells have hundreds of different biochemical components, but cells and organelles share many similar features in respect to their macromolecular structure. The majority of these macromolecules are large carbohydrates, which can roughly be classified as structural, functional, regulatory, storage and secondary molecules. Some of them, for example, lipids and sugars, contain only carbon, oxygen and hydrogen, whereas others, for example, proteins, have significant quantities of also other molecules such as nitrogen or sulphur. These components are important as functional groups of molecules.

Functional Substances

Functional substances enable, by definition, processes in living organisms. All metabolic pathways consist of several interlinked processes, involving numerous functional and regulating substances.

Most functional substances are *proteins*, some of which are composed of several chemical entities that are functional only when they are together. Proteins form about one-tenth of the total dry matter in living plant cells, and they can be either soluble or bound to various membraneous structures. Proteins contain most of the plant nitrogen. Proteins function as enzymes, signalling molecules, structural compounds and membrane pumps (see membrane composition). Proteins have molecular masses ranging from 1 to 10^3 kDa, and they may be composed of several subunits. All proteins are formed of 20 different amino acids, which all contain at least one amine group ($-NH_2$) and a carboxylic acid group (–COOH), attached to a carbon chain. Many amino acids also contain sulphur (e.g. cysteine) or additional nitrogen atoms (e.g. arginine, glutamine). Amino acids are organised into long polypeptide chains, which in turn are assembled into protein molecules. The nitrogen content of amino acids is 15–17%.

The vast majority of proteins are *enzymes*, which act as biological catalysts in all metabolic processes that occur in living tissues. Enzymes are highly specific to their substrates and end products, which partially explains why there are so many different enzymes involved even in the simplest metabolic chains. They also have rather specific temperature and pH requirements where they can operate. Their catalytic capacity makes the enzymatically catalysed reactions typically in the order of 10^8–10^{12} times faster than the corresponding uncatalysed reactions. The amounts of enzymes in the cell compartments are determined by the relative rates of their synthesis and degradation, that is, turnover rates of the enzymes. Most enzymatic reactions show an exponential increase in rate with increasing temperature, up to the point where the enzymes are denatured and a rapid and irreversible decline in activity results.

All cellular proteins are also constantly degraded and resynthesised. Natural catabolism by protease enzymes leads to breakdown of proteins into amino acids and other simple molecules, which then can be recycled for build-up of other proteins. Protein catalysis is thus forming a complex system for controlled turnover of proteins and is vital for proper cellular functions in all living cells. Individual proteins turn over at different rates. Their half-lives can vary from a few minutes to several weeks, the fastest growing cells having most rapid turnover rates (Plotkin 2011). Rapid protein turnover ensures that functional substances are able to respond to constantly changing conditions.

Of the total soluble protein in green leaves, ca. 50% is in the chloroplast enzyme ribulose bisphosphate carboxylase/oxygenase (Rubisco), making it the most abundant single protein in plants. Rubisco is unique in that the same enzyme catalyses two important reactions, namely, carboxylation and oxygenation. Rubisco is found in the chloroplasts. Typically its concentration per leaf area is ca. 8 g m^{-2} (Warren et al. 2003).

A plant can control the activity of its *light-absorbing pigments* and thereby photosynthesis. Photosynthetic light absorption occurs in thylakoid membranes of chloroplasts (see photosynthesis in Sect. 4.2.2). Light quanta are absorbed by chlorophylls *a* and *b* and carotenoid molecules in photosynthetic light reactions (e.g. Bendall 2006). Chlorophylls are lipid-like molecules which form a tetrapyrrole ring with a Mg atom in the centre. Carotenoids and xanthophylls are products of the isoprenoid pathway and can be rapidly converted to each other by changes in light intensity (see Sect. 4.2.2.5). They are important in protecting the photosynthetic apparatus from excess light. Photosynthetic reaction centres are integral thylakoid membrane protein complexes, containing both chlorophyll molecules and several electron acceptor molecules acting as converters of light energy into chemical energy in a highly organised manner.

Regulating Substances

The functional substances are under strict control by the biochemical regulation system which provides homeostasis and enables the annual cycle of metabolism and acclimation to prevailing conditions. This control is vital for proper operation of functional substances and achieved by regulating substances, which participate in synthesis, decomposition, activation or deactivation of the functional substances.

The regulating substances are a very heterogeneous group of compounds, including:

- Many metabolites and proteins, taking part in the activation and turnover of enzymes or expression of genes

- Hormones and other signalling molecules, transmitting information between distant plant parts
- Inorganic molecules such as NO_3^-, S, Ca or P

Storage Compounds

The main form of carbon storage in coniferous trees is *starch* (Fig. 4.2), whereas the most important transportable carbohydrate is *sucrose*. The syntheses of starch and sucrose are competing processes depending, for example, on the inorganic phosphorus level in cytoplasm. Sucrose is a disaccharide and the major soluble end product of photosynthesis in green leaves.

Starch synthesis occurs in chloroplasts where it is also stored as large granules. Specialised starch storages serve as energy reserves in rapidly developing tissues such as seeds or buds, or in woody tissues during winter. Starch grains vary in size from less than 1 μm in leaves to larger than 100 μm in seeds and can grow in multiple layers. The amount of starch varies greatly within season and plant organ.

ATP (adenosine triphosphate) is the universal energy carrier which is readily useful for energy to numerous cellular reactions. It is formed in both photosynthetic light reactions and in the respiratory pathways and consumed, for example, in the Calvin-Benson cycle for reducing CO_2 into carbohydrates and in almost all anabolic reaction chains thereafter. ATP is hydrolysed to liberate the energy and yield ADP (adenosine diphosphate) and inorganic phosphate, which then can be recycled. Other forms of chemical energy (e.g. NADH (nicotinamide adenine dinucleotide) or $FADH_2$ (flavin adenine dinucleotide)) need to be converted to ATP before they can be used as energy sources.

All energy-demanding or energy-binding biochemical reactions involve the formation or participation of cofactors NAD^+ (nicotinamide adenine dinucleotide), FAD^+ (flavin adenine dinucleotide) or $NADP^+$ (nicotinamide dinucleotide phosphate). These cofactors are reduced by accepting electrons and one or two protons to produce the corresponding reduced compounds NADH, $FADH_2$ and NADPH.

Secondary Compounds

In addition to the primary carbon metabolites, plant cells include numerous compounds performing specialised functions. Although they commonly are known as secondary metabolites, the group of compounds included in this category is very heterogeneous, and many of their functions are vital for plant survival and growth. One important feature of secondary compounds is that the same compounds can be either constitutive (permanent storage pool in specialised cells) or induced (temporary pool which is produced after activation of metabolic pathways).

Most Important Nutrients

Carbon, oxygen and hydrogen are the most abundant elements in plant tissues. The share of other macronutrients, nitrogen, phosphorus and sulphur, ranges from 0.1 to 4% of plant dry weight, whereas micronutrient concentrations are only millionths of dry weight. Yet their deficiencies can produce growth abnormalities or lead to plant death, and thus, maintaining balanced uptake rates is essential for the plant survival.

From plant dry matter, *nitrogen* (N) forms ca. 0.1–4%. Since leaves contain most of the functional substances that are rich in nitrogen, the N content in leaves can be considerably higher, especially in some broad-leaved trees. In conifer needles, however, the N content is normally 1–1.5% (Helmisaari 1992; Pensa and Sellin 2002). An average of 15–17% of the molecular weight of proteins is nitrogen. A sufficient supply of nitrogen is a major determinant for plant growth because it is needed for functional substances.

Phosphorus (P) concentration in plant tissue is ca. 0.2% (dry weight) involving the main energy storing and transport components ATP, ADP, NAPD and NADPH. Additionally, a majority of biosynthetic pathways, such as starch synthesis and degradation, nitrogen and sulphur assimilation and biosynthesis of cell wall components and isoprenoids, involve phosphorylated intermediates. P has strong chemical bonds to soil clay minerals, and therefore, it is the least available macronutrient from the soil for root uptake. However, the uptake of phosphorus is greatly enhanced by mycorrhizal associations and also by root exudates.

Sulphur (S) plays an important role in many macromolecules. Sulphur controls inter alia protein activity by forming disulfide bridges in molecules and participating in catalytic site formation. Sulphur content in conifer needles ranges between 0.7 and 1.7 mg (g dry weight)$^{-1}$ (e.g. Manninen et al. 1997; Rautio et al. 1998).

Magnesium (Mg) is the central atom in chlorophyll molecules and is required in many enzymes involved in phosphate transfer. *Calcium* (Ca) is a constituent in the middle lamella of cell walls, required as a cofactor in some enzymes and vitally important in signalling of a variety of physical and chemical factors. *Potassium* (K) is required by many enzymes as a cofactor, and it is essential in establishing and maintaining cell turgor. It is the most abundant cation with millimolar concentrations and needed, for example, in photosynthesis, oxidative metabolism and protein synthesis.

Also boron (B), chlorine (Cl), zinc (Zn), manganese (Mn), copper (Cu), iron (Fe), molybdenum (Mo) and nickel (Ni) are essential for normal plant growth and reproduction. Their physiological roles involve participation in, for example, protein synthesis and electron transport reactions.

Isoprenoids are a large group of carbohydrates varying both in size and in chemical structure. The smallest molecule is isoprene (C_5H_8), which is the basic structural component in all larger isoprenoids. Their synthesis occurs in both plastids and cytoplasm, and two partially independent pathways are involved (e.g. Kesselmeier and Staudt 1999; Pichersky and Gershenzon 2002). Isoprene and many other small isoprenoids such as mono- and sesquiterpenes are volatile at ambient temperatures and thus tend to vapourise, producing the characteristic smells associated with vegetation (e.g. pine forests smell of pinene). Many isoprenoids are specifically synthesised for, for example, defence purposes and stored in specialised storage tissues, although not all of their precise functions are fully elucidated to date. Recently, a protective role against high temperatures and reactive oxygen species has also been postulated for many isoprenoids. Plants produce also copious numbers of non-isoprenoid volatile compounds, which are by-products of other major biological processes such as pectin biosynthesis (methanol), fermentation (acetaldehyde) or decarboxylation of carbon compounds in cells (acetone) (Fall 2003).

4.1.2 Tissues

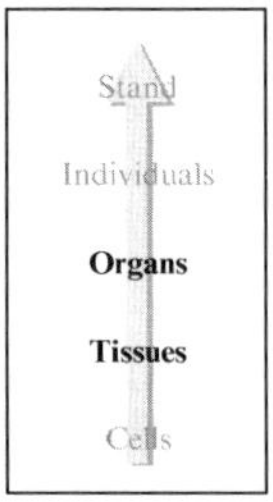

In a tissue, cells act in concert in response to external or internal cues. Tissues consist of functionally similar cells and in plant organs; they can be separated into dermal, parenchyma, meristematic and vascular tissues.

Dermal tissues are the outermost layer of cells facing the atmosphere or soil. They protect the inner tissues from extreme environmental conditions but also selectively exchange compounds between the plant and the environment. Stomata are highly specialised dermal structures, involved in gas-phase diffusion of many substances.

Ground parenchyma in leaves is formed from actively metabolising cells involved in photosynthesis and storage. Most leaves have two or more layers of parenchyma cells called mesophyll cells, rich in plastids involved in light capture and CO_2 assimilation. In roots, cortical parenchyma allows the diffusion of water, nutrients and oxygen from the root hairs inwards and also serves as starch reserve for root metabolism.

The vascular tissues can be divided into xylem (conducting water) and phloem (conducting assimilates). In conifer needles and roots, the vascular tissue is surrounded by an endodermal cell layer, which controls the movement of water, ions and hormones into and out of the vascular system.

Meristematic tissues in buds and vascular cambium are the sites where primary or secondary growth occurs. Meristematic cells are undifferentiated and do not possess secondary cell walls; thus, they can divide rapidly, producing new tissues and organs during the non-dormancy period.

Stomata

Gas-phase diffusion of many substances occurs through stomata, which are highly specialised type of dermal tissues in leaves, stems and other aboveground organs (Fig. 4.3). Stomata actively control the exchange of water and gaseous compounds (CO_2, O_3, VOCs, etc.) between the plant and environment. They are formed by pairs of specialised guard cells, which control the opening of a small stomatal pore. Stomatal apertures typically vary in response to changes in light intensity, saturation deficit of ambient water vapour and soil moisture availability. The stomatal pore is approximately 5 μm wide and 10 μm long when it is fully open.

Opening and closing of stomata are physiologically controlled by accumulation of solutes, especially potassium, in the guard cells. As a result, both the osmotic potential and water potential of the guard cells are lowered, a water potential gradient is created between the guard cells and the neighbouring cells, and water moves into the guard cells. This induces swelling in the guard cells and results in opening of stoma. With a decline in guard cell solute concentrations, water moves out of the guard cells, resulting in closing of stoma.

Stomata are in many leaves located in the lower (abaxial) leaf surface, whereas hypostomatous leaves such as conifer needles have stomata in both surfaces. A leaf can have hundreds to thousands of stomata, and their density is mostly controlled by the environmental conditions during development. Stomatal density is highest in leaves developed at high light, low CO_2 and moist environments. Leaf surfaces and the stomatal antechamber are covered with epicuticular waxes, protecting the surface and providing a barrier for uncontrolled water flow through the leaf surface.

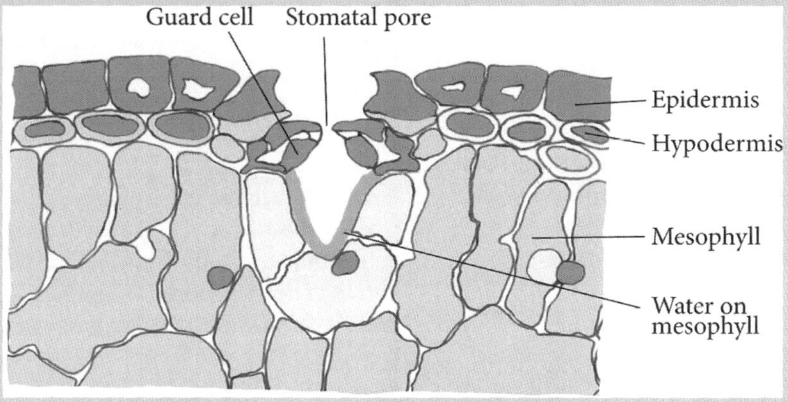

Fig. 4.3 A schematic presentation of the structure of a stoma and the underlying mesophyll tissue

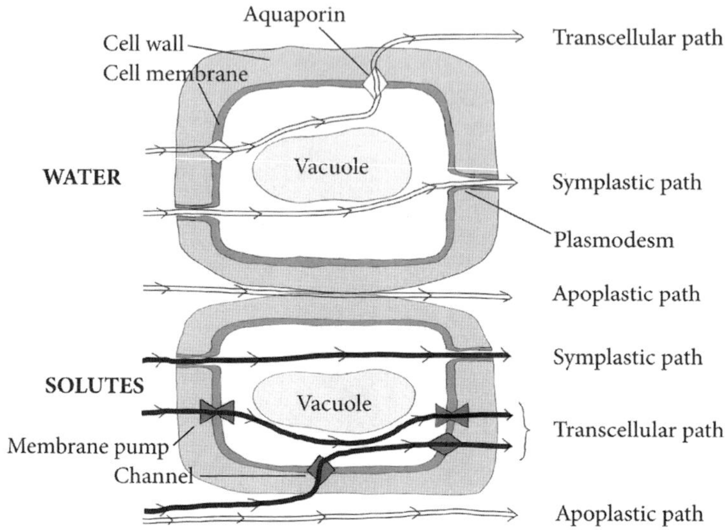

Fig. 4.4 Schematic illustration of assimilate and water transport via symplastic, apoplastic and transcellular pathways. Symplastic flow connects the adjacent cells within a tissue through plasmodesmata, transcellular water flow goes via aquaporins, transcellular solute flow requires membrane pumps or carriers for membrane penetration and the apoplastic flow occurs in the cell walls and intercellular spaces

4.1.2.1 Structures for Transcellular Water and Sugar Transport

Plant tissues contain two parallel pathways for aqueous transport between cells (Fig. 4.4). Water and solutes can move either in the symplast, that is, between the cytoplasms of neighbouring cells, passing via plasmodesmata from one cell to another, or in the apoplast, that is, in the cell walls and extracellular spaces, without passing the plasma membrane (Fig. 4.5). Solutes can move from one cell to another along the osmotic pressure gradient within the symplast through plasmodesmata, which is an opening in the cell wall. Apoplast is formed by the continuum of extracellular spaces and cell walls of adjacent cells, but interrupted by the Casparian bands in roots, endodermis surrounding the vascular tissue in, for example, conifer needles and the cuticle on the outer surface of the plant. Xylem tissue is part of the apoplast, whereas the living phloem cells belong to the symplast, and this has important consequences to the control of solute and water flows.

The apoplast is an important set of channels for the plant's interaction with its environment. Most components (CO_2, O_2, hormones, volatile organic compounds) entering or leaving the cell are passing through the apoplast, where dissolving or evaporation of molecules occurs. Symplast and apoplast are connected via structural membrane proteins, which control the molecular traffic between the transport pathways. The most important membrane proteins are aquaporins, which

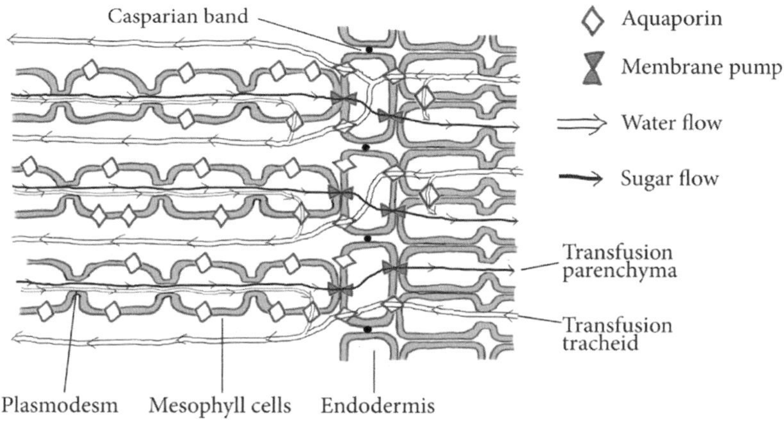

Fig. 4.5 Schematic illustration of assimilate and water transport pathways within and between tissues in conifer needles. Water is following the apoplastic path from xylem to the endodermis. The Casparian band closes the apoplastic pathway, and water molecules have to pass the cell membranes through aquaporins and continue mainly in apoplast into the mesophyll. Assimilates are transported via the symplastic path through plasmodesmata all the way until the phloem sieve element

control the water movement, and membrane pumps and ion channels regulating the selective movement of sugars and ions.

Assimilate and water transport and phloem-loading pathways have been intensively studied for decades, but they still remain enigmatic, and fundamental differences have been reported between plant families in particular in assimilate transport modes (e.g. Rennie and Turgeon 2009; Turgeon 2010; Liesche et al. 2011). Here, we describe the current view of transport in coniferous needle tissues.

Water transport from roots towards leaves is occurring in xylem vessels, which are part of the apoplast. The transpiration stream pulling water and nutrients from soil up to the leaves is driving the apoplastic water flow in xylem and transfusion tracheids. However, the impermeable Casparian bands in radial cell walls of the endodermis block the apoplastic water transport. In order to continue towards mesophyll, water is passing the plasma membrane in endodermal cells via aquaporins and transported to mesophyll apoplast through the symplast of endodermis. After entering mesophyll tissue, water is mainly flowing in the apoplastic space again (Fig. 4.5).

The assimilate transport pathway involves three separate compartments:

1. Mesophyll where sucrose is formed by photosynthesis
2. Phloem companion cells and sieve elements which are the collection compartment
3. Sieve tubes which are the export compartment

The transport pathway between these compartments differs between species, in particular between woody and herbaceous species.

Evidently a symplastic passage of sucrose molecules through the mesophyll cells and endodermis occurs in all species (Fig. 4.5). In herbaceous plants, an active apoplastic loading into sieve element symplast is a general rule, whereas sugar diffusion through symplastic connections prevails in angiosperm trees (Turgeon 2010). In some cases, an energy-demanding polymer formation (sucrose is polymerised to raffinose) may precede the loading of sieve elements. Based on the high number of plasmodesmata in all cell types, including sieve elements and transfusion parenchyma, conifers are also supposed to use a symplastic phloem-loading pathway (Liesche et al. 2011). This implies transport through the plasmodesmata between mesophyll cells and endodermis and all the way to transfusion parenchyma cells, companion cells and sieve elements (see Fig. 4.5). The phloem sieve elements are connected to each other with sieve plates and form sieve tube elements, where sugar transport occurs from source organs towards the different sink organs.

Unloading of sucrose in the meristematic and elongating root tips occurs primarily via symplastic pathway, although in some other sinks, sucrose unloading may involve also partial apoplastic steps (Patrick 1997).

4.1.3 Organs

The traditional characterisation of vascular plant body involves the differentiation of plant parts as organs that are more or less specialised to certain functions. The main organs are stem, leaf and root and flowers. The stem, leaf and flowers form a shoot (above-ground organs) in correspondence to the root system (below-ground organs). Leaves are specialised for photosynthesis, roots for nutrient and water uptake and anchorage, whereas the stem tissues support the above-ground plant parts and form a continuum between the below-ground and above-ground plant parts, acting as pipelines for transport of water and assimilates. Structurally these organs mainly differ in the relative distribution and organisation of the vascular and ground tissues.

Leaves are the organs where light is captured for providing the energy to drive the chemical CO_2 fixation and other vital functions in plant. Leaves also participate in gas exchange between the plant and atmosphere through the stomata. Leaves can vary greatly in shape and size, but a general phenomenon is that they have a flat and rather thin leaf blade (broad-leaved species) or a thicker, cylindrical structure (needle- or scale-like leaves, Fig. 4.6).

The leaves are usually grouped together in a species-specific manner. In many species (e.g. in the genus *Pinus*), the needles form a cluster (fascicle) consisting of two or more needles which can be from 2 to 15 cm in length. Each fascicle is produced from a small bud on a dwarf shoot in the axil of a scale leaf. The fascicles are combined to form a unit, which consists of a group of needles of same age, that is, they all are formed during the same growing season (or during one flush in those

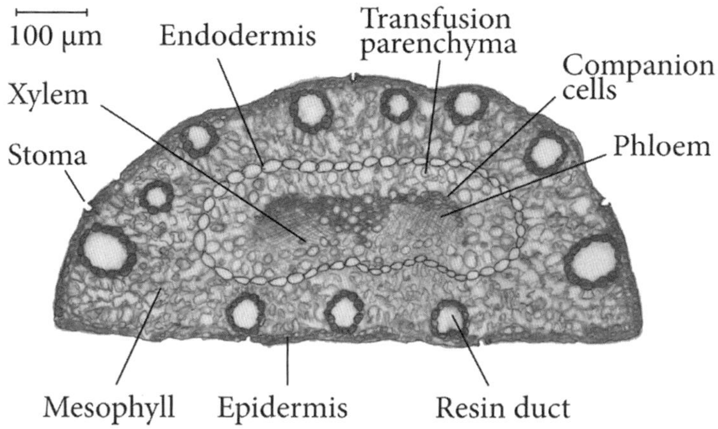

Fig. 4.6 Cross section of a Scots pine needle

Fig. 4.7 Schematic illustration of radial components of a mature coniferous tree stem. Photo: K. Kabiri Koupaei

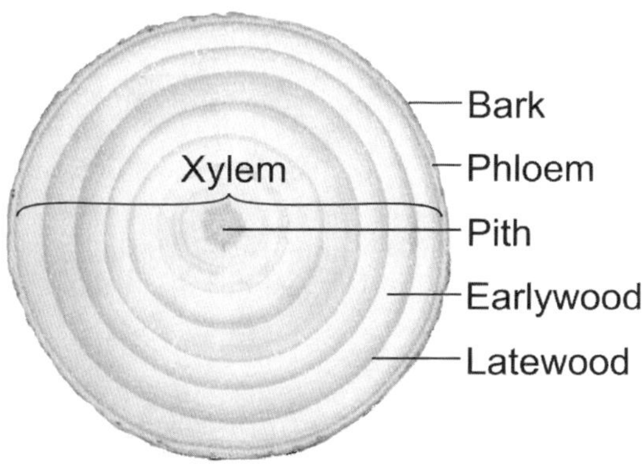

species having several flushes per growing season). Leaves can be shed annually (in deciduous trees) or sustain and maintain their vitality for 2–10 years (even 40 years in some evergreen trees).

The *stem* of a vascular plant is specialised in structural support and conducting functions between the above-ground parts (shoot) and below-ground parts (roots). Therefore, the main proportion of a stem is formed of thick-walled, water-conducting tissue (xylem), participating in both of these functions (Fig. 4.7). Assimilates are transported in the living cells of the phloem tissue, which form a minor proportion of the cross section in a stem. Stem also forms a significant storage compartment for water and reserve carbohydrates during dormancy period.

In conifers and in many hardwoods, the living parenchyma cells in the central part of the stem die and may release secondary substances in the process of changing the chemical composition of the wood. This part of the stem is called heartwood, whereas the water-conducting part of the stem is called sapwood. The xylem cells in wood are arranged in cylinders parallel to the length axis of the stem, and the growth rings form concentric layers with alternating lumen and wall dimensions. The early wood consists of large lumen cells that grow in early summer and late wood has thick cell walls and smaller lumen.

Roots absorb water and nutrients from the soil, produce many functional substances such as growth regulators and can also store sugars and starch. Efficient acquisition of soil resources depends on both the quantity and functioning of the below-ground structures. Both nonwoody and woody roots with their mycorrhizal hyphae participate in water and nutrient uptake.

Most boreal forest trees possess a tap root system, where the primary root grows vertically down into the soil. Later growing lateral or secondary roots and other branches form the branching root system capable of supporting large trunks and acquiring water and nutrients from a large area surrounding the tree trunk.

Roots are divided into both functionally and structurally different coarse (transport) and fine roots. Generally, fine roots (diameter <1 mm) located in the uppermost soil layers are the principal sites for nutrient and water absorption, and they can account for >90% of the whole root length (Finér et al. 1997). Under drought conditions, the tap root system and sinker roots penetrating deeper into soil will be of greater importance for water uptake.

The roots are formed by an apical meristem at root tips. This meristem is protected by a root cap surrounding the root apex. On ageing, the root tissues differentiate into three anatomically different longitudinal zones (Peterson et al. 1999; Hishi 2007):

1. The white zone of active cells nearest to the root apex
2. The brown-coloured condensed tannin zone with dead cortical cells
3. The cork zone characterised by a secondary growth by vascular and cork cambia

The nonwoody white root tips with their mycorrhizal hyphae are the main part of the absorption of nutrients in most species. The elongation zone behind the apical meristem is important as the main site for root exudate liberation to the surrounding soil. In the root hair zone, fragile epidermal structures have developed to facilitate interaction between the root and soil surrounding it. Root hairs greatly increase the absorptive area and contact with soil particles and microorganisms.

In mature root cells of the lateral (woody) root zone, the cellular differentiation has been completed, and this has great effects on the functions of root tissue. Endodermal cells divide the root into the cortex (outside) and stele (inside), which contains the phloem and xylem tissues. The endodermal cell walls contain a thin suberin layer called a Casparian band, which acts as a hydrophobic barrier preventing the uncontrolled apoplastic movement of water and solutes across the

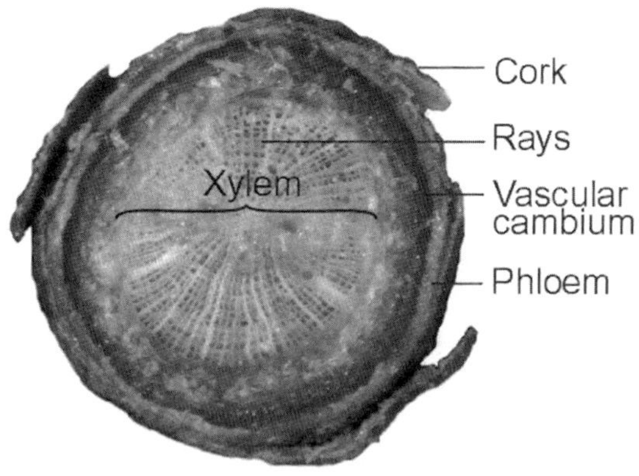

Fig. 4.8 Cross section of a transport root in Norway spruce (*Picea abies*). Photo: K. Kabiri Koupaei

root. The cork layer is produced inside the endodermis by a cell layer called pericycle, which has the capacity to produce lateral roots and cork. Newly formed cork layer is completely impermeable against water and nutrients; cracking, dead lateral roots and lenticells may however allow some movement of water and solutes through older cork zones (Peterson et al. 1999).

Most of the boreal woody species, including Scots pine and Norway spruce (*Picea abies*), are associated with an ectomycorrhizal symbiont, which greatly increases the effective-absorbing surface area of roots and improves water and nutrient (particularly nitrogen and phosphorus) availability for the plant, especially in nonhomogeneous environments such as forests. It is commonly accepted that there is mutual benefit for the mycorrhizal partners, due to the exchange of plant-derived soluble carbohydrates for amino acids and mineral nutrients taken up from soil by the fungal associate (Nehls et al. 2001).

The ectotrophic mycorrhizal hyphae form a thick sheath around the root tips and partially penetrate between the root live cortical cells, forming a netlike structure called a Hartig net. Outside of the roots, the mycorrhizal hyphae (diameter few micrometres) extend to the surrounding soil layers and form large structures inside the soil and fruiting bodies above the soil. Woody plants are probably largely dependent on the symbiotic fungi to supply mineral nutrients, as there is little root surface area capable of nutrient absorption outside the fungal mantle and hyphae (Taylor and Peterson 2005).

Rather thin transport roots connect the fine roots and their mycorrhiza with the tree stem. The structure of transport roots is to some extend similar to that in the stem (Fig. 4.8). However, the annual rings are less clear, and the tracheids are larger. The woody xylem transports water and nutrients upwards and phloem sugars towards fine roots.

4.1.4 Individual

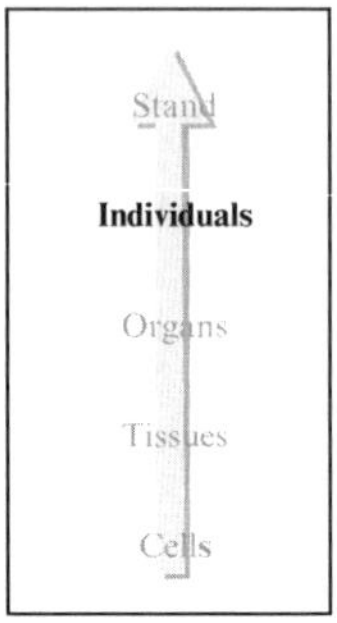

The individual organism is the central level of the hierarchy of biological organisms since the evolutionary selection mechanisms in general act upon it. The success of the individual determines its capacity for producing offspring. A critical component of the success of a single, individual tree is its structure. Structure determines how well trees are able to reach favourable positions in the vegetation in order to capture the resources it needs, how well they are able to withstand high mechanical stress that wind or heavy snow loads can cause and how well they balance their allocation of resources between growth and reproduction over their lifespan.

The central driving force for the structural development of plants is photosynthetic production. Photosynthesis takes place predominantly by leaves using the solar radiation that they intercept. However, photosynthesis also requires a continuous supply of water from other parts of the tree to the leaves, as carbon dioxide intake is associated with loss of water vapour through stomata. The detailed biological machinery responsible for photosynthesis in leaves is built up of resources that originate from soil. Thus, the vascular system that supplies the crown with water and nutrients and that supplying the root system with energy and structural carbohydrates have to be developed in concert.

The particular feature of trees that distinguishes them from other plants is their large crowns and stems. An important aspect of tree structure is the above- and below-ground architecture as it determines the extent over which trees extract resources. Height and size are critical structural features as they determine the proportion that the individual captures from the total flux reaching the site. The architecture of the crown and root system determines the spatial distribution of the exchange sites. The closer the sites are, the more they interact with each other. From the whole-plant point of view, this is unimportant since the resources captured are used by the same organism. However, different spatial arrangements of the exchange sites can affect the amount of resources used to bring about the uptake (e.g. Givnish 1985; Küppers 1989). Closely clumped structures require lower woody growth in branches and stem, and even if they are not able to intercept as much light energy, the saving in structural investments may make such architectural design more efficient than a more widely spread crown.

At crown level, the structure has a central role in determining the intercepted light at a certain point (e.g. Kellomäki and Oker-Blom 1983; Kuuluvainen and Pukkala 1987, 1989; Oker-Blom and Smolander 1988). Closely clumped foliage will have strong internal shading, while widely spread foliage does not interfere as much. Below ground, the heterogeneity of soil imposes specific demands on the root system structure of a tree. Nutrient availability can deviate by an order of magnitude between the locations occupied by different roots of single plant

(Jackson and Caldwell 1993; Gross et al. 1995; Farley and Fitter 1999), and the rate of nutrient mineralisation varies, for example, due to weather (Grime 2002). The branching angles of root branches determine the time at which an emerging lateral root branch escapes from the depletion zone around its parent (Fitter 1991). The scale and precision of the soil foraging determine water and nutrient uptake efficiency of a total root system (Campbell et al. 1991).

Tree, canopy and root structure can also be viewed from the point of view of the developmental processes that brings it about or from the functional significance that the structure has. From the developmental point of view, a tree is simply a collection of elementary units that form larger structures (branches, etc.) over passage of time (Prusinkiewicz and Lindenmayer 1990). Analyses of plant architecture have suggested a number of possible elementary units, such as the metamer and growth unit (Room et al. 1994; Caraglio and Barthélémy 1997). A metamer is defined as an internode with axillary bud(s) and leaf (leaves) in its upper end, but without any shoots resulting from growth of the axillary buds (Caraglio and Barthélémy 1997). A growth unit, initially called a unit of extension by Hallé et al. (1978), is the part of the shoot resulting from uninterrupted extension growth. Room et al. (1994) describe it as 'extension of the contents of a previously dormant apical bud followed by growth of neoformed leaves (if any) and formation of a new, dormant, apical bud'.

The architecture of a root system can be described with the aid of multiple variables such as the branching frequency of the root branches, length of root segments, branching angles with respect to the previous branch, the mortality of apices and axes and topology, that is, how the different parts of the root system are related to each other irrespective of their dimensions (Fitter 1991; Harper et al. 1991; Lynch 1995; Hodge 2004). Below ground, the basic unit could be determined to comprise a coarse woody root segment with fine roots, root hairs and possible nodules (Sievänen et al. 2000). Ectomycorrhiza-forming fungi could also be included in below-ground basic unit since fungal hyphae of ectomycorrhiza-forming fungi form a sheath or mantle around the root, essentially replacing the root hairs. This effectively means that nutrients are taken up primarily via the fungus (Smith and Read 1997).

The crown structure has a dual role in a tree's life. On the one hand, it influences the energy capture for photosynthetic production, and on the other hand, it influences the way in which a tree can exploit space in the future. Tree crown is formed as a result of reiteration of bud formation, their activation and flushing, axis growth and new bud formation. Successful tree development seems to presuppose that only a few specific buds grow and the rest either become dormant or abort (Stafsform 1995). Depending on the competitive environment, different height growth strategies and thus also bud formation and activation are favoured (Givnish 1995). Shoots grown in the upper and lower parts of the tree crown play different roles in crown dynamics. Because young, first-order shoots in the upper crown are unlikely to be immediately shaded, they play an important role in supporting tree structure for many years. They will be incorporated in the structures that will provide material for the active crown above for many years. On the other hand, shoots in the lower crown, even if at first in a strong light environment, become shaded and are less able to support other shoots and are more likely to be structures that a tree can shed.

Similarly also the structure of a tree root system determines the success of soil foraging, and different root growth strategies could be defined depending on the environment. Large root surface area and relative continuous readjustment of the spatial distribution of adsorptive surfaces ensure the rapid establishment and high use of soil borne resources, while small number of root branches and low turnover rate may be more cost-efficient for the exploitation of temporary pulses of higher resource availability in infertile sites. Trees with many short lateral roots may be strong below-ground competitors with the nearest neighbours, while more sparse systems with long lateral root spread may enable plants to search over longer distances and locate pockets where nutrients are abundant or competing roots absent (Casper et al. 2003).

Tree stems and crowns reflect their past developmental history; in that sense, they are unique. However, a closer look at the stems reveals repeating patterns in different species. This is quite natural as tree stems have certain functional requirements that they need to fulfil. Already in his notes, Leonardo da Vinci (MacCurdy 2002) puts forwards the concept that was later called the pipe model theory (Shinozaki et al. 1964a, b). This observation has shown that there is a constant relationship between the sapwood area and foliage area above in many species. Therefore, stem thickness increases from the tip of the tree crown downwards, in a saturating manner, as the foliage area above accumulates. Below the living crown, the sapwood area should not increase anymore according to the pipe model principle, but the whole stem thickness should increase. This is due to accumulation of disused pipes into the stem, in conifers very often into distinctly distinguishable heartwood.

This structural linking between the wood and foliage produces, at least qualitatively, the well-known tapering of tree stems. However, quantitatively the assumptions of pipe model do not exactly hold below the crown or even within the crown. Below the living crown, it has often been observed that the sapwood thickness, rather than area, seems to remain constant (Longeutaud et al. 2006), or even within the crown, there seems to be variation in the relationship (Berninger and Nikinmaa 1994). However, in many coniferous species, such as Scots pine, the relationship seems to be fairly constant when measured at the same reference point (e.g. below crown) in trees grown in very different growing conditions and positions within the same climatic zone.

A number of structural changes occur in the longitudinal direction in conifer stems and also from the core of the stem outwards. In the upper part of the stem, the annual tree rings are wider, with considerably thinner sections of latewood than at the base of the stem. In wood technology, the wood within the crowns of fairly young trees is commonly referred as the juvenile wood. It has also been observed that the average size (length and thickness) of stem tracheids in the conifers becomes larger as the cambial age, that is, the number of years when the cambium was initiated, increases. This observation is in agreement on theoretical derivations of structural scaling of all living creatures where the transport of substances equal to different tissues is fundamental to maintain metabolic activities (West et al. 1997).

Water-conducting cells in roots are usually larger in diameter and longer than in stem, and the size of conduits increases as a function of distance from the tree base, root wood thus becoming increasingly less stem-like (Fayle 1968; Tyree and Zimmermann 2002). Also variable patterns in a change of root conduit size from the core to the outer wood have been observed (Bannan 1941; Peterson et al. 2007; Christensen-Dalsgaard et al. 2008).

Also the mechanical properties of xylem tissue change as a function of cambial age. The elasticity of wood rapidly decreases as the cambium ages until a rather constant value is reached (Mencuccini et al. 1997). The wood dimensions seem to follow also the requirement to withstand mechanical loads. Critical factors are the diameter reference point and its distance from the centre of the loading point. A common feature in many tree species is the formation of so-called reaction wood at places which are under strong mechanical stress. In reaction wood, the cell wall thickness normally increases versus the lumen size in comparison to normal wood.

4.1.5 *Stand*

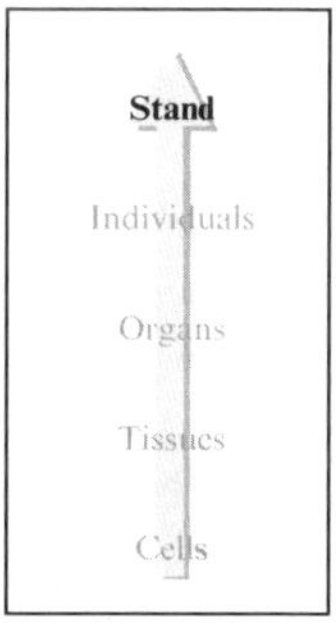

A population consists of all the organisms that belong to the same species and live in the same geographical area. Several coexisting species form a community. Groups of rather even-aged trees growing on similar soil are known as stands, and they can be easily recognised in boreal forests. A stand is usually dominated by one species, although often other species are present in low numbers. During the stand development, the trees grow in height and diameter, but simultaneously the difference between big trees and small trees becomes larger due to the competition. Eventually the smallest trees start to die because of insufficient light for photosynthesis and lack of nutrients or water. This kind of orderly changes in the composition or structure of an ecosystem is called succession.

Tree stands are structurally clearly distinguishable units in the landscape. The structural features that set the stands apart are the species composition and the tree size distribution. The reasons for the spatial differences of these features lie in the edaphic-growing conditions of trees and the disturbance history of the region. A common feature of the boreal landscape is the alternation of lakes, wetlands and uplands that form a mosaic of site-growing conditions. Superimposed on these site-dependent differences is the impact of the disturbance history. Forest fires of different intensity, storms and large-scale insect outbreaks create areas with different number and size of remaining trees of different species. These sites are gradually further stocked with newly born and often fast-growing trees. More protected areas remain uninfluenced by the large-scale disturbances, and there only internal dynamics of gap establishment, growth, site dominance and tree mortality create vertically very heterogeneous stand structures with a small number of very large trees and increasing number of smaller trees. Due to the variation of factors causing the stand properties to vary, also their size variation in the landscape in the natural forests is large.

Most well-stocked stands, in terms of tree stem biomass, are the result of strong and large-scale disturbance that has created fully stocked rather even-aged stands where all the trees have established approximately at the same time. The proportion of these types of stands have increased due to silvicultural practices in the boreal region that try to mimic this type of stand development dynamics so that they are by far the most common stand structure presently, particularly in Fennoscandia (Kuuluvainen 2002).

Within a tree stand, trees interact with each other via shading or via competition for nutrients and water in the soil. Competition for limited resources starts when the domains of resource uptake overlap (Walker et al. 1989). In structurally heterogeneous stands, there are clear differences between the trees both in terms of their height and biomass which quite clearly determine their competitive status. The development of a small understorey sapling into a canopy tree requires several incidents of gap opening (Canham 1985). In the case of even-aged stand development, the trees in the stand develop together, and the differentiation into size classes results from the competition. Simultaneously with the size increase, the number of trees decreases due to self-thinning that can be described with a power relationship between the tree number and their average size (Zeide 1987).

The competition for light drives the differentiation between trees in closed canopies so that small differences in height are quickly reflected on the captured energy and growth that further increase the size differences leading into vicious circle and eventual death of the most suppressed individuals. Particularly light-demanding tree species allocate more of their growth into height extension than canopy width, foliage biomass and stem thickness when shaded (Ilomäki et al. 2003; Vanninen and Mäkelä 2005) that maintains the differences between trees in the stand always. This competition is a continuous but slowing down process during the first 100 years of stand development in boreal forests as the individual size of trees increases. Because of the continued change, there are always a number of different-sized trees present in the canopy layer that are the same age despite their size differences. These size differences have been used in silviculture to divide the canopy into dominance classes and as the basis for different silvicultural treatments.

Leaf area in the canopy changes over the growing season as new leaves and needles emerge, and old ones are senescing and dropped down. Figure 4.9 demonstrates a needle mass distribution within the canopy in a 40-year-old Scots pine in southern Finland. Leaf area index (LAI) is generally defined as the ratio of

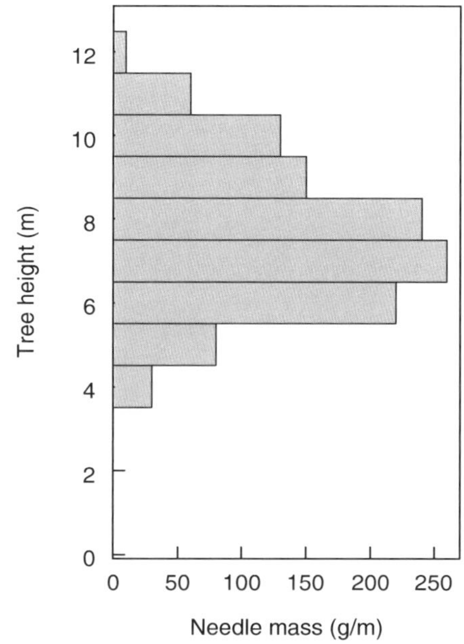

Fig. 4.9 Average needle mass distribution at SMEAR II in a 40-year-old Scots pine (Redrawn from Ilvesniemi and Liu 2001)

total upper leaf surface of vegetation divided by the surface area of the ground. The average LAI in boreal needle-leaved forests is 3.5 (standard deviation 3.3) $m^2\ m^{-2}$ (Scurlock et al. 2001, global leaf area index[2]). Crown architecture is a major determinant for the radiation intensity distribution within the crown, and therefore, leaf area is a very commonly used characteristic to describe the light environment inside and below canopy. As the light response of photosynthesis is saturating, different shading patterns within the crown may bring about different productivity per leaf area. Narrow leaves far apart will have large penumbra area, that is, shading only part of the Sun disc, while wide leaves close to each other will have very patchy light environment with bright and deeply shaded areas (Horn 1971, see Sect. 3.4). In the former case, the light is more evenly distributed within the crown, enhancing the production efficiency in comparison to the latter case.

As a typical example, an even-aged pine stand has grown around the SMEAR II measuring station (Chap. 9). Its age was estimated to be 40 years when it was measured in autumn of 2002. Trees were classified to five size classes according to their diameters. The annual height and radial increments were measured back to the age of 3 years. The height developments in each size class were very similar. The biggest trees at the age of 40 years were also the biggest at the age of 3 years and retained their ratios, respectively, to other size classes (Fig. 4.10a).

[2]Global Leaf Area Index Data from Field Measurements, 1932–2000 http://daac.ornl.gov/VEGETATION/lai_des.html

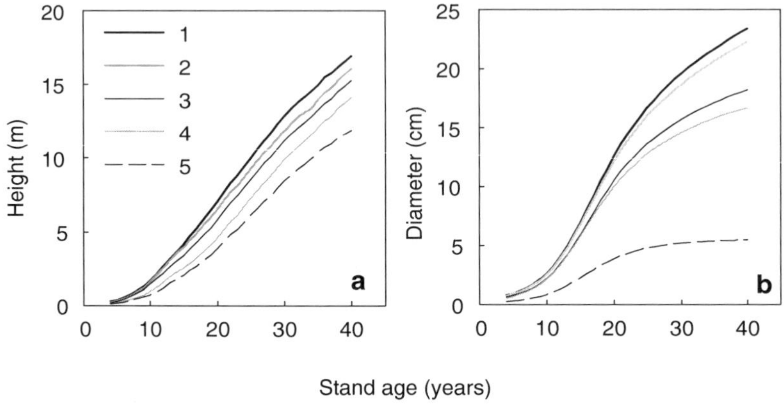

Fig. 4.10 The height (**a**) and diameter (**b**) development in five size classes in the Scots pine stand around SMEAR II

The diameter development was similar to that of height development. However, the differences between size classes were larger (Fig. 4.10b). Read more of the test from Sect. 7.4.

A tree population forms a community with other species in the stand. Ground vegetation typically covers the forest floor level. The species composition and density usually changes during a succession due to interspecific competition and changes in environmental factors generated by surrounding trees. In the early phases of succession, the radiation environment is unshaded, and fast-growing and opportunistic species with rapidly reproducing new tissues dominate. In boreal forests, such opportunist, pioneer species are, for example, *Deschampsia flexuosa*, *Calamagrostis* spp., *Epilobium angustifolium* and *Rubus* spp. After 10–15 years, seedlings start to shade the ground level. Then evergreen and slowly growing species are more competitive than the earlier pioneer species. Such evergreen species are, for example, *Vaccinium* spp., *Calluna vulgaris*, *Ledum palustre* and mosses (e.g. *Pleurozium schreberi*, *Polytrichum* spp., *Dicranum* spp.). After the canopy closure, composition of ground vegetation is still characterised by temporal variation. Small disturbances like thinning, windfall or insect damage generate living space and locally increase radiation intensities in the environment at ground level and alter the dominating species.

The species composition and density of ground vegetation are spatially very heterogeneous in all phases of succession. Below a homogeneous tree canopy, ground vegetation type can vary and consist of small areas of diverse types of vegetation depending on small-scaled differences in water-holding capacity and radiation environment. Mosses and lichens grow on stones; unlike vascular plants, they do not require soil as a substrate. Small differences in soil characters, for example, affect ground vegetation more notably than trees which are usually dominated by only one tree species and therefore have no interspecific competition. In addition, their coarse root system is more extensive than that of ground vegetation.

The plant leaves together with fine roots form the surface between the ecosystem and the atmosphere. The material and energy fluxes conveying the interactions take place through this surface.

4.2 Vegetation Processes

Cell walls form a stable and inert compartment in which the metabolic action, that is, plant processes such as photosynthesis and respiration, takes place. Each process has its specific biologically determined functional substances, that is, enzymes, membrane pumps and pigments under the control of the biochemical regulation system, and they act in a very coherent way.

Large organic molecules form the plant structure, and metabolism synthesises these macromolecules. Plants convert light energy into chemical energy during photosynthesis (Sect. 4.2.2), and the resulting sugars can be converted to insoluble forms (e.g. starch) and stored. The required inflow of carbon dioxide (CO_2) into a leaf occurs via stomatal pores. The release of the stored and captured energy is called respiration (Sect. 4.2.3). Transpiration (Sect. 4.2.2.2) is the loss of water vapour via stomata. Photosynthesis, respiration and transpiration are strongly connected with each other since respiration produces CO_2 that can be utilised in photosynthesis, whereas diffusion of CO_2 from surrounding air into a stoma is impossible without outflow of water vapour, that is, transpiration. In this chapter, we consider these interlinked processes separately and refer to the other processes when needed. Other introduced key processes in the flows of material and energy are senescence (Sect. 4.2.4), nutrient and water uptake (Sect. 4.2.5), bud burst (Sect. 4.2.6), growth (Sect. 4.2.7) and NO_x exchange of needles (Sect. 4.2.8).

The response of a process to the prevailing environment (Fig. 4.11) and the amount of functional substances changes slowly, often on a timescale of days. The two timescales, instantaneous and delayed, open versatile possibilities for research. We can assume that the amount of functional substances is stable in a plant if we generate rapid changes in the environment and study the instantaneous responses. The same separation can also be done when analysing field data. This opportunity is widely utilised later in this chapter.

Cold, dark winters and rather warm irradiation-rich summers are characteristic for the climate of northern forests (Chap. 3). Dormant plants can tolerate very low temperatures, often as low as below –40°C or –50°C. Biochemical regulation systems have developed during evolution to manage the very regular alternation of favourable and unfavourable periods during a year (see basic idea 6, Chap. 2).

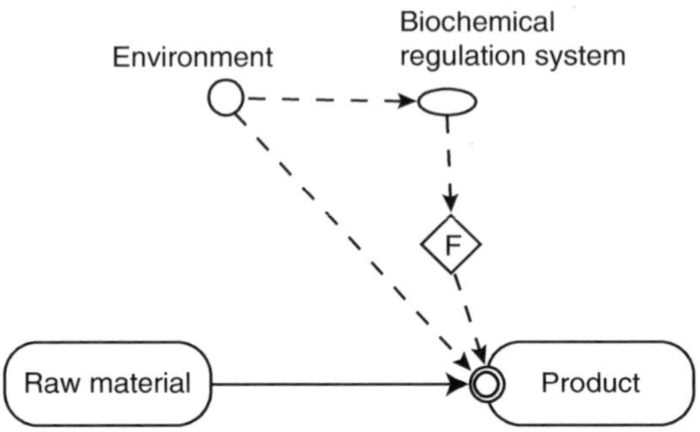

Fig. 4.11 The effect of environment and of functional substances (*F*) on process during a prolonged period when the action of the biochemical regulation system changes the functional substances of the process and the change is reflected into the response of the process to the environment (The symbols are introduced in Fig. 2.3)

The plants have to be active during the favourable period, and they have to tolerate hard winter conditions. Proper timing of active and inactive periods has been crucial for the survival and success of plats (see basic idea 7, Chap. 2). Figure 4.12 shows that in early spring, photosynthesis can be almost totally blocked. The daytime CO_2 exchange of a Scots pine shoot in northern Finland is near zero or even negative during a cool cloudy day in April. Evident photosynthesis can be seen later in May in similar conditions. This chapter introduces also the annual cycles of plant processes generated by the action of the biochemical regulation system.

The biochemical regulation system reacts also to several irregular changes in the environment, resulting in acclimations of the processes. Photosynthesis and its stomatal control change during drought (Sect. 4.2.2.4). Energy utilisation in light reactions is reduced at high irradiance (Sect. 4.2.2.6), and the increase of atmospheric CO_2 concentration changes light and carbon reactions of photosynthesis (Sect. 4.2.2.5). Acclimations are common, and they evidently strongly contribute to the success of plants because they have been thoroughly tested during evolution (see basic idea 8, Chap. 2).

A complex web of regulation substances forms the biochemical regulation system. The genetic code determines the structure of the web. The regulation system makes operational the information gained during evolution. For example, the regulation system controls the functional substances involved into the annual cycle and acclimations.

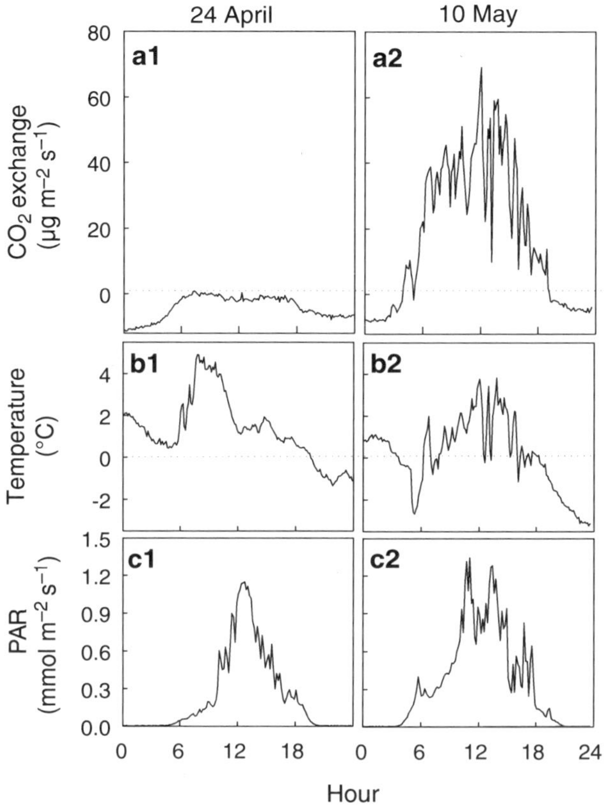

Fig. 4.12 Measured CO_2 exchange of a Scots pine shoot (**a**) and air temperature (**b**) and photosynthetically active radiation, PAR, (**c**) at SMEAR I in northern Finland on 24 April and 10 May 2001. Photosynthesis is almost missing in April, despite the high photosynthetically active radiation: Photosynthetic CO_2 uptake is lower than release of CO_2 by respiration resulting in negative daytime CO_2 exchange. At similar air temperature in May, there is strong CO_2 uptake by the shoot

4.2.1 Annual Cycle of Processes

Pertti Hari, Risto Häkkinen, Tapio Linkosalo, and Heikki Hänninen

Strong seasonal variation characterises the climate on the globe, especially at high latitudes (Chap. 3). In the most continental parts of the boreal zone, the minimum

---

P. Hari (✉)
Department of Forest Sciences, University of Helsinki, P.O. Box 27, 00014 University of Helsinki, Helsinki, Finland
e-mail: pertti.hari@helsinki.fi

air temperatures may drop below −70°C during winter, whereas the summertime maxima are often above +30°C. In these extreme conditions, the annual amplitude of air temperature thus exceeds 100°C. In less continental areas, the amplitude is smaller, but still well above 50°C. Unlike ground vegetation overwintering below a sheltering snow cover, trees experience – with the exception of roots – all these temperature extremes (Sakai and Larcher 1987). Trees are adapted to this extreme climatic seasonality with their biochemical regulation system of the annual cycle so that the alternation of the frost hardy dormant state and the susceptible growth state are synchronised with the annual climatic cycle (Sarvas 1972, 1974; Fuchigami et al. 1982; Koski and Sievänen 1985; Hänninen and Kramer 2007).

The biochemical regulation system of the annual cycle takes its signal from the annual cycle of environmental factors. Subsequently the amounts and activities of several functional substances, such as enzymes, are regulated in such a way that the timing of the growth and dormancy of the trees is synchronised with the climatic cycle prevailing at the growing site (Fig. 4.11). This requires a proper orchestration of the functional substances involved. The functional substances in turn affect the synthesis of several other macromolecules, such as proteins, cellulose and lignin. Finally, all of these complicated physiological and morphological phenomena taking place at molecular, cellular and tissue levels are manifested in the annual cycle of phenological events which is readily visible to the naked eye at the whole tree level. This phenological cycle consists of such events as bud burst and onset of elongation in the spring and growth cessation and bud set during early autumn (Sarvas 1972, 1974; Fuchigami et al. 1982; Koski and Sievänen 1985; Hänninen and Kramer 2007).

The activation of photosynthesis of evergreen coniferous trees is initiated during early spring, mainly as a result of rising air temperature. During this time of the year, a lot of incoming solar radiation is available, but its utilisation is impaired (Pelkonen 1980; Pelkonen and Hari 1980; Hari and Mäkelä 2003; Mäkelä et al. 2004). Impaired water uptake by frozen soil may also slow down photosynthesis. The photosynthetic ability is restored due to prolonged exposure to relatively high air temperatures. A simultaneous gradual dehardening, that is, loss of frost hardiness, of the needles takes place (Repo et al. 2006). These metabolic changes are to some extent reversible so that the needles may reharden and their photosynthetic ability may decrease during a cold spell occurring after the dehardening and activation of the needles during earlier periods of relatively high air temperatures (fluctuating development, Hänninen and Kramer 2007). Later during spring, formation of new tissues starts, that is, cambial and elongation growth are initiated (fixed sequence development, Hänninen and Kramer 2007).

Early summer is the season of high metabolic activity in trees. At this time, resource acquisition and use for tissue formation take place at a high rate. Towards the latter part of the summer, the metabolic rates are slowed down. Generally, the formation of new tissues stops earlier than the processes of resource acquisition, that is, photosynthesis and uptake of nutrients. For instance, in central Finnish conditions, the elongation growth of mature Scots pine trees stops in late June to early July, and the growth of needles and cambium in late July to early August

(Kanninen et al. 1982), whereas photosynthesis continues until late autumn (Hari and Mäkelä 2003; Mäkelä et al. 2004; Kolari et al. 2007). Thus, the latter part of the growing season is characterised by resource acquisition and collecting carbohydrate reserves for the next growing season. During this time of the year, also the overwintering buds are formed. The environmental conditions of the growing season affect bud formation. In the case of species with the so-called fixed growth habit, the number of needle primordia in the bud, and thus the number of needles per shoot during the next summer, is determined on the basis of the environmental conditions prevailing during the summer when the bud is formed (Garrett and Zahner 1973).

Intracellular ice formation is always lethal to plant cells, and the trees are no exception to this rule. Frost survival of boreal trees is based on two mechanisms: freezing avoidance and freezing tolerance (Sakai and Larcher 1987). *Freezing avoidance* is based on the phenomenon of supercooling where the temperature of the cell sap drops below the freezing point without formation of any ice crystals. With supercooling, the tree tissues can tolerate slight short-term frosts even during the active growing season. Depending on the species and tissue, supercooling is an important survival mechanism also during overwintering, but it is not alone sufficient. Rather, during winter, the mechanism of *freezing tolerance* is also required. Freezing tolerance has two components: firstly, tolerance to extracellular ice formation and secondly, tolerance to a drastic dehydration of the cell sap. As active growth processes are not possible in severely dehydrated tissues, the freezing tolerance is accompanied by a deep dormancy of the meristematic tissues (Sakai and Larcher 1987).

During late summer, approximately at the time of cessation of tissue formation, the process of frost hardening is initiated, that is, the two components of freezing tolerance start to develop in the tree tissues. Several functional substances, such as dehydrins, are involved in this process (Sutinen et al. 2001). Sugars lower the freezing temperature, and transformation of starch into soluble sugars is an essential part of the frost-hardening process. For this reason, the collection of carbohydrate reserves during the later part of the growing season is essential also with respect to the frost survival of the trees. When air temperature drops sufficiently below zero, ice crystals start to be formed in the extracellular spaces. This leads to a flow of water molecules from the cells to the extracellular spaces. In this way, the size of the ice crystals in the extracellular spaces is increased, and the cell sap is gradually dehydrated. As a result of the dehydration, the freezing point of the cell sap is lowered so that no intracellular freezing takes place even during the lowest temperatures prevailing during winter (Sutinen et al. 2001).

According to the classical theory, the frost hardening takes place in two phases so that increase of night length is essential during the first and declining air temperatures during the second phase (Weiser 1970). According to a contrasting theory, the effect of these two is additive during the entire hardening phase (Chen and Li 1978; Greer 1983; Leinonen 1996). During spring, dehardening of the tree tissues is driven by rising air temperatures (Repo and Pelkonen 1986). During the dehardening, the cells are gradually rehydrated, so the process of dehardening is closely related to recovery of the metabolic activity of the tissues. Thus, in general, the development taking place during spring is opposite to that taking place during autumn. Repo et al. (2006) found, however, that the relationship between frost hardiness and photosynthetic capacity is not identical during autumn and spring. This finding suggests that partially different metabolic mechanisms with changes in their, different functional substances take place during these two phases of the annual cycle.

Overall, the biochemical regulation of various aspects of the annual cycle of metabolism takes place as an interaction of genetic and environmental factors. The production of functional substances is controlled by genes whose functioning responds to environmental factors. During the course of adaptation, different genetic properties have been selected under different climatic conditions. For this reason, there is much genetic variation among tree species and provenances in their response to the environmental factors controlling the annual cycle (Aitken and Hannerz 2001; Clapham et al. 2001). In other words, the amount and activity of functional substances respond to environmental factors in a different way in different provenances.

Each process in the vegetation has its own specific web of functional substances, namely, enzymes, membrane pumps and pigments. The alternation of active and passive metabolic periods reflects changes in the functional substances, they are missing or in an inactive state during passive periods and they are abundant during active spells. This alternation of passive and active functional substances is a very demanding task for the biochemical regulation system. The short lifetime of the functional substances is an additional complication in the maintaining of proper concentrations.

The biochemical regulation system synthesises, decomposes, activates and deactivates the functional substances and synchronises the metabolism with the annual cycle of light and temperature. The genetic code controls the action of the biochemical regulation system and carries information on the alternation of the favourable and critical periods on the site.

Long chains of steps form each metabolic process, and each step is based on a specific enzyme, membrane pump or pigment. The proper ratios in the concentrations of the functional substances in a chain of reactions are important for the smooth running of the process. The biochemical regulation system is able to maintain the needed balance between the steps in the metabolic process. The structure of the reaction chain determines the dependence of process rates on environmental factors. However, the process rates depend also on the activity of the functional substances involved. The action of the biochemical regulation system changes the activities of the functional substances in such a way that the reactions in the chain are balanced with each other. This generates an emergent property in the functional substances called *efficiency of the functional substances*.

The action of the biochemical regulation system changes the amounts and activities of the functional substances during the annual cycle. This generates emergent regularities in the concentrations of active functional substances. We call these emergent regularities in the functional substances as the annual state of the functional substances of the specified process.

The structure of the reaction chain generates the dependence of the process rate on environmental factors, and the amounts and activities of the functional substances determine the amplitude of the process rate. The biochemical regulation system changes the amounts and activities of the functional substances, and consequently, the annual state of the functional substances responses to the annual cycle in environment. The efficiency of the functional substances reflects the annual state of the functional substances of the process, and it changes along the annual cycle.

4.2.2 Photosynthesis and Transpiration

4.2.2.1 Photosynthesis

Pertti Hari, Eija Juurola, Pasi Kolari, and Jaana Bäck

The Role of Photosynthesis in Forest Ecosystems

Photosynthesis is a fundamental process in plants providing raw material for growth and storing energy in chemical form to be used later in energy-consuming processes such as biosynthesis and transport. In photosynthesis, radiation energy is converted into chemical form as sugars using atmospheric CO_2 and water originating from soil. Sugars formed in photosynthesis are the dominant source of raw material for synthesis of macromolecules, as well as the source of energy for metabolism, obtained via respiration. Thus, understanding the effect of environment on the photosynthetic process is the key to link climate with the growth and development of ecosystems as well as to understand the effects of climate change on forests (Sect. 8.3).

CO_2 is taken up by chloroplasts in the light, thus creating a low concentration of CO_2 in the leaf. This creates a concentration gradient between the outside and inside of the leaf, and so CO_2 diffuses from the atmosphere to the photosynthesising tissues through the open stoma. The gas molecules are drawn towards the chloroplasts in intercellular air spaces due to simple physical diffusion from higher concentration to a lower one (see Appendix 4: "Transport Phenomena in Ecosystems"). After dissolving in the water film that covers the mesophyll cells, diffusion of CO_2 occurs in water. Eventually, the dissolved CO_2 passes through the chloroplast membrane and is bound to RuBP in chloroplast via the action of a very important functional substance, Rubisco (see Sect. 4.2.1). The route of CO_2 from surrounding air to mesophyll creates opening for water emission, transpiration, from intercellular space to atmosphere, and therefore, the transpiration is linked to the photosynthesis (Fig. 4.13).

P. Hari (✉)
Department of Forest Sciences, University of Helsinki, P.O. Box 27,
00014 University of Helsinki, Helsinki, Finland
e-mail: pertti.hari@helsinki.fi

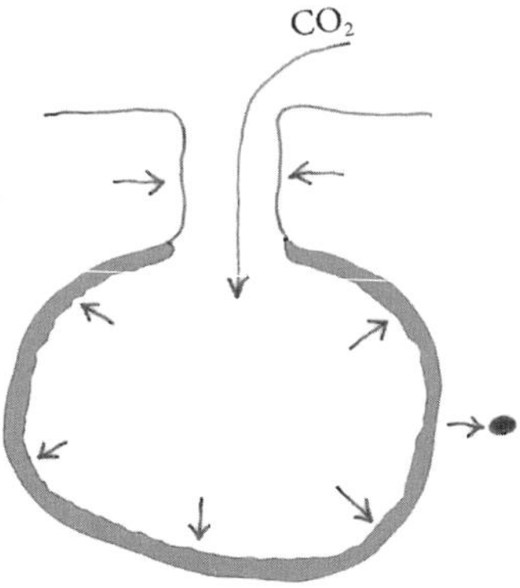

Fig. 4.13 CO_2 molecules flow from air into leaf mesophyll through a stoma. The *black dot* is chloroplast and grey line on the mesophyll surface indicates water

Stomata (see Fig. 4.3) are highly specialised dermal tissues in leaves, controlling actively the exchange of gaseous compounds (CO_2, H_2O, O_3, VOCs, etc.) between the plant and surrounding air. They are formed by pairs of specialised guard cells which make a port of entry. These cells perceive environmental signals and control the opening of a small stomatal pore. The stomatal pore is approximately 5-μm wide when it is fully open. Stomatal density can acclimate to the prevailing environment (e.g. Woodward 1987; Woodward and Kelly 1995; Brownlee 2001), and it varies a lot between species, and in many species, it also depends on the surface of the leaf. The Scots pine needle stomatal density lies between 40 and 140 stoma mm^{-2} (Turunen and Huttunen 1996; Lin et al. 2001; Luomala et al. 2005).

Synthesis of Sugars

There are two main reaction complexes in photosynthesis. The light reactions occurring in thylakoids of chloroplasts convert solar radiation energy into chemical form as ATP and NADPH. In addition, water is split, and oxygen is released. Carbon reactions (Calvin-Benson cycle) in the chloroplast stroma join one carbon atom to a five-carbon sugar utilising energy released from ATP and NADPH. They were traditionally called *dark reactions* to distinguish them from the photosynthetic light capture, although the reactions actually are occurring in light and are directly affected by light.

Figure 4.14 illustrates the hierarchy of structure in higher plant leaves. Gases diffuse through open stomata to the mesophyll (CO_2) or to surrounding air (H_2O) (Fig. 4.14a). Mesophyll tissue is composed of irregularly shaped photosynthesising cells (Fig. 4.14b) forming a porous matrix where diffusion of gases and light absorption is possible throughout the entire tissue. Each mesophyll cell contains several tens to several hundreds of chloroplasts arranged in a manner maximising their light and CO_2 availability, that is, generally lining the cell walls as a single-plastid layer (Fig. 4.14c).

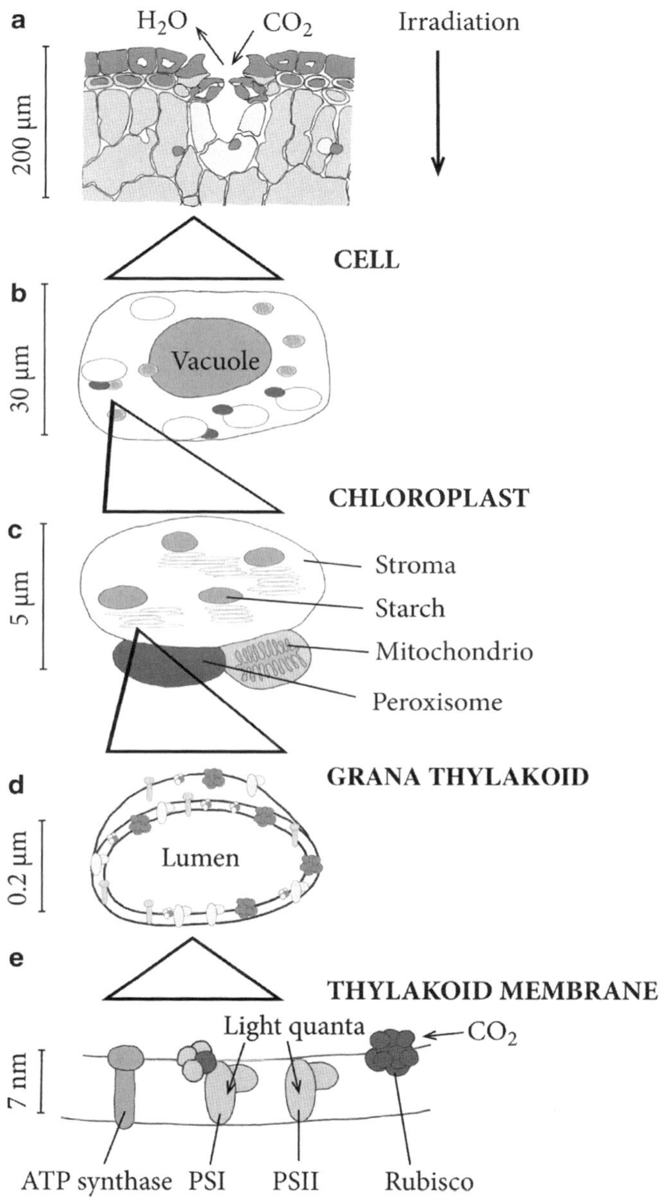

Fig. 4.14 Structural hierarchy in photosynthesising tissues. Cross section of photosynthesising tissue in a conifer needle (mesophyll) (**a**), a photosynthesising mesophyll cell (**b**), chloroplast (**c**), grana thylakoid and the intrathylakoid space (lumen) (**d**) and thylakoid membrane where the photosynthetic machinery, that is, PSI, PSII, ATP synthase and Rubisco molecules, is embedded (**e**)

Chloroplasts are small, ellipsoidal organelles, separated from the cell interior by an envelope membrane. The chloroplast matrix (stroma) is rich in proteins and houses the machinery for binding carbon into large organic molecules, as well as the reserve carbohydrates stored as starch grains. The inner, complex membranes are forming highly organised thylakoid structures, grana stacks (Fig. 4.14d), where light capturing and conversion into chemical energy occurs.

Thylakoid membranes contain the photosynthetic pigments, mainly chlorophylls a and b and carotenoids. Thylakoids are the site of light absorption by the action of chlorophyll pigments. The majority of pigments are situated in two closely linked photosystems (PSI and PSII, Fig. 4.14e), which act in a synchronous manner to convert the incoming light quanta into a chemical form of energy, namely, ATP (adenosine triphosphate) and NADPH (nicotinamide adenine dinucleotide phosphate).

Light Reactions. Photosystems are composed of several large pigment and protein complexes, which become excited with light energy. The molecules transport the excitation energy in a highly organised process, electron transport chain and eventually split the water molecules into hydrogen and oxygen. This creates a proton concentration gradient across the thylakoid membrane and a low pH value in the lumen (intrathylakoid space, Fig. 4.14d). The membrane-bound ATP synthase enzyme uses the energy released by the protons moving across the membrane and creates chemical energy as ATP. The process of using the proton gradient between lumen and stroma to fuel an energy-requiring process is known as chemiosmosis. Excitation energy is passed to PSI, and finally NADPH is formed. Thus, both the two photosystems and the ATP synthase have their unique role in energy capture, and their synchronised operation is a prerequisite for an active photosynthesis.

Carbon Reactions. After the light energy has been converted into chemical energy in thylakoids, it can be used in carbon reactions (Calvin-Benson cycle) in plastid stroma to reduce carbon dioxide into small carbohydrate molecules. In the Calvin-Benson cycle, a five-carbon molecule (RuBP, ribulose 1,5-bisphosphate) is bound with CO_2 in a reaction catalysed by Rubisco enzyme to form two molecules of three-carbon intermediates, triose phosphates. These molecules are phosphorylated by ATP and reduced by NADPH to glyceraldehyde 3-phosphate (G3P). After several enzymatically catalysed reactions, the RuBP molecule is formed, and carbon binding can start over again. The whole cycle consumes nine molecules of ATP and six molecules of NADPH in the formation of one triose phosphate molecule. Calvin-Benson cycle intermediates are either used for storage starch biosynthesis in chloroplasts or exported into cytoplasm in exchange to inorganic phosphorus by a triose phosphate pump and converted into sucrose, which is the main soluble sugar compound in photosynthesising cells.

The biochemical regulation system of annual cycle changes the concentrations and activities of the functional substances involved in photosynthesis. There is, however, a period in the summer when the regulation system of the annual cycle exerts only small changes to the functional substances. First, we consider the situation in summer and assume that the functional substances are stable.

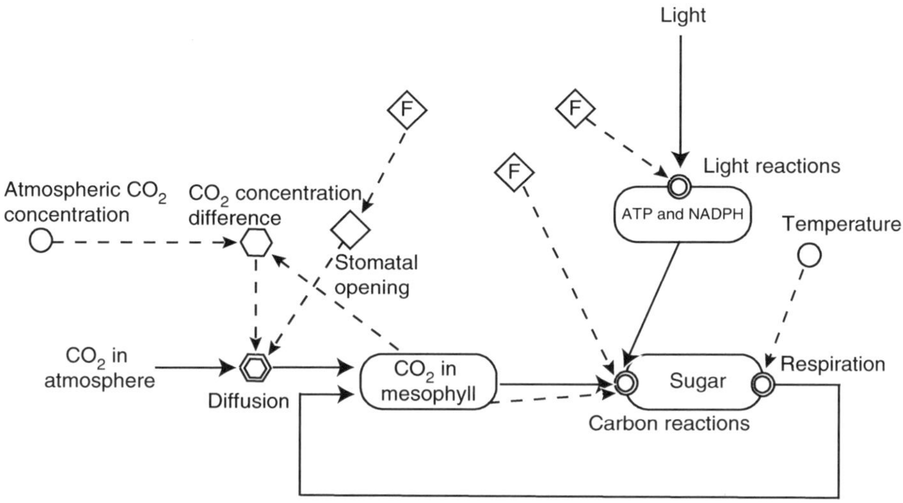

Fig. 4.15 The role of biochemical processes and transport in photosynthesis: ATP and NADPH are produced in light reactions. The captured energy in ATP and NADPH is used in carbon reactions to form sugar utilising atmospheric CO_2. Diffusion transports CO_2 into mesophyll, and stomatal action controls the carbon flow (The symbols are introduced in Fig. 2.3)

A thin water layer covers the mesophyll surfaces around the stomatal cavity, and the air in the stomatal cavity is water saturated having higher water vapour concentration than the ambient air. Diffusion generates water flow through the stomatal pore into the ambient air. The long transport distance from fine roots to leaves and low availability of water in the soil call for a need to reduce transpiration with partial closure of stomata. The stomatal control of transpiration and photosynthetic material flows has been important during evolution since water and carbon fluxes are the two dominating fluxes in living trees.

The driving flux of photosynthesis is solar energy in the form of light quanta. They excite the chlorophyll which begins the chain of light reactions resulting in formation of ATP and NADPH that carries the energy to the carbon reactions of photosynthesis. The carbon reactions of photosynthesis consume CO_2 resulting in decrease in the carbon dioxide concentration in the mesophyll and in the stomatal cavity. Diffusion generates CO_2 inflow from the atmosphere into the stomatal cavity due to the concentration difference. The energy and CO_2 flows associated with photosynthesis are visualised in Fig. 4.15.

Theory of Optimal Stomatal Control of Photosynthesis

Specified Basic Concepts and Ideas

Photosynthesis can be quantified either by the amounts of sugars produced or by the CO_2 consumed. In principle, the quantification by the sugars produced is more justified, but the CO_2 approach is more practical from the measurement point of view. The transition between the two different units can be done by multiplication with a factor, rather close to one. We define the *specific photosynthetic rate* at a point and moment as the CO_2 consumed in a small volume element around the point and during short time interval divided by the product of the leaf mass in the volume of the space element and of the length of the interval. Analogous concepts are commonly used in physics; thus, physical knowledge is used to concretise the cover theory to the treatment of photosynthesis.

In physics, the energy of solar radiation is studied in units W m^{-2}. However, in photosynthetic studies, the number of light quanta in the visible range of spectrum is used because it refers directly to the capturing of light quanta in the light reactions of photosynthesis. Nearly half of solar radiation at the wavelengths of infrared radiation is useless in photosynthesis and not used for the light reactions.

The *flux of photosynthetically active radiation* (PAR) is defined as the number of visible light quanta falling on a small area element during short time interval divided by the area of the element and by the length of the interval. The area element and the time interval have to be so small that the light intensity is stable. Its units are usually moles of photons per area of leaf per second.

The theoretical starting point of our theory of optimal control of photosynthesis consists four main basic ideas.

Photosynthesis is a complicated reaction chain starting with capturing energy from light, thereafter formation of ATP and NADPH and finally synthesis of sugars. Numerous steps are involved in photosynthesis, and each step requires its own functional substance. These pigments, membrane pumps and enzymes are non-stable compounds. However, the regulation system of photosynthesis maintains their concentrations and activities resulting in smooth photosynthesis although the underlying biochemical system is non-stable.

The action of the photosynthetic regulation system generates an emergent property in the involved functional substances. This emergent property is called as the *efficiency of photosynthetic functional substances*. It characterises the performance of the whole chain of functional substances. This new concept enables the aggregation of the very detailed physiological knowledge of single steps to whole process level.

Metabolism, namely, the action of pigments, membrane pumps and enzymes, generates the energy flow from solar radiation into chemical form and carbon flow from the atmosphere into sugars.

Basic idea P1: The efficiency of the functional substances affects photosynthetic rate.

Photosynthetically active radiation (PAR) is the source of energy for light reactions, and therefore, photosynthesis depends on it. Synthesis of sugars is located deep inside the leaf, and carbon dioxide has to travel through the stomatal pore and cavity into mesophyll. CO_2 molecules are transported by diffusion, which is rather inefficient. Thus, carbon dioxide concentration is rather low in the chloroplasts.

Basic idea P2: The photosynthetic rate depends on the availability of light energy and carbon dioxide in the chloroplasts.

Diffusion transports CO_2 into and simultaneously water vapour out of stomata. The concentration difference, the driving force of diffusion, is often nearly hundredfold for water vapour when compared with that of carbon dioxide. Thus, large amounts of water are needed for photosynthesis to allow the CO_2 diffusion into chloroplasts. However, the water transport capacity in coarse roots, stems and branches may be insufficient, and plants have to reduce the diffusion out of the stomatal pores.

Basic idea P3: Effective stomatal control of photosynthetic and transpiration rates has developed in evolution.

Cowan (1977) and Cowan and Farquhar (1977) introduced the idea of optimal stomatal control. They were, however, unable to solve the proposed optimisation problem, but the solution has been introduced later (Hari et al. 1986).

Basic idea P4: The functioning principle of stomata is to maximise the photosynthetic production minus transpiration costs.

Derivation of the Theoretical Model

Diffusion transports CO_2 into and out from stomatal cavity depending on whether concentration is lower or higher than in the outside air. Let g_{CO_2} denote the diffusive flux through a stomatal pore, $C^{s}_{CO_2}$ the CO_2 concentration in the stomatal cavity and $C^{a}_{CO_2}$ the ambient CO_2 concentration. The diffusive flux is proportional to the concentration difference; thus,

$$g_{\mathrm{CO}_2} = g_{\mathrm{s}}\left(C^{a}_{CO_2} - C^{s}_{CO_2}\right), \tag{4.1}$$

where g_{s} is a parameter, frequently called stomatal conductance (Gaastra 1959).

The energy captured in light reactions is used to form sugars utilising CO_2 from the stomatal cavity as raw material. Let p denote the photosynthetic rate and I_{PAR} the photosynthetically active radiation flux. The efficiency of photosynthetic functional substances, $E_{\mathrm{P},}$ characterises the performance of the functional substances to carry out photosynthesis. Further, the photosynthesis depends on the availability of raw materials for carbon reactions, that is, energy and CO_2. We introduce simple model for the photosynthetic rate:

$$p = E_{\mathrm{P}} f\left(I_{\mathrm{PAR}}\right) C^{s}_{CO_2}, \tag{4.2}$$

where the function f is

$$f\left(I_{\mathrm{PAR}}\right) = \frac{I_{\mathrm{PAR}}}{I_{\mathrm{PAR}} + \gamma}. \tag{4.3}$$

The parameter γ introduces saturation of light reactions.

Diffusive flux between the stomatal cavity and ambient air, photosynthesis and respiration change the CO_2 concentration inside the leaf. Now we can apply the general principle of conservation of mass to construct a dynamic model of a mass in a container, as treated in Chap. 2 (Eq. 2.3). The amount of CO_2 in the stomatal cavity is concentration multiplied by the volume of the cavity, V_i. The differential equation for the CO_2 in the cavity is

$$\frac{\mathrm{d}\left(V_i\,C^s_{CO_2}\right)}{\mathrm{d}t} = A_s\left(g_s\left(C^a_{CO_2} - C^s_{CO_2}\right) - E_P\,f\left(I_{PAR}\right)C^s_{CO_2} + r\right), \tag{4.4}$$

where A_s is the leaf area that a stoma feeds with CO_2. The terms on the right-hand side in the above equation are inflow by diffusion, outflow by photosynthesis and inflow by respiration, r.

Assume that the CO_2 concentration in the cavity is in steady state, which means that the concentration does not change. Then

$$\frac{\mathrm{d}\left(V_i C^s_{CO_2}\right)}{\mathrm{d}t} = 0. \tag{4.5}$$

Thus,

$$A_s\left(g_s\left(C^a_{CO_2} - C^s_{CO_2}\right) - E_P f\left(I_{PAR}\right)C^s_{CO_2} + r\right) = 0. \tag{4.6}$$

When the stomatal carbon dioxide concentration, $C^s_{CO_2}$, is solved from the above equation, we get

$$C^s_{CO_2} = \frac{g_s\,C^a_{CO_2} + r}{g_s + E_P f\left(I_{PAR}\right)}. \tag{4.7}$$

When the obtained dependence of $C^s_{CO_2}$ on light is introduced in Eq. 4.2, we get a model for photosynthesis:

$$p = E_P\,f\left(I_{PAR}\right)\frac{g_s\,C^a_{CO_2} + r}{g_s + E_P\,f\left(I_{PAR}\right)}. \tag{4.8}$$

CO_2 is dissolved into the water on mesophyll cells in the stomatal cavity. This generates a close coupling between photosynthesis and transpiration, since water molecules evaporate from the water film on mesophyll cells and the stomatal cavity is saturated with water vapour. The water vapour concentration is higher in the cavity than in the surrounding air giving rise to diffusion. The diffusion of CO_2 and water vapour through the stomatal pores generates a close coupling between photosynthesis and transpiration. Thus, photosynthesis is impossible without transpiration.

The above model assumes that diffusivity from the atmosphere through stomatal pores into the stomatal cavity does not change. The stomatal action that reduces diffusion is, however, evident during high evaporative demand (basic idea P4).

Stomatal closure reduces the diffusion of CO_2 and water vapour into and out from a leaf. The parameter g_s, called stomatal conductance for CO_2, takes this reducing effect in consideration. We introduce the concept 'degree of stomatal opening', u, to describe the reduction in diffusion caused by the stomatal action. We assume

$$g_s = u\, g_{max}, \tag{4.9}$$

where g_{max} is stomatal conductance for CO_2 when the stomata are fully open. The degree of stomatal opening can take values only between 0 and 1. When u is introduced into Eq. 4.8, we get

$$p = E_P\, f\,(I_{PAR})\, \frac{u\, g_{max}\, C^a_{CO_2} + r}{u\, g_{max} + E_P\, f\,(I_{PAR})}. \tag{4.10}$$

Light intensity and the degree of stomatal opening vary during a day, and photosynthesis reflects this variation.

The stomatal action has two conflicting goals: (1) to maximise photosynthetic production and (2) minimise amount of transpiration. These goals can be combined by assuming that transpiration of 1 kg of water requires λ kg CO_2 for formation and maintenance of water uptake and transport capacity. The optimal stomatal control can now be formulated as a maximisation problem where the photosynthetic production is maximised regarding to transpiration costs in CO_2.

Let g_{H_2O} denote transpiration rate and D_{H_2O} the water vapour concentration difference between stomatal cavity and ambient air and $g_S^{H_2O}$ the stomatal conductance for water vapour. The diffusion out from the leaf is analogous to the diffusion of CO_2 into the cavity. As water molecules are lighter than CO_2 molecules, diffusivity of water vapour is higher than diffusivity of CO_2. Stomatal conductances for H_2O and CO_2 are related to each other by factor $a = 1.6$. Thus,

$$g_S^{H_2O} = a\, g_s. \tag{4.11}$$

When we introduce the changing stomatal regulation and concentration difference, we get

$$g_{H_2O}(t) = a\, u(t)\, g_{max}\, D_{H_2O}(t). \tag{4.12}$$

The optimisation problem is to find the best possible pattern of the degree of stomatal opening $u(t)$ that maximises the photosynthetic production minus transpiration costs during a prolonged period from t_1 to t_2. The solution can be found with calculus of variations using rather old mathematical tools.

Let $P_P(t_1,t_2)$ denote photosynthetic production during the interval from t_1 to t_2 and $G_{H_2O}(t_1,t_2)$ the amount of transpiration during the same interval. They are obtained by integration:

$$P_P\,(t_1, t_2) = \int_{t_1}^{t_2} p(t)\, dt, \tag{4.13}$$

and

$$G_{H_2O}(t_1, t_2) = \int_{t_1}^{t_2} g_{H_2O}(t)\, dt. \tag{4.14}$$

The optimisation problem can now be formulated in an exact form:

$$\underset{u}{\text{Max}} \left\{ P_P(t_1, t_2) - \lambda\ G_{H_20}(t_1, t_2) \right\}. \tag{4.15}$$

The above optimisation problem can be solved with the Lagrange method, which is rather commonly used in systems analysis and economics. As a first step, the Lagrange function, L, is formed:

$$L = E_P\ f(I_{PAR})\ \frac{u\ g_{max}\ C_a + r}{u\ g_{max} + E_P\ f(I_{PAR})} - \lambda\ u\ g_{max}\ D_{H_2O}. \tag{4.16}$$

The next step is to differentiate the Lagrange function with respect to u. The optimal stomatal opening during the interval from t_1 to t_2 is obtained as a zero point of the derivative of the Lagrange function.

The derivative of the Lagrange function with respect to u includes u as first and second power. Thus, it can be solved resulting in

$$u^* = \left(\sqrt{\frac{C^a_{CO_2} - (r/(E_P\ f(I_{PAR})))}{\lambda\ a\ D_{H_2O}}} - 1 \right) \frac{E_P\ f(I_{PAR})}{g_{max}}. \tag{4.17}$$

The above solution yields values that are greater than 1 or negative which are outside the range of degree of a stomatal opening. In these cases, u takes value 1 or 0. The final solution is

$$u = \begin{cases} 0, & \text{if } u^* \le 0 \\ u^* & \text{if } 0 < u^* < 1 \\ 1 & \text{if } u^* \ge 1 \end{cases}. \tag{4.18}$$

The light intensity and water vapour concentration difference between the stomatal cavity and surrounding air vary during a day, and the degree of stomatal opening, u, responds to this variation as well as to light intensity. Since the water vapour concentration difference is low in the mornings, the stomata are usually fully open after sunrise, and they partially close if the day is sunny and the water vapour concentration difference is large. These variations in the light and water vapour concentration difference generate strong daily patterns in photosynthesis.

When we combine Eqs. 4.10, 4.17 and 4.18, we obtain the theoretical model for photosynthesis.

Field Measurements

We utilise the field measurements at SMEAR I in northern Finland to test the optimal stomatal control theory of photosynthesis. We describe the measurements in more detail in Chap. 9.

As we have no leaf temperature measurements, we have to use a proxy obtained from air temperature and solar radiation. We assume that solar radiation warms the leaf surface where the photosynthesis takes place. Let T_L denote leaf temperature, T air temperature and I_{PAR} photosynthetically active radiation. The proxy for leaf temperature is

$$T_L = c_L \, I_{PAR}, \tag{4.19}$$

where c_L is an empirical constant having an approximate value 1.5°C $mmol^{-1}$ m^2 s^1.

Test of the Theory with Field Measurements

Testing of the theory includes four phases that play very different roles in the evaluation. We have to determine the values of the parameters to obtain clear predictions with the theoretical model. The measured data repeats, to great extent, the same pattern, and the possibilities to utilise the measurements are rather limited. There are, however, interesting phenomena and periods of measurements that carry useful information for testing the theory of optimal stomatal control of photosynthesis. We focus the testing on these characteristic periods. Next, we evaluate the adequacy of the theoretical model, its theoretical basis, simplifying and idealising assumptions. Thereafter, we consider the explaining power of the theory, especially focusing on the characteristic behaviour. The comparison with other theories closes the evaluation.

Estimation of Parameter Values. We use the normal statistical least squares method to estimate the values of the model parameters. If the explaining factors are independent of each other and if the model parameters do not compensate each other, then the estimation of the parameter values is a straightforward task. However, in the case of our model, the explaining variables (solar radiation, leaf temperature and water vapour deficit) are causally connected with each other, and in addition, the structure of the model is such that the parameters also compensate each other. This means that there are an infinite number of different combinations of estimated parameter values that all give the sum of squared residuals close to the global minimum. This leads to instability in numerical estimation indicating that the model may be over parameterised.

We selected gas exchange measurements on one Scots pine shoot at SMEAR I in the summer of 2006 for the first test of the optimal stomatal control model. First, we selected a period of 5 days from the middle of July for test data to estimate the values of the parameter λ and the efficiency of the functional substances. The estimation procedure was somewhat unstable. A more stable parameter set could be found when transpiration measurements were also introduced. The problem

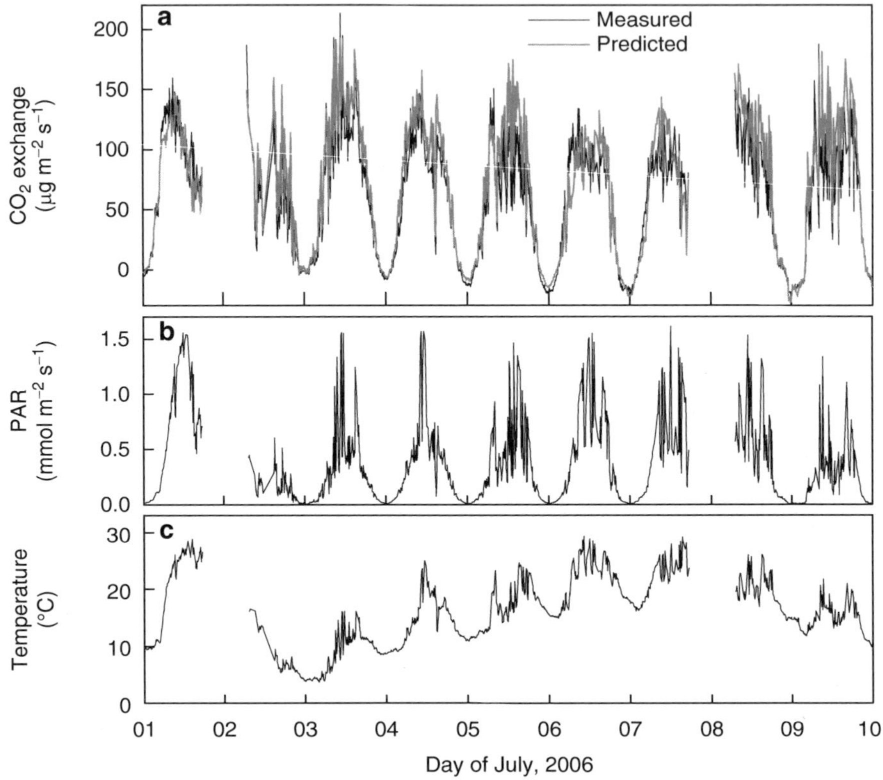

Fig. 4.16 The measured and predicted CO_2 exchange (**a**), measured photosynthetically active radiation (*PAR*) (**b**) and measured temperature (**c**) at SMEAR I, northern Finland, during 1–10 July 2006. The *gaps* in data are caused by thunderstorms since the measuring system is coupled off to prevent lightning damage

in transpiration is the systematic error in the measurement when the air is moist and that the data must be rejected if air humidity is high. The obtained values are $E_P = 0.001$ m s^{-1}, $\gamma = 1.2$ mmol m^{-2} s^{-1}, $\lambda = 0.012$ g (CO_2)/g (H_2O) and $g_{max} = 0.001$ m s^{-1}.

Prediction of Characteristic Features of Photosynthesis. We based our analysis of the performance of the theory and the theoretical model on test data that was not utilised in the estimation of the parameter values. The first characteristic feature to be analysed is the strong daily patterns in CO_2 exchange (Fig. 4.16a). The inflow of CO_2 is strong during day, and rather small outflow of CO_2 takes place in the night. The model predicts the same daily patterns very clearly.

The typical summer weather at our measuring station is sunny, and temperature is about 20°C. During these days, there are clouds travelling across the sky generating strong variation in the photosynthetically active radiation. The daily pattern of CO_2 exchange is clear, but the clouds generate variation also in the CO_2 flux

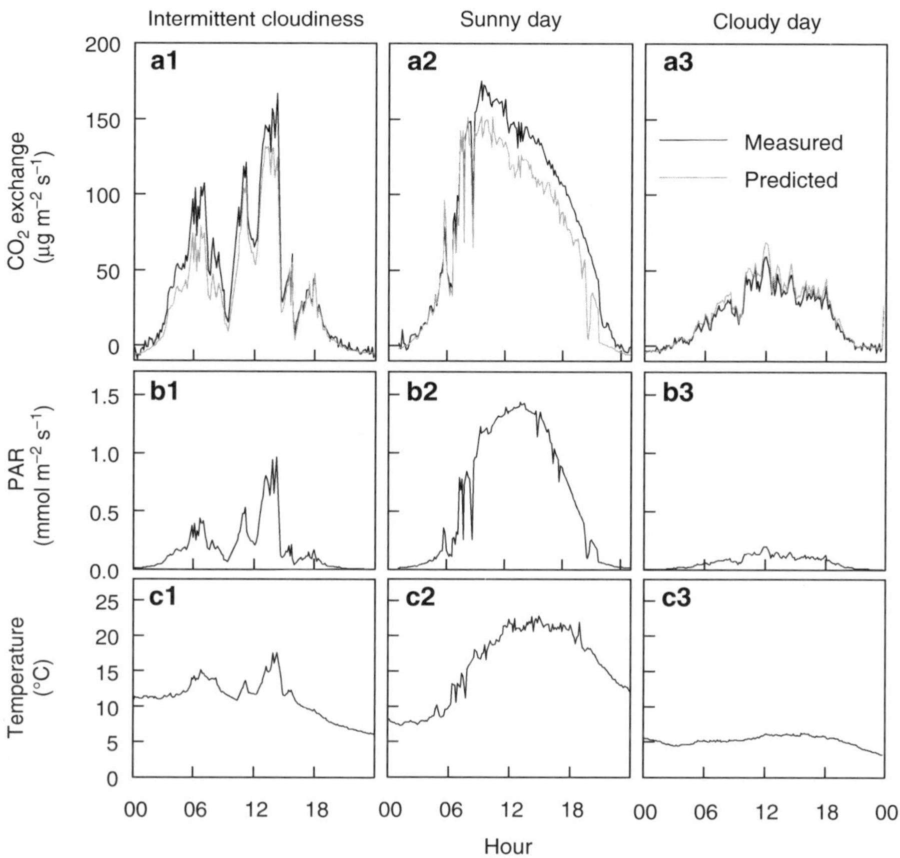

Fig. 4.17 The measured and predicted CO_2 exchange (**a1**–**a3**), measured PAR (**b1**–**b3**) and measured temperature (**c1**–**c3**) at SMEAR I, northern Finland, during a day of intermittent cloudiness (26 June), a sunny day (2 July) and a cloudy day (6 August) in 2006

(Fig. 4.17a1). The prediction with the optimal stomatal control theory and model also results in strong variation during intermittent cloudiness.

There are two extreme weather conditions that are informative for the testing of our theory, that is, totally cloudless and totally cloudy days. These days are not common, and they occur during a couple of days in the summer.

The daily pattern in CO_2 exchange during sunny days is clear, rapid increase in the morning, saturation around 10–12 o'clock and clear decline in the afternoon (Fig. 4.17a2). The optimal stomatal control model predicts the same daily pattern, but the level is a bit low.

The totally cloudy days are often also rainy, and the temperature is low, around 10–15°C. The daily pattern of measured CO_2 flux during cloudy days differs clearly from that during sunny weather (Fig. 4.17a3). The increase in the morning is delayed to some extent, and a rather large and irregular variation is characteristic for the

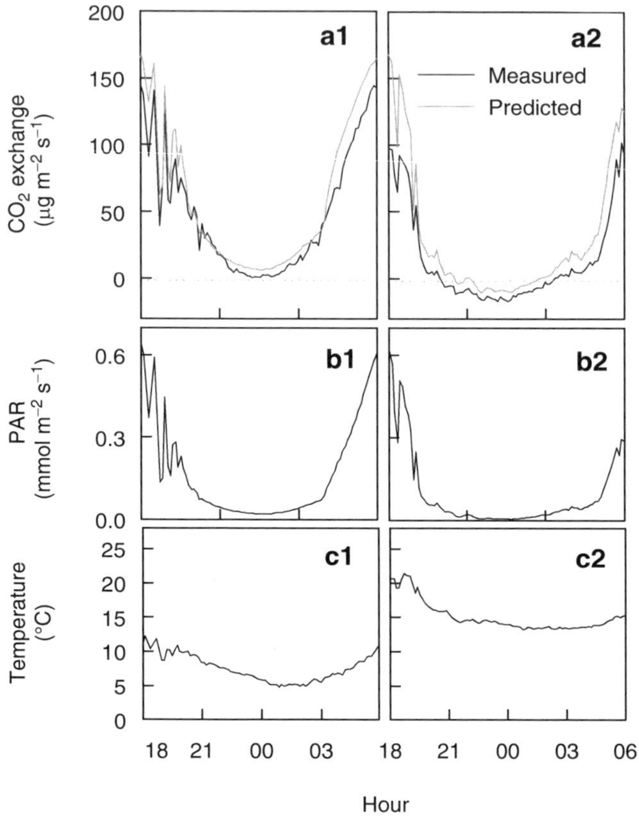

Fig. 4.18 The measured and predicted CO_2 exchange (**a1–a2**), measured PAR (**b1–b2**) and measured temperature (**c1–c2**) at SMEAR I, northern Finland, during the polar day, 27–28 June (*left panels*) and during 14–15 July in 2006 (*right panels*)

daily pattern, and the CO_2 uptake is rather low. The optimal stomatal control model of photosynthesis is again able to predict the daily pattern, the variation between measurements rather closely and also the low CO_2 uptake.

Our measuring station is located about 200 km north of the polar circle, and the Sun never sets during a month around midsummer. This fact generates an additional characteristic feature in our measurements of the CO_2 exchange during polar day; then photosynthesis is able to compensate the respiration during night (Fig. 4.18a1). The optimal stomatal control model predicts this compensation during polar night.

Our theory of optimal control of photosynthesis together with the optimal stomatal model was able to predict the observed CO_2 exchange during all characteristic periods we had detected in our data.

Adequacy of the Model Structure. The theory of optimal stomatal control of photosynthesis is based on four basic ideas that give rather simplified picture of the complicated web of interconnected reactions. For example, we assume implicitly

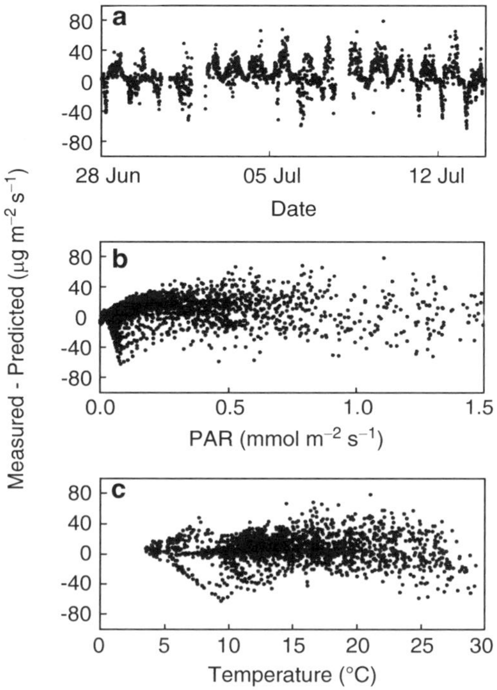

Fig. 4.19 The difference between measured and predicted CO_2 exchange against date (**a**), PAR (**b**) and measured temperature (**c**) at SMEAR I during 28 June–14 July 2006

that temperature has no effect on photosynthesis, although all biochemical reactions and the dissolution of CO_2 in water are strongly temperature dependent. The temperature increase accelerates the biochemical reactions and decrease solubility of CO_2 in water. However, these opposite effects do not compensate each other totally. Also the derivation of the equations for the fluxes includes simplifications and idealisations, such as that CO_2 concentration in the stomatal cavity equals with that in chloroplasts. Thus, we should be able to improve our theory with more detailed description of photosynthesis in the chloroplasts.

The chosen periods of characteristic features cover only a rather small part of the available data; thus, the information in the data is rather poorly utilised. In addition, the analysis is so far rather qualitative. The analysis of residuals enables utilisation of large data sets, and the result is condensed into clear pictures. We studied the behaviour of the residuals as function of time, temperature and photosynthetically active radiation (Fig. 4.19). The residuals as function of time and PAR showed some systematic behaviour, but the absolute values were always very small.

Explaining Power of the Theory of Optimal Stomatal Control of Photosynthesis. The optimal stomatal control model is able to predict the characteristic features of the observed patterns of CO_2 exchange. Photosynthetically active radiation played a dominating role in the explanation of the variation in the observed CO_2 exchange. However, also the stomatal action reduces the CO_2 inflow during afternoons. Our theory of optimal control of photosynthesis explains the great variation in

CO_2 exchange during intermittent cloudiness with variation in photosynthetically active radiation and stomatal action is of minor importance. The behaviour of CO_2 exchange during cloudless days can be explained by the combined action of light and stomata over the course of the day. While the increasing CO_2 uptake follows the increasing light in the morning, the stomatal action clearly reduced photosynthesis later in the day. Low radiation energy for photosynthesis explains the low level of photosynthesis during cloudy days. Also the strange photosynthesis in the middle of night during the polar day is connected to light; since then, the Sun never sets, and the low photosynthetically active radiation in the night is sufficient for carbon fixation. Thus, our theory is capable to explain all observed characteristic features in CO_2 exchange during midsummer.

We have a large data set of more than 10 years, but the study of characteristic behaviour covers only small part of that data set. However, the proportion of explained variance tells us effectively the ability of the model to predict and explain the measurements with the theory. The proportion of explained variance when the CO_2 exchange is predicted with our model is high, usually 90%, sometimes clearly higher but also lower in some cases.

Comparison with Other Approaches. In the past, simple empirical light response models were commonly used in the analysis of photosynthesis in field conditions. The fit of the models was often rather poor, in part due to the low quality of data. Statistical methods have commonly been used in the analysis of photosynthesis under field conditions. The relationship between light and photosynthesis has been described using simple functions, such as the Michaelis-Menten-type saturation response. Our model is also Michaelis-Menten-type model when the stomata are fully open; however, the stomatal action reduces frequently photosynthesis, and then the saturating light dependence is insufficient.

The 'analogy modelling' tradition that follows the ideas presented by H. T. Odum (1983) has been strong in the analysis of the behaviour of photosynthesis. Then the diffusion is treated with normal conductance model, and the analogue is thereafter expanded to the photosynthetic biochemical reactions with mesophyll conductance (Linder and Troeng 1980; Collatz et al. 1991, 1992; Jarvis et al. 1999). When mesophyll conductance is used, then it is implicitly assumed that the CO_2 concentration in chloroplasts is zero, but however, that is not justified.

Farquhar et al. (1980) presented a model of the biochemistry of photosynthesis, which can be combined with statistical models of stomatal action. The theoretical basis of Farquhar model is rather close to our formulation of the basic ideas P1, P2 and P3. Farquhar model is more detailed than the optimal stomatal control model and includes more parameters. The estimation of the values of the parameters in Farquhar model is problematic when confronted with field data. A comparison between the Farquhar model and the optimal stomatal control model showed rather similar performance with shoot CO_2 exchange data measured at SMEAR I (Aalto et al. 2002). The difference in the number of parameters to be estimated hampered the comparison. Although the theoretical basis of the optimal stomatal control model is completely different from the biochemical Farquhar model, the behaviour of those two models is remarkably similar.

Noe and Giersch (2004) developed a dynamic model for photosynthesis that describes the carbon flow from the atmosphere into sugars in leaves. The model construction is based on conservation of carbon atoms. Thus, it utilises the modelling principle presented in Sect. 2 Eq. 2.3. Three differential equations describe the behaviour of photosynthetic intermediates, CO_2 concentration in the stomatal cavity and action of stomata. The treatment of photosynthesis is rather similar with our approach. They test their model with field measurements of oak photosynthesis during 5 days. The fit between measured and modelled CO_2 exchange is rather good.

Conclusion: The theory of optimal stomatal control of photosynthesis passed successfully all the severe tests, and our theory and model gained evident corroboration.

Theory of Annual Cycle of Photosynthesis

Observed Annual Cycle of Photosynthesis. The annual cycle in photosynthesis is strong in most forests. There is intensive sugar formation in summer and very small, if any, photosynthetic activity in winter. The measured mean CO_2 exchange has a strong annual pattern (Fig. 4.20) that is generated by the annual pattern of environmental factors, especially light, and by the annual cycle of photosynthetic functional substances. The variation from the mean behaviour is rather considerable, especially during spring (see later Fig. 4.24).

The dotted lines in Fig. 4.21 indicate predicted CO_2 exchange assuming that the relationship between photosynthesis and light and stomatal action obtained in the middle of the summer is valid also in the spring and in the autumn. This prediction results in too high photosynthesis at the ends of the photosynthetically active period. The annual cycle in photosynthetic functional substances explains this discrepancy, because the efficiency of the functional substances of photosynthesis is low early in the spring and late in the autumn.

Background of Regulation of Annual Cycle. The annual cycle is strong in the functional substances involved in photosynthesis due to the action of the biochemical regulation system. Photosynthesis is based on a complicated chain of reactions, both biochemical and physical. Each step in the metabolic chain requires its own functional substances, namely, pigments, enzymes or membrane pumps. The number of functional substances involved in photosynthesis is large, evidently over one hundred. The regulation system responds to the annual cycle of environmental factors such as light and temperature and changes the concentrations and activities of photosynthetic functional substances, which further influence the efficiency of the photosynthesis process itself.

Photosynthetic rate is at its maximum in summer, and the functional substances involved in photosynthesis can be considered to be in a fairly stable state in midsummer. The introduced optimal stomatal control theory and model predicted CO_2 exchange rather well during the midsummer when the variation of irradiance was the most important factor behind the short-time fluctuations in photosynthetic rate. If the model parameter values obtained in mid-July are used for predicting

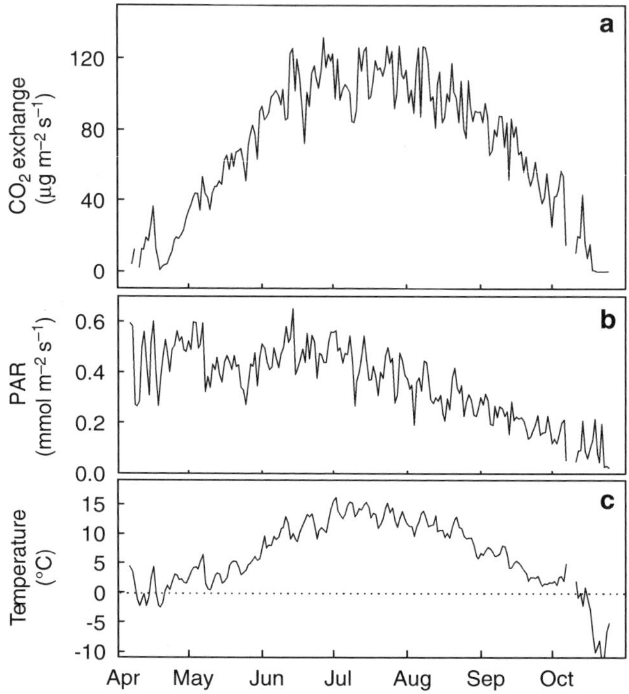

Fig. 4.20 Measured averages of the seasonal courses of daily mean midday CO_2 exchange of Scots pine shoots (**a**), photosynthetically active radiation (**b**) and air temperature (**c**) from April to October during 2001–2007 at SMEAR I, northern Finland. The fluxes were averaged daily over all shoots that were under monitoring simultaneously (2–4 shoots) and between 11:00 and 14:00 solar time

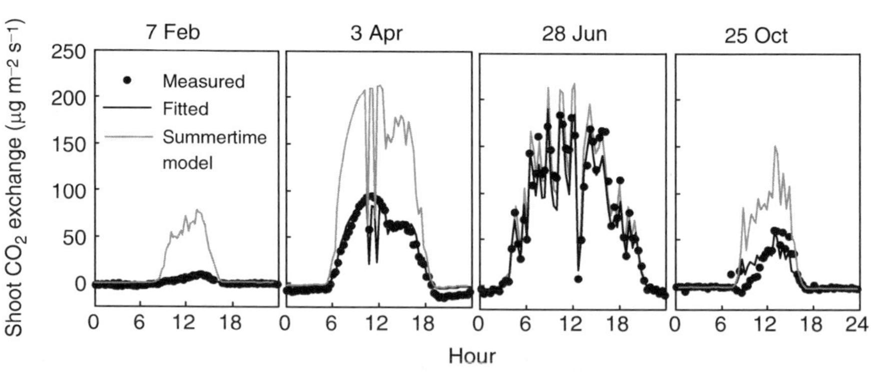

Fig. 4.21 Examples of diurnal courses of Scots pine shoot CO_2 exchange at SMEAR II in Hyytiälä in winter, spring, summer and autumn 2005. The *dots* represent measured CO_2 exchange, *gray lines* CO_2 exchange predicted with the optimum model and midsummer parameters' values and *black lines* the model when the efficiency of photosynthetic functional substances is estimated daily

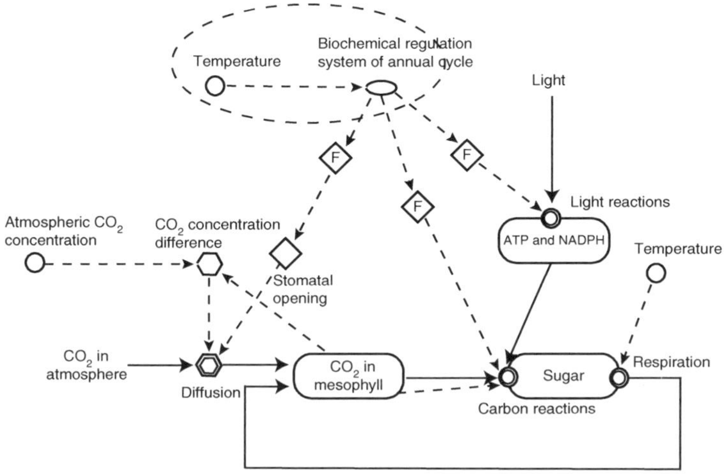

Fig. 4.22 Visualisation of the action of the biochemical regulation system of annual cycle on photosynthesis. The carbon and energy flow part of the figure is the same as in Fig. 4.15. The biochemical regulation system is added to introduce the annual cycle in the functional substances. Temperature affects the regulation system, and in this way, it synchronises the annual cycle in photosynthesis with the environmental cycle. The *dash line circle* stresses the biochemical regulation in the figure (The symbols are introduced in Fig. 2.3)

CO_2 exchange over the whole year, the discrepancy between the predicted and the observed CO_2 exchange is evident, especially in spring (Fig. 4.21). The daily patterns seem to be rather similar, but the level of photosynthesis varies greatly over the year.

The derivation of the optimal stomatal control model is valid whenever the functional substances do not change during the period under consideration. The regulation system, however, changes the concentrations and activities of functional substances during recovery of photosynthesis in spring and also in autumn (Fig. 4.22), but we can assume that these changes are so slow that the efficiency of functional substances is constant during a day. Thus, we can analyse within-day data with the optimal stomatal control model and study day-to-day changes in the efficiencies of the photosynthetic functional substances. Previously it has been shown that although the absolute level of CO_2 exchange of Scots pine shoots varies during the year, the diurnal patterns and the relationships of CO_2 exchange with environmental driving factors remain similar qualitatively (Pelkonen and Hari 1980; Hari and Mäkelä 2003; Mäkelä et al. 2004; Kolari et al. 2007).

The synthesis, decomposition, activation and deactivation change the efficiency of the functional substances slowly, in the timescale of several days. We first estimated the values of the efficiency of the functional substances (E_P, Eq. 4.2) for each day, and we kept the values of the other parameters as constant. Then we used the obtained value of E_P to predict the CO_2 exchange of the next day to determine the predictive power of the optimal stomatal control model without fitting the model. The model was able to predict the daily patterns of CO_2 exchange during the whole active period as well as the within-day rapid variation caused by movements of clouds. There are, however, two exceptions. After freezing nights, CO_2 uptake is strongly reduced before noon. This effect seems to be related to daily minimum temperature (Kolari et al. 2007) and is normally most obvious in spring. On sunny days in spring, stomatal action is sometimes stronger than the model predicts, and using the summertime value of the parameter cost of water overestimates CO_2 exchange and transpiration. We omit these small effects in our analysis because they do not disturb the big picture of the annual cycle in photosynthesis.

The average predicting power in the CO_2 exchange data for the whole summer was 93.9% when using the efficiencies of the functional substances obtained from the measurements during previous day. The explaining power of the optimal stomatal control model was close to that obtained in predicting midsummer CO_2 exchange, and the model could produce the daily patterns as well as the short-term variation caused by clouds.

The regulation system of the annual cycle in photosynthesis changes the functional substances involved in photosynthesis; it synthesises and activates in the spring, keeps as stable in the summer, deactivates and decomposes in the autumn and finally keeps stable in the winter. The biochemical regulation system responds to the annual cycle of temperature. The annual cycle of photosynthesis has two components: the annual cycle of environmental factors and the annual cycle of photosynthetic functional substances (Fig. 4.22).

Basic Concepts and Ideas

The numerous functional substances involved in photosynthesis change in a balanced way and the photosynthetically active radiation effects in a similar way the photosynthesis during the whole year, but the level of photosynthesis changes greatly from spring to summer. The biochemical regulation system of photosynthesis is able to maintain a good balance between the steps in photosynthesis starting from the light reactions until the final step in the synthesis of sugars. Thus, the efficiency of the photosynthetic functional substances changes during the annual cycle. Photosynthetically active radiation and availability of CO_2 are the primary drivers of photosynthesis.

Basic idea AP1: The photosynthetic rate during the annual cycle depends on the efficiency of the functional substances, on the availability of light energy and carbon dioxide in the chloroplasts.

The action of the biochemical regulation system of photosynthesis changes smoothly the concentrations of active photosynthetic substances resulting in a new emergent property that is involved in the annual cycle of metabolism. *The annual state of photosynthetic functional substances* describes, by definition, the emergent regularities in the photosynthetic functional substances generated by the action of the biochemical regulation system during the annual cycle.

The good agreement between the measured fluxes and CO_2 exchange predicted with previous-day efficiency of the functional substances supports the idea that the biochemical regulation system changes slowly the annual state of photosynthetic functional substances. Over the year, however, the changes are substantial, and a fixed set of summertime parameter values poorly predicts the observed seasonal pattern in CO_2 exchange.

Basic idea AP2: The efficiency of the functional substances depends on the annual state of photosynthetic functional substances.

The changes in photosynthesis are connected with the annual cycle of temperature. Our starting point is, as in the case of bud development (see Sect. 4.2.6), that the regulation system slowly changes the annual state of the photosynthetic functional substances to reflect the annual cycle of environmental factors, especially temperature.

Basic idea AP3: The biochemical regulation system changes the annual state of the photosynthetic functional substances in such way that high temperatures activate and low ones deactivate photosynthesis.

These general ideas AP1–AP3 can be formulated into precise model of photosynthesis.

Derivation of the Theoretical Model

We can now simplify the basic idea AP3 into the model of the annual state, S_P, of the photosynthetic functional substances as follows:

$$\frac{dS_p}{dt} = \frac{T - S_p}{\tau}, \tag{4.20}$$

where the parameter τ is frequently called time constant. In this formulation, S_P will always move towards the prevailing temperature. The time constant τ describes the speed of the changes in the state of photosynthetic functional substances, large values of τ indicating slower change than small ones.

There is a connection between the functional substances and photosynthesis as stated in basic idea AP2. This idea can be expressed with mathematical notations as follows:

$$E_P = E_P(S_P). \tag{4.21}$$

There is no prior information on the metabolic mechanism behind the dependence of the efficiency of photosynthetic functional substances on the annual state of photosynthetic functional substances, that is, $E_p(S_p)$. Thus, we have to use the simplicity criterion and start with the linear relationship proposed by Mäkelä et al. (2004). Assume

$$E_P(S_P) = \text{Max}\{0, a_S(S_P - S_0)\}, \tag{4.22}$$

where a_S and S_0 are parameters. The threshold state of functional substances S_0 describes the onset of photosynthesis in the spring.

The optimal stomatal control model of photosynthesis describes phenomena in the timescale of a day, and Eqs. 4.20 and 4.22 introduce the slow changes in the values in the efficiency of the functional substances $E_p(S_p)$. When these models are combined, we get a model, called PhenPhoto, covering the whole year. We collected the equations into Box 4.1.

Box 4.1 The Model PhenPhoto Includes Seven Equations. They Are:

$$p = E_P f\,(I_{PAR})\,\frac{u\,g_{max} C^{a}_{CO_2} + r}{u\,g_{max} + E_P f\,(I_{PAR})} \tag{4.23}$$

$$f\,(I_{PAR}) = \frac{I_{PAR}}{I_{PAR} + \gamma} \tag{4.24}$$

$$g_s = u\,g_{max} \tag{4.25}$$

$$u* = \left(\sqrt{\frac{C^{a}_{CO_2} - (r/\,(E_P\,f\,(I_{PAR})))}{\lambda\ a\ D_{H_2O}}} - 1\right)\frac{E_P\,f\,(I_{PAR})}{g_{max}} \tag{4.26}$$

$$u = \begin{cases} 0, & \text{if } u* \leqslant 0 \\ u* & \text{if } 0 < u* < 1 \\ 1 & \text{if } u* \geqslant 1 \end{cases} \tag{4.27}$$

$$\frac{dS_p}{dt} = \frac{T - S_p}{\tau} \tag{4.28}$$

$$E_P\,(S_P) = \text{Max}\,\{0, a_S\,(S_P - S_0)\} \tag{4.29}$$

Variables in the optimal stomatal control model and in PhenPhoto:

$C^{a}_{CO_2}$	Atmospheric CO_2 concentration
D	Water vapour concentration difference between stomatal cavity and ambient air
E_P	Efficiency of photosynthetic functional substances
P	Photosynthetic rate
S_P	Annual state of photosynthetic functional substances
r	Respiration rate
u	Degree of stomatal opening

Parameters in the optimal stomatal control model and in PhenPhoto:

g_{max}	Stomatal conductance when the stoma are fully open
γ	Introduces saturation of light reactions
λ	Cost of water in transpiration
S_o	Annual state of photosynthetic functional substances at which photosynthesis is inhibited
a_S	Level parameter

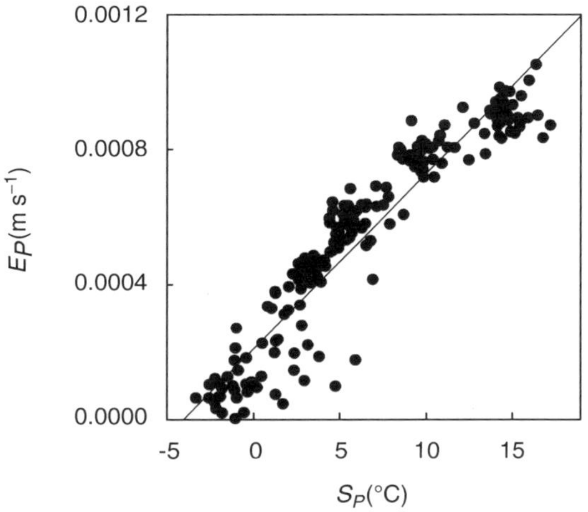

Fig. 4.23 The relationship between the efficiency of the functional substances (E_P) and the annual state of the photosynthetic functional substances (S_P) in Scots pine (mean of three shoots) in 2004 at SMEAR I, northern Finland

Test of the Theory with Field Measurements

Estimation of Parameter Values. We selected data measured during the year 2004 for studying the relationship between the efficiency of photosynthetic functional substances (E_P) and the annual state of photosynthetic functional substances, S_p (Fig. 4.23). The estimation resulted $\tau = 200$ h, $a_s = 0.000052$ m s^{-1} °C^{-1} and $S_o = -4$°C (Eq. 4.22). The other parameters of the photosynthesis model seem to change less during the year. Cost of transpiration (λ) typically shows also a slightly decreasing trend from the early spring to the beginning of summer and again an increase in the autumn (Sect. 4.2.2.2). We omit these small changes and assume that the values of all other parameters in the equations in Box 4.1 remain constant.

Prediction of Characteristic Features of Photosynthesis During the Annual Cycle. There is shoot-specific variation in photosynthesis that has to be introduced into the predictions. This is why we estimated the value of the parameter a_s in Eq. 4.22 for each chamber and year using estimation data sets consisting of measurements during 5-day period in the middle of the summer. The data used in the estimation were rejected from the test data set.

The variation between years is large: Photosynthesis exceeds respiration during an early spring about a month earlier than during a late one (Fig. 4.24). Our theory predicts the same behaviour. In the year 2005, there was first a warm spell and thereafter a cold one. The CO_2 exchange reacted clearly to the warm and cold spells. During warm weather, the photosynthesis recovered rapidly, and during cold one, the trees went back towards the slow photosynthesis earlier in the spring (Fig. 4.24). The prediction with PhenPhoto model produces the same pattern as the measured CO_2 exchange.

Cold nights occur every spring during the recovery of photosynthesis; then the temperature can fall even below −5°C. Photosynthesis is strongly reduced after these cold nights, but the PhenPhoto model is unable to predict photosynthesis after cold nights.

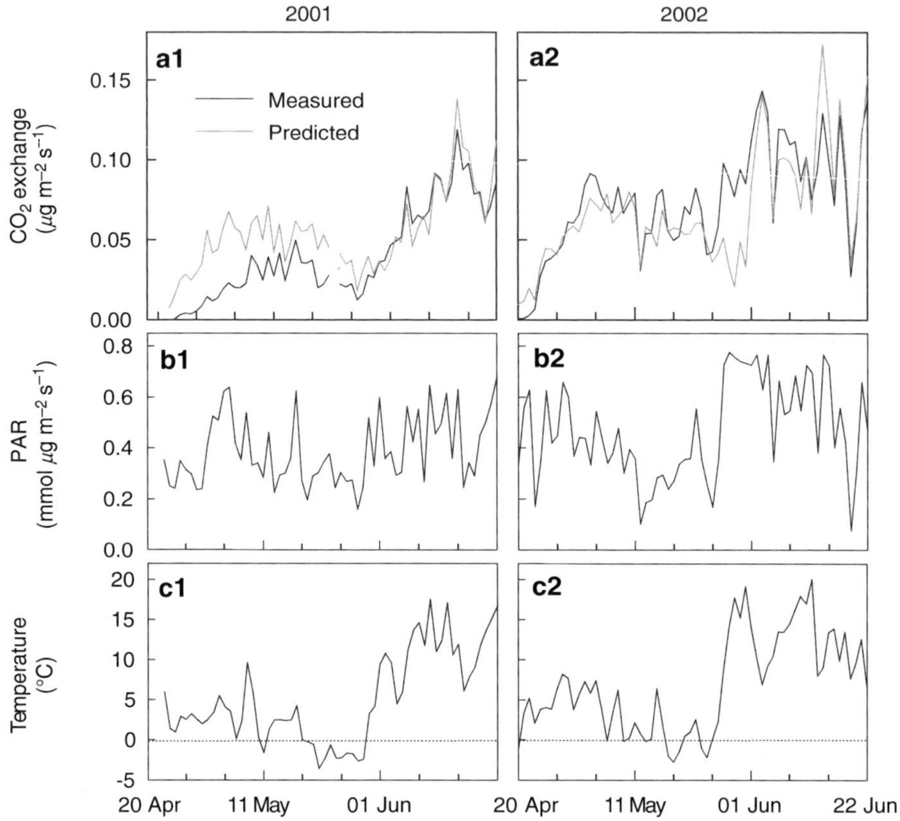

Fig. 4.24 The observed and predicted midday (11:00–14:00) mean CO_2 exchange (**a1**–**a2**), PAR (**b1**–**b2**) and temperature (**c1**–**c2**) at SMEAR I, northern Finland, in the springs 2001 and 2002. The prediction is based on the PhenPhoto model

The CO_2 has a strong annual pattern: slow recovery in the spring, peak at late summer and decline towards winter. The weather during each year causes variation from the general annual trend. Our theory and the PhenPhoto model are able to predict the daily photosynthesis utilising the development of the efficiency of the photosynthetic functional substances and prevailing light and temperature (Fig. 4.25).

Adequacy of the Model Structure. The annual cycle of photosynthesis is caused by the changes in the functional substances. Unfortunately, the knowledge on the operation of the biochemical regulation system on photosynthetic functional substances has hardly ever been incorporated in models of photosynthesis. Instead, the seasonal changes in the state of the photosynthetic machinery are often completely omitted in the models. We have rather limited knowledge on the operation of the biochemical regulation system on photosynthetic functional substances.

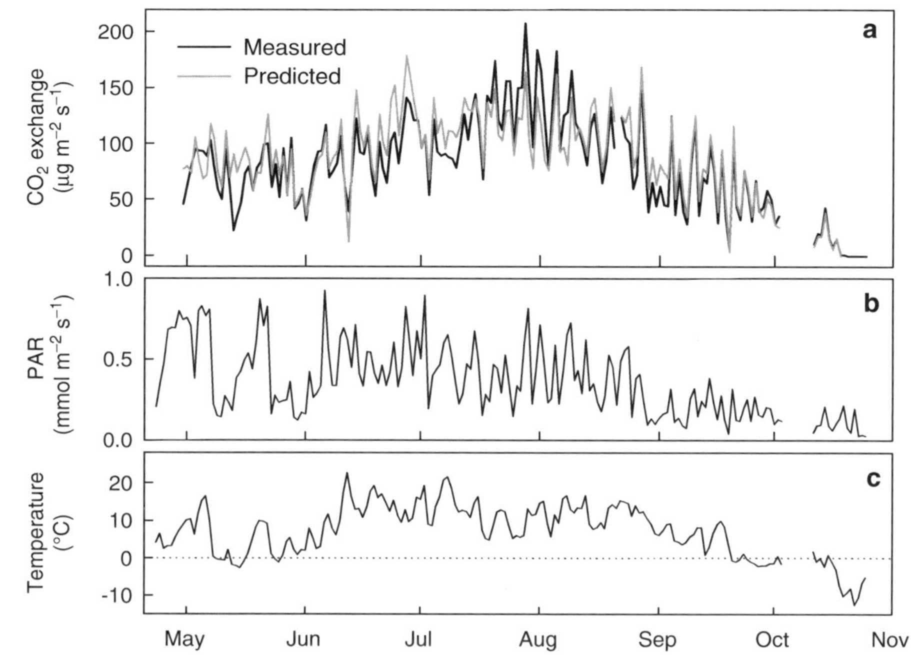

Fig. 4.25 Measured and predicted midday (11:00–14:00) mean CO_2 exchange of Scots pine shoots (**a**), photosynthetically active radiation (*PAR*) (**b**) and temperature (**c**) at SMEAR I, northern Finland, at noon in 2006

Research technical argumentation dominates in the field, and we should utilise more physiological knowledge in the analyses of the annual state of the photosynthetic functional substances. Lloyd et al. (2002) present interesting results from Siberia dealing with the changes in photosynthesis during growing season. The research of the action of the biochemical regulations system of annual cycle on photosynthesis will be important in the future.

Also the discrepancy between the prediction and the measured fluxes after cold nights indicates shortcomings in the applied model structures. Evidently combination of strong light and freezing temperatures in clear spring mornings leads to further downregulation of photosynthesis (Sect. 4.2.2.6; Ensminger et al. 2004; Porcar-Castell et al. 2008a). The residuals as function of time showed some systematic behaviour (Fig. 4.26) indicating need of improvement in the temperature effect on the development of the annual state of the photosynthetic functional substances.

Explaining Power of the Theory of Annual Cycle of Photosynthesis. The action of the biochemical regulation system explains the recovery of photosynthesis in the spring and the decrease in the autumn in our theory of the annual cycle of photosynthesis. The regulation system synthesises new substances and converts inactivated ones to active state in the spring and decomposes and deactivates them

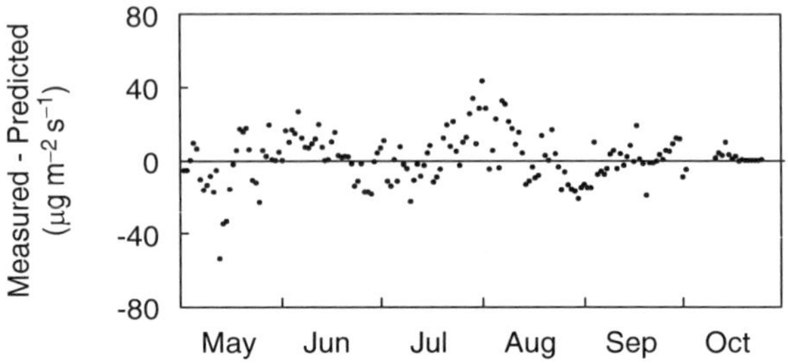

Fig. 4.26 The residuals of CO_2 exchange (measured-predicted) against date at SMEAR I in northern Finland in 2006

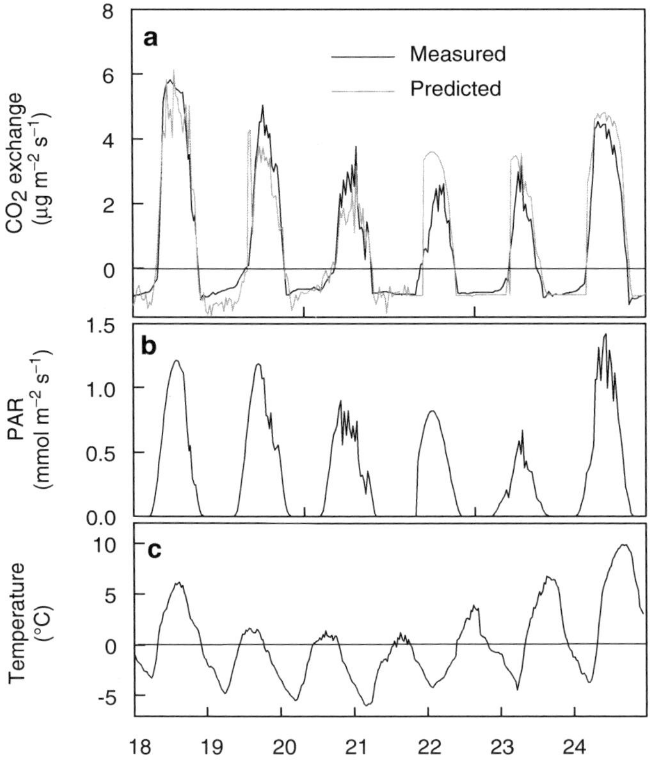

Fig. 4.27 Observed and predicted CO_2 exchange (**a**), photosynthetically active radiation (*PAR*) (**b**) and temperature (**c**) at SMEAR I during 17–24 April in 2005

in the autumn. These changes increase the efficiency of photosynthetic functional substances in the spring and decrease it in the autumn resulting in increasing and decreasing photosynthesis as seen in Fig. 4.25.

Temperature affects the recovery of the biochemical regulation system. High temperatures accelerate the action of the system, and low temperatures can even result in decomposition and deactivation of the photosynthetic functional substances (Fig. 4.27). The proportions of the explained variance in the prediction of measured CO_2 were high as in the case of the predictions in midsummer around 90%.

Comparison with Other Approaches. Several empirical and modelling studies, however, have indicated a strong relationship between temperature and the seasonal changes in light-saturated photosynthesis and in the efficiency of light reactions (e.g. Leverenz and Öquist 1987; Bergh et al. 1998; Lundmark et al. 1998). The seasonal changes in photosynthesis are also linked with frost tolerance (e.g. Vogg et al. 1998; Repo et al. 2006).

We have not found in the literature any theory of the annual cycle of leaf or shoot photosynthesis over several active periods that utilise physiological knowledge as background. This omitting of the physiological knowledge has evidently deep roots in the applied philosophy of science.

Conclusion: We utilised about 700,000 measurements in the test of the theory of annual cycle of photosynthesis. The theory was able to explain the characteristic features in annual photosynthesis. However, our theory was unable to explain the behaviour of photosynthesis during cold mornings when the temperature is below zero. The systematic features in residuals were rather small. The explaining power of our theory is high, about 90%. We have not found any theory of the recovery of photosynthesis in the spring that utilises physiological knowledge as background. Thus, our theory of annual cycle in photosynthesis gained strong corroboration in the severe test.

4.2.2.2 Transpiration

Pertti Hari, Kourosh Kabiri Koupaei, and Pasi Kolari

Structural and Physical Background

Water can be in either gas, liquid or solid form, and the phase transitions between those forms consume or release energy. The metabolic actions of vegetation take place in liquid water; thus, water is necessary for life. Within the temperature range from 0 to 100°C and under normal pressure, water can be in either liquid or gas form. The maximum water vapour concentration in air is temperature dependent, and air carrying the maximum water vapour concentration is called water-vapour-saturated air, often shortly saturated air.

Inside leaves, a thin water layer covers the surface of the mesophyll cells surrounding the stomatal cavity, and water can evaporate to form saturated air in the substomatal cavity (Fig. 4.28). The water vapour concentration in the ambient air is usually clearly lower than that in the stomatal cavity.

P. Hari (✉)
Department of Forest Sciences, University of Helsinki, P.O. Box 27,
00014 University of Helsinki, Helsinki, Finland
e-mail: pertti.hari@helsinki.fi

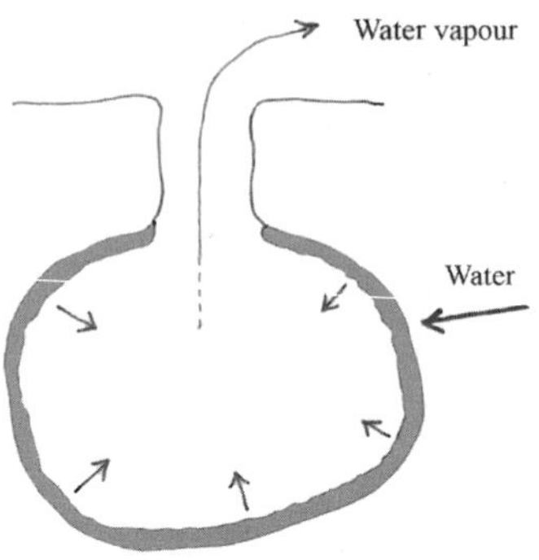

Fig. 4.28 Flow of water molecules by diffusion from water film on mesophyll cells via stomatal pore into the atmosphere. The grey line on the mesophyll surface indicates water

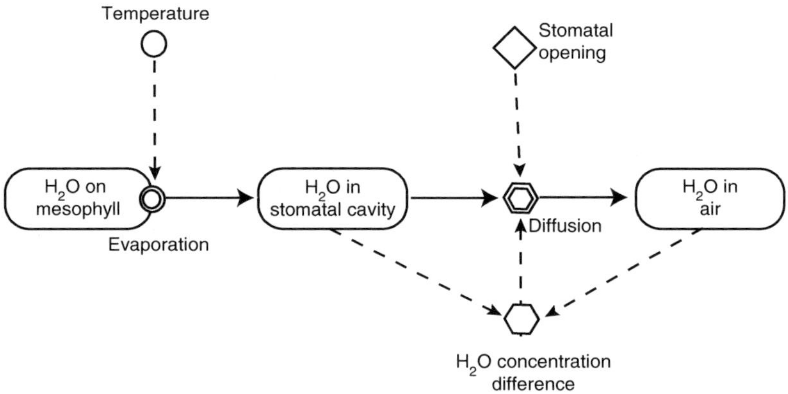

Fig. 4.29 Transpiration as combination of evaporation and diffusive transport. The water evaporates from the surface of the mesophyll cells after which diffusion transports it out through the stomatal pore. The stomatal cavity air is water vapour saturated. Stomata control the flow (The symbols are introduced in Fig. 2.3)

The stomatal pores enable diffusion of water vapour into the atmosphere and simultaneous diffusion of CO_2 into the mesophyll. CO_2 molecules diffused into the stomatal cavity dissolve into the water laying on the cell surfaces and further continue their journey deeper into the cells. Simultaneously water vapour evaporates from the water surface on the mesophyll cells.

Theory of Optimal Stomatal Control of Transpiration. The action of the stomatal pore can control the water flux from the leave into the atmosphere. Partial closure of the pore reduces transpiration and inflow of CO_2 generating a strong coupling between transpiration and photosynthesis. Effective stomatal control has developed during evolution to match the transpiration and photosynthesis with each other and with the availability of water.

The interplay of evaporation, stomatal control and diffusion generating the water flux from leaves into the atmosphere is visualised in Fig. 4.29.

Specification of Basic Concepts and Ideas

Temperature mainly determines the saturated water vapour concentration in the stomatal cavity. The ambient concentration is usually clearly lower than inside the leaf. The water vapour saturation deficit is defined as the difference between the water vapour concentrations inside stomata and in the ambient air.

The treatment of transpiration flow of water out from the leaf requires an exact concept. The *specific transpiration rate* is defined, analogously to the definitions of photosynthetic and respiration rates, as the ratio of the amount of transpired water from a small tissue element during a short interval and divided by the leaf mass in the tissue element and the length of the interval. Thermal movement transports molecules in the air, also water molecules in the stomatal cavity, through the pore into the ambient air.

Basic idea TR1: Transpiration rate is driven by diffusion of water vapour molecules.

The diffusion transports water vapour molecules in and out through the stomatal pores. The concentrations in the stomatal cavity and in the ambient air determine the in and out fluxes.

Basic idea TR2: Transpiration rate is proportional to water vapour concentration difference between stomatal cavity and ambient air.

Transpiration requires effective water transport structures in coarse roots, stems and branches. Transport is expensive: Large amounts of sugars have to be used in the construction of the water pipes to obtain enough water transport capacity. The stomatal control of transpiration evidently reduces transpiration during high saturation deficit and in this way reduces the sugar consumption to the water transport system. The stomatal action reduces inflow of CO_2 and in this way also photosynthesis responses simultaneously.

Basic idea TR3: Effective stomatal control of transpiration and photosynthetic rates has emerged in evolution.

The combination of transpiration loss and photosynthetic gain can be done with the optimality hypothesis proposed by Cowan (1977) and Cowan and Farquhar (1977).

Basic idea TR4: The functioning principle of stomata is to maximise the photosynthetic production minus transpiration costs.

Theoretical Model and Measurements

The core of the theoretical model describing the relationship between transpiration rate and environmental factors is rather simple. The transpiration flux driven by diffusion is proportional to the concentration difference between stomatal cavity and ambient air. The treatment of the stomatal action is more complicated, but it was considered already in the connection of photosynthesis (Eqs. 4.10, 4.17 and 4.18).

We utilise the field measurements at SMEAR I in eastern Lapland, Finland, in the test of the optimal stomatal control theory of transpiration. We describe the measurements in more detail in Chap. 9.

Tests of the Theory

Estimation of Parameter Values. We estimated the values of the parameters already in the connection of photosynthesis when we formed the combined residual sum of squares for CO_2 exchange and transpiration. We utilise the results obtained in the analysis of photosynthesis for transpiration.

Prediction of Characteristic Features of Transpiration. The transpiration is more variable from day to day than photosynthesis. There are days of high and low transpiration, but in photosynthesis, the days are more similar with each other (Figs. 4.30a and 4.16). The theory of optimal stomatal control of transpiration is able to predict this great variability between the days.

During typical summer day of intermittent cloudiness, the transpiration pattern is clear, showing slow increase until late afternoon and thereafter rather rapid decline. The theory of optimal stomatal control of transpiration predicts great variation during the day although the pattern is a bit different (Fig. 4.31a1).

Transpiration has rather characteristic pattern on sunny days: rapid increase in the morning, saturation before midday and thereafter slow decline. The theory of optimal stomatal control of transpiration is able to predict this pattern (Fig. 4.31a2).

During cloudy and often rainy days, the transpiration is nearly missing; only very small fluxes can be measured, if any. The theory of optimal stomatal control of transpiration predicts also this behaviour (Fig. 4.31a3).

Adequacy of the Model Structure. The diffusion of water vapour through stomatal pores and the action of stomata form the model structure. The diffusion submodel follows the modelling tradition proposed by H. T. Odum (1983). In this case, this model is rather adequate description of the diffusion of water vapour molecules. We introduce the action of stomata with optimal stomatal control theory that is used rather seldom, although it has sound evolutionary basis and it has gained strong corroboration in field tests.

The residuals for the optimal stomatal control model of transpiration show some systematic behaviour (Fig. 4.32). This can reflect three different problems: (1) systematic measuring errors, (2) problems in the determination of the parameter values and (3) shortcomings in model structure. Evidently, the systematic measuring errors dominate the residuals.

Explaining Power of the Theory of Transpiration. The great variation in the difference in water vapour concentration between the stomatal pore and the ambient air is the dominating factor in explaining the observed variation of transpiration. The stomatal action reduces, however, transpiration in the afternoons, especially during sunny and warm days.

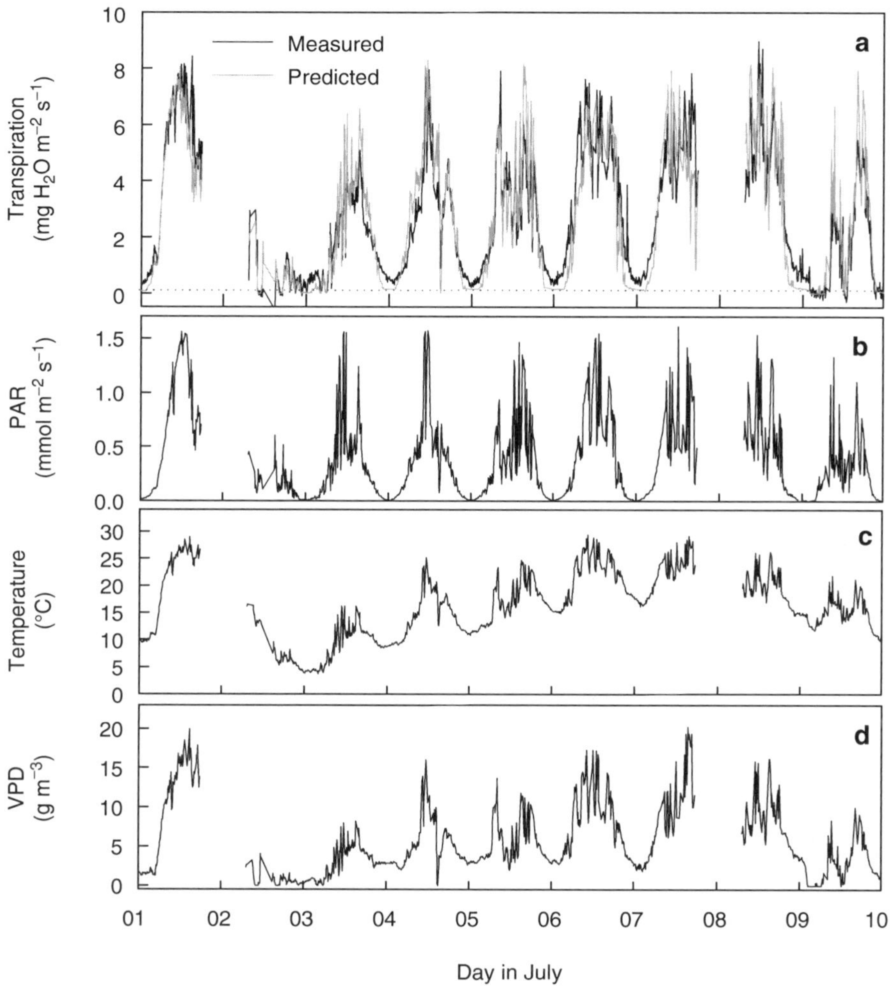

Fig. 4.30 The measured and predicted transpiration (**a**), measured photosynthetically active radiation (*PAR*) (**b**), temperature (**c**) and VPD (**d**) at SMEAR I, northern Finland, during 1–10 July in 2006. VPD is the difference in the water vapour concentrations between the stomata interior and the ambient air. The *gaps* in data are caused by thunderstorms since the measuring system is coupled off to prevent lightning damage

The explaining power of the theoretical model of transpiration varies between chambers and years but is usually around 70%. This result is mainly caused by the small variation in the accepted measurements since night-time values are rejected as biased due to condensation on the chamber walls. In addition, the systematic measuring errors reduce the explained proportion of variance power.

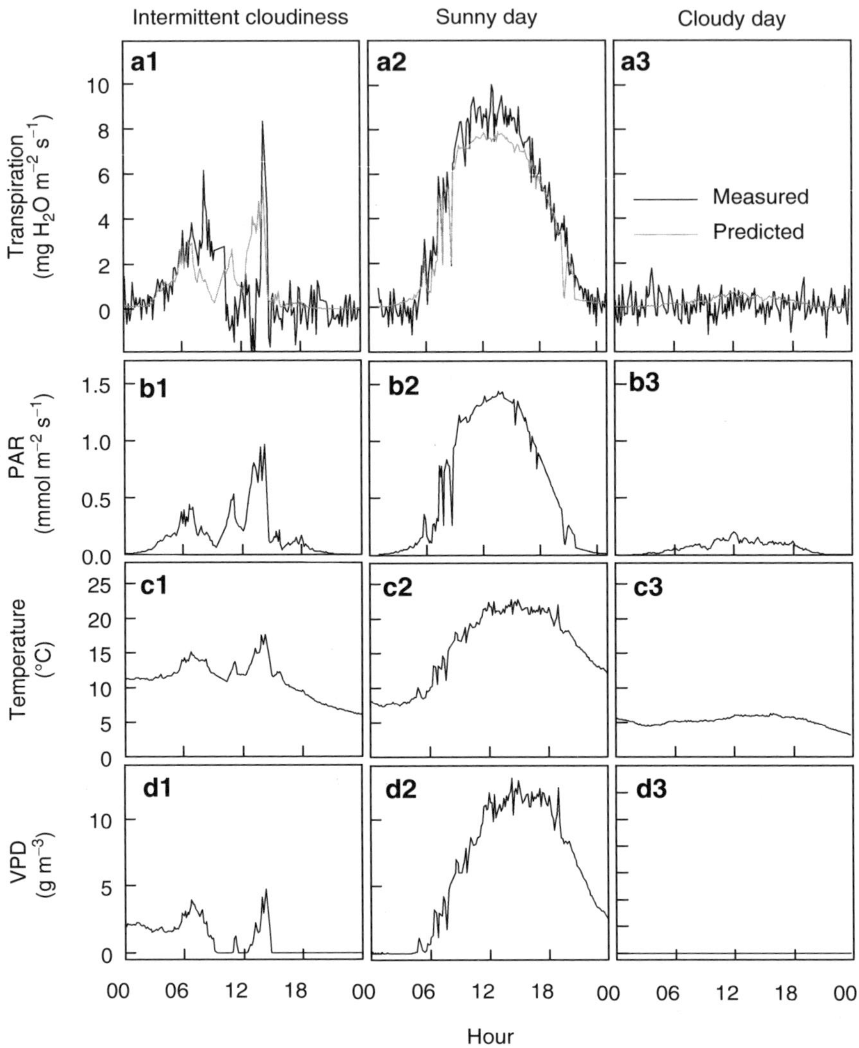

Fig. 4.31 The measured and predicted transpiration (**a1**–**a3**), measured PAR (**b1**–**b3**), temperature (**c1**–**c3**) and VPD (**d1**–**d3**) at SMEAR I, northern Finland, during a day of intermittent cloudiness (26 June), a sunny day (2 July) and a cloudy day (6 August) in 2005

Comparison with Other Approaches. Empirical regressions (e.g. Ball et al. 1987; Leuning 1995) dominate the research of the effect of stomatal action on transpiration and photosynthesis. Important results have been gained, but the theoretical understanding and explanation are missing.

The optimal stomatal control hypothesis by Cowan (1977) and Cowan and Farquhar (1977) opened new theoretical argumentation into the problem. The

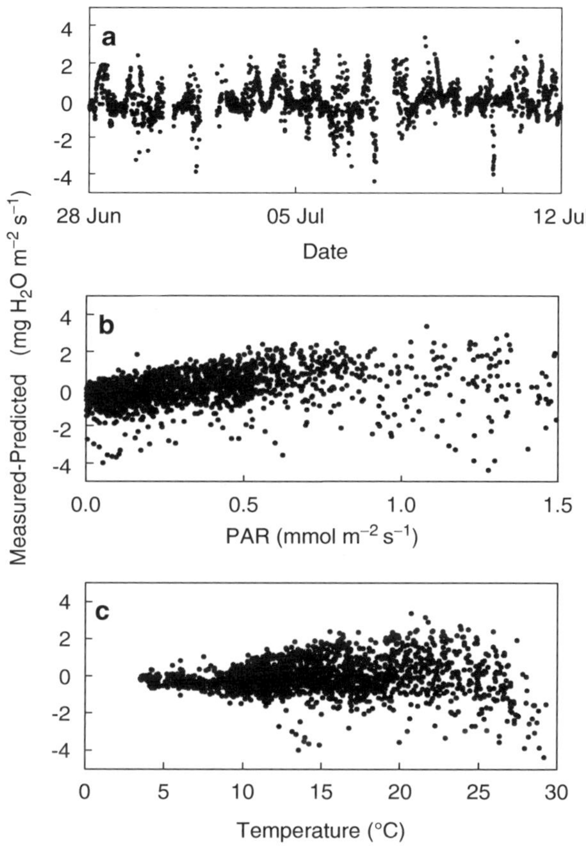

Fig. 4.32 The difference between measured and predicted transpiration against date (**a**), PAR (**b**) and VPD (**c**) at SMEAR I, northern Finland, in midsummer 2006

solution of the optimality hypothesis with additional assumptions by Hari et al. (1986) enables the testing of the hypothesis. This possibility is not yet fully utilised.

Conclusion: The interpretation of the test of the optimal stomatal control of transpiration is problematic, since the evident systematic measuring errors disturb conclusions. However, we have to live with the measuring problems and draw limited conclusions. Thus, the theory of optimal control of transpiration gained weak corroboration in the test.

Annual Cycle of Transpiration

Structural, Metabolic and Physical Background

Water vapour deficit in the air is high in the spring when the soil is often frozen and availability of water for roots is low. In addition, the frozen stems are

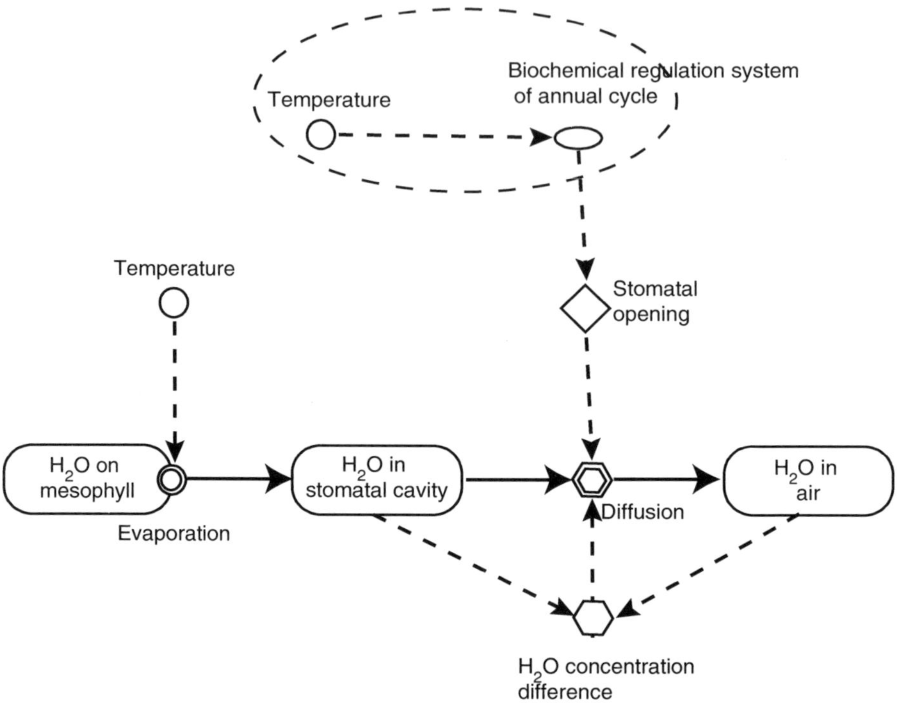

Fig. 4.33 Visualisation of the effect of the annual biochemical regulation system on transpiration. The water flow is as in Fig. 4.29. The annual biochemical regulation system changes the functional substances involved in the stomatal action, and in this way, it generates the annual cycle in transpiration

ineffective water transport systems. In these conditions, the trees have to reduce their transpiration and close, at least partially, their stomata. Thus, annual cycle in transpiration is a necessity for trees to overcome the winter.

The biochemical regulation system of the annual cycle changes the concentrations and activities of the functional substances involved in the stomatal action and in this way reduce transpiration (Fig. 4.33), and the daily transpirations have a clear annual pattern (Fig. 4.34).

Specification of Basic Concepts and Ideas

The stomatal actions are involved in the diffusion of water vapour and carbon dioxide out and in from the stomatal cavity. Consequently, both transpiration and photosynthesis react to changes in the stomatal action.

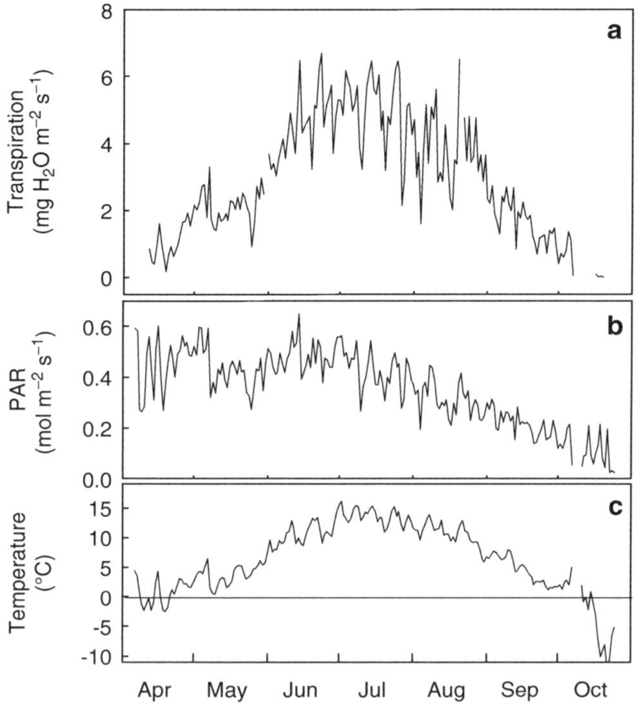

Fig. 4.34 The measured midday mean transpiration (**a**), photosynthetically active radiation (*PAR*) (**b**) and temperature (**c**) at SMEAR I, northern Finland during 2001–2007. The weather in April in Lapland is usually very sunny as can be seen also in the photosynthetically active radiation

Basic idea TRA1: The annual cycle in stomatal action derived for photosynthesis is valid also for transpiration.

Theoretical Model and Field Measurements

We introduced the annual cycle of photosynthesis with the annual state of photosynthetic functional substances and with the relationship between the efficiency and the state of photosynthetic functional substances. The optimality idea connects transpiration and photosynthesis via stomatal action with each other. The combination of the transpiration equations with annual cycle of stomatal action results into five equations (Box 4.2).

We utilise the field measurements at SMEAR I in eastern Lapland, Finland, in the test of the optimal stomatal control theory of transpiration. We describe the measurements in more detail in Chap. 9.

Box 4.2 The Equations Describing Transpiration Including the Effects of Air Saturation Deficit, Stomatal Action and Annual Cycle:

$$g_{H_2O}(t) = a\,u(t)g_{max}D_{H_2O}(t) \tag{4.30}$$

$$u^* = \left(\sqrt{\frac{C^a_{CO_2} - (r/(E_P\,f\,(I_{PAR})))}{\lambda\ a\ D_{H_2O}}} - 1\right)\frac{E_P\,f\,(I_{PAR})}{g_{max}} \tag{4.31}$$

$$u = \begin{cases} 0, & \text{if } u^* \leqslant 0 \\ u^*, & \text{if } 0 < u^* < 1 \\ 1, & \text{if } u^* \geqslant 1 \end{cases} \tag{4.32}$$

$$\frac{dS_p}{dt} = \frac{T - S_p}{\tau} \tag{4.33}$$

$$E_P(S_P) = \text{Max}\,\{0, a_S\,(S_P - S_0)\} \tag{4.34}$$

Test of the Annual Theory of Transpiration

Parameter Estimation. No additional parameter estimation is needed since we have already determined the values of all parameters in the treatment of CO_2 exchange.

Prediction of the Characteristic Annual Features in Transpiration. The transpiration has a clear annual pattern: no or very little transpiration in winter, slow recovery during spring, high transpiration during summer and finally declining transpiration in autumn. The annual theory of transpiration is able to predict the annual pattern (Fig. 4.35).

The variation in transpiration between days is large. There are days with high transpiration and very low transpiration throughout the whole growing season. The annual theory of transpiration predicts the high variability in transpiration.

Adequacy of the Model Structure. The possible problems in the model structure are the same as in the case of the theory of the annual cycle of photosynthesis. The residuals indicate some systematic behaviour (Fig. 4.36), especially in the spring.

Explaining Power of the Annual Theory of Transpiration. The dominating reason for the strong annual cycle in transpiration is the annual cycle in the water vapour concentration difference between the stomatal cavity and ambient air. In addition, the action of the stomata makes the annual cycle of transpiration more pronounced.

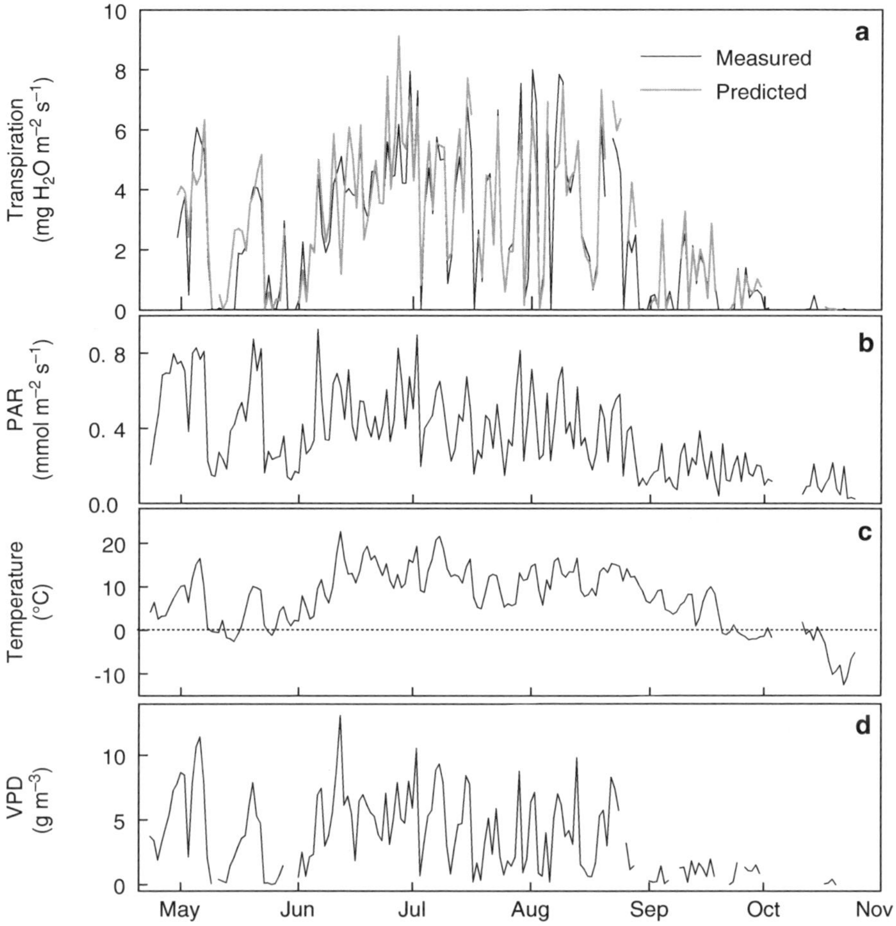

Fig. 4.35 Measured and predicted midday average transpiration (**a**), PAR (**b**), temperature (**c**) and VPD (**d**) at SMEAR I station, northern Finland during 15 April–1 November in 2006

The proportion of the explained variance varied between years and chambers; 70% is the typical value. The systematic measuring errors and problems in the estimation of the parameter values contribute considerably to the poor fit between prediction and measurements.

Comparison with Other Theories. The annual cycle of transpiration has received rather little attention in the research. The empirical approach has dominated the field. We know no other theories that include the action of the biochemical regulation system on stomata.

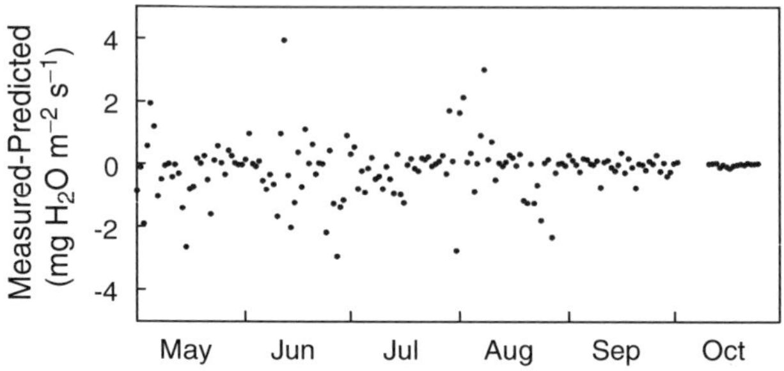

Fig. 4.36 The midday residuals of measured and predicted mean daily transpirations as function of date at SMEAR I in 2006

Conclusion: Our theory passed successfully the tests. Thus, the theory of optimal stomatal control of annual transpiration gained corroboration, but the evident systematic measuring errors and problems in estimation reduce the corroboration.

4.2.2.3 Theory of Optimal Stomatal Control During Drought

Pertti Hari, Pasi Kolari, and Eero Nikinmaa

Drought is very common in forests at the global scale. Only in northern forests, the water pool in the soil is sufficiently large to provide enough water during most of the growing season. Vegetation can reduce transpiration by partially closing the stomata that also slows down photosynthesis. In this way, the water pool, originating from the snow of winter or from previous rainier periods, can last until the rains increase again the soil water pool.

Water is compulsory for all metabolism, and drying of leaves is lethal. Acclimations to drought have developed in evolution to overcome the drought periods. The regulation system of acclimations to drought is activated, when the water pool in soil is small enough. It changes the functional substances involved into the stomatal action resulting in stronger stomatal closing and reduced transpiration and photosynthetic fluxes between leaf element and the atmosphere (Figs. 4.37 and 4.38).

In the fairly moist conditions of the northern forests, there are seldom warm and dry periods long enough to dry out the soil. We have not been able to observe clearly drought-induced decline in shoot CO_2 uptake at SMEAR I near the arctic-alpine timberline. The forest ecosystems in the continental and the southern regions

P. Hari (✉)
Department of Forest Sciences, University of Helsinki, P.O. Box 27,
00014 University of Helsinki, Helsinki, Finland
e-mail: pertti.hari@helsinki.fi

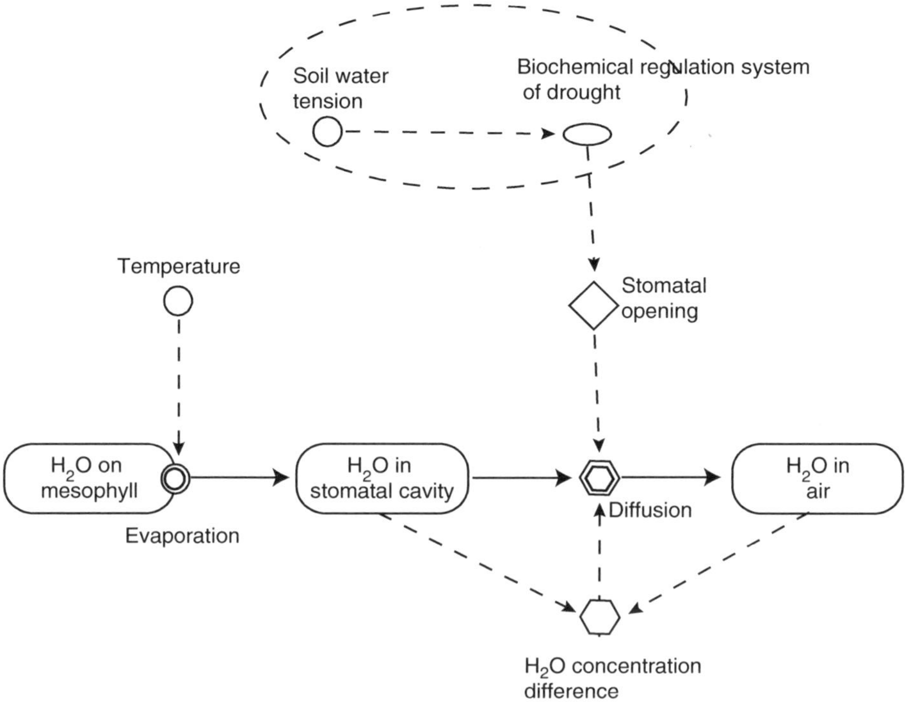

Fig. 4.37 The effect of drought on transpiration via regulation of functional substances. The biochemical regulation system responses to drought resulting in increasing closure of stomata. Otherwise, the transpiration is the same as illustrated in Fig. 4.33

are more prone to suffer from soil water deficit. In boreal conditions of SMEAR II, severe drought is still rare, but in few years, there have been short periods in late summer when the effect of soil water deficit has been obvious in the gas exchange data. In the warming climate, the frequency of drought periods and the ability of forest ecosystems to withstand drought may become a more important question. Therefore, it is important that we are also able to introduce the effects of dry soil in predicting photosynthesis into our theory.

Basic Ideas

Droughts are common in the global scale, and the acclimations to droughts strongly contribute to the success of trees and ground vegetation.

Basic idea PD1: When soil water storage decreases, the regulation system changes the functional substances involved in stomatal action in such a way that the stomata become more active to reduce transpiration.

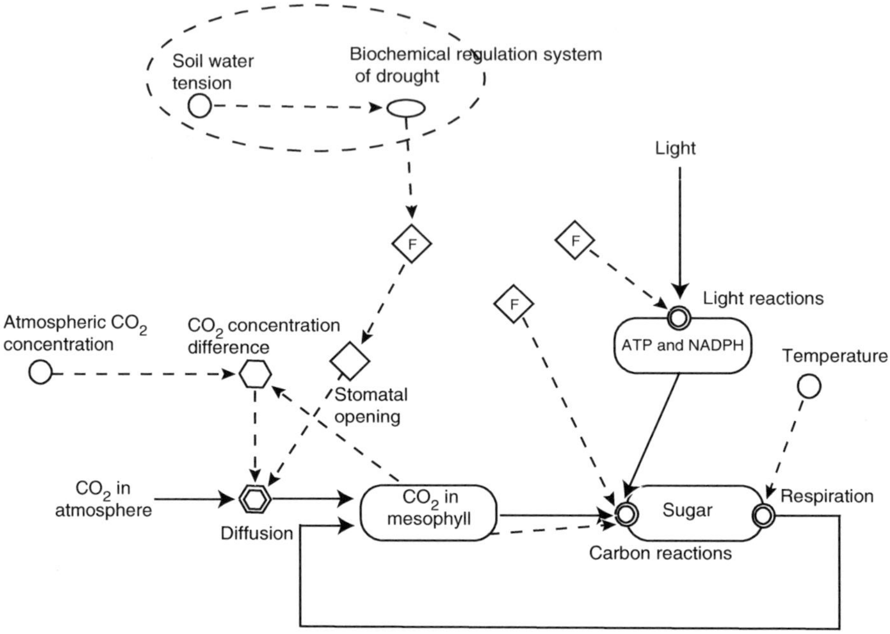

Fig. 4.38 The effect of drought on photosynthesis via action of functional substances involved in stomatal closure. The biochemical regulation system responses to drought by increasing the stomatal control of water vapour and carbon dioxide flows; otherwise, the figure is rather similar with Fig. 4.22

The changes in the functional substances are slow, but during longer periods, they change plant metabolism and gas exchange. The soil water tension increases during drought, and high tensions hinder water uptake.

Basic idea PD2: The cost of transpiration in the theory of optimal stomatal control of photosynthesis increases when the soil water tensions increase.

Theoretical Model

During conditions of sufficient water pool in the soil, the previously used optimal stomatal control model is able to predict well the daily behaviour of gas exchange over the summer with a fixed parameter set, as seen in Sect. 4.2.2. The derivation of the optimal stomatal control model is valid also during drought. However, the cost of water increases when the water pool in the soil is depleted according the basic idea PD2.

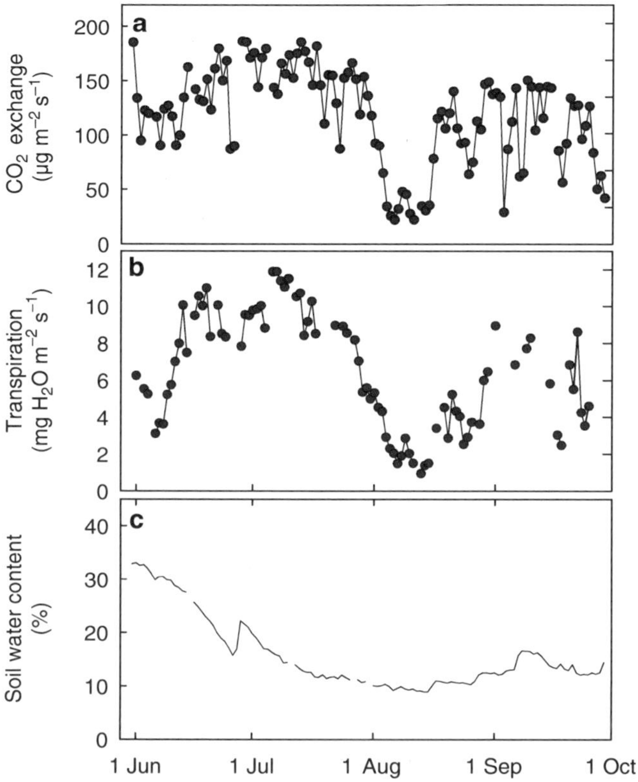

Fig. 4.39 The daily midday (from 11:00 to 14:00) mean CO_2 exchange (**a**), transpiration of Scots pine shoots (**b**) and soil water content (**c**) at SMEAR II in southern Finland in 2006

Test with Field Measurements

The summer of 2006 was extremely dry in southern Finland. There was very little rain from early July to mid-August. During late July and the first half of August, there was a clear reduction in CO_2 uptake and transpiration in Scots pine shoots at SMEAR II (Figs. 4.39 and 4.40). The decline could be first seen in the afternoon and later, when the soil kept drying out, over the whole day. Eventually transpiration and CO_2 exchange dropped to about 10% of their pre-drought levels. The rate of respiration (night-time CO_2 exchange in Fig. 4.40) also declined with photosynthetic CO_2 uptake. The role of stomatal action in the reduction of gas exchange can be concluded from the measured transpiration that indicates almost complete closure of stomata in the last few days of the drought period. When the soil water storage was replenished in mid-August, gas exchange recovered to nearly normal level in few days.

We estimated the daily values of the cost of transpiration using both CO_2 exchange and transpiration measurements. The efficiency of photosynthetic functional

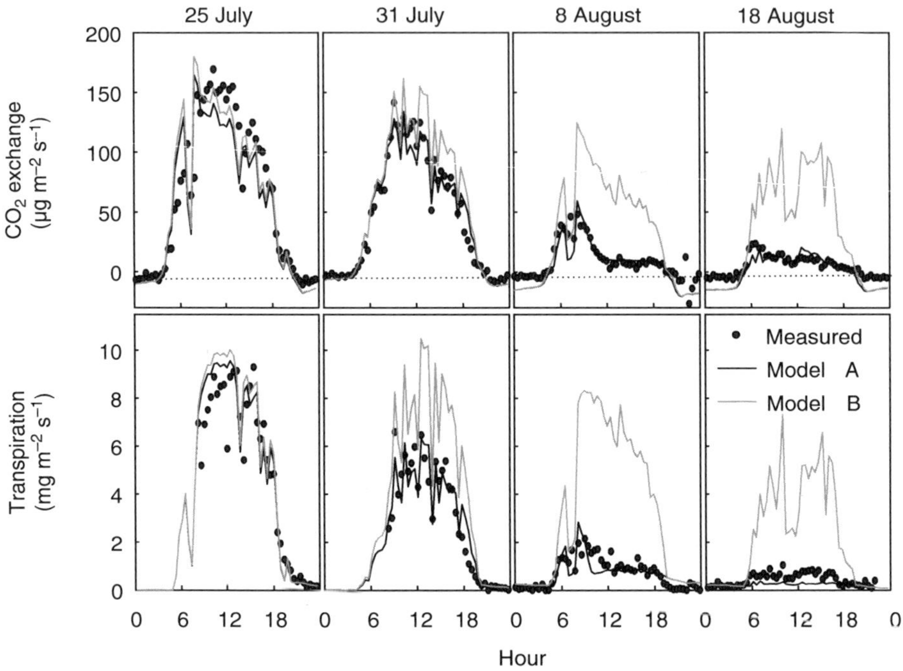

Fig. 4.40 Modelled with daily cost of water (A, *black line*), modelled with constant cost of water (B, *grey line*) and measured (*dots*) diurnal course of CO_2 exchange (*upper panels*) and transpiration (*lower panels*) of a Scots pine shoot at SMEAR II on four selected days during progressive drought in 2006. The *grey line* indicates the gas exchange predicted with the optimal stomatal control model and constant cost of water (λ), and the *black line* modelled gas exchange when λ was estimated daily

substances E_P was obtained from the annual cycle model (Box 4.2) and calibrated for each shoot using values estimated for the period of 15 June–15 July when there was still ample water in the soil. The other photosynthetic parameters were fixed.

The partial closing of stomata reduces transpiration and simultaneously also photosynthesis. When the drought becomes more severe, the stomata get more closed. We cannot predict the stomata behaviour during drought since we have no test data, and consequently, we have to use all data in the estimation. However, the optimal stomatal control model when the value of the parameter λ is estimated daily can describe the daily transpiration and photosynthesis patterns rather well.

The stomatal control of transpiration becomes more pronounced when the drought comes more severe as can be seen from the declining trend in transpiration. This behaviour can be seen in the parameter values of the optimal stomatal control model, since the cost of water increases when water pool in the soil is depleted (Fig. 4.41). The relationship between the daily values of λ and soil water tension is rather clear (Fig. 4.42).

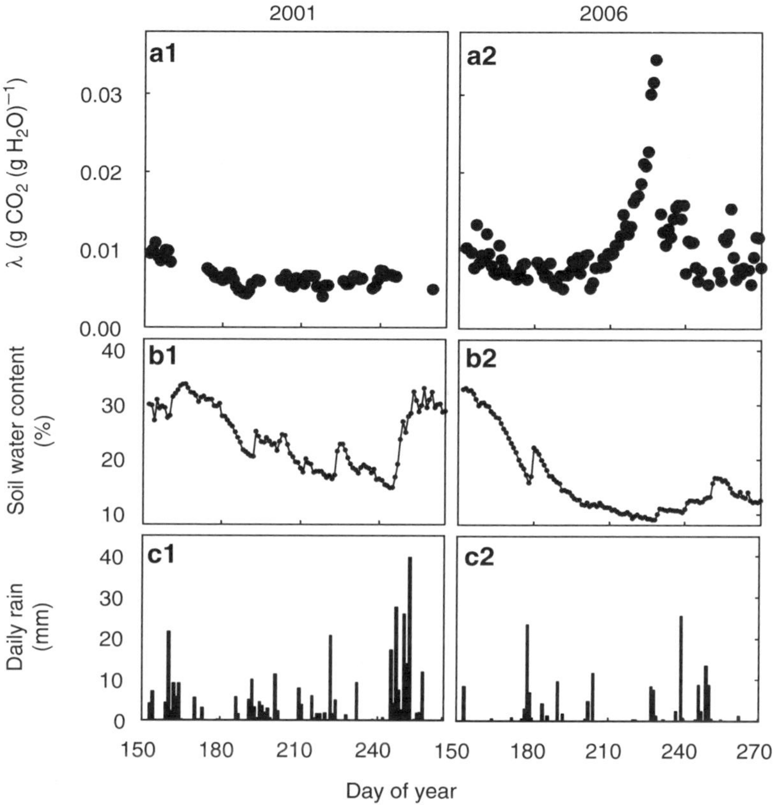

Fig. 4.41 Estimated daily values of cost of water (λ) (**a1**), volumetric water content in the uppermost 5-cm layer of mineral soil (**b1**) and daily precipitation (**c1**) in a typical summer (2001) and in a summer when there was severe drought (**a2**, **b2**, **c2**) at SMEAR II (2006)

Fig. 4.42 The relationship between the daily values of cost of water λ and soil water tension at SMEAR II southern Finland, in the summer 2006

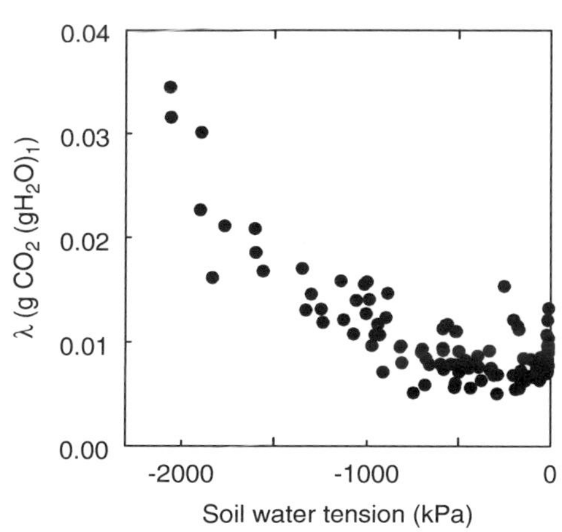

The optimal stomatal control provides a good framework to study the effects of drought on transpiration and photosynthesis. Its theoretical basis is rather non-problematic, and it has passed successfully sever tests. However, our present data are rather limited for proper analysis of residuals.

The increasing cost of transpiration during drought gives satisfactory explanation for the changes in gas exchange during drought. The proportion of explained variance is rather meaningless because the limited data available exclude predictions needed in the tests.

The empirical regressions dominate the research of the effect of drought on photosynthesis and transpiration. Valuable results have been obtained, but the theoretical explanation is missing, and applications, without theoretical understanding, are problematic.

Conclusion: Our theory of optimal stomatal control passed successfully the tests. However, the limited data available reduced strongly the severity of the test, and this is why the theory gained only weak corroboration.

4.2.2.4 Increasing CO_2 and Acclimation of Photosynthetic Processes and Structure

Eija Juurola and Pertti Hari

Background

Besides the annual cycle, there are slow (i.e. during the lifetime of individuals) and irregular changes in the environment. In Chap. 2, we defined acclimations of processes as changes in the biochemical regulation system and in the functional substances matching processes in vegetation with slow and irregular changes in environment. Also we defined acclimations of structure as slow changes in structure matching the structure with irregular changes in the environment. In contrast to adaptations, which occur over evolutionary timescales, acclimations occur over days, weeks and months and do not involve alterations in the genetic make-up (Chap. 2).

Atmospheric CO_2 concentration has varied considerably in the timescale of million or milliard years. During the last 600,000 years, it has fluctuated between 200 and 300 ppm (see Fig. 8.11). The fine structures of leaves as well as the functional substances that carry out photosynthesis have been subjected to this

E. Juurola (✉)
Department of Forest Sciences, University of Helsinki, P.O. Box 27, 00014 University of Helsinki, Helsinki, Finland

Department of Physics, University of Helsinki, Helsinki, Finland
e-mail: eija.juurola@helsinki.fi

fluctuation in atmospheric CO_2 concentration. Still, during the lifetime of an individual tree, atmospheric CO_2 concentration has been rather stable due to the large atmospheric carbon pool. Current large emissions of CO_2 have, however, caused very rapid change that has never occurred before – at least not in millions of years. Hence, the availability of CO_2 now changes even during the lifetime of a single tree.

As CO_2 is the key substrate for photosynthesis, its availability largely determines the speed of the process. Initially, with increasing CO_2 concentration, photosynthesis is accelerated due to the increased availability of main substrate and the biochemical properties of the key enzyme in CO_2 fixation process, Rubisco. Thus, the primary acclimation originates from the CO_2-fixing enzyme Rubisco, and all the other acclimations follow from this primary effect. As a consequence, for example, the water-use efficiency (WUE, CO_2 fixed per water lost) increases. In the long term, if the CO_2 concentration stays high, complicated acclimations arise leading to alterations in concentrations or activities of functional substances involved in photosynthesis and finally to reallocation of resources within the photosynthetic apparatus and changes in leaf structure (e.g. Woodrow 1994; Drake et al. 1997; Luo et al. 1998).

Rubisco is an enzyme that catalyses both the fixation of CO_2 (carboxylation) and O_2 (oxygenation) (Woodrow and Berry 1988). The ratio of carboxylation to oxygenation primarily depends on the concentrations of CO_2 and O_2 at the carboxylation site, temperature and the Rubisco specificity factor of the species (Viil and Pärnik 1995).

With increasing CO_2 concentration, the catalysing reaction of Rubisco turns in favour of carboxylation, and thus, the fixation of O_2 and the release of CO_2 in photorespiration diminish (Jordan and Ogren 1984; Brooks and Farquhar 1985; Ghashghaie and Cornic 1994). As the photorespiration is relatively more enhanced with increasing temperature at current atmospheric CO_2 concentration, the temperature dependence of photosynthesis changes at elevated CO_2 concentration (e.g. Juurola 2005) leading to a higher optimum temperature.

During the last 250 years, the atmospheric CO_2 concentration has increased from near 280 ppm in the pre-industrial era (i.e. before 1750, Houghton et al. 2001; Keeling and Whorf 2005) to near 400 ppm, and it is expected to reach 700–1,000 ppm by the end of the twenty-first century (Houghton et al. 2001). Such a rapid increase in atmospheric CO_2 concentration is unprecedented in the genetic history of tree species. The acclimation of structure and function of trees to such a rapid change in CO_2 concentration has not been tested during their evolutionary history, and even unfavourable acclimations may occur. Indeed, despite extensive research on long-term effects of elevated CO_2 concentration on plants, it has remained unclear why large differences in photosynthetic response to elevated CO_2 concentration exist among (Tjoelker et al. 1998) and within species (Pettersson et al. 1993; Gunderson and Wullschleger 1994; Rey and Jarvis 1998).

Recent results show that coniferous and broad-leaved tree species have different strategies in investing nitrogen to CO_2-fixing enzyme Rubisco (Warren and Adams 2004), which in turn may lead to different acclimation patterns. On top of all

Fig. 4.43 The temperature dependence of CO_2 exchange in silver birch (*Betula pendula*) seedlings grown in different CO_2 concentrations and measured at the growth CO_2 concentration. Symbols are as follows: 350 ppm (*black circles*), 600 ppm (*white circles*), 1,500 ppm (*grey circles*) and for comparison, a temperature dependence of silver birch seedling grown at 350 ppm but measured at 2,000 ppm of CO_2 (*crosses*)

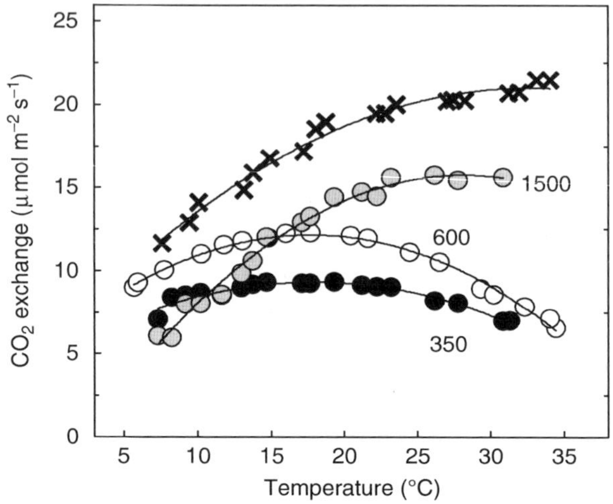

this, the Rubisco reaction is a temperature-dependent process, and increasing CO_2 concentration may affect the temperature dependence of CO_2 assimilation that is shown to vary between species and according to growth conditions (Björkman 1981a, b; Bernacchi et al. 2003).

Acclimation of Processes

In acclimation of processes in C3 plants to elevated atmospheric CO_2 concentration, Rubisco plays a key role (e.g. Eamus and Jarvis 1989; Stitt 1991). In the long term, increasing CO_2 concentration may lead to reduction in Rubisco activity or concentration because the amount of Rubisco required for maintaining the same assimilation rate decreases (Woodrow 1994; Drake et al. 1997). Despite this decrease, the increase in CO_2 concentration may still allow for a higher CO_2 assimilation rate (e.g. Hikosaka and Hirose 1998). Furthermore, because of the general kinetics of the Rubisco-catalysed enzymatic reactions, that is, saturation at high substrate concentrations and two competitive substrates (CO_2 and O_2), the acclimations mediated through Rubisco should be nonlinear in response to increasing CO_2 concentration (Woodrow 1994; Luo et al. 1998; Juurola 2003). Finally, due to the general kinetics of Rubisco, the prolonged effect of increased CO_2 concentration on photosynthesis should be temperature dependent leading to altered optimum temperature (e.g. Long 1991; Drake et al. 1997, Fig. 4.43).

The acclimation of Rubisco is not, however, the only change in the photosynthetic functional substances we should expect since there are strong connections between the functional substances. The increased performance of dark reactions is reflected also in light reactions and stomatal action. Thus, we can assume that concentrations of all functional substances involved in photosynthesis react to the increasing CO_2 concentration as visualised in Fig. 4.44.

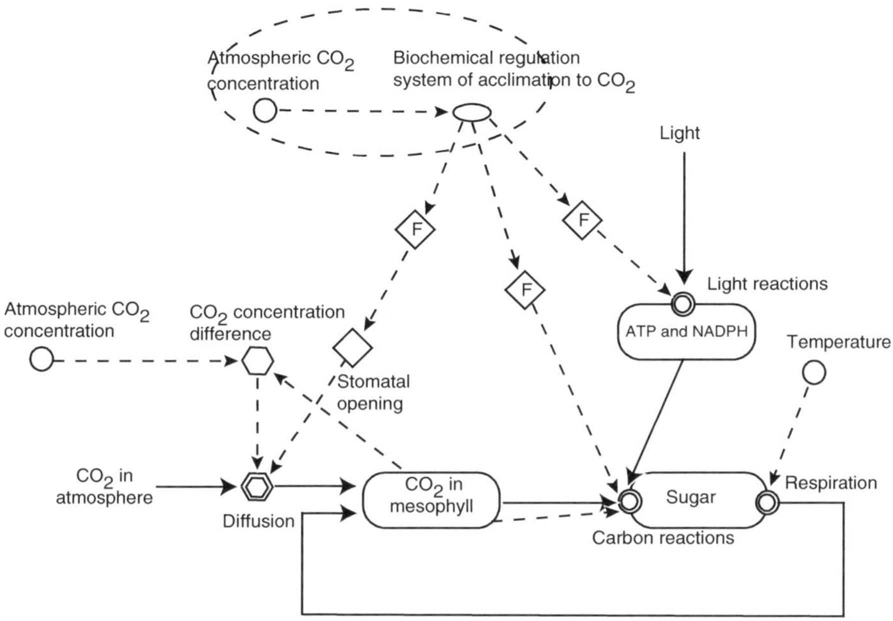

Fig. 4.44 Visualisation of the acclimation of photosynthesis to increasing atmospheric CO_2 concentration. The biochemical regulation system reacts to the atmospheric CO_2 concentration (The symbols are introduced in Fig. 2.3)

Acclimations of Structure

In the long term, changing CO_2 concentration induces alterations in leaf structure as well as in leaf development (Pritchard et al. 1999 and references therein). The leaf dry mass to fresh mass ratio can change due to, for example, accumulation of carbohydrates into the leaf. Due to the accumulation of carbohydrates, the chloroplast structure and organisation may change (Pritchard et al. 1997). Also leaf thickness might change reflecting, for example, the adjustment of altered ratio of penetrating irradiance and absorbed CO_2 as well as increased water-use efficiency (Ceulemans and Mousseau 1994).

Importantly, the increased water-use efficiency can lead to decreased stomatal density of leaves (Wynn 2003; Ainsworth and Rogers 2007). Indeed, it has been found that around natural CO_2 sources, the stomatal density is generally decreased (Tognetti et al. 2000). Also, Fernandez et al. (1998) reported that close to the CO_2 sources, the stomatal density is decreased but the size of the stomata is increased.

However, it is also stated that although stomatal density is well known to decline in response to increasing CO_2 concentration during growth, the response appears to saturate close to present-day ambient levels and FACE experiments typically show little or no change in stomatal density (Buckley 2008 and references therein).

The Responses to Increasing CO_2 Concentration Are Nonlinear

When the effects of increasing CO_2 concentration are studied in plants grown at one or two elevated CO_2 concentrations, it is implicitly assumed that the responses are linear, although it was suggested already in 1990s by Bowes (1991) and Woodrow (1994) that this was unlikely. Thus, to reveal the pattern of acclimation, it is essential to understand changes in plant physiology in response to a wide range of treatments. This kind of analysis is difficult in so-called FACE experiments. We studied the acclimation of young silver birch (*Betula pendula*) and Scots pine seedlings to a series of increased CO_2 concentrations from 350 to 2,000 ppm for two growing seasons (Juurola 2003). We measured chlorophyll fluorescence, Rubisco and chlorophyll contents, C/N ratios and contents and the gas exchange properties at the end of both growing seasons. All gas exchange measurements were performed in the laboratory with a dynamic system for measuring gas exchange (described in Aalto and Juurola 2001).

While there is evidence for a decreasing Rubisco concentration and activity at elevated CO_2 concentration, it is not always the case for the light reactions: Both an unchanged and decreased chlorophyll concentrations as well as PSII photochemistry have been observed (Wilkins et al. 1994; Lawlor et al. 1995; Lewis et al. 1996; Scarascia-Mugnozza et al. 1996; Rey and Jarvis 1998; Gielen et al. 2000; Jach and Ceulemans 2000; Riikonen et al. 2005). In Juurola (2003), however, a clear decreasing trend was observed also in chlorophyll concentrations in silver birch and Scots pine leading to unchanged balance between light and dark reactions in seedlings grown at elevated CO_2 concentrations (Fig. 4.45). This implies that the acclimation is a complicated phenomenon including changes both in functions and structure.

There exist several possible reasons for the diverse results on gas exchange in different species. First, due to the higher availability of CO_2, plants can more efficiently control the transpiration stream by changing stomatal aperture. The response may be acclimatory or a direct effect of CO_2 on stomatal control. In fact, a doubling of CO_2 concentration has often been reported to have no effect on stomatal conductance in conifers in contrast to deciduous species, reflecting perhaps the differences in leaf structure and in the strategy for water use (Eamus and Jarvis 1989; Roberntz and Stockfors 1998; Juurola 2003). Decrease in the density of stomata might be a favourable structural acclimation because it utilises the increased CO_2 availability to reduce water loss.

The temperature dependence of CO_2 assimilation is also an important attribute in acclimation to increasing CO_2 concentration. Therefore, the acclimation patterns observed in CO_2 exchange may also reflect changes in the temperature response of photosynthesis originating from the combined effect of the CO_2/O_2 specificity of Rubisco (Ghashghaie and Cornic 1994; Hikosaka and Hirose 1998) and structural differences between species. Indeed, it is often found that the temperature optimum increases at elevated CO_2 concentration (Fig. 4.43) (e.g. Juurola 2005).

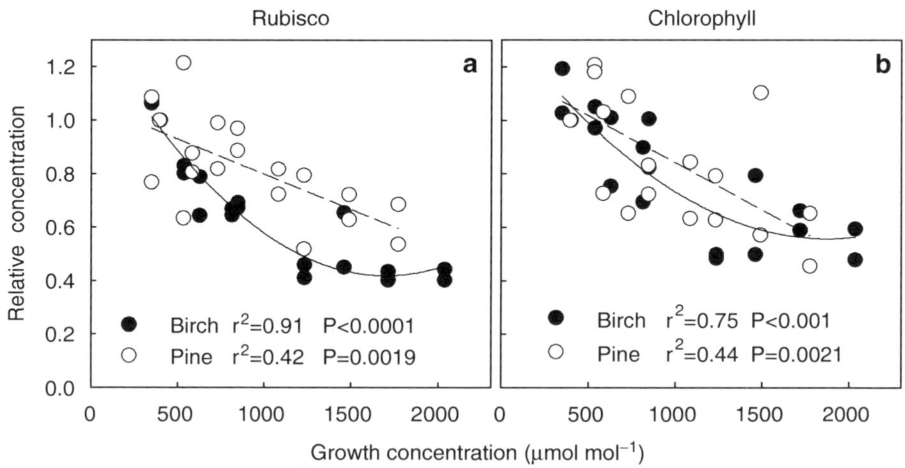

Fig. 4.45 Relative Rubisco concentration (**a**) and relative chlorophyll concentration (**b**) as function of growth CO_2 concentration in silver birch (*black circles*) and Scots pine (*white circles*). The data from 2 years were pooled and expressed as relative values with respect to seedlings grown at 400 μmol mol^{-1}. Nonlinear or linear regressions were applied separately for both species. Second-order regressions were applied only when the nonlinear term in the model was statistically significant

Changes in the temperature response of CO_2 assimilation reflect the properties of the Rubisco-catalysed reaction but also acclimation of the photosynthetic functional substances to increasing CO_2 concentration (e.g. Bunce 2000). This acclamatory effect was highlighted when the traditional Farquhar model (Farquhar et al. 1980) was compared to the measured temperature responses of birch seedlings acclimated to different CO_2 concentrations (Fig. 4.46). The ratio of CO_2 exchange rate at growth concentration to that at current atmospheric CO_2 concentration (A_{growth}/A_{350}) was calculated both for the birch seedlings acclimated to elevated CO_2 concentrations and by utilising the temperature response from the Farquhar model that was parameterised for birch (Aalto and Juurola 2001). In the model analysis, the measured respiration rates and intercellular CO_2 concentrations were taken into account to ensure that the possible differences would not originate from differences in the respiration or stomatal conductance. The results at different temperatures show that the observed changes in temperature response are not only direct effect resulting from the inherent properties of Rubisco. The changes are also due to the acclamatory effect on other functional substances, which is shown as a discrepancy in the calculated ratio especially at high temperatures.

Although there is an increasing amount of information on the response of photosynthetic functional substances to increasing CO_2 concentration, the picture is not clear. There are observed differences between species and even within species in the acclimations to elevated CO_2 concentrations. Some of the results may be caused by experimental techniques, since proper experiments are difficult

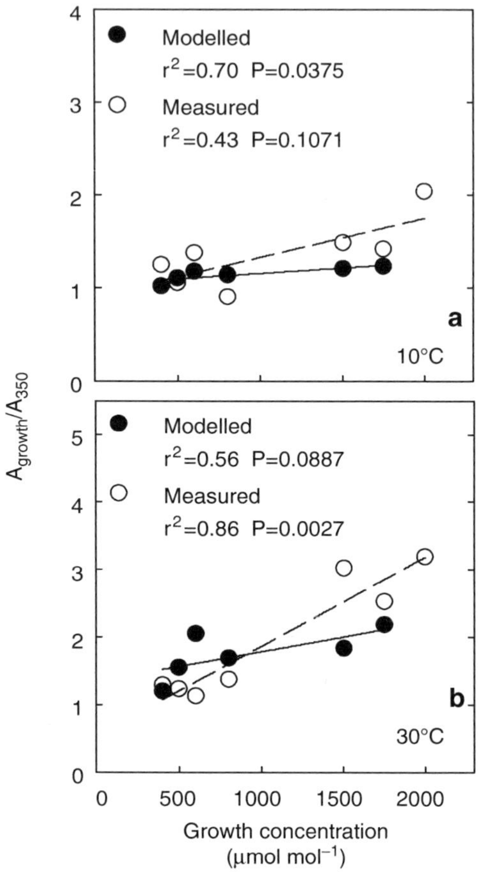

Fig. 4.46 The relative CO_2 exchange as function of growth CO_2 concentration at two temperatures. The modelled (*black circles*) and measured (*white circles*) ratios of CO_2 exchange at growth CO_2 concentration to current ambient CO_2 concentration (A_{growth}/A_{350}) in silver birch leaves at 10°C (**a**) and at 30°C (**b**). Linear regressions were applied separately for both modelled and measured A_{growth}/A_{350}. Note the different scales in A_{growth}/A_{350}

to implement. However, these differences can be expected, since acclimation to present atmospheric concentrations has not been tested in evolution, at least during the last one million years in which the atmospheric CO_2 concentration has been below 300 ppm.

The controlled environment studies give some idea about the acclamatory responses, but it makes a big leap of fate to generalise them to field conditions with a gradually changing climate (Körner 2006). Recent results from long-term FACE experiments, as well as chamber experiments, show that the response to increasing CO_2 at natural ecosystems and in mature trees may be temporal and less than expected (Morgan et al. 2005; Körner 2006), although there are results showing a higher photosynthetic rate even after 7 years of FACE exposure (Springer and Thomas 2006, see also Liberloo et al. 2007 for similar results). Furthermore, it seems that higher rates of photosynthesis will not necessarily produce 1:1 increase in growth (Körner 2006). However, Kimball et al. (2007) showed a clear increase in growth with a decreasing effect on photosynthesis. It seems that the soil properties, nutrition and underground processes play an important role in photosynthetic responses to increasing CO_2 in the long run.

Conclusion: Photosynthetic functional substances and, evidently, also the fine structure of leaves acclimate to increasing atmospheric CO_2 concentration, but the acclimations vary. This is to be expected since the acclimations have not been tested in evolution during the last one million years.

4.2.2.5 Acclimation of Photosynthesis to the Environment: An Optical Perspective

Albert Porcar-Castell, Beñat Olascoaga Gracia, Eija Juurola, and Pertti Hari

Photosynthesis is constantly acclimating to the environment. The efficiency to which incident solar energy can be converted into chemical energy via photosynthesis varies in a very dynamic way during the course of the day or the passing of the seasons. As a result, if we want to understand the complex spatiotemporal dynamics in plant gross primary productivity and how this is affected by the environment, it is essential to measure the acclimation of photosynthesis.

Fortunately, many mechanisms involved in the acclimation of the photosynthesis are directly interacting with light or involve changes in the concentration of specific pigments that do so. Therefore, the acclimation of photosynthesis generates an optical signal that can be measured with optical sensors. Tracking the acclimation of photosynthesis using optical means has a number of advantages:

1. Measurements can be done from a distance without interference with the sample, thus facilitate the study of how plants respond to their environment *in situ*.
2. The signal can be scaled up by moving the sensor away from the sample.
3. Measurements are remote-sensing friendly, allowing the acquisition of data from sensors onboard aircrafts or satellites.

In this chapter, we present the basic theory behind the acclimation of the light reactions of photosynthesis and the optical signal it generates. We end up discussing some important considerations that need to be taken when implementing these approaches from a physiological point of view.

A. Porcar-Castell (✉)
Department of Forest Sciences, University of Helsinki, P.O. Box 27,
00014 University of Helsinki, Helsinki, Finland
e-mail: joan.porcar@helsinki.fi

Photosynthesis and the Need for Acclimation

Light is essential for photosynthesis to take place, but it can also become a hazard if its energy is not rapidly used or dissipated. Light absorbed in excess promotes the formation of reactive oxygen species which may lead to photodamage of the photosynthetic machinery (Barber and Andersson 1992). To minimise the effects of photodamage plants, use temporally flexible acclimation mechanisms that are able to adjust the photosynthetic energy balance under random and non-random variation in the environment.

Photosynthesis can be regarded as a complex set of chemical reactions where reactants (water and carbon dioxide) are supplied with light energy to yield products (sugars and oxygen) that have captured part of the energy carried by light into their chemical bonds. If plants could perform photosynthesis in a laboratory, they could mix the desired amount of chemicals under a given set of conditions, supply the required amount of light energy and obtain the necessary sugars, without risk of photodamage caused by excess (light) energy. However, because plants cannot directly control environmental factors such (as) incident radiation, air temperature or water availability, they require flexible acclimation mechanisms that keep adjusting the energy input to the prevailing energy requirements of photosynthesis (Demmig-Adams et al. 2006).

Light energy (in the form of photons) is captured in a first step by photosynthetic pigments that are organised into photosystems in the thylakoid membranes of chloroplasts. Once a photon has been absorbed, the excitation energy (in the form of excitons) is partitioned between different processes in the photosystem:

1. Excitation energy can be used to split a water molecule and initiate electron transport.
2. Excitation energy can be lost as thermal energy.
3. Excitation energy can be reemitted as fluorescence.
4. Excitation energy can lead to the formation of chlorophyll triplet states with the associated risks of photodamage mentioned above.

These processes compete for excitation energy, following first-order kinetics (Butler and Kitajima 1975), that is, the rate of the process depends on the concentration of excitons and a process rate constant. Acclimation mechanisms are needed to keep the formation of chlorophyll triplet states in check (Horton and Ruban 2005).

In a second step, the energy captured by the photochemical pathway in photosystem II (PSII) is used to split the water molecule releasing two protons in the thylakoid lumen and initiating a linear electron transport chain between photosystem II and photosystem I (PSI) (Fig. 4.47). The remaining energy is used during linear electron transport to pump additional protons from the chloroplast stroma to the thylakoid lumen. Together, the proton pumping and the splitting of water result in a transmembrane proton gradient that is used later on to generate ATP, when protons exit the lumen via ATP synthase. Simultaneously, the excitation energy captured by the reaction centre in PSI is used to add further chemical-reducing

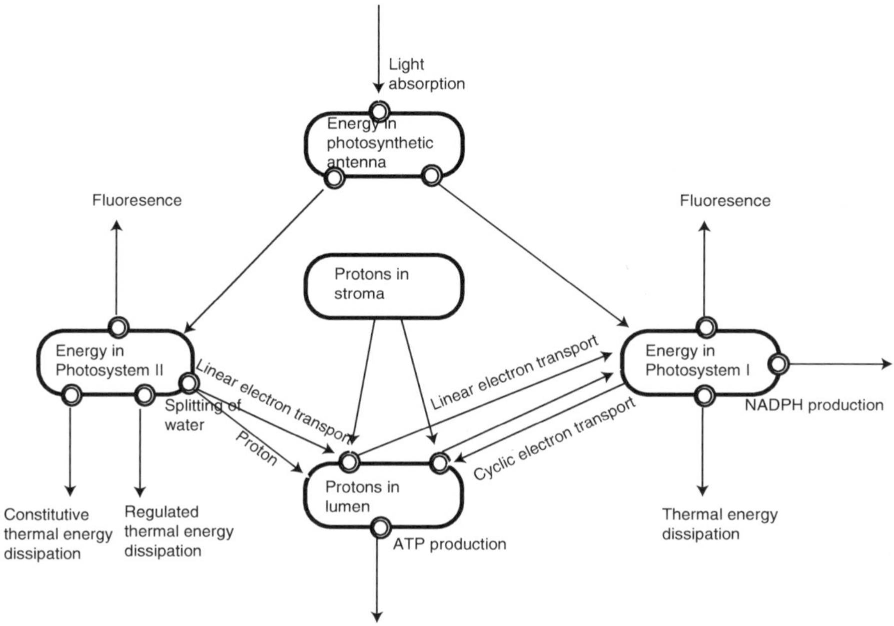

Fig. 4.47 Partitioning of absorbed light energy in the light reactions of photosynthesis. Absorbed light energy may be captured by pigments associated to two different photosystems, photosystem II (*PSII*) and photosystem I (*PSI*). Part of the energy captured by PSII is used to split water, a process that generates two protons and the initial electrons for the linear electron transport. Linear electron transport results in pumping of protons from the chloroplast stroma to the thylakoid lumen and provides electrons to PSI. In PSI, part of the absorbed excitation energy is used to add further reducing power to the electron carriers and eventually used to produce NADPH from NADP and H^+. In addition, the accumulation of protons in the lumen is used to produce ATP from ADP + P. Alternatively, cyclic electron transport routes also occur which function around PSI to pump protons yielding ATP only

power to the carriers of the electron transport chain and eventually bound by the conversion of $NADP^+$ into NADPH. In a final phase, the energy stored in ATP and NADPH is used by the carbon reactions of photosynthesis to synthesise sugars using atmospheric carbon dioxide in a multistep enzymatic process known as the Calvin-Benson cycle (Sect. 4.2.2.1).

Light and carbon reactions of photosynthesis are energywise intimately interlinked (Fig. 4.48). In order to optimise resources while keeping the risks of photodamage low, there will always be a tendency to adjust the energy production in the light reactions of photosynthesis to the prevailing consumption levels of the carbon reactions (Huner and Öquist 2003). Photosynthetic energy consumption by the carbon reactions may undergo different types of fluctuations such as:

- Seasonal: Photosynthetic energy consumption is high during the growing season when environmental conditions are favourable for the normal functioning of the carbon reactions and low during suboptimal environmental conditions (e.g. drought, low temperatures) (Kolari et al. 2007).

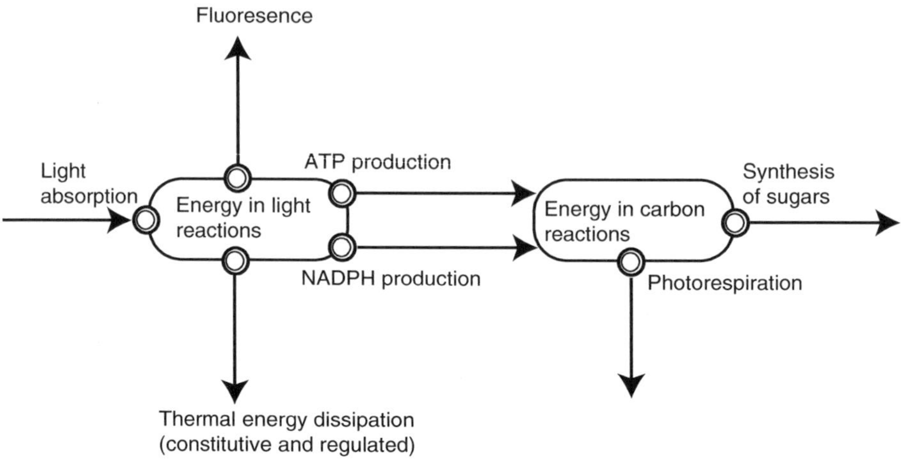

Fig. 4.48 Partitioning of absorbed light energy during photosynthesis. Absorbed light energy in the light reactions is mainly partitioned between thermal energy dissipation, emission of chlorophyll fluorescence and photochemistry leading to the production of ATP and NADPH. In turn, the ATP and NADPH produced in the light reactions is used by the carbon reactions to synthesise sugars or used by alternative energy routes such as photorespiration

- Diurnal: Photosynthetic energy consumption is constrained by diurnal fluctuations in vapour pressure deficit (VPD) or temperature; for example, stomata may close at noon to minimise water loss in leaves where water is a valuable asset (Hari et al. 1986; Gamon et al. 1997; Flexas et al. 2000).

In turn, photosynthetic energy input in the light reactions is controlled by the photosynthetic photon flux density incident on the leaf, which also undergoes fluctuations at different timescales: seasonal, diurnal and fast random variation (e.g. sunflecks or clouds). Thus, the dynamics of photosynthetic energy input and consumption take place to a large extent independently of each other. Accordingly, the energy balance has to be maintained under the highly dynamic variation in factors that independently control energy input and consumption (Fig. 4.49a). If energy production fails to meet the demand by the carbon reactions, the light reactions will tend to increase production capacity by up-regulating the light-harvesting machinery (Fig. 4.49b), as, for example, during spring recovery of photosynthesis in boreal evergreen foliage (Ensminger et al. 2004; Porcar-Castell et al. 2008b, c). Otherwise, if energy production exceeds its consumption, the light reactions will respond by downregulating photosynthetic energy production capacity (Fig. 4.49c) to reduce the probability of photodamage (Huner and Öquist 2003). Importantly, these imbalances may appear at very diverse temporal scales, for example, from a short-lived sunfleck lasting a few seconds to the seasonal imbalance caused by low winter temperatures. Therefore, acclimation mechanisms need to be equally flexible with a well-integrated set of mechanisms that operate at equivalent temporal scales.

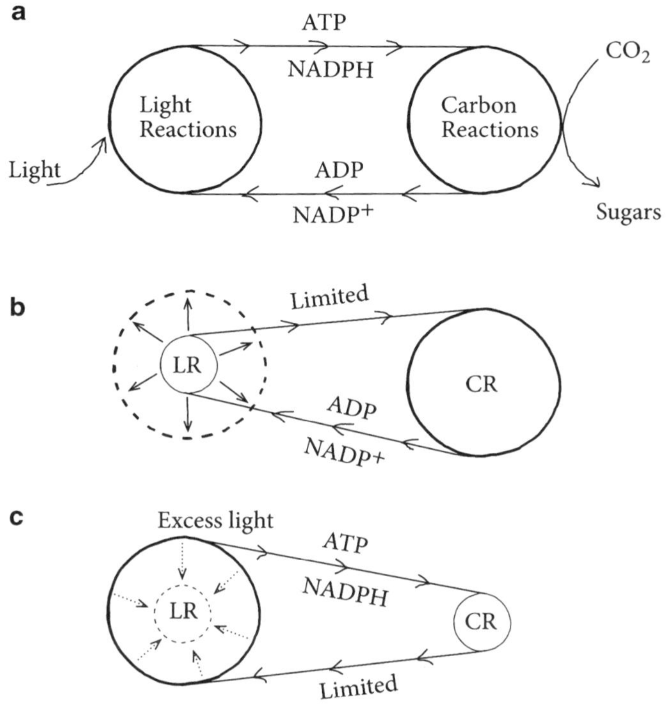

Fig. 4.49 Types of energy imbalances between light and carbon reactions of photosynthesis and the way acclimation of the light reactions operates. Energy balance between light and carbon reactions: No acclimation needed (**a**); energy production by the light reactions is lower than consumption by carbon reactions: Acclimation mechanisms upregulate the light reactions (**b**); and energy production by the light reactions is higher than consumption by carbon reactions: Acclimation mechanisms downregulate the light reactions (**c**)

Mechanisms of Acclimation

Acclimation in Photosynthetic Energy Input

Plants make use of different mechanisms to regulate the flux of absorbed light energy entering photosynthesis at different temporal and spatial scales (Table 4.1). At the temporal scale of minutes, some species are able to regulate light energy input by adjusting the leaf angle in respect to the incident light (Yu and Berg 1994; Arena et al. 2008) or at the cellular level via chloroplast movement (Terashima and Hikosaka 1995; Wada et al. 2003). At the seasonal temporal scale (days), plants are able to adjust the concentration of light-absorbing pigments in the leaf (chiefly chlorophyll-*a* and chlorophyll-*b*), regulating in this way the photosynthetic energy input (Huner and Öquist 2003; Porcar-Castell et al. 2008b, c). Similarly, at the whole-plant level, the concentration of light-absorbing pigments and energy input is regulated by adjusting the total leaf area, for example, deciduous plant species

Table 4.1 Spatiotemporal distribution of mechanisms to acclimate light absorption capacity in plants

	Subcellular	Leaf	Canopy
Seconds, minutes	Chloroplast movement	Leaf angle adjustments	
Days		Acclimation in pigment contents, epicuticular wax dynamics	Acclimation in number and distribution of leaves, leaf area index (LAI)
Years			Adjustments in crown morphology

decrease light energy input to zero during the dormant season by shedding all their leaves. Other mechanisms affecting photosynthetic energy input are related to the structural characteristics of the leaf surface, for instance, the presence of trichomes or epicuticular waxes (Holmes and Keiller 2002). Waxes are able to reflect variable percentages of the incident photosynthetically active radiation (PAR) energy that may go up to 50% (Mulroy 1979; Ehleringer 1981; Johnson et al. 1983) and which appear to vary in response not only to weathering processes but also to seasonal environmental factors (Olascoaga et al. manuscript). Finally, at a temporal scale of years, woody plants adjust their crown morphology to the long-term photosynthetic energy input requirements (Beaudet and Messier 1998).

Acclimation in Energy Partitioning Between Photochemical and Non-photochemical Pathways

Once excitation energy has been absorbed by pigment molecules in PSII, this energy is partitioned into different processes: splitting of water molecule and electron transport(photochemical utilisation) (*p*), emission of fluorescence – or chlorophyll-*a* fluorescence in higher plants (*f*), dissipated thermally as heat either via constitutive (*d*) and regulated thermal energy dissipation (*n*), the later widely addressed as non-photochemical quenching (NPQ) or experience intersystem crossing and form a potentially hazardous chlorophyll triplet state (*i*) (Parson and Nagarajan 2003). The yield of chlorophyll triplet formation is usually very small; however, this can increase under conditions of excess light, enhancing the probability of photodamage (Huner and Öquist 2003; Takahashi and Badger 2010). When ATP and NADPH production by the light reactions exceed its consumption (excess light) (Fig. 4.49c), accumulation of ATP and NADPH in the chloroplast stroma will slow down the linear electron transport rate between photosystems (Fig. 4.47), leading to the accumulation of protons in the thylakoid lumen and consequent decrease in pH (Müller et al. 2001). Simultaneously, because the photochemical utilisation of excitation energy cannot deal with all excitation energy, the excitation density in the antenna of PSII will increase, with a corresponding increase in energy available to fluorescence, constitutive heat dissipation and hazardous chlorophyll triplet-state formation. In order to keep the formation of chlorophyll triplet states under control, plants need to compensate for the insufficient photochemical energy utilisation by increasing the capacity of thermal dissipation of excitation energy.

When lumen pH decreases under excess light due to saturation of the carbon reactions, specific proteins from PSII are protonated, and the xanthophyll-cycle pigment violaxanthin is released from the thylakoid membrane and de-epoxidised to antheraxanthin and zeaxanthin by the enzyme violaxanthin de-epoxidase (VDE) (Niyogi 1999; Müller et al. 2001). This process is known to be responsible for the formation of a quenching complex that favours the thermal dissipation of excitation energy, thus enhancing the capacity for thermal energy dissipation in the photosystem (Müller et al. 2001; Horton and Ruban 2005). This process has been widely addressed as non-photochemical quenching (of the chlorophyll fluorescence signal) or NPQ (Schreiber et al. 1986; Bilger and Bjorkman 1990; Krause and Weis 1991; Müller et al. 2001; Demmig-Adams and Adams 2006). Regulated thermal dissipation or NPQ is a reversible process that operates typically in a timescale of seconds to minutes (Krause and Weis 1991; Müller et al. 2001). Nevertheless, under certain conditions, NPQ may be retained during cold weather not relaxing during night (Adams and Barker 1998; Porcar-Castell 2011; Porcar-Castell et al. 2008b). Conditions of excess light may appear in response to fast variation in illumination (e.g. sunflecks), diurnal changes in illumination (day-night fluctuations) or seasonal changes in light, temperature or water availability (e.g. excess light induced by low-temperature metabolic limitations). Similarly, NPQ presents a wide range of temporal components that cooperate in an additive fashion to respond to the multitemporal dimension in the energy imbalance (Demmig-Adams et al. 2006; Porcar-Castell 2011).

The partitioning of excitation energy in photosystem II can be described, for the sake of simplicity, taking into account only the most energy relevant processes and including also the formation of chlorophyll triplet states given its important implications. Variations in the excitation energy density (E) (μmol m^{-2}) in a population of N PSII units distributed in a given leaf area, with $N \rightarrow \infty$, depend on the incident photosynthetic photon flux density ($PPFD$) absorbed by the N PSII units, the fluorescence flux (F), the electron transport rate (P), the rates of constitutive (D) and regulated (NPQ) thermal energy dissipation and the rate of formation of chlorophyll triplet states (TS), all in (μmol m^{-2} s^{-1}), as

$$\frac{\mathrm{d}E}{\mathrm{d}t} = PPFD\ A\ a_{II} - F - P - D - NPQ - TS, \tag{4.35}$$

where A is the leaf absorptance (adimensional) and a_{II} the relative absorption cross section of PSII (relative to the total absorption PSII + PSI). In turn, each process rate can be related to its first-order rate constant k_{i} (s^{-1}) and the current excitation density (E) as

$$\frac{\mathrm{d}E}{\mathrm{d}t} = PPFD\ A\ a_{\mathrm{II}} - k_{\mathrm{f}}E - k_{\mathrm{P}}E - k_{\mathrm{D}}E - k_{\mathrm{NPQ}}E - k_{\mathrm{TS}}E \tag{4.36}$$

and k_f, k_P, k_D, k_{NPQ} *and* k_{TS} are the first-order rate constants associated to the emission of fluorescence, constitutive thermal energy dissipation, regulated thermal energy dissipation or NPQ and formation of chlorophyll triplet states.

Acclimation in Alternative Electron Transport Routes and Energy Sinks

In addition to the linear electron transport between PSII and PSI that produces both ATP and NADPH, a number of alternative electron transport routes also exist that lead to the production of ATP alone. Cyclic and pseudocyclic electron transport routes around PSI are coupled to ATP sunthesis but do not produce, in contrast to linear electron transport, any NADPH (Allen 2003). Cyclic electron transport routes may play a double functional role by adjusting the ratio of ATP to NADPH or by protecting the photosystems under stress conditions promoting the fast acidification of the thylakoid lumen and formation of NPQ in PSII (Turpin and Bruce 1990; Endo et al. 1999; Kramer et al. 2004a).

In addition, the ATP and NADPH produced by the light reactions of photosynthesis may be used in other processes apart from fixation of carbon dioxide. Photorespiration, for example, is an alternative pathway that consumes ATP and NADPH from the light reactions of photosynthesis, reducing the rates of CO_2 assimilation. Rubisco is an enzyme that catalyses both the fixations of CO_2 (carboxylation) and of O_2 (oxygenation) (Woodrow and Berry 1988), where the affinity for O_2 despite of CO_2 increases as temperature rises up or stomata close during droughts. Photorespiration may reduce potential photosynthetic carbon assimilation up to 75% under extreme environmental conditions (Sharkey 1988) in plants with C_3 metabolism, while it is very low in plants with C_4 metabolism, where the carboxylation reactions are promoted relative to oxygenation in specialised bundle sheath cells. Although photorespiration may appear at a first sight as a futile energy loss in C_3 plants, it may play an important photoprotective role by acting as safety valve by keeping up the regeneration of ADP and NADP under conditions that otherwise would restrain the carbon reactions (e.g. noon stomatal closure or drought) (e.g. Kozaki and Takeba 1996). In addition, photorespiration plays a role in leaf nitrogen cycle and protein synthesis, thereby potentially being an important process in plant metabolism (e.g. Betsche 1983; Oaks 1994; Wingler et al. 2000).

Acclimation in the Carbon Reaction Capacity

Environment exerts direct and indirect effects regulating the capacity of the carbon reactions of photosynthesis. Direct effects are, for example, the direct effect of temperature on the enzymatic reactions involved in the Calvin-Benson cycle or the direct effect of environmental variables on stomatal conductance, where stomata tend to close at high vapour pressure deficit (VPD) with the subsequent reduction in leaf internal CO_2 and thus in assimilation capacity (Farquhar et al. 1980; Hari et al. 1986). In turn, environmental factors control also the slow acclimation in the capacity of the carbon reactions by modulating plant growth. In boreal conifers, the whole tree carbon balance and carbohydrate sink activity are modulated according to the phases of growth. The synthesis of primary carbon compounds and their use are intimately linked. Changes in whole tree source-sink balance as such can induce raffinose synthesis and change the carbohydrate balance in tree. For example, the cessation of growth in autumn is considered to be driven by changes in temperature and photoperiod that decrease sink demand of sugars (Öquist and Huner 2003). There is also ample evidence on downregulation of photosynthetic carbon reactions in response to external feeding of glucose or sucrose (e.g. Stitt and Schulze 1994).

The Optics of Photosynthetic Acclimation

Acclimation in the light reactions of photosynthesis generates a series of optical signals that can be remotely measured with optical sensors placed at variable distances from the target (ranging from a few mm to km in satellite platforms). These optical signals can be used as proxies of changes in the photosynthetic capacity of vegetation across temporal and spatial scales in a non-destructive way. Optical signals of photosynthesis include the spectra of the radiation reflected by the leaf as well as the spectra and properties of the emitted chlorophyll fluorescence.

Spectral Reflectance

Electromagnetic energy incident on a leaf may be absorbed, reflected or transmitted. Importantly, this energy partitioning is wavelength dependent, and different wavelengths are absorbed, reflected or transmitted through the leaf to a different extent depending on interaction with the biochemical and structural properties of the leaf. The most important pigments responsible for the absorption of PAR in higher plant leaves are chlorophyll-*a* and chlorophyll-*b*, which have strong absorption bands in the blue (400–500 nm) and red (600–700 nm), rapidly decreasing in the far red (>700 nm). Therefore, differences in absorption in the red compared to the far red/near infrared will be larger in samples with higher chlorophyll content. Because higher chlorophyll content involves higher light absorption capacity, the previous difference has been widely exploited by the so-called greenness indexes, where different properties of the shift in reflectance between the red and near infrared are exploited to follow the adjustments in light absorption capacity of a sample or target area (Peñuelas and Filella 1998). The most widely used greenness index has been probably the normalised difference vegetation index NDVI (Tucker 1979), estimated as:

$$\mathrm{NDVI} = \frac{\rho_{\mathrm{NIR}} - \rho_{\mathrm{Red}}}{\rho_{\mathrm{NIR}} + \rho_{\mathrm{Red}}}, \qquad (4.37)$$

where ρ_i represents the reflectance in the near-infrared (NIR) or red regions of the spectra. Based on the correlation between seasonal changes in chlorophyll concentrations and seasonal changes in photosynthetic capacity, the NDVI index has been widely used to follow changes in green biomass at large spatial scales. However, other greenness indexes have been also used: such as the enhanced vegetation index (EVI), simple ratios (SR) or red edge (Peñuelas and Filella 1998; Hilker et al. 2008a).

Another important feature in the vegetation reflectance spectra linked to the acclimation of photosynthesis is found in the reflectance around 531 nm. Changes in reflectance around 531 nm have been associated with the operation of the xanthophyll cycle and NPQ and used to construct the photochemical (or physiological) reflectance index, PRI (Gamon et al. 1992), which is generally estimated as:

$$\mathrm{PRI} = \frac{\rho_{531} - \rho_{570}}{\rho_{531} + \rho_{570}}, \tag{4.38}$$

where ρ_i represents the reflectance in a particular wavelength (in nm). Based on the correlation between PRI and NPQ, the PRI index has been successfully used to follow both the rapid and slow adjustments in photochemical capacity and overall light-use efficiency (LUE) (in mol CO_2/mol PAR absorbed) at the leaf, canopy and landscape levels (Gamon et al. 1997; Nichol et al. 2002; Rahman et al. 2004; Garbulsky et al. 2008). The general advantage of PRI over greenness indexes is that the signal behind PRI is generated in response to changes in excitation energy partitioning in the light reactions (Gamon et al. 1992); therefore, PRI is not restricted to the detection of changes in photosynthetic capacity occurring alongside vegetation greenness and has the potential to track both the seasonal and diurnal changes in photosynthetic light-use efficiency of evergreen vegetation. However, because other physiological factors such as seasonal changes in leaf chlorophyll and carotenoid contents also affect the PRI (Filella et al. 2009; Porcar-Castell et al. 2012), a general relationship between PRI and NPQ or LUE cannot be obtained.

Radiation reflected from foliage can be measured from a few mm distance to several km from satellite platforms. Importantly, while changes in spectral reflectance at the leaf level can be associated to physiological or morphological properties of the sample, changes in spectral reflectance at the canopy or ecosystem level are affected also by physical factors such as geometry between light source (i.e. Sun), target and sensor or wavelength-dependent scattering in the atmosphere or within the canopy (Barton and North 2001; Hilker et al. 2008b). Therefore, depending on the spatiotemporal domain, a different set of factors needs to be considered for interpreting the physiological information carried by optical data.

Chlorophyll Fluorescence

A small proportion of the excitation energy absorbed by photosynthetic pigments in higher plant leaves is eventually lost as chlorophyll-*a* fluorescence. At ambient temperatures, chlorophyll fluorescence is predominantly emitted from photosystem II (Krause and Weis 1991). The intensity of the chlorophyll fluorescence emission will depend on the rate of light absorption by the photosynthetic pigments in the sample or target area, as well as on the acclimation in energy partitioning of the light reactions (Eqs. 4.35 and 4.36). To exclude the effect of potential changes in ambient illumination from the chlorophyll fluorescence signal, a pulse-amplitude-modulated (PAM) method is used in field fluorometers (Schreiber et al. 1986). PAM fluorometers supply the leaf with a pulsed measuring light of very low but constant intensity; this pulsed light generates a pulsed fluorescence signal that can be later on separated from the total emitted fluorescence. Therefore, while total fluorescence emitted from the leaf depends on both physiological and illumination factors, the pulsed fluorescence depends on physiological factors only facilitating the study of PSII under field conditions or under variable illumination.

Because excitons equilibrate in the antenna within a few picoseconds (Krause and Weis 1991), it can be assumed that the excitation density in PSII is at the steady state and that $dE/dt = 0$ in Eqs. 4.35 and 4.36. Consequently, the amount of modulated fluorescence emission emanating from a sample (F) can be expressed as (Porcar-Castell 2011; Porcar-Castell et al. 2008b):

$$F = \mathrm{PPFD_{ML}}\ A\ a_{\mathrm{II}} \frac{k_{\mathrm{f}}}{k_{\mathrm{f}} + k_{\mathrm{P}} + k_{\mathrm{D}} + k_{\mathrm{NPQ}}}, \tag{4.39}$$

where $\mathrm{PPFD_{ML}}$ is the constant light intensity of the measuring light. Furthermore, the rate constants of fluorescence (k_{f}), basal thermal energy dissipation (k_{D}), the leaf absorptance (A) and the PSII absorption cross section (a_{II}) are assumed to remain constant, and the rate constant associated to the formation of triplet states (k_{TS}) is considered to be insignificant relative to the other rate constants and omitted. Therefore, an increase in fluorescence (F) will denote a decrease in either photochemical or non-photochemical thermal energy dissipation capacities, represented by their respective rate constants k_{P} or k_{NPQ}, respectively, and *vice versa* in the case of a decrease in F. Yet, how can we discriminate between photochemical and non-photochemical processes from the observed variations in F?

The technique currently in use consists in shutting down the photochemical pathway by supplying the sample with a saturating light pulse that momentarily reduces all the primary electron acceptors so that the photochemical capacity tends to zero ($k_{\mathrm{P}} = 0$). Under these conditions, the fluorescence yield increases to a maximum level or maximal fluorescence (F_{m}). To evaluate how much energy was actually being used by photochemistry, the fluorescence before the pulse (F) and the maximal fluorescence F_{m} can be compared. In summary, if shutting down photochemistry with a saturating pulse has little effect on fluorescence and F_{m} remains almost equal to F, it reflects that photochemistry was using up a very little fraction of the excitation energy available. In contrast, if the difference between F_m and F is large, it reflects that photochemistry was performing at a high efficiency. The operating quantum yield of photochemistry φ_P (Butler and Kitajima 1975) can be thus derived as:

$$\varphi_P = \frac{F_m - F}{F_m}. \tag{4.40}$$

Subsequently, if we know the intensity of the illumination (PPFD, μmol photons m^2 s^{-1}), the absorptance of the leaf (A) and the relative cross-sectional area of photosystem II (a_{II}) and the current electron transport rate (ETR, μmol e^- m^2 s^{-1}) can be estimated as (Genty et al. 1989):

$$\text{ETR} = \text{PPFD}\ A\ a_{II}\ \varphi_P. \tag{4.41}$$

This is typically done assuming $A = 0.84$ and $a_{II} = 0.5$.

In the absence of cyclic electron transport routes, ETR will be proportional to the formation of ATP and NADPH, which in the absence of photorespiratory and other alternative energy sinks will correlate well with gross photosynthetic carbon assimilation. As expected, a good correlation between ETR obtained from ChlF using Eq. 4.41 and net CO_2 assimilation can be obtained in C_4 plant species with low photorespiration rates, while different curvilinear relationships are obtained in C_3 species (Genty et al. 1989; Edwards and Baker 1993).

In addition, because photochemistry is always zero during the saturating light pulse used to obtain the maximal fluorescence F_m ($k_P = 0$), changes in F_m through time for a given sample will reflect changes in the capacity of thermal energy deactivation, denoted by NPQ (with the rate constant k_{NPQ}). Typically, a relatively high F_m will be measured in dark-acclimated foliage where NPQ is absent ($k_{NPQ} = 0$) or minimum (Porcar-Castell 2011), while a smaller F'_m is obtained under illumination when part of the excitation energy is thermally dissipated. The parameter NPQ (Bilger and Bjorkman 1990) has been derived to follow variations in the capacity for regulated thermal energy dissipation, as:

$$\text{NPQ} = \frac{F_m}{F'_m} - 1. \tag{4.42}$$

Similarly, parameters associated with the rate constants of photochemical and non-photochemical processes, the yields of each of the energy-consuming processes in the photosystem and the different temporal components of these acclimation processes can be estimated through analysis of PAM fluorescence data (Hendrickson et al. 2004; Kramer et al. 2004b; Porcar-Castell 2011; Porcar-Castell et al. 2008b).

Chlorophyll fluorescence emitted from leaves can be measured from a few mm to several metres with active fluorometers (i.e. fluorometers that make use a saturating light pulse to shut down photochemistry and obtain F_m) (Schreiber et al. 1986; Flexas et al. 2000). However, because the use of saturating light pulses is not practical at distances over several metres, this technique cannot be used from airborne or satellite platforms. In contrast, passive fluorescence measurements (i.e. measurements of the natural fluorescence induced by sunlight) have been proved feasible at distances ranging from several metres to km in spaceborne platforms using a line-filling principle (Moya et al. 2004; Meroni et al. 2010). A few absorption bands in the Earth's atmosphere and the Sun's photosphere (Fraunhofer lines) fall within the emission spectra of chlorophyll fluorescence. Inside these narrow bands, the light incident on the Earth's surface is drastically attenuated by the properties of the Earth's atmosphere or Sun's photosphere, and fluorescence can be estimated based on the difference in intensities between a point located within the absorption line and a nearby point located outside this line.

Limitations and Potential of Optical Measurements

Carbon reactions are the critical step of photosynthesis, where the atmospheric CO_2 is converted into sugars. Therefore, if we are able to estimate the performance of the carbon reactions, we can estimate the fluxes of CO_2. However, because the carbon reactions do not generate any optical signal, we need to interpret photosynthesis from a light reaction point of view or more specifically from a photosystem II (PSII) point of view. Extrapolation of optical data to the estimation of processes such as light-use efficiency (LUE) or gross primary production (GPP) is based on the assumption that the acclimation in the light reactions of photosynthesis is closely following that of the carbon reactions. Although this is to some extent the case, a couple of important considerations need to be taken into account and examined when interpreting the dynamics of optical data in connection with photosynthesis: (1) Is the optical signal under consideration varying at the same timescale than that of property of photosynthesis we are interested in estimating? For example, it may be reasonable to pick up a greenness index to evaluate the seasonal changes in GPP in a grassland, but not necessarily in a thick and evergreen spruce forest, and (2) are processes not visible to the optical signal likely to affect the property of photosynthesis we want to estimate? For example, photorespiration may significantly decouple electron transport rate from gross photosynthesis.

The study of the optics of photosynthesis has a large potential to facilitate the measurement and understanding of the spatiotemporal variation of photosynthesis. In particular, recent advances in the remote sensing of indexes such as the photochemical reflectance index (PRI) or chlorophyll fluorescence are very promising. However, technical limitations aside, the physiological interpretation of the resulting data is yet on an infant state. Assimilation of optical data into mechanistic models that consider the key physiological processes will certainly help advancing in this area.

4.2.3 Respiration

Pertti Hari, Pasi Kolari, and Jaana Bäck

In living cells, there are several processes, such as synthesis of new molecules (see Growth), maintaining vital functions and transport through membranes, which need energy. This energy is released from oxidation of energy-rich carbon molecules, such as sugars, and transiently stored in ATP. This process, called respiration, produces many important carbon precursors for cellular metabolism and releases CO_2 into mesophyll. In general, living cells need always energy provided by ATP, and formation of ATP mainly occurs through respiration. Thus, respiration is a characteristic feature of living organisms.

Structural and Metabolic Background

Mitochondria house the essential machinery that generates ATP in aerobic respiration, and they are most abundant in actively metabolising cells. During aerobic respiration in mitochondria, the reduced carbon compounds (glucose, sucrose, hexose, phosphates as well as lipids, organic acids or even proteins) are oxidised. In this process, free energy is released and transiently stored in ATP, which then can be used for maintenance and growth of the cells. In the course of this, O_2 is consumed, and CO_2 is released. ATP is needed for numerous biosynthetic pathways, for example, in root N uptake and nitrate reduction, phloem loading, protein turnover, macromolecule biosynthesis and maintenance of ion concentration gradient.

Respiration is relatively more sensitive to temperature changes than photosynthesis (Amthor 1994). In cold environments, the properties of respiratory enzymes are the most important factors limiting respiration, whereas under warmer conditions, the substrate and adenylate concentrations determine the respiration (Atkin and Tjoelker 2003).

In daytime, cells in leaves are simultaneously photosynthesising and respiring. The leaf respiration in light is, however, considerably smaller than in the dark, although exact measurements are impossible and controversy regarding the magnitude still exists (e.g. Hoefnagel et al. 1998; Pinelli and Loreto 2003). Although the activity of, for example, pyruvate dehydrogenase, one of the key enzymes in the citric acid cycle, decreases considerably in light, it is clear that even in illuminated leaves, the mitochondrial respiration is a major supplier of ATP to the cytosolic processes.

Respiration generates ATP flow from mitochondria to other parts of a cell providing energy for metabolism and release of CO_2 into the atmosphere (Fig. 4.50).

P. Hari (✉)
Department of Forest Sciences, University of Helsinki, P.O. Box 27,
00014 University of Helsinki, Helsinki, Finland
e-mail: pertti.hari@helsinki.fi

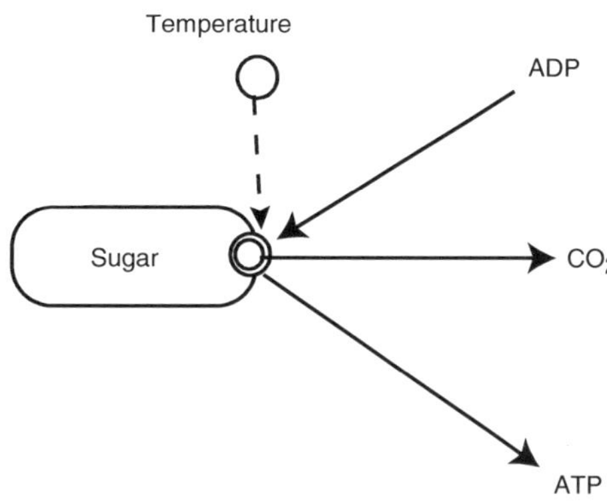

Fig. 4.50 Visualisation of respiration. ADP is converted to energy-rich ATP utilising chemical energy in sugar. The carbon in sugar is released as CO_2

Basic Concepts and Ideas

The *specific respiration rate* is defined, analogously to the definition of specific photosynthetic rate, as the CO_2 produced in small tissue element during a short time interval divided with the mass of tissues in the element and with the length of the interval. Respiration is an enzymatic chemical reaction, and temperature affects enzymatic reactions.

Basic idea R1. Temperature has an effect on specific respiration rate.

A long chain of biochemical reactions and membrane penetrations form the respiration process, and each step in the chain has its specific functional substance that is a non-stable compound. The biochemical regulation system maintains proper relations in the concentrations of the involved functional substances.

The action of the biochemical regulation system generates emergent property *the efficiency of the respiratory functional substances.*

Basic idea R2. The efficiency of the respiratory functional substances characterises the performance of the chain of the functional substances in respiration.

Theoretical Model for Respiration

Let r denote specific respiration rate and T temperature. Strong exponential relationship between reaction rate and temperature is characteristic for enzymatic reactions, and therefore, the rate of respiration in plant tissues depends exponentially on temperature. The efficiency of the respiratory functional substances, E_R, determines together with temperature the rate of specific respiration:

$$r = E_R Q_{10}^{T/10}. \tag{4.43}$$

The temperature sensitivity Q_{10} describes the temperature response of the functional substances. In biological processes, Q_{10} is often about two at ambient

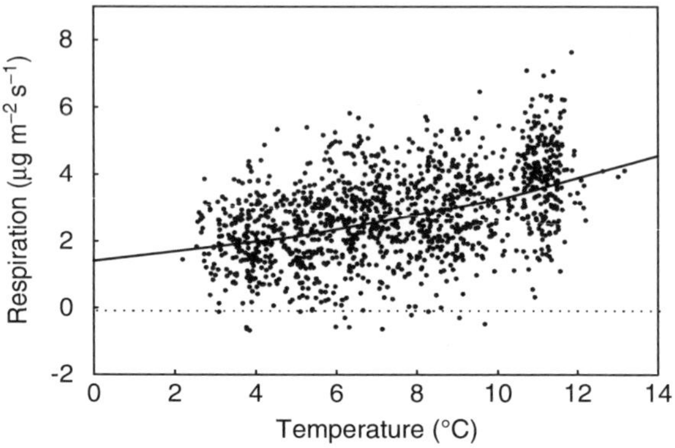

Fig. 4.51 The observed and modelled relationship between dark respiration and temperature. We determined the measured dark respirations by selecting measurements of CO_2 exchange when the photosynthetic radiation is below 5 μmol m^{-2} s^{-1} in chamber 3 during the period from 29 July to 29 August 2011 at SMEAR I

temperatures between 0 and 30°C (reviewed by Atkin and Tjoelker 2003). Thus, respiration is doubled when temperature increases 10°C.

Field Measurements. We describe the measurements in Chap. 9. The measurement is CO_2 exchange, but in the darkness when photosynthesis is missing, the measured value can be interprcted as respiration. The obtained signal is rather small when compared with the measuring noise.

Parameter Estimation. CO_2 exchange during nights gives the rate of respiration, but respiration in light is problematic. We estimated the values of the efficiency of respiration E_R and the temperature sensitivity Q_{10} during summertime with the normal statistical principle by minimising the residual sum of squares utilising data measured during darkness in chamber 3 from 29 July to 29 August 2011 at SMEAR I. The obtained values are $E_R = 1.4$ μg m^{-2} s^{-1} and $Q_{10} = 2.3$. The signal-to-noise ratio is rather small, and in addition, the variation in night temperatures is rather small (Fig. 4.51). These technical problems make meaningless to analyse further the summertime respiration.

Conclusion: The rather large measuring noise hinders clear conclusions.

Annual Cycle of Respiration

The energy consumption in living cells has a clear annual pattern. Trees are dormant during winter, and their metabolic activity and need of energy are low; thus, respiration should be reduced. The metabolism recovers during spring, and the functional substances needed for the recovering processes are synthesised or activated. This requires energy in the form of ATP. The functional substances are non-stable compounds; thus, their synthesis and activation is needed also in summer. In addition, the transport of sugars away is based on the action of membrane pumps that consumes energy.

As the rate of ATP formation, that is, respiration, is related to temperature, respiration over a year in general follows the seasonal course of temperature (Fig. 4.52). The biochemical regulation system of the respiratory annual cycle

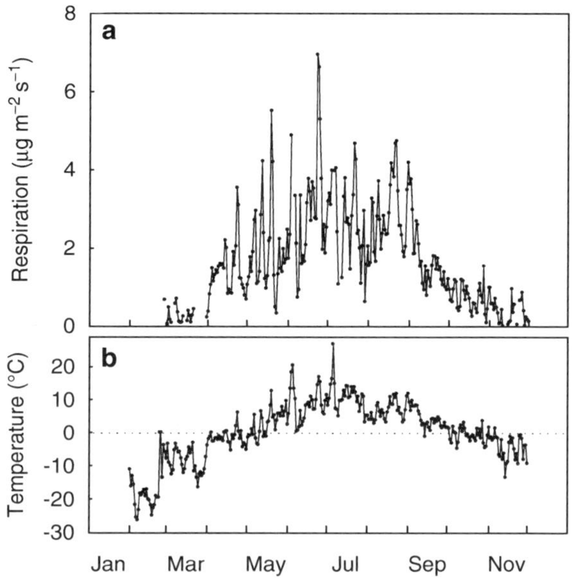

Fig. 4.52 Seasonal pattern of shoot respiration, that is, daily mean night-time CO_2 exchange (**a**) and night-time temperature (**b**) at SMEAR I in 2011

changes the concentrations and activities of functional substances. Respiration reflects these changes; thus, the efficiency of the respiratory functional substances changes. The annual patterns in temperature and in the respiratory functional substances are connected with the observed annual pattern in respiration.

Basic Ideas

The need of energy for metabolism in leaves has a strong annual pattern and respiration response to it.

Basic idea AR1: The efficiency of respiratory functional substances reflects the annual pattern of need of energy for metabolism.

Estimation with Field Measurements

We estimated the daily values of the efficiency of respiratory functional substances from the data measured at SMEAR I in the year 2011. Outside the period of polar day, we used CO_2 measurements when the photosynthetic radiation was less than 5 μmol s^{-1} m^{-2}. During polar day, we used CO_2 exchange measurements when photosynthetically active radiation was less than 20 μmol s^{-1} m^{-2}.We subtracted from the measured CO_2 exchange photosynthesis obtained from the optimal stomatal control model for photosynthesis. This procedure caused evidently additional noise in the obtained values.

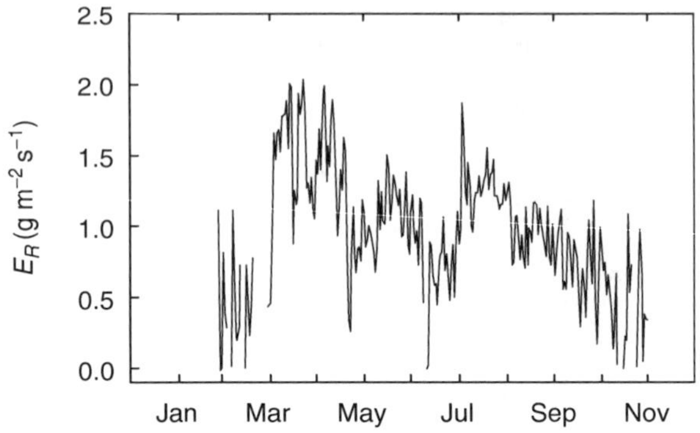

Fig. 4.53 The annual pattern of the efficiency of respiratory functional substances in 2011 in chamber 1 at SMEAR I. The measured respiratory signal is below −10°C so small that we cannot determine the efficiency of the respiratory functional substances. This is why quite many points are missing in winter

The annual pattern in the efficiency of respiratory functional substances in 2011 at SMEAR I is rather clear (Fig. 4.53). During midwinter, the small signal and low temperatures make the determination of the efficiency of the respiratory functional substances problematic. We can determine the efficiency only when the temperature is above −10°C. The efficiency of the respiratory functional substances is very low during dormancy in winter. Recovery of the efficiency takes place when temperatures increase in late March and early April. At midsummer, the efficiency declines and recovers again in late summer. During autumn, a slow and steady decline takes place towards the low values during dormancy.

Conclusion: The annual cycle in respiration is evident, and it is generated by the annual cycle of the efficiency of the functional substances and of temperature, and it reflects the annual pattern of need of energy for metabolism, namely, low during dormancy, moderate during active summer season and high during spring recovery. The low signal-to-noise ratio hampers the analysis.

4.2.4 Senescence

Jaana Bäck and Pertti Hari

The lifetime of living cells is rather short when compared to that of individual plants, especially of trees. The leaves of annual plants and deciduous perennials grow in the spring, photosynthesise in the summer and die in the autumn. The leaves of coniferous trees and of other evergreen plants live longer but usually only a couple of years. The death of a cell or organ is a slow phenomenon and occurs in a predetermined and well-coordinated manner under the control of the biochemical regulation system.

J. Bäck (✉)
Department of Forest Sciences, University of Helsinki, P.O. Box 27,
00014 University of Helsinki, Helsinki, Finland
e-mail: jaana.back@helsinki.fi

As the first step of senescence, the biochemical regulation system controlling the proper functioning of the cell starts to slow down the metabolism, and then the activity and concentrations of the functional substances in the cell are reduced. Major part of the proteins in enzymes, membrane pumps and pigment complexes are broken down into amino acids and transported away from the cell to be used for synthesis of new proteins in remaining organs. Simultaneously, most of starch is broken to sugars and transported away for later use. The cellulose and lignin in the cell walls and lipids in the membranes remain in the senescent cells.

Autumn leaf colours are one example of the action of the biochemical regulation system: Protease enzymes dismantle the chlorophyll molecules in the photosystems of chloroplast thylakoids, whereby the yellow xanthophylls and orange B-carotene pigments become visible. The released amino acids are stored in tree branches and roots and recycled to new leaves emerging in the following spring.

When the senescing process continues further, the petioles of leaves develop a specialised abscission zone. Cell walls in abscission zone are weak and loosely connected and allow the leaves that have become inefficient to be shed easily without a damage to the remaining branch.

The chemical composition of needle litter, that is, senescent needles, differs clearly from that of active needles. The relative concentrations of proteins and sugars and starch in litter are clearly lower than in the green needles (Simojoki et al. 2008) reflecting the action of the biochemical regulation system and reuse of proteins, starch and sugars. The relative concentrations of cellulose, lignin and lipids in the litter are higher than in needles because considerable amounts of proteins and lipids have been translocated. Leaf senescence offers the perennials a way to recycle valuable nutrients, especially nitrogen in amino acids.

4.2.5 Uptake of Water and Nutrients by Roots

Asko Simojoki, Jussi Heinonsalo, Sari Timonen, Kari Heliövaara, and Pertti Hari

Flow Paths of Water and Nutrients in the Soil-Plant-Atmosphere Continuum

Most of the water and elements required for plant growth are taken up from the soil by roots. The main exceptions are carbon and part of oxygen taken up as carbon dioxide by plant leaves from the atmosphere for the photosynthesis. The nutrient uptake of trees is generally assisted by the symbiosis between roots and mycorrhizal

A. Simojoki (✉)
Department of Food and Environmental Sciences, University of Helsinki, P.O. Box 27, 00014 University of Helsinki, Helsinki, Finland
e-mail: asko.simojoki@helsinki.fi

Fig. 4.54 A pine seedling with a mycorrhizal root system. Mycorrhiza (*M*), rhizomorph (*R*) and hyphal front (*H*). Seedling was grown on 5-mm-thick organic layer soil between two Perspex® plastic sheets with dimensions 20 × 20 cm for 4 months (Photo: Sari Timonen)

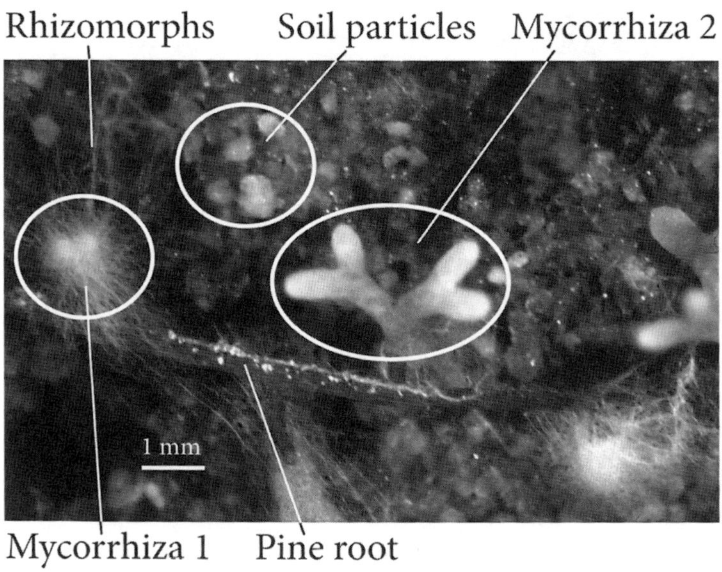

Fig. 4.55 A pine root with two different kinds of mycorrhizal root tips. The first mycorrhiza has a mantle with a furry outer layer and rhizomorphs, while the other has a smooth, thin hyphal mantle. Note the size of mineral soil particles (Photo: Jussi Heinonsalo)

fungi (Smith and Read 2008). Mycorrhizas are important for the nutrient uptake of woody plants (Figs. 4.54 and 4.55), as they may cover most of the root surface area capable of efficient nutrient absorption (Smith and Read 2008; Cairney 2011;

Lehto and Zwiazek 2011). Soil in the immediate vicinity of roots (rhizosphere) or mycorrhiza and its hyphae (mycorrhizosphere) is the most important for the uptake of water and nutrients by plants.

Soil water flows through the mycorrhizosphere into root cells, and through the plant to the atmosphere, and carries solutes along. Nearly all water is eventually transpired into the atmosphere, whereas the nutrients are mostly incorporated into plant tissues. Uptake of water and nutrients by plant requires adequate amount of continuous water-filled pores in the surrounding soil and a good contact with them. The fluxes of water and nutrients to plants are driven mainly by the spatial gradients of water tension (or pressure) and nutrient concentrations created by the transpiration of water, and the selective uptake and metabolic reactions of solutes by cells. Transmembrane pathways provide sites for the active control of water and nutrient transport by root cells (Figs. 4.4 and 4.5). At a larger scale, soil factors play an important role, as the acquisition of water and nutrients by plant roots is determined by the soil volume accessible for the growth of roots and mycorrhizal fungi. Each of these aspects of root water and nutrient uptake is discussed in more detail below.

Growth of the Root System into Soil Containing Available Water and Nutrients

The prerequisite for the uptake of water and nutrients by roots or mycorrhizal fungi is that water is not too tightly bound to soil pores and particle surfaces and that soil water contains enough nutrients in dissolved form. The uptake of water and nutrients increases the water tension and decreases the nutrient concentrations locally in soil and creates spatial gradients that drive water and nutrient fluxes from the surrounding soil to meet the uptake and replenish the decreased concentrations. However, as the flux rates and transport distances are generally much limited by soil properties, the ability of plants to take up water and nutrient from soil also depends on the spatial distribution of the root system (plant roots or external hyphae of mycorrhiza) that needs to grow and form contact with adequate volumes of soil containing available water and nutrients. As coarse-textured soils typical of boreal forests are weak in retaining moisture, it is important that the root-hyphal system can reach large soil volumes. Otherwise, the supply of water to plants would be fluctuating a lot and be more dependent on rainfalls.

The ability of roots and fungal hyphae to grow into various parts of soil is determined mainly by pre-existing soil properties, as the possibilities of growing roots or fungi to change soil properties are limited. For instance, plant-available reserves of water and nutrients in subsoil horizons in fact remain inaccessible to plant, if oxygen deficiency (poor aeration), high acidity or high mechanical impedance prevents root or fungal growth into these soil layers. On the other hand, localised favourable conditions like high concentrations of easily available nitrogen in topsoil may concentrate ample root and hyphal growth and cause less growth into less fertile but wetter subsoil.

The responses of plants and fungi to soil conditions vary in a species-dependent manner. Plant roots can to some extent overcome soil mechanical stresses, transport gases, water and nutrients internally within the plant from one location to another and show some acclimation to varying soil conditions by changing the anatomy and pattern of root and mycelial growth. In grass species, root width has a special role, as fine roots make it possible for plant to construct a longer root system and higher absorbing surface area with a given amount of photosynthesised carbon and nitrogen. In trees, the allocation of plant-synthesised carbon to mycorrhizal symbionts is in comparison even more cost-effective for constructing a large root system, as the width of fungal hyphae ($\sim$10 μm) is typically more than one order of magnitude smaller compared with plant roots. In oxygen-deficient wet soils, however, the ability of roots for shallow rooting and internal aeration would be more important in ensuring sufficient oxygen supply to roots, especially if the roots are thick. Mycorrhizal fungi generally do not like to grow in anaerobic conditions in wet soils. Tips of roots and hyphae are the major organs for sensing soil conditions around the roots. Branching increases their number and allows them to respond in a more sensitive way to variations in soil conditions.

Membrane Transport Proteins and the Selectivity of Water and Solute Uptake

Soil water carries numerous solutes in various chemical forms, such as inorganic and organic ions and neutral molecules. Their amounts vary, and their chemical forms depend on their exchange and retention by soil solid phase as well as on the acid-base and oxidation-reduction reactions in soil. However, plant roots and fungal hyphae are remarkably selective in their uptake of elements. As a consequence, the cytosolic solute concentrations generally deviate from those in the soil solution.

Within plants and fungi, water and nutrients move through (1) cell membranes and vacuolar membranes (tonoplasts, transmembrane pathway), (2) the network of cytoplasm interconnected by pores (plasmodesms) in the walls of neighbouring cells (symplastic pathway inside the cell membrane of living cells) and (3) the continuum of extracellular space (apoplastic pathway outside the cell membranes, outside living cells) (Figs. 4.4 and 4.5). The selective control for inflow and outflow of water and solutes at the cell membrane is the main mechanism by which the roots and fungi maintain the concentrations in the cytosol often very far from equilibrium with those in the surrounding solution. The metabolic control of cell-to-cell transport through plasmodesmata (Ehlers and Kollmann 2001; Burch-Smith et al. 2011) is not directly involved in the uptake of nutrient ions by cells.

The high selectivity of solute uptake by root cells is attributed to highly substrate-specific membrane transport proteins that facilitate most of the transport of water and solutes through biological membranes (Figs. 4.4 and 4.5). Similar mechanisms operate in the membranes of fungal hyphae. Only nonpolar and small uncharged molecules permeate rapidly through the phospholipid part of the membranes. Besides cell membrane, similar transport proteins operate across other biological membranes as well, such as tonoplastic and mitochondrial membranes. In addition to nitrogen that is needed for synthesis of these functional transport substances ("Functional Substances"), several other elements, like potassium, phosphorus, magnesium and calcium, are necessary for their proper functioning. The membrane transporters are commonly classified as channels, carriers and pumps according to their type of functioning.

Channels are transport proteins that cross the whole membrane and facilitate passive transport along the decreasing electrochemical gradient (Taiz and Zeiger 2006). They are essentially transmembrane selective pores that open and close in response to cellular signals ('gating'). The substances diffuse very rapidly through channels, if they are open (10^8 ions s^{-1} per channel). Specific channels exist for numerous cations and anions. Aquaporins are channels specific for water molecules (Luu and Maurel 2005). They also allow transport of small neutral solutes and/or gases.

Carriers do not have pores crossing the whole membrane. Instead, they bind the molecule on one side and release it on another (Taiz and Zeiger 2006). The slow rate of required conformational changes in the protein make the carrier transport 10^6 times slower compared with the channel transport. Carrier proteins mediate both passive and active transports. Active carrier-mediated transport uses energy stored in the electrochemical potential gradient of protons (higher concentration of protons outside than inside the cells) to drive the transport of other ions in the same direction (symport, coport) or the opposite direction (antiport) compared with that of protons.

Pumps are membrane transport proteins that use energy obtained from direct ATP hydrolysis to carry out active energetically uphill transport of substances, such as that of protons or calcium ions out of cytosol (Taiz and Zeiger 2006). This is called the primary active transport, in contrast to the secondary active transport by carriers for which the energy is provided by the proton motive force or other ion gradients created indirectly by pumps rather than by direct ATP hydrolysis. Most pumps are ion pumps. Proton is the most important ion transported by this way in the membranes of plants. Symporters of sucrose and glucose (Sauer 2007; Kühn and Grof 2010; Geiger 2011) are examples of secondary active transport.

Due to the membrane transport proteins, practically all substrates have their own specific uptake mechanism (see a nice animation on the functioning of transport proteins at www.teachersdomain.org/resources/tdc02/sci/life/cell/membraneweb, WGBH 2003), although the specificity may not be absolute. The transport proteins may increase the permeability of membrane for ions and water by one or more orders of magnitude compared with that in a pure phospholipid bilayer, but they have no effect on nonpolar gases and small uncharged molecules (Taiz and Zeiger 2006). The molecular biology of different transport proteins is currently an active field of research (e.g. Miller and Cramer 2004; Luu and Maurel 2005; Raghothama and Karthikeyan 2005; Demidchik and Maathuis 2007; Isayenkov et al. 2010; Kollist et al. 2011; Schulz 2011).

Passive and Active Transport

Water and nutrients are transported in the soil-fungus-plant-atmosphere continuum by convection and diffusion mechanisms (see Appendix 4). Passive transport occurs spontaneously along a decreasing potential gradient so that the free energy of system decreases in accordance with second law of thermodynamics. On the other hand, the transport of species actively against their potential gradients requires external work. For details on how the concentrations, temperature, pressure, electrical forces and gravitation affect the energy status of solutes and water, and the free energy of whole system, we refer to standard textbooks in soil physics and plant physiology (e.g. Nobel 1991; Hillel 1998; Taiz and Zeiger 2006). In short, water tends to flow spontaneously towards lower elevations, temperatures and pressures (or higher tensions), as well as higher solute concentrations, whereas the solutes carried by water tend to distribute themselves towards lower temperatures and concentrations (with all other things being equal). Electric forces influence the movement of solute ions, but not that of water and other uncharged species.

The movements of particularly nutrients and other solutes occur frequently against the directions of spontaneous transport within plant, and this requires metabolic energy. Even if water itself is not transported actively, water transport is often under indirect metabolic control by solute transport. Membrane transport proteins capable of pumping ions actively against their gradients are the most important in this respect. The energy for maintaining the concentration gradients and the transport protein system is provided by the hydrolysis of ATP molecules produced by cellular respiration (see Sect. 4.2.3). In this way, root metabolism controls both the driving force (concentration gradients) and material properties (transport proteins) determining the flux of water and nutrients across the plasma membrane.

The uptake of water and nutrients by plants involves several passive and active transport mechanisms and gradients that often influence and oppose each other in a coupled way. For this reason, the full description of material fluxes in the systems with multiple constituents requires explicit accounting for all transport mechanisms for the species involved (see, e.g. Corey and Logsdon 2005). Below, we outline the mechanisms involved in the uptake of water and nutrients by roots and mycorrhizal fungi.

Uptake of Water by Roots

Plant roots have specialised structures for water and nutrient uptake. The outermost cell layers in a non-mycorrhizal root are called the rhizodermis (epidermis) and the

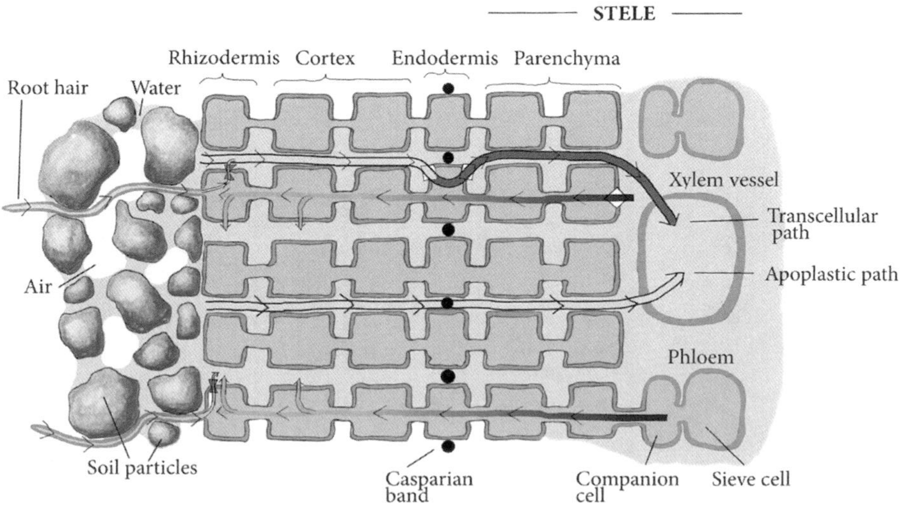

Fig. 4.56 A trans-section of non-mycorrhizal root in contact with moist soil, including the different flow paths of water and solutes from soil to xylem and the flow path of sugars from phloem to root surface and surrounding soil

cortex (Fig. 4.56). Root hairs are elongated extensions of rhizodermis that increase the contact area of roots with soil. The inner boundary of cortex surrounding the stelar tissues is called the endodermis. Endodermal cell walls become lignified and suberised with waxlike materials (Casparian bands) with ageing of cells. Inside the endodermis, parenchyma cells connect the endodermis to xylem and phloem.

Plants need both large-scale transport and small-scale transport of water. Large-scale transport such as flow of water from roots to leaves does not usually cross cell membranes, whereas small-scale transport such as the uptake of water into cells generally does. Water is driven from soil to plants by the gradients of both gravitational, pressure and osmotic (solute concentration) components of water potential (Nobel 1991; Hillel 1998), but the relative importance of different components varies depending on the flow path. In soil, only the convection (bulk flow) of water driven by the gradients in pressure (or tension) and gravity is relevant. In plants, however, both convection and osmosis (diffusion) are important: The hydrostatic pressure gradient is the main driving force for the convective water flow along the apoplastic and symplastic pathways (Figs. 4.56, 4.58 and 4.59) between plant tissues, whereas the osmotic uptake of water into root cells (transmembrane pathway) depends on the gradients of solute concentration and water tension between the apoplast and root cell (Fig. 4.57), and the gravity can be neglected. The solute concentration and the symplastic flow in root cells are tightly coupled to the sugar supply from the phloem (Figs. 4.56 and 4.59).

At a cellular scale, the concentration gradients driving the osmotic uptake of water by cells are created mainly by the selective transport of sugars, ions and other solutes as across cell membranes and tonoplasts and their metabolic reactions within

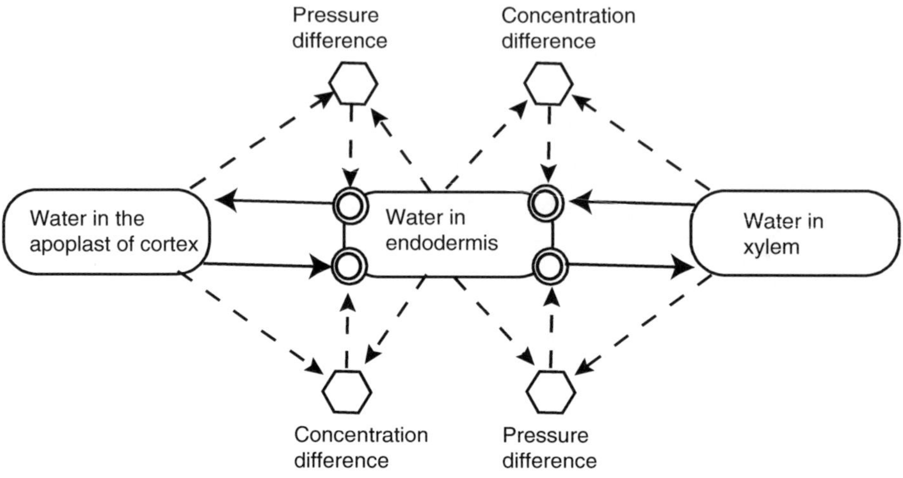

Fig. 4.57 Water flow from the apoplast of cortex via a transmembrane pathway through endodermis cells to xylem. Water flow into and out of the cell is driven by the total solute concentration and pressure differences between endodermis cells and the apoplast (in cortex and xylem), respectively

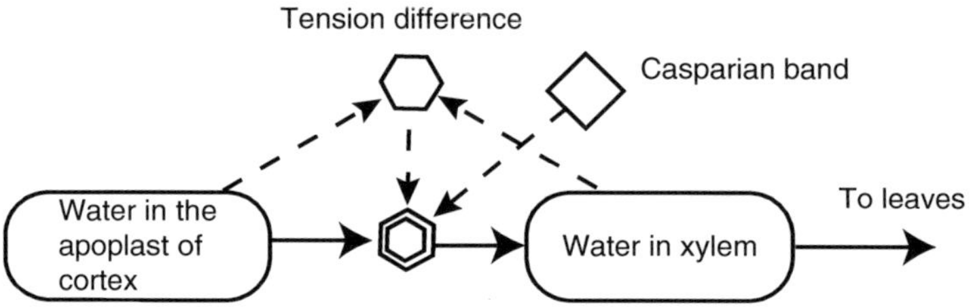

Fig. 4.58 Water flow in the cortical apoplast to the xylem via an apoplastic (extracellular) pathway through the endodermis as driven by a tension difference

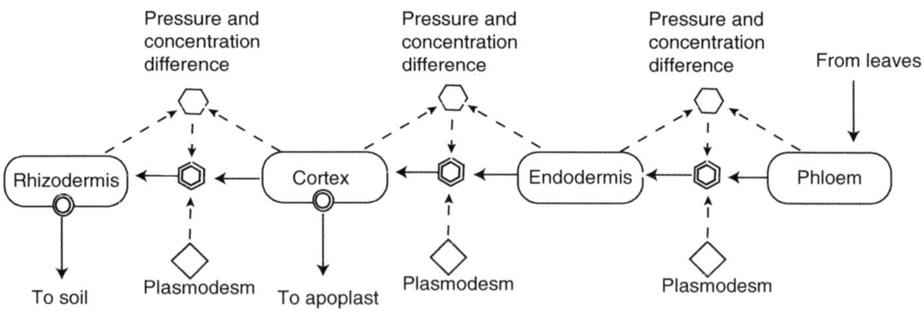

Fig. 4.59 The flux of photosynthates (sugars) from leaves to roots and surrounding soil. Sugars are translocated from leaves to roots by the pressure flow of phloem sap. Within roots, sugars are transported symplastically (through plasmodesmata) driven by pressure and solute concentration differences towards outer root surface. Part of sugars stored or metabolised by the cell (not shown), and part is exuded by root cells into the apoplast and soil

cells. The reactions include, for example, the synthesis and decomposition of starch (Taiz and Zeiger 2006). Plant roots can adjust themselves osmotically to water stress more rapidly than leaves (Hsiao and Xu 2000). In particular, the metabolic control of solute concentrations at desired levels requires a continuous supply of sugars from the phloem (Figs. 4.56 and 4.59), not only for solutes, but also for a source of energy. The high cellular concentrations of sugars and other solutes (decreased water potential) drive the osmotic water flow from the apoplast through aquaporins into the cells, which causes the build-up of positive hydrostatic pressure (turgor) within cells at a rate depending on the rate of water flow and the mechanical properties of cell wall (Mengel and Kirkby 2001; Taiz and Zeiger 2006). The build-up of turgor, in turn, drives the flow of water out of cells until an equilibrium or steady state is reached at which the inflow and outflow of water balance each other (Fig. 4.57).

Positive turgor is important for cells not only because of its role in the control of water uptake but also because it provides the physical pressure for the expansion of growing cells (Greacen and Oh 1972; Bengough et al. 2006). If the plant loses its turgor, the cells cannot grow and the plants wilt. Empirically, it has been confirmed that permanent wilting occurs for many plant species at soil moisture contents corresponding to greater than about 1.5 MPa soil matric suction.

At a whole-plant scale, the tension differences driving the convection of water are caused by the transpirational pull of water transmitted from the leaves to the root system. The prerequisite of water flow from soil to plant leaves is that the water tension (underpressure) in xylem is higher than in soil. The water tension generally increases on the way from soil to the cortical apoplast and the xylem because the transpiration of water by leaves creates a tension gradient that drives water from soil to leaves by the cohesion-tension mechanism (Steudle and Petersen 1998; Steudle 2001; Angeles et al. 2004). Only if soil water tensions and transpiration rates are low, small positive pressures (typically <0.1 MPa) may build up in xylem due to accumulation of solutes (Steudle and Petersen 1998; Taiz and Zeiger 2006).

The apoplast allows the most rapid water flow from soil to leaf. At the endodermis, however, the flow is partly blocked by Casparian bands, and at least part of the flow will pass in and out of endodermal cells (transmembrane path) through aquaporins (Fig. 4.56). It is also worth noting that the symplast probably does not contribute significantly to the inward transport of water in root, because the convective flow in the symplast occurs in the opposite direction transporting sugars out of the phloem to other root tissues (Figs. 4.56 and 4.59). This limits the inward symplastic water transport only to slow diffusive fluxes at rather small distances (Fig. 4.56).

The Role of Endodermis for Water Transport Within Roots

Endodermis is a major control site for water and solute transport in roots by at least three different mechanisms (Peterson et al. 1999; Enstone et al. 2003). Although water and solutes may enter the symplast anywhere in the cortex, 1) the partial blocking of apoplastic pathway by Casparian bands is particularly efficient in forcing water and solute flows through the membrane transport proteins into the root cells, which allows a high degree of metabolic control of solute transport in roots by endodermal cells. Casparian bands also 2) prevent the back diffusion of solutes from xylem to cortex and 3) prevent the growth of mycorrhiza into the stele. Casparian bands may also develop, although at a more slowly rate, to some extent in the tissue just beneath rhizodermis (exodermis). However, this is plant specific, and some studies even suggest that most gymnosperms do not develop exodermis (Hose et al. 2001). The impact of exodermis on water flow is small in young roots but may become large in older roots with mature exodermis (Hose et al. 2001; Enstone et al. 2003).

According to the composite model of water flow (Steudle and Petersen 1998), water and small molecular solutes pass the endodermis to variable extent both along the apoplastic (through Casparian band) and transcellular paths. This is true even in mature roots with well-developed Casparian bands that restrict the apoplastic path.

The transcellular flow of water into and out of endodermal cells is driven by the concentration and pressure differences between the endodermal cells and apoplast (Fig. 4.57). The biochemical regulation system maintains high sugar concentration in the endodermis cells relative to the apoplast, which gives rise to the osmotic inflow of water from the cortical apoplast into the endodermal cells through aquaporins (water-specific transport channels). Simultaneously, the higher water tension in the xylem relative to the endodermis pulls water out of endodermal cells into the stelar apoplast. Together these mechanisms drive water flow from the cortex to the xylem in a transpiring plant. On the other hand, the higher solute concentration of endodermal cells relative to xylem and the higher pressure relative to cortical apoplast can drive the transcellular water flow also in the opposite direction, from the stele to the cortex (Fig. 4.57), but this can only occur at conditions of minimal transpiration. The resistance to flow along the transcellular path is high, as water must pass twice through the cell membrane. The contribution of transcellular path to the total water flow across endodermis depends also on the functioning of apoplastic path.

The apoplastic flow of water through the Casparian band is driven by the water tension difference between the xylem and the cortical apoplast (Fig. 4.58). The rate of apoplastic water flow through endodermic is much influenced by the properties of Casparian band. In older roots, the rate is generally much reduced by the suberisation of Casparian band, which decreases the relative importance of apoplastic pathway with root age (Peterson et al. 1999; Steudle 2001; Enstone et al. 2003). However, root branching initiated at the tissue just beneath the endodermis (pericycle) temporarily breaks down the endodermis and Casparian band and enhances apoplastic flow.

In boreal forests, ectomycorrhizas cover most of tree roots. They form a dense layer (mantle) on the root surface and densely packed hyphal tissue between cortical cells (Hartig net) (Fig. 4.60). The external hyphae growing in soil supply water to roots from larger distances by enhancing the contact of root system with soil and expanding the soil volume utilised for the uptake of water and nutrients

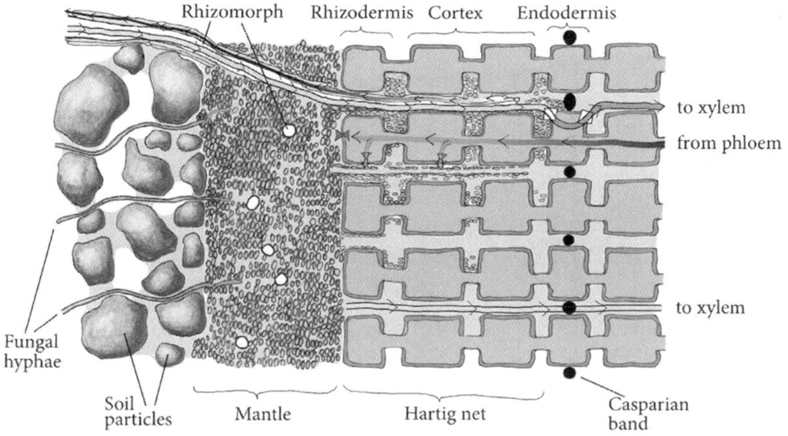

Fig. 4.60 A trans-section of ectomycorrhizal root in contact with moist soil, including the different flow paths of water and solutes from soil to Hartig net and xylem and the flow path of sugars from phloem to Hartig net and external mycelia

(Smith and Read 2008). In a sense, the mycorrhiza performs a similar function as root hairs in a non-mycorrhizal roots, yet with much larger amount and length. The structural features of different ectomycorrhizas are discussed in detail by Agerer (2006).

The most rapid ectomycorrhizal pathways for convective water flow over large distances are provided by large vessel-like empty spaces (fungal apoplast) within the aggregated bundles of hyphae called rhizomorphs (mycelial strands, mycelial cords). The spaces are formed by the decomposition of one or more hyphae within rhizomorphs and have typical diameters between 6 and 20 μm (Lehto and Zwiazek 2011). Such spaces extend and provide rapid water flow from soil through the mantle into the cortex of root, where the fungal and root apoplasts form a continuum (Fig. 4.60). In the cortex, water and nutrients may be subsequently taken up by root cells or transported via xylem to the leaves as described in Figs. 4.57 and 4.58.

A positive pressure (turgor) within cells provides the physical force for the hyphal tip growth of fungi, just like for root growth (Steinberg 2007; Lew 2011). A regular supply of sugars from the host plant makes this possible. Even if the optimal moisture contents for both plant roots and fungi correspond to about -10-kPa water suction, a notable difference is, however, that fungi can tolerate more than one order of magnitude higher matric water suction (Coyne 1999, p. 177) than the typical wilting point for plants (about 1.5 MPa). This gives an advantage to mycorrhizal roots compared with non-mycorrhizal roots in dry soil, as fungal hyphae may continue growing, reach deeper soil layers with available water and provide the host plant with water and nutrients even when the lack of water impedes the non-mycorrhizal root growth in topsoil (Lehto and Zwiazek 2011). This also helps the host plant to recover more rapidly after dryness.

Transport of Sugars Within Roots and Mycorrhiza

The sugars play important roles in the penetration of water across cell membranes by osmosis, in the action of membrane pumps and in maintaining the turgor in living cells. The biochemical regulation system has a demanding task in controlling the proper sugar concentrations in the various cells of roots. Plasmodesmata connect the plant cells and allow the symplastic sugar from the phloem to the rhizodermis. Membrane pumps or specific transport channels transport the sugars between apoplast and cells (Fig. 4.59).

Energy-demanding metabolic processes, such as the nutrient uptake (see below) and growth of roots and fungi, require a regular flow of sugars photosynthesised in the leaves. According to the pressure-flow model by Ernst Münch (Hölttä et al. 2006, 2009; Taiz and Zeiger 2006), the increase of turgor pressure resulting from the active loading of sugars and consequent intake of water into the leaf phloem drives the solute flow through the phloem to the roots. In roots, the sugars in the phloem are symplastically and/or apoplastically unloaded by neighbouring cells (Patrick 1997; Turgeon and Wolf 2009). The positive pressure in the phloem contributes to the symplastic mass flow (incl. sugars) towards active sinks in outer root tissues and growing root tips, and the decreasing sugar concentrations and turgor pressures towards the root surface drive the flow to the same direction towards active sinks.

While the symplastic mass flow from phloem provides the cells with sugars and other essential solutes for cellular growth and energy, it simultaneously prevents the symplastic water flow to the opposite direction (from the root surface inwards towards stele). The symplastic mass flow from phloem is quite large and may account for about one-third up to two-thirds of the water needed for root growth in well-aerated moist soil (Boyer et al. 2010). For this reason, only a small diffusive flow of water is possible in the symplast to the countercurrent direction, as noted earlier (Fig. 4.56). Nevertheless, even if diffusion is a slow mechanism of flow, it can be relevant at small distances, such as for transporting water from root surface cells to outer cortical cells, particularly as the rate of diffusive flow from cell to cell is known to be increased by cytoplasmic streaming that decreases the concentration gradients within cells. In addition, the direction and speed of symplastic flows in roots are not constant, but may vary locally and temporally in response to the rate of phytosynthesis and transpiration in leaves and metabolic reactions in root cells.

Part of the sugars are exuded into soil or apoplast, where they are taken up by other root cells, mycorrhizal fungi or other microbes. Alternatively, they may be carried with water into xylem and transported back to leaves. However, roots are efficient in recycling sugars into use before they enter the xylem. Even if elevated concentrations sugars and other organic substances in the xylem sap (Satoh 2006) indicate some exudates do enter the transpiration stream, part of the sugars are taken up before and recycled back to the symplastic flow by stelar parenchyma cells.

The sugars produced and exuded by the host plant enable the growth of mycorrhizal fungi. Because the symplasts of roots and mycorrhizal fungi do not form a continuum and the growth of mycorrhiza is limited outside endodermis by

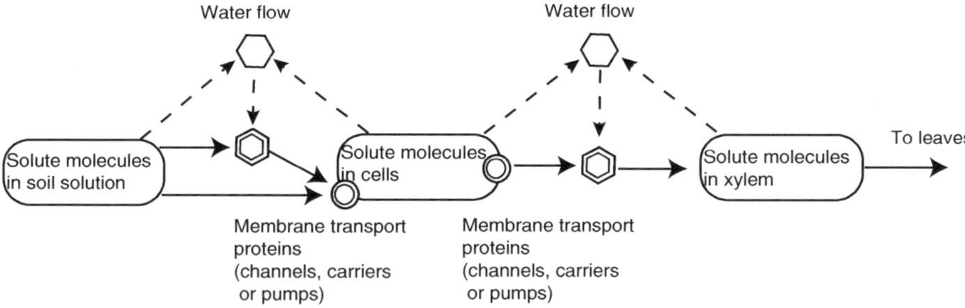

Fig. 4.61 The flux of solutes from soil solution into root cells and xylem to leaves. Solutes are carried by water to root apoplast, where they can be taken up into root cells through specific transport proteins. Likewise, some nutrients may be transported out of the cell and carried apoplastically by water to xylem and leaves

the Casparian band, the fungi are dependent on the sugars exuded by the root cells into the apoplastic continuum in cortex (apoplast shared by both roots and fungi), where the sugars can be taken up into the hyphae.

The long-distance symplastic transport of sugars within fungal hyphae to the growing hyphal tip is probably mostly driven by solute concentration and pressure gradients rather similarly as in plant roots. Plasmodesmata-like pores in the septa separating the individual cells of hyphae allow the transport of sugars in the symplastic pathway in the mycorrhiza. The elongate cells of hyphae probably cause less resistance to flow compared with root cells, as they have a less complex internal structure. In addition, active forms of transport operate within fungal cells, but their efficiency for long-distance transport is unclear (Steinberg 2007; Lew 2011).

Uptake of Nutrients

Solutes in soil, such as nutrient ions and amino acids, are carried by water flow or diffusion to the vicinity of root cell membranes (apoplast), where they can be taken up into root cells through compound-specific membrane transport proteins (channels, carriers and pumps). Similar membrane transport proteins also transport ions and molecules out of cell. For instance, the outflow of ions from endodermal or stelar parenchymal cells to the stelar apoplast allows the transport of ions to the leaves via the xylem (Fig. 4.61). The combination of water tension and osmosis drives water through the aquaporins (water-specific transport channel) into and out of cells. This enables a very strict control of ions and molecules taken up from the soil since the transport of both water and solutes across cell membrane is based on highly specific membrane transport proteins.

Living cells exhibit a net negative charge. The voltage difference across the cell membrane ranges between about −60 and −300 mV. It is caused by the unequal flux rates and active transport of ions into and out of the cells (Ritchie 2006; Taiz and Zeiger 2006). This has major consequences for the nutrient uptake by cells. In order to maintain the net electrical balance on cation uptake, the cells emit an equivalent amount of protons to soil (Mengel and Kirkby 2001). Similarly, the uptake of anions is balanced by the emission of organic anions. On the other hand, the net negative charge impedes the uptake of anions and enhances the uptake of cations by roots. Accordingly, the theoretical predictions show the uptake of anions is active (energy-demanding), whereas the uptake of most cations is passive, with only potassium taken up actively at low external concentrations. In contrast, such cations as sodium, calcium and protons are pumped actively out of cytosol into extracellular space and vacuole to maintain low cytosolic concentrations (Taiz and Zeiger 2006).

Following the nutrient uptake by roots, the nutrient ions must ultimately be transported from root to the xylem and upper parts of plant. This occurs most rapidly by the water flowing in the apoplast, as the symplastic diffusion of nutrients is efficient only at small distances. However, at the endodermis, the apoplastic flow is much impeded. Water, ions and small molecules may go partially through the Casparian band in the endodermis of young roots. At the endodermis of mature roots, the Casparian band impedes the apoplastic pathway and forces the nutrients to the transcellular pathway through endodermal cells (Figs. 4.56, 4.57 and 4.58).

The nutrient uptake into mycorrhizal hyphae involves similar mechanisms as described for roots (Smith and Read 2008). Following the nutrient uptake from soil by fungi, the nutrient ions must be transported to the root cortex of host plant, where they can be taken up into the root cells. For long distances within fungal hyphae, the nutrients are most rapidly carried by the water flowing in the apoplastic empty spaces within rhizomorphs (Fig. 4.60).

The secretion of organic acids, protons and other ions, along with that of sugars, polysaccharides and enzymes (see Sects. 4.4, 5.4 and 5.6) by roots and mycorrhizal fungi into the surrounding soil, contributes to the modified physico-chemical and microbiological properties of mycorrhizosphere soil compared with bulk soil (reviewed recently by Hinsinger et al. 2005, 2009; Gregory 2006). Such biological activity contributes to the increased rates plant nutrients are released from soil minerals by mineral weathering (reviewed in detail by Gadd 2007; van Schöll et al. 2009; Taylor et al. 2009; Martino and Perotto 2010; Sokolova 2011).

The capability of roots for nutrient uptake varies according to the anatomy of roots and architecture of the root system (Peterson et al. 1999; Hishi 2007). Most of the nutrients are taken up by the active white zones near the tips of apical first-order roots and some through the pigmented zones of condensed tannin, whereas the pigmented cork zones form a barrier for nutrient uptake (Hishi 2007). The external hyphae of mycorrhizas enhance the uptake of all plant nutrients by increasing the effective-absorbing surface of roots and by enhancing the enzymatic hydrolysis of organic matter (Chalot et al. 2002; Miller and Cramer 2004; Pritsch and Garbaye 2011). In particular, the uptake of ammonium is enhanced by the enzymatic hydrolysis of N-containing organic macromolecules in soil. On the other hand, the phosphorus nutrition of mycorrhizal trees is enhanced by the release of both inorganic phosphorus by low-molecular-weight organic acids and organic phosphorus by phosphatase enzymes, respectively, as reviewed by Plassard and Dell (2010).

Ammonium and nitrate are traditionally considered the major nitrogen forms taken up directly by roots. The concentrations of nitrate and ammonium in soil solution of natural ecosystems vary widely ranging from several tens or hundreds of mmol m^{-3} up to several mol m^{-3} (Britto and Kronzucker 2002; Miller and Cramer 2004). These values are typically much less than the estimates for their cytosolic concentrations ranging from several mol m^{-3} to several tens of mol m^{-3}, with the values in the lower range being more probable (Mengel and Kirkby 2001; Loque and von Wirén 2004; Ritchie 2006). For this reason, active uptake of nitrate anions by plant roots is required almost in the whole range of concentrations encountered in the soil (Miller and Cramer 2004), whereas the active uptake of ammonium cations is required at low concentrations, as their uptake becomes passive only at concentrations higher than about 500 mmol m^{-3} (Britto et al. 2001).

Most N in soil is in organic form. The enzymatic hydrolysis of proteins in litter and soil organic matter releases amino acids into soil. The uptake of amino acids by roots is an active energy-requiring process as well. Many plant species can take up amino acids by roots, but the importance of this as a major pathway for nitrogen acquisition in the field has not been confirmed (Jones et al. 2005; Persson et al. 2006). While the roots and mycorrhiza do take up organic N forms and the uptake mechanisms of amino acid by roots and mycorrhiza are well established, in lack of direct evidence, it remains an open question, whether organic N really contributes significantly to plant nutrition in any ecosystem, although several lines of indirect evidence support the important role of organic N forms in some ecosystems (Näsholm et al. 2009). In forest ecosystems, the complexation of protein with polyphenols limits the direct access of ectomycorrhizal fungi to organic N sources (Wu 2011). Nevertheless, as the nitrate concentrations are low, and many plant species including conifers take up ammonium preferentially over nitrate (Kronzucker et al. 1996; Mengel and Kirkby 2001; Persson et al. 2006), ammonium ion and amino acids may be considered the major chemical forms of nitrogen taken up by plants in boreal forests.

The mechanisms involved in the uptake of nitrate and ammonium by plant roots have been recently reviewed by Glass et al. (2002), Loque and von Wirén (2004), Miller and Cramer (2004) and Näsholm et al. (2009). Several high- and low-affinity transport systems have been indicated for both nitrate and ammonium. The cytosolic concentrations are determined by the net influx and efflux, as well as the cellular compartmentation and assimilation of ions, all of which are controlled largely by plant metabolism at the transcriptional, protein, cellular and whole-plant levels. Nitrate is generally considered to be taken up by a secondary active $2H^+/NO_3^-$ coport or by a primary active ATP-driven pump (Miller and Cramer 2004; Ritchie 2006), whereas anion channels are the most obvious route for nitrate efflux (Miller and Cramer 2004). Ammonium ions enter the cells by an active $H^+/NH_4{}^+$ coport or through cation-selective channels or carriers for $NH_4{}^+$ and K^+ as chiefly driven by the negative membrane potential of plant cell (Mengel and Kirkby 2001; Miller and Cramer 2004).

Gaseous ammonia (NH_3) is a weak base ($pK_a = 9.3$) that accompanies ammonium ions (NH_4^+) in aqueous solutions at high pH. The concentrations of ammonia increase with increasing pH, but this does not influence the uptake of ammonium by plant roots at the normal apoplastic pH range. Nevertheless, ammonia does contribute to the overall ammonium efflux (Loque and von Wirén 2004; Britto et al. 2001). The high cytosolic pH favours the passive efflux of ammonia. It may permeate the cell membranes through the phospholipid bilayer or through aquaporin-like transport proteins. The extent to which plants control the transmembrane ammonia fluxes by gating of such channels is not known (Loque and von Wirén 2004). In addition, the energy-demanding active efflux of ammonium ions has been implicated as a possible cause for ammonium toxicity at high concentrations (Britto and Kronzucker 2002).

The energy requirements and mechanisms of ion transport across biological membranes through transport proteins are broadly similar in mycorrhizal fungi and plant roots. Ammonium ions, amino acids and sugars pass the fungal and plant membranes through specific transport proteins. For instance, fungal cells may take up ammonium ions through the transport proteins in the fungal cell membrane. Ammonium ions may then be assimilated into amino acids, and amino acids and ammonium ions be transported to the host plant. This requires ion transport through two biological membranes: through the fungal cell membrane into the cortical apoplast and from there into root cell across the cell membrane of root. The energy for fungal metabolism is provided by the flow of sugars from the host plant to the fungus. Chalot et al. (2002) summarised the membrane transport proteins involved, of which part are fully characterised, whereas some remain hypothetic. Recent papers by Govindarajulu et al. (2005), Chalot et al. (2006), Schüßler et al. (2006), Martin et al. (2007), Müller et al. (2007), Parrent and Vilgalys (2009) and Nehls et al. (2010) provide a more detailed discussion of the molecular mechanisms of the transfer of nitrogen and carbohydrates in the mycorrhizal symbiosis. Acioli-Santos et al. (2011) provides a comprehensive review of the main genes involved in the symbiosis grouped according to their physiological functions.

4.2.6 Bud Burst Phenology

Pertti Hari, Tapio Linkosalo, Risto Häkkinen, and Heikki Hänninen

Physiological Basis of Bud Development

Nearly all plants have a clear annual cycle in their metabolism, although some plants in the humid tropics may be an exception. Trees and ground vegetation lay down

P. Hari (✉)
Department of Forest Sciences, University of Helsinki, P.O. Box 27, 00014 University of Helsinki, Helsinki, Finland
e-mail: pertti.hari@helsinki.fi

buds in late summer, and the growth during the coming summer is programmed to a great extent in the buds since there are primordia of new leaves waiting for growth in the spring. The buds can tolerate the hard conditions during winter dormancy, but they are vulnerable to low temperatures during growth. For example, the young shoots of Norway spruce (*Picea abies*) suffer lethal damage if heavy frost occurs in the spring. Thus, too early growth may cause severe damage. On the other hand, too late onset of growth may imply loss of metabolic production, especially photosynthesis, during favourable weather. Consequently, the proper timing of bud burst in the spring is of vital importance for the success of a plant.

Bud development during early spring starts with formation of macromolecules, cellulose, lignin, lipids, etc. (Sutinen et al. 2009), to construct cells in the new tissues. Each macromolecule has its specific enzymes for the biochemical steps in the synthesis. There is a clear order in the construction, for example, synthesis of cellulose in the primary cell walls must take place before formation of lignin. Thus, the biochemical regulation system synthesises, activates, decomposes and deactivates the enzymes involved in the bud developed in a very synchronised way.

The buds are dormant in the late autumn and early winter, so that they do not respond to favourable environmental factors, especially to temperature. This dormancy disappears during the winter, and the buds start to response to elevated temperatures.

Formation of new cells and their growth lead into irreversible ontogenetic development, that is, to the microscopic anatomical changes within the bud which finally lead to the visible bud burst (Sutinen et al. 2009). The anatomical changes are tightly synchronised with the changes in the functional substances, since construction and growth of the new cells in the buds have very specific requirements for raw material, that is, for cellulose and lignin for cell walls and proteins for functional substances, etc. The important role of the biochemical regulation system and of the functional substances in the formation of new cells and finally in the bud break is visualised in Fig. 4.11.

Basic Concepts

The biochemical regulation system of the annual cycle changes the functional substances in a synchronous way so that the interconnected synthesis of macro-molecules proceeds in a reasonable sequence. Thus, there are regularities in the functional substances, and we introduce the *annual state of functional substances of bud development* to describe these emergent regularities.

The state of functional substances changes during the annual cycle, and the synthesis of new molecules and cell formation respond to the changes in the functional substances. The *rate of change of the annual state of functional substances* describes the changes taking place in the functional substances during bud development, and it is defined as the time derivative of the annual state of the functional substances.

The biochemical regulation system of the annual cycle can change the state of the functional substances towards summer state or backwards towards winter, since it characterises the concentrations and activities of the functional substances. In contrast, the formation of new cells is irreversible, since existing cell structures cannot be removed (apart from senescence where whole organs are inactivated). Thus, there are two different aspects in the bud development: the state of the functional substances and the state of bud development.

The concept of *state of ontogenetic development* describes the often invisible anatomic features within the bud (Hari et al. 1970; Hari 1972; Hari and Häkkinen 1991; Hänninen 1995; Hänninen and Kramer 2007). The *rate of ontogenetic development* describes the small anatomic changes within the bud, and it is by definition the time derivative of the state of bud development. The scale of ontogenetic development is arbitrary, depending on the specific approach used.

Observation of the slow development inside the buds like developmental phases and growth of primordial shoot is problematic because it destroys the bud; instead we can make point observation of its ending. Bud burst occurs, by definition, when first time in spring clear changes in the bud is observed.

Basic Ideas

The role of the state of the functional substances in the ontogenetic development is to start the growth in buds in proper time early in the spring:

Basic idea PH1: The state of functional substances and bud development are synchronised. The ontogenetic development starts when the annual state of the functional substances has reached capacity to synthesise the needed macromolecules.

The construction of new cells and their growth require large number of enzymes. The biochemical regulation system synthesises or converts enzymes into active form before onset of bud development. The buds are dormant when the key enzymes are missing. The dormancy disappears slowly during winter, and the buds become able to react to elevated temperatures.

Basic idea PH2: Temperature and light affect the rate of change of the annual state of the functional substances.

The ontogenetic development involves enzymatic reactions in the synthesis of new macromolecules that are temperature dependent.

Basic idea PH3: The rate of ontogenetic development depends on temperature and on the annual state of the functional substances.

The bud development is a slow accumulating process, and only its final state can be observed without destroying the bud.

Basic idea PH4: Bud burst takes place when biochemical reactions have produced sufficiently raw material for the formation of new cells and growth has proceeded long enough.

Theoretical Model

The state of ontogenetic development is in a key role in the quantitative analysis of the timing of bud burst. Assume that the state of the ontogenetic development can be described with one number, and assume further that the state of the functional substances can be described with one number. Both these quantities are immeasurable without destruction of the bud; only the state of ontogenetic development can be observed at the bud burst.

Integration over time enables the transition from the rates to states. Thus, if we know the rates as a function of time, we can obtain also the states. The basic idea PH1 states that the ontogenetic development does not proceed during dormancy, and thereafter, it depends on temperature and on the annual state of the functional substances (basic idea PH3). If we assume that the temperature dependence is exponential as in all enzymatic reactions, we get a model for the state of ontogenetic development.

The annual state of functional substances, S_F, changes during the annual cycle, and the synthesis of new molecules and cell formation respond to the changes in S_F. The rate of change of the annual state of functional substances, s_F, is the time derivative of the state of the functional substances, that is,

$$s_F = \frac{dS_F}{dt}. \tag{4.44}$$

The biochemical regulation system of the annual cycle takes its signal from temperatures or from light, and it changes the functional substances towards summer state according temperature and light. This can be introduced into the analysis by assuming that the rate of change of the annual state of the functional substances depends on temperature and light:

$$s_F = s_F(T, I). \tag{4.45}$$

Each step in the long chain of biochemical reactions in the synthesis of macromolecules has its characteristic functional substances. The biochemical regulation system controls the balance between the steps, and this results in stable relationship between ontogenetic development and temperature. The annual state of bud development, S_B, describes the ontogenetic state of the bud, and the rate of ontogenetic development, s_B, is the time derivative of the state of bud development:

$$s_B = \frac{dS_B}{dt}. \tag{4.46}$$

During dormancy, the rate of ontogenetic development is zero and very rapid in late spring. The above argumentation results in model structure that is analogous to that we have used for photosynthesis, respiration and daily height increments:

$$s_B = E_B\,(S_F)\,s_o(T), \tag{4.47}$$

where $E_{\mathrm{B}}(S_{\mathrm{F}})$ is the dependence of the efficiency of the functional substances of ontogenetic development and s_{O} is the stable relationship between ontogenetic development and temperature:

$$s_{\mathrm{o}}(T) = \frac{28.31}{1 + e^{-0.185(T-18.431)}}. \tag{4.48}$$

The state of bud development is obtained as integral of the rate of ontogenetic development:

$$S_{\mathrm{B}}(t) = \int_{t_0}^{t} E_{\mathrm{B}}\left(S_{\mathrm{F}}(t)\right)\; s_{\mathrm{o}}\left(T(t)\right) \mathrm{d}t, \tag{4.49}$$

where t_{o} is a moment during dormancy.

Bud burst is predicted to occur when the state of bud development attains a critical value, that is, the high temperature requirement of bud burst (or flowering), H_{crit}. Thus, the moment of bud burst, t_{B}, is the solution of the following equation:

$$H_{\mathrm{crit}} = \int_{t_0}^{t} E_{\mathrm{B}}\left(S_{\mathrm{F}}(t)\right)\; s_{\mathrm{o}}\left(T(t)\right) \mathrm{d}t. \tag{4.50}$$

Measurements

The phenological observations at Oulainen-Ohineva (64°13′N, 24°53′E) from 1953 to 2002 were made by Aarne Juhonsalo and his family. These data were used for the estimation of model parameters. There were 49 observations for the bud burst for birch (*Betula*), 41 for the flowering of bird cherry (*Prunus padus*) and 50 for flowering of rowan (*Sorbus aucuparia*). The average timing of bud burst of *Betula* was defined as when the buds were opened having leaf length about 10 mm, and the flowering on *Prunus* and *Sorbus* were recorded when the flower buds were clearly open and the general appearance of the trees had turned to white.

A larger data set collected by the Finnish Society of Sciences and Letters, and more recently by the Zoological Museum at the University of Helsinki, was used to test the models. This data set consists of observations within a circle of 250 km in diameter around the city of Jyväskylä (62°14′N, 25°20′E) in Central Finland. The total number of observations was for *Betula* $n = 4321$, *Prunus* $n = 4600$ and *Sorbus* $n = 4055$. These data were combined into a single time series that represents the conditions in Jyväskylä according to an averaging method described by Häkkinen et al. (1995). The data covered the years from 1896 to 2002, with the annual number of observations varying up to 190.

The old time series are always problematic to use since we have no means to check and control the measurements and we have to accept the data as they are. In addition, our data are combined from rather sporadic observations in various places in southern and Central Finland. The noise of each annual observation is couple of days, and we must hope that the systematic measuring errors are missing.

Test with Field Observations

Estimation of Parameter Values. We tried alternative relationships between the rate of change in the annual state of the functional substances and temperature and light. In addition, we varied the relationship between the efficiency of the functional substances and the state of the functional substances. This analysis suggested that trees evidently take the signal to start the ontogenetic development from light, and the efficiency of the functional substances turn suddenly from 0 to 1. As the starting date and bud burst threshold parameters in the temperature sum model are strongly correlated (e.g. Linkosalo et al. 2008), estimating the precise date when the signal from light conditions activates the functional substances is problematic. For different species, the estimated onset dates varied from early January to mid-April. Such variability between species is unlikely in nature, but direct measures of the activation of functional substances would be needed to pinpoint the precise date of activation of bud development. For modelling purposes, however, the variation of onset date can be compensated with proper selection of bud burst threshold (Linkosalo et al. 2008).

We used as estimation data the phenological observations done in Revonlahti during the years 1953–2002. The combined phenological observations during the years 1896–2002 and temperature recordings in Jyväskylä served as test data.

Characteristic Features of Timing of Bud Burst. The great variability between years is characteristic for the timing of the bud burst in the spring. The difference between earliest and latest bud burst dates in Fig. 4.62 is over 1 month. The theory and the model based on the theoretical ideas are able to predict this great variation.

The moment of bud break has slowly become earlier; it takes place about 10 days earlier than 100 years ago, although the great variation hampers the detection of the shift in bud brake time. Our theory and the model predict the same advance in the spring development (Figs. 4.62 and 4.63).

Adequacy of the Theoretical Model. The construction of theories of phenology is rather problematic since the measurements are rather limited, only one single observation at the end on the phenological development and daily temperature records. Additional information is needed, and physiology, especially studies of enzymes and of small anatomic changes in the buds, can provide it. At the moment, these measurements are insufficient (Sutinen et al. 2012), and we have to rely on theoretical argumentation and to test these models. Our approach is an attempt in this line.

Explaining Power of the Theory. Our theory explains the great annual variability in the bud burst dates with the variability in spring temperatures. If the spring is warmer

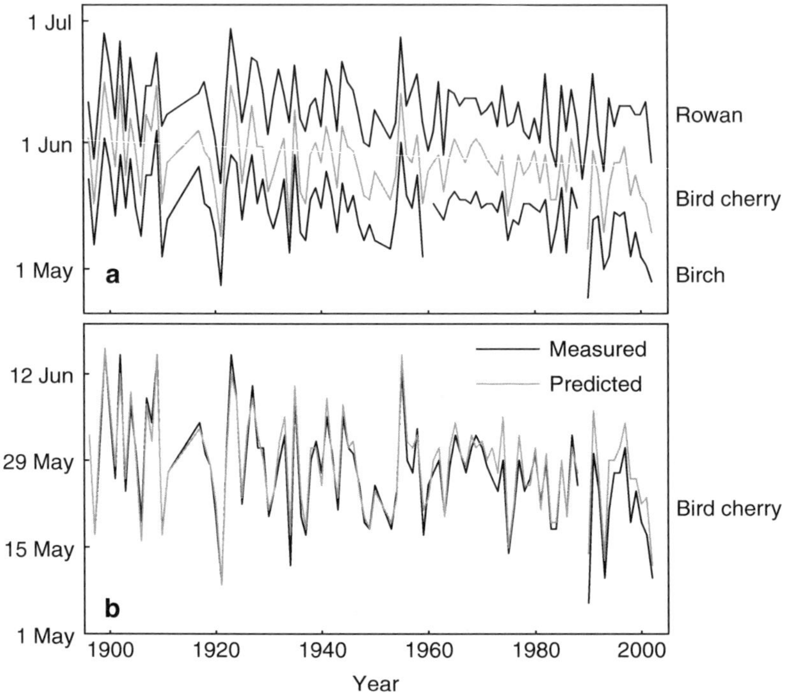

Fig. 4.62 Measured time of bud burst of birch, bird cherry and rowan during 1896–2002 in Jyväskylä (**a**) and measured (*black line*) and predicted (*gray line*) of the bud burst of bird cherry. (**b**) The model parameters are obtained from the measurements during 1953–2002 in Revonlahti

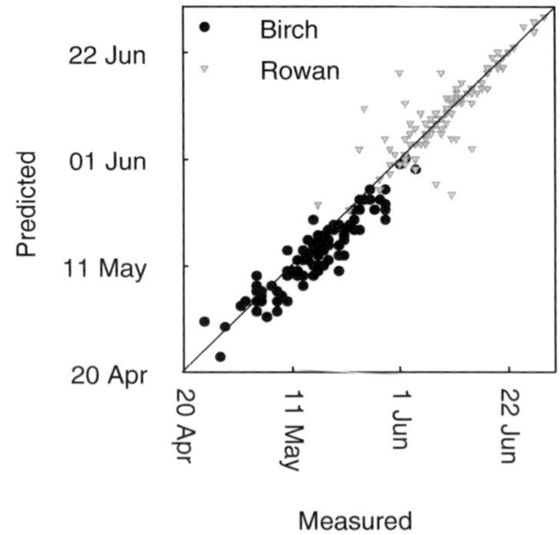

Fig. 4.63 Comparison of measured and predicted time of bud burst of birch and rowan during 1896–2002 in Jyväskylä. The model parameters are obtained from the measurements during 1953–2002 in Revonlahti

than a normal one, then the bud burst takes place early and during cold springs late, correspondingly. Reamur in the eighteenth century was the first to express this fact. The springs have been warmer, especially after 2000, and this explains according our theory the earlier dates of bud burst.

The theory was able to explain annual variation in bud burst dates rather satisfactorily. The proportion of explained variance, PEV, was 87% for birch, 92% for bird cherry and 73% for rowan.

Comparison with Other Approaches. Phenological research is an old tradition; the systematic annual observations started over 100 years ago. Hari (1972) introduced the concept state of development and the rate of maturation to deal with the growth potential of vegetation in the spring. Sarvas (1972, 1974) expanded the theoretical thinking to the annual cycle of plants.

Ecophysiological models for dormancy release (or rest break) and growth onset and bud burst have been presented for boreal trees since the 1970s (Sarvas 1972, 1974, for reviews, see Hänninen 1990; Häkkinen 1999; Chuine et al. 2003; Hänninen and Kramer 2007; Linkosalo et al. 2008). Here, we used the original temperature dependence of the ontogenetic development/activity of active substances as tabulated by Sarvas (1972). However, our interpretation of the phenomenon is partially different in comparison with the original papers. This is reflected in our terminology so that we introduce the key concept functional substances used throughout the present volume into the name of the model variable describing the dormancy (or rest) status of the buds. Thus, we use the concept of annual state of functional substances of bud development, instead of the previous names of the model variable, which refer either to accumulation of chilling (Landsberg 1974; Richardson et al. 1974; Sarvas 1974; Hänninen 1990; Kramer 1994a, b; Chuine et al. 1998, 1999; Linkosalo et al. 2008) or to the progress of the process of dormancy release (or rest break), that is, the removal of the growth arresting physiological conditions in the bud (Hänninen 1995, 2006; Häkkinen et al. 1998; Hänninen and Kramer 2007).

Conclusion: The theory of timing of bud burst predicted satisfactorily the annual variation in the bud burst date and also the observed earlier dates after the year 2000. The residuals did not detect any shortcomings in the model. The theory of timing of bud burst explained a high proportion of the variance in observed bud burst dates. Our theory gained corroboration in the test.

4.2.7 Shoot Elongation

Pertti Hari and Eero Nikinmaa

The lifetime of leaves is rather short: in deciduous trees, herbs and grasses, only one growing season and in coniferous trees, a few years. To maintain the green leaves for

P. Hari (✉)
Department of Forest Sciences, University of Helsinki, P.O. Box 27, 00014 University of Helsinki, Helsinki, Finland
e-mail: pertti.hari@helsinki.fi

photosynthesis, plants must grow new leaves annually. Coniferous trees lay down buds in the late summer for the growth during the coming summer. In Scots pine, the primordials of all needles can be found in the buds. The growth begins with shoot elongation, and the needles grow thereafter. The shoot elongation is a prerequisite for the needle growth and renewal of the photosynthetic system.

Shoot elongation is a vital component of the annual cycle of trees and other vegetation. Then new shoots and needles (leaves) in them are formed using sugars from photosynthesis and nitrogen from nutrient uptake by roots and from reuse in internal circulation as raw material for construction of macromolecules. The biochemical regulation system of the annual cycle takes care of the timing of different phenomena in proper order in respect to the annual environmental cycle.

Physiological Aspects of Shoot Elongation

The meristems in buds and cambium become active in the spring when the conditions for the synthesis of new macromolecules and formation of cells become favourable and these processes begin. The growth phenomena are very intensive in early summer, and they gradually slow down during summer and end well before autumn. In trees with determinate growth pattern, such as Scots pine, the cessation of growth is associated with fulfilment of the maximum size of the organ. During favourable year, this takes place earlier, while during, for example, cold summers, the development may still be incomplete when the growth stops due to the starting of the hardening processes before winter. All the growth processes, be it shoot elongation, growth of leaves or wood in stems, follow rather similar patterns, but their timing during summer may vary (Pietarinen et al. 1982). For example in pine, shoot extension growth gets ready first, followed by the needle growth. The stem thickness growth extends well into late summer, and the root growth in the soil may continue as long as the soil temperature allows.

The formation of new cells into the growing shoot consists of several steps, each of them having its specific enzymes. First, the cellulose is synthesised for the cell walls, thereafter, lignin to strengthen the walls and finally, the lipids form the membranes and proteins and the enzymes and membrane pumps. The synthesis of macromolecules requires sugars as raw material and energy from ATP, produced in respiration. Especially lignin and lipids are energy rich, thus their synthesis generates respiration.

The growth is based on a large number of functional substances, such as enzymes and membrane pumps. The composition and amounts of the functional substances change under the control of the biochemical regulation system during the growing season since the needs of new structures develop along the season. The action of the regulation system generates regularities in the functional substances.

Basic Concepts

The lengths of the growing shoot increase slowly, but the temperature is varying rather rapidly, in the timescale of hours. To solve the discrepancy between the timescales, we need a new concept that is able to cope with the rapid temperature variation. The *height growth rate* is defined as the time derivative of the shoot length.

The action of the biochemical regulation system of growth changes the functional substances involved in height growth. This action generates regularities in the functional substances, and we call these emergent properties in the composition of the functional substances as *the state of the functional substance of height growth*. The *rate of development* is defined as the time derivative of the state of the functional substances.

The synthesis of macromolecules for height growth includes several steps, each of them having specific functional substance. The biochemical regulation system changes the concentrations and activities of the functional substances involved into shoot growth in a balanced way and resulting in an emergent property, the *efficiency of the functional substances in growth*. It characterises the ability of the shoot to grow.

Basic Ideas

The synthesis of new macromolecules into the growing shoot is based on sugars originating from photosynthesis. The sugar pool in growing tissues is large, and evidently there are always enough sugars for the synthesis. In the case of sufficient raw material availability, the shoot elongation is driven by enzymatic reactions, that is, synthesis of new macromolecules, and because the rates of all enzymatic reactions depend on temperature, we arrive to the following basic ideas:

Basic idea HG1: Height growth rate depends exponentially on temperature.

The regulation system changes the concentrations of enzymes, and thus, the growth also reacts to these changes.

Basic idea HG2: The changes in the state of the growth functional substances change the efficiency of the functional substances of growth in such way that there is slow increase, steady response and slow decline in height growth rate.

When the growth proceeds, new cell layers are laid down in the shoot. Thus, growth is an irreversible phenomenon, and the state of functional substances should also go towards summer state after shoot elongation has started.

Basic idea HG3: The rate of development depends on temperature, and it is always positive.

Theoretical Model

Conservation of mass or energy is frequently used in our book to derive the theoretical model from basic ideas. However, we cannot apply this general principle in the connection of shoot elongation; instead we have to use more heuristic approach. The growth includes two types of phenomena, the instant response of the enzymatic reactions and slow changes in the state of the functional substances. Thus, there are two timescales involved, simultaneous and delayed one.

The state of functional substance changes slowly, and the efficiency of the functional substances changes, and the relationship between temperature and growth responds to the state of the functional substances. Let g_{Shoot} (mm s^{-1}) denote the growth rate. It depends on temperature T and the state of functional substances S_{G}:

$$g_{\text{Shoot}}(t) = g\left(T(t), S_{\text{G}}(t)\right), \tag{4.51}$$

where t denotes time.

Assume that the effects of temperature and state of functional substances of growth, E_{G}, are multiplicative:

$$g_{\text{Shoot}}(t) = g_{\text{T}}\left(T(t)\right) E_{\text{G}}\left(S_{\text{G}}(t)\right), \tag{4.52}$$

where g_{T} describes the effect of temperature and E_{G} is the efficiency of the functional substances.

The measurements of continuous growth rate are difficult to do both due to small changes of growing dimensions in time, especially in the boreal environment, and also since changes in the water status of plant reflect also in the dimensional changes. However, daily shoot elongations can be determined rather easily. The transition from process rate to daily values is done by integration. We get

$$G_{\text{Shoot}}\left(t_{i+1}\right) - G_{\text{Shoot}}\left(t_i\right) = \int_{t_i}^{t_{i+1}} g_{\text{Shoot}}(t)\,\mathrm{d}t = \int_{t_i}^{t_{i+1}} g_{\text{T}}\left(T(t)\right)\; E_{\text{G}}\left(S_{\text{G}}(t)\right)\,\mathrm{d}t, \tag{4.53}$$

where $G_{\text{shoot}}(t_i)$ is the length of the shoot at the moment t_i.

The state of functional substances changes slowly, so let's assume that it remains constant during a day. We get

$$G_{\text{Shoot}}\left(t_{i+1}\right) - G_{\text{Shoot}}\left(t_i\right) = E_{\text{G}}\left(S_{\text{G}}\left(t_i\right)\right) \int_{t_i}^{t_{i+1}} g_{\text{T}}\left(T(t)\right)\,\mathrm{d}t. \tag{4.54}$$

The identification of this type of a model is feasible since the daily shoot elongations and diameters are rather non-problematic to measure as well as temperature.

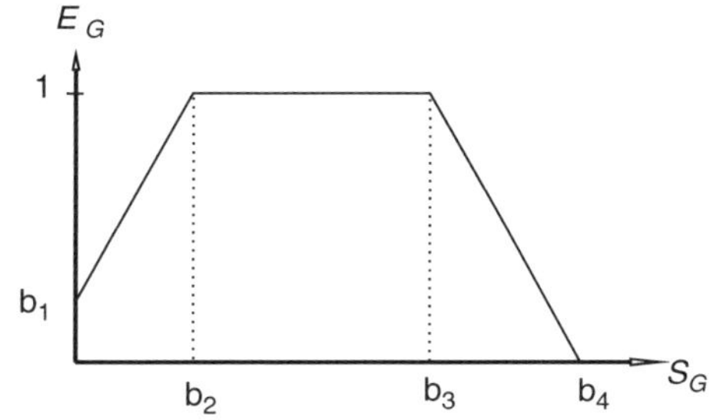

Fig. 4.64 The relationship between the state of the functional substances of growth, $S_{G,}$ and the efficiency of growth functional substances E_G. The parameters b_1, b_2, b_3 and b_4 define the beginning and end points of the linear sections

If the availability of sugars does not restrict the height growth, then the simultaneous response is caused by enzymatic reactions depending exponentially on temperature. To operationalise the model, we need the dependence of growth on temperature that, in principle, it can be obtained with measurements of shoot elongation at short interval, less than 1 h. This measurement is rather problematic to arrange, and the accuracy and precision of the result are poor. We need to find temperature dependence of some quantity that would reflect the effect of temperature in a similar way than growth responses to temperature, and we use the obtained relationship as proxy. The synthesis of new macromolecules requires energy in the form of ATP, which is formed in respiration. The respiratory release of CO_2 is easy to measure, and it enables approximation of the temperature dependence of growth. Let g_T denote the relationship between respiration and temperature; thus, assume

$$g_T = r_0\, r(T), \tag{4.55}$$

where r_o is a parameter.

Assume further that the rate of development depends similarly on temperature as respiration. We obtain the state of the growth functional substances by integration over time:

$$S_G(t) = \int_{t_0}^{t} r\,(T(t))\ \mathrm{d}t, \tag{4.56}$$

where t_0 is the beginning moment of the growing season.

According the basic idea HG2, there are three phases in height growth, increasing, steady and declining ones. Thus, we can introduce the effect of the state of growth functional substances by linear parts as shown in Fig. 4.64.

Measurements

We measured 35 years ago daily shoot elongations and temperature in Hyytiälä, rather close to the present SMEAR II (Pietarinen et al. 1982). A pin was placed under the ten buds, and the lengths of the shoots were measured daily in the morning with the accuracy of 0.1 mm. Temperature was measured with thermo-hydrograph 24 times a day.

The measuring accuracy was rather good, and it introduced only minor random variation in the data. Also temperature measurements with rather primitive thermo-hydrograph were sufficiently accurate, but they are vulnerable to systematic errors if the paper is not properly placed on the rotating cylinder.

Test with Field Measurements

Estimation of Parameter Values. The direct temperature responses were derived from published respiration measurements (Dahl and Mork 1959). The parameters in the model were estimated annually numerically with the normal least squares principle.

Characteristic Features of Height Growth. The daily height increments of Scots pine have three characteristic features: (1) slow increase, steady growth and slow decline, (2) high peak values, even several centimetres during a day, and (3) great variability between days (Fig. 4.65, especially in the year 1971). We introduced the slow beginning, steady growth and slow decline with the dependence of the efficiency of the growth functional substances on the state of the functional substances, and this produced the characteristic timing rather well. The big temperature variation between days generated the daily variation in height increments. The large height increments during the maximum height growth phase are generated by warm days and high efficiency of the functional substances. The theoretical model was able to describe all characteristic features in daily height increments.

Adequacy of the Model Structure. We developed the height growth model already about 40 years ago, although then we did not understand the metabolic basis of the model as well as is presented above. This new formulation of the model makes the rather sound physiological background clear. The rather limited number of measuring points and old data made the proper analysis of residuals impossible.

Explaining Power of the Theory. The theory explained the variation in daily height increments with the changing efficiency of the functional substances of growth and with temperature differences between days. The proportion of explained variance was high about 90%. The rather large number of parameter values estimated from the same data generates bias in the proportion of explained variance.

We have applied our theory to the analysis of height increments of several boreal species from fern to herbs (Vuokko et al. 1977) and for needle and radial growth of Scots pine (Pietarinen et al. 1982). The agreement between the theory modelled and observed has been within measuring accuracy. Thus, our approach can explain the rapid growth in early summer in the northern climate, and the other factors have rather small effect.

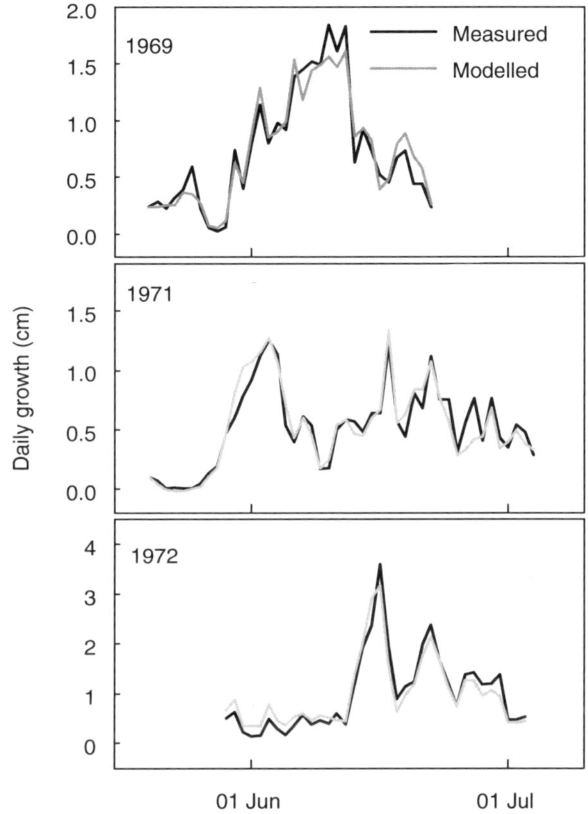

Fig. 4.65 Measured and modelled daily shoot elongation of Scots pine in southern Finland in the summers 1969, 1971 and 1972

Comparison with Other Approaches. Although we published these results several decades ago, we know no other similar analysis of daily height growths.

Conclusion: Our theory of the height growth of plants gained support in the tests; however, the missing test data reduce the power of the evidence.

4.2.8 NO_x Exchange of Needles

Maarit Raivonen, Johanna Joensuu, and Pertti Hari

Nitrogen oxides (here: $NO_x = NO + NO_2$) are atmospheric trace gases that play an important role in tropospheric chemistry. Nitrogen oxides are reactive gases,

M. Raivonen (✉)
Department of Physics, University of Helsinki, P.O. Box 48, 00014 University of Helsinki, Helsinki, Finland
e-mail: maarit.raivonen@helsinki.fi

and their lifetime in the atmosphere is rather short (less than 1 or 2 days in the summer). The nitrogen oxides in the atmosphere can originate from fluxes of ecosystems, and they also generate fluxes from the atmosphere to ecosystems. The major sources of the primary oxidised nitrogen compounds, NO (nitric oxide) and NO_2 (nitrogen dioxide), are fuel combustion, biomass burning, lightning and some microbial processes in the soil, with anthropogenic emissions dominating those from natural sources. Typical NO_x concentrations in rural areas are only a few parts per billion (ppb), being well below 1 ppb in the most remote areas. In urban regions, the concentrations are generally at tens of ppb, and the peak values can approach 1,000 ppb (Seinfeld and Pandis 1998). Nitrogen oxides disappear from the atmosphere mostly by wet deposition (through rain or snow) or dry deposition. Nevertheless, vegetation plays a role in the atmospheric balance of NO_x. It has been shown that plants absorb NO_2 and to a lesser extent NO from the air when the ambient concentration of these gases is high enough (Rondón et al. 1993; Sparks et al. 2001). Plants can even utilise the nitrogen of absorbed NO_x in their metabolism (e.g. Segschneider et al. 1995). Some observations exist also on NO_x emission from plants at ambient concentrations close to zero (Wildt et al. 1997; Sparks et al. 2001; Teklemariam and Sparks 2006), but the issue is controversial (e.g. Lerdau et al. 2000; Chaparro-Suarez et al. 2011). The critical concentration at which net NO_2 emission turns to net absorption, also called the compensation point, has usually been below 2 ppb. This means that if plants were able to emit NO_x, vegetation would more likely to be a source than a sink of NO_x in rural and remote areas.

The absorption rate of NO_2 has been observed to depend on the degree of stomatal opening and on the concentration difference between plant interior and ambient air (e.g. Rondón and Granat 1994; Geßler et al. 2002; Chaparro-Suarez et al. 2011). It is still somewhat unclear whether NO_2 dissolves in the cell wall water fast enough so that the internal concentration could be assumed zero (e.g. Rondón and Granat 1994) or if the dissolution is slower and a non-zero internal concentration should be taken into account (e.g. Thoene et al. 1996). Atmospheric NO_2 deposits also onto plant surfaces, but this non-stomatal absorption has been found to be small compared to stomatal absorption (Geßler et al. 2002).

Emissions of NO_x seem to be connected with nitrite metabolism of plant leaves. It has been shown that NO can be produced from nitrite and NADH by nitrate reductase (Yamasaki and Sakihama 2000). High internal concentrations of nitrite are necessary for significant NO emission to occur, and this apparently requires that nitrate, not ammonium, is the nitrogen source for the plant (Wildt et al. 1997). Also NO_2 emission rates were found to correlate positively with leaf nitrogen content, which suggested a positive correlation between nitrate reductase activity and NO_2 emission (Sparks et al. 2001).

Fluxes of NO_2 and NO into a plant canopy were first studied by Hill (1971). When the measuring station SMEAR II was planned and implemented, we established a project of measurements of NO_y flux from the atmosphere into Scots pine needles to quantify the nitrogen flux and to evaluate its importance. Our original measurement system detected the flux of total reactive nitrogen NO_y, which besides

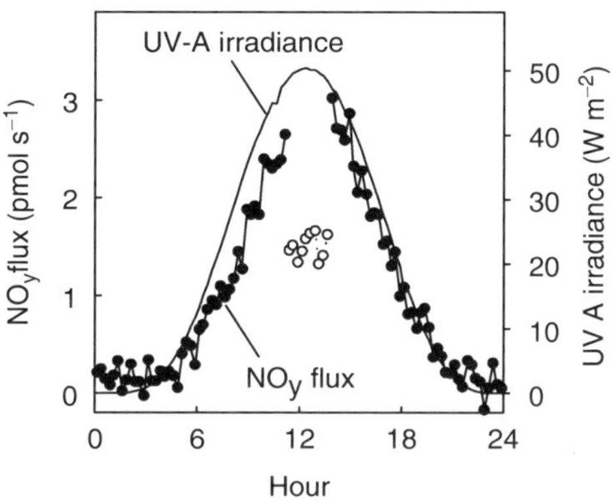

Fig. 4.66 Measured NO_y fluxes of one clear-sky day in a branch chamber (*black circles*). The branch was removed from the chamber for a few hours in the middle of the day (*white circles*)

NO_x includes, for example, nitrous acid (HONO), nitric acid (HNO_3) and peroxy acetyl nitrates (PAN). Later, the system has been improved to measure specifically NO_x (NO and NO_2).

When the measurements were started, it turned out that there was a NO_y flux *from* the needles *to* the atmosphere. In addition, the flux seemed to be connected with UV radiation (Hari et al. 2003). The UV-driven NO_y emission most likely originates from the surface of the needles. However, the mechanism is not yet fully understood. We have suggested that the NO_y emissions consist mainly of NO_2 and that they originate from photolysis of nitric acid that has deposited on the needle surface from the air (Raivonen et al. 2006). A similar phenomenon has earlier been observed to produce NO_2, NO and HONO on snow and glass surfaces (Dibb et al. 2002; Zhou et al. 2002) and also from organic sources (Twigg et al. 2011; Stemmler et al. 2006). The NO_x concentrations in air are very low at SMEAR II (about 1–3 $\mu g\ m^{-3}$), which explains why we have not constantly seen deposition on needles. However, there are spells when the air mass is coming from urban areas and its NO_x concentration is higher (about 10 $\mu g\ m^{-3}$). During these periods, we have observed also some deposition.

Measuring NO_y fluxes is problematic, since the reactive NO_y interact with the chamber and slightly also with the tubing needed for the transport of gas (Raivonen et al. 2003). Figure 4.66 shows an example of the chamber effects. On a sunny day, the NO_y emissions in a branch chamber followed solar irradiance. When the branch was removed, the emission level dropped – not to zero but to roughly half of the previous value: The chamber was emitting NO_y as much as the pine branch. We reduced the systematic measuring error caused by interactions of NO_y with the instrumentation by a correction method. Alongside the chambers enclosing pine branches, we monitored simultaneously an empty chamber. We considered the chamber blank of the branch chambers to be similar to the empty one.

NO_y emission can be seen especially on sunny days when the ambient NO_y concentrations are low. The NO_y emissions of the branch had a linear relationship

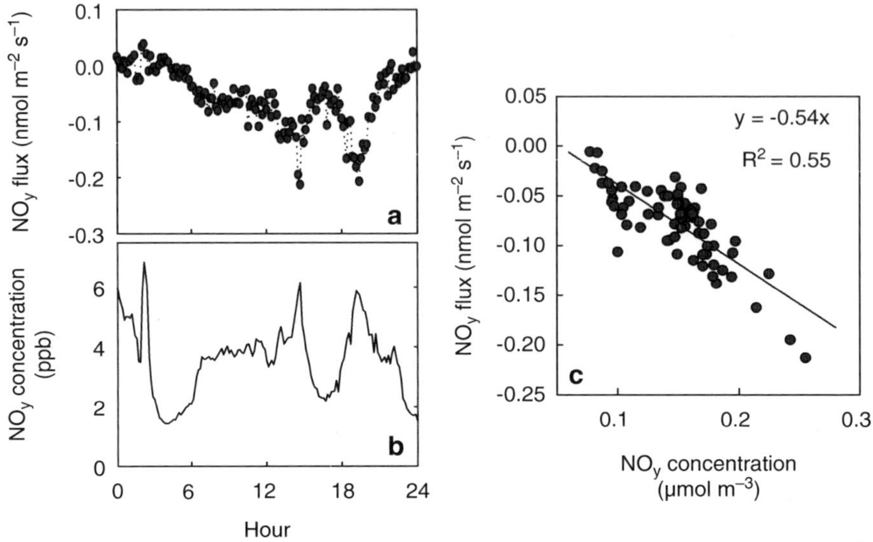

Fig. 4.67 Measured NO_y fluxes (**a**) on a cloudy day with high ambient NO_y concentrations (**b**) and the relationship between the blank corrected NO_y flux and the NO_y concentration (**c**). The line is fitted using the least squares method

with UV-A irradiance and with total solar irradiance. Uptake of NO_y was clear on a day when the air mass contained unusually high concentrations of NO_y. The degree of stomatal opening apparently affects the uptake rate, too. Because of this, we excluded measurements with low PAR (below 100 μmol s^{-1} m^{-2}), suggesting small stomatal aperture, from the analysis. The relationship between ambient NO_y concentrations and the uptake rates with open stomata was roughly linear.

The absorption of NO_x has been demonstrated frequently: One can relatively safely assume that vegetation acts as a NO_x sink at high ambient concentrations. However, measurements of small fluxes at near-zero concentrations are demanding because reactions on the chamber surfaces disturb the analysis. So, the universality and the underlying processes of NO_x emissions remain unsolved. In case there really is a UV-induced surface process of nitric acid that generates NO_x, it could occur always when nitrate/nitric acid and sunlight are present, recycling deposited nitrate back to the atmosphere.

Flux of nitrogen oxides in vegetation has been suggested to be bidirectional: Plants can both absorb and emit NO_x, though the emission of NO_x is still somewhat controversial. Our results indicate that at low ambient NO_x concentration and high UV-A irradiance, emissions dominate the NO_y balance and at high ambient NO_x concentrations, deposition occurs (Fig. 4.67). The fluxes are, however, very small when compared with nitrogen deposition from the atmosphere. The annual emission of NO_y is less than 1 mg m^{-2} $year^{-1}$.

4.3 Structure of Forest Soil

Asko Simojoki, Mari Pihlatie, Jukka Pumpanen, Jukka Kurola, Mirja Salkinoja-Salonen, and Pertti Hari

Mineral Soil

Soil denotes the loose upper layer of the Earth's crust covering the solid bedrock. It is a heterogeneous and porous mixture of mineral and organic matter, where the voids (or pores) between solid particles are filled with water or air. In the boreal zone, the bedrock is composed mainly of magmatic and metamorphic rocks, such as granite and gneiss. Soils are classified according to particle size into textural classes. Coarser mineral soil particles include silt and sand fractions that are 20–200 μm and 200–2,000 μm in diameter, respectively. Particles larger than 2,000 μm in diameter are classified as gravel or stones. Coarse particles are composed mainly of such primary silicates as quartz, feldspars and mica, reflecting the mineralogy of parent rocks. The smallest soil particles with <2 μm in diameter are called the clay fraction. Clay particles are dominated by secondary minerals formed by modification of original rock minerals or recrystallisation of their decomposition products during soil formation. These include the secondary layer silicates called clay minerals as well as hydrous oxides (hydroxyoxides) of Al, Fe and Mn. The oxides are partly poorly crystalline. In contrast to the more or less spherical shape of coarser soil particles, the clay particles are platelike, reflecting the shape of layer-silicate minerals.

Soil *chemical properties* are largely dominated by the reactions occurring at the surfaces of small particles. This is due to the large specific surface area and polar nature of clay minerals and oxides in the smallest particle size fractions. Most clay minerals have a permanent negative electrical charge due to isomorphic substitution of cations within the crystals. These substitutions occur during the crystallisation of minerals and remain unchanged thereafter. On the other hand, the oxides may be negative, neutral or positive depending on soil acidity that determines the protonation and deprotonation of their hydroxyl groups. Soils generally have a net negative charge and thus adsorb cations on their surfaces by electrostatic forces and retain anions on the hydroxyl groups of clay mineral edges and oxide surfaces by specific mechanisms. These processes are discussed in more detail later (Sect. 4.4).

Boreal forest soils are typically coarse-textured soils that often contain considerable amounts of gravel and stones. However, these larger fractions are chemically

A. Simojoki (✉)
Department of Food and Environmental Sciences, University of Helsinki, P.O. Box 27, 00014 University of Helsinki, Helsinki, Finland
e-mail: asko.simojoki@helsinki.fi

inert and mainly occupy space from more fertile and porous soil composed of finer fractions. In practice, soil texture refers to the relative proportions of size fractions finer than 2 mm in diameter. Moreover, the conventional pretreatments for the determination of particle size distribution include the removal of organic matter and oxides that bind the silicate particles together.

Soil Organic Matter

Soil organic matter (SOM) denotes all living and dead biologically derived organic material at various stages of decomposition in the soil or on the soil surface, excluding the above-ground portion of living plants (Baldock and Nelson 2000). In practice, the concept has often a narrower scope referring to materials passing through a 2-mm sieve (Nelson and Sommers 1996) excluding undecayed plant and animal residues.

Origin and Molecular Structures of SOM Compounds

Soil organic matter includes all organic compounds entering the soil through the root growth, the litter-fall from above-ground plant parts and root exudates as well as the products of their microbial decomposition and resynthesis. Microbial decomposition of organic macromolecules produces a range of small molecules as well as the decay-resistant, dark, amorphous organic matter called humus.

The chemical composition of soil organic matter is similar to that of litter unless not changed by microbial decomposition. Biological tissues are composed of various organic macromolecules and a lesser amount of small molecules. The chemical composition of organic residues depends on the tissue (reviewed by Baldock and Nelson 2000; Derenne and Largeau 2001; Kögel-Knabner 2002; Blume et al. 2010a). Within the plasma membrane, the cells contain phospholipids, carbohydrates, lignin, nucleic acids and proteins, as well as such smaller molecules as sugars, amino acids and dissolved inorganic ions. The cell walls of plants are composed mainly of cellulose, hemicellulose and lignin. Cellulose is a carbohydrate polymer of 1–4 β-linked glucose units, whereas hemicellulose contains other sugars as well. In contrast, lignin is not a carbohydrate and has a complex structure composed of linked phenyl propane units. The nitrogenous compounds constitute a small but important group, as nitrogen is a structural component of enzymes and other proteins as well as nucleic acids, ATP and such pigments as heme and chlorophyll.

The chemical composition of microbial cells resembles plant cells. However, the cell walls of fungi contain mainly chitin, a polymer of 1–4 β-linked *N*-acetyl-glucosamine units, and those of bacteria peptidoglycans and/or lipopolysaccharides. As a consequence, the composition of soil microbes is dominated by phospholipids and protein-like chemical species. The chemical composition and functional role of biological macromolecules in plant and microbial tissues is discussed in more detail in the preceding chapter.

Humus is traditionally considered to be composed of large molecules with a complex structure not amenable for detailed structural characterisation. The classical fractionation of humic substances is based on their acidity-dependent solubility characteristics. The humic substances that are extracted by an alkaline solution and remain soluble after subsequent acidification are fulvic acids, whereas those in the acid-precipitated fraction are humic acids. Still, large amounts of humic substances called humins may remain unextracted by the alkali. Humins are mostly humic acids that have very high molecular weight or are protected by mineral soil particles. The classical fractionation of humus has some practical merits, as it separates the fulvic acids that are soluble and potentially mobile in a broad pH range in soil, in contrast to the humic acids that are immobile in acid conditions. Nevertheless, the molecular structure of humus has long been and still is a topic of active research and discussion (Schnitzer 2000; Burdon 2001; Piccolo 2001; Sutton and Sposito 2005). Current views question the macromolecular nature of humus. Instead, the humic substances are best described as collections of diverse, relatively low molecular mass components that form supramolecular associations stabilised by hydrophobic interactions and hydrogen bonds (Sutton and Sposito 2005).

Humus contains numerous acidic groups that can dissociate in a broad pH range which gives rise to pH-dependent negative charge of soil particles. For this reason, it is the dominant soil constituent responsible for the cation exchange capacity in coarse-textured soils, such as those in boreal forest.

Analysis of Soil Organic Matter (SOM)

The amount and composition of soil organic matter can be analysed by various methods. The crudest approaches involve the determination of organic matter based on the loss-on-ignition or the organic carbon content. However, the treatment of soil organic matter according to its chemical structure allows efficient use of quantitative methods, since all soil components can, at least in principle, be directly measured. Quantitative data on the different chemical groups of biological macromolecules can be obtained by combinations of selective extractions, such as the proximate methods used in the food chemistry, or more generally, by thermochemolytic degradation of macromolecules, followed by compound-specific analysis of the products (see Kögel-Knabner 2000). Non-destructive spectroscopic methods allow the characterisation of various groups of atoms, but the quantification of various molecular groups is not yet reliable (Kögel-Knabner 2000), although, for example, Nelson and Baldock (2005) made an attempt to apply a solid-state ^{13}C NMR technique towards this aim.

Traditionally, different empirical wet chemical methods have been used for the analysis of litter and humus. Nonetheless, the strict separation of soil organic matter into litter and humus fractions is not feasible with conventional extractions (Beyer

1996). For this reason, Beyer et al. (1996) recommended applying litter compound analysis to soil as well. Moreover, application of the same analytical procedures for both plant and soil samples provides methodological advantages.

For the above reasons, we planned and carried out a uniform analytical procedure for the determination of chemical composition of plant tissues, litter and soil organic matter by combining dry and wet chemical methods of food and soil chemistry (outlined by Simojoki et al. 2008). The analyses take advantage of the fact that biological macromolecules are composed of specific monomers that can be determined by compound-specific analysis, following acid hydrolysis. This allows the estimation of main organic macromolecules, including (1) lipids, (2) sugars and starch, (3) hemicellulose + cellulose, (4) lignin and (5) proteins and other nitrogenous compounds. The lignin is determined as undissolved Klason lignin residue. In soil samples, this residue contains humus as well. Nitrogenous organic compounds fall into several fractions (see Fig. 4.77). For simplicity, however, the protein content in this study was estimated according to the convention of food chemistry as a crude protein by multiplying the content of total hydrolysable N by 6.25 (assuming 16% N in the proteins). 'Humus N' was obtained as a difference between the total and hydrolysable N. This approach is in line with studies confirming that acid hydrolysis methods can provide meaningful estimates of bioavailable organic C and N pools in soil (Xu et al. 1997; Silveira et al. 2008).

Soil Structure, Water and Air

Forest soil is a mixture consisting of soil particles, water and air. Soil structure generally denotes the spatial arrangement of soil particles and pore space between them. Natural structural development of soil involves spatial heterogenisation of soil. In fine-textured soils that contain much clay, the individual mineral and organic particles are clustered together into larger units called aggregates or peds of various shapes and sizes. The aggregates in soils containing even modest amounts of organic matter show a hierarchical structure (Tisdall and Oades 1982; Oades 1993): The particles form microaggregates that are joined together into macroaggregates. Clay, humus and oxides often coat the coarse particles and aggregate surfaces. This kind of secondary soil structure is not well developed in boreal forest soils that are mostly coarse in texture. The structure of coarse-textured soils is predominantly single-grained, with roots, fungal hyphae, organic biopolymers and oxides providing only limited binding of particles together. However, the channels of old plant roots and earthworms or other burrowing animals often provide a stable system of continuous large pores in both fine- and coarse-textured soils (Lee and Foster 1991; Oades 1993; Angers and Caron 1998).

Besides the description of soil aggregates, soil structure can be characterised by the amount, shape, size and distribution of pores. The total porosity of soil (ε_T) depends on the bulk and particle densities of soil and can be calculated as:

$$\varepsilon_T = 1 - \rho_b / \rho_p, \tag{4.57}$$

where ρ_b is the bulk density of soil (the dry mass per unit volume of soil) and ρ_p is the density of soil particles (the dry mass per unit volume of solid phase). The particle density is usually close to that of quartz (2.65 g cm^{-3}). In contrast, the bulk density can vary much in the soil profile depending on texture and structure, organic matter content and soil depth. It is determined by taking volumetric samples from soil.

The pore space between soil particles or aggregates is filled with water or air and is a site for chemical reactions and a habitat for soil organisms. Most reactions and microbial activity occur in the water-filled pore space. Water retention by soil depends on the properties of both soil (pore size, wettability of particle surfaces) and water (surface tension). Smaller pores tend to retain water more strongly than larger ones for reasons outlined below.

The facts that soil particles have charged surfaces and that water molecules are permanent dipoles, that is, water molecules have electrically positive and negative areas, explain the behaviour of water in the soil (Hillel 1998). Electric forces bind the water molecules to particle surfaces and with each other. Water forms layers on soil particles in which the first layer of water molecules is bound electrically to the particle surface (adhesion) and thereafter the water molecules are bound electrically with each other (cohesion).

At the border (interface) between water and air, the neighbouring water molecules interact with each other unequally at different directions, as the molecules inside the water phase pull the molecules from the surface inwards, while the pull towards the air phase is negligible (Hillel 1998). This net inward force causes a flow of water molecules from the surface deeper into the water until an equilibrium is reached at which the electric forces in the interface between water and air are strong enough to resist further flow and the surface area of water is minimised. As a result, the water molecules at the interface are bound electrically more strongly with each other than with the molecules of bulk water and form a clear film (meniscus) separating the water and air phases. The film acts like an elastic membrane between water and air phases resisting any forces that tend to increase its surface area. The magnitude of phenomenon is quantified by surface tension (unit N m^{-1}) or equivalently by surface energy (unit J m^{-2}).

The shape of water-air interface at equilibrium with external forces neglecting gravity satisfies the Young-Laplace equation:

$$\Delta P = \gamma \left(\frac{1}{r_1} + \frac{1}{r_2} \right) \tag{4.58}$$

or

$$\Delta P = \gamma \frac{2}{r}, \text{ with } r = \frac{2\, r_1\, r_2}{r_1 + r_2}, \tag{4.59}$$

where P ($=P_{\text{water}} - P_{\text{air}}$) is the pressure difference across the water-air interfacial membrane, γ the surface tension, r the equivalent radius of curvature of the surface and r_1 and r_2 the radii of curvature lying in planes normal to each other at a point

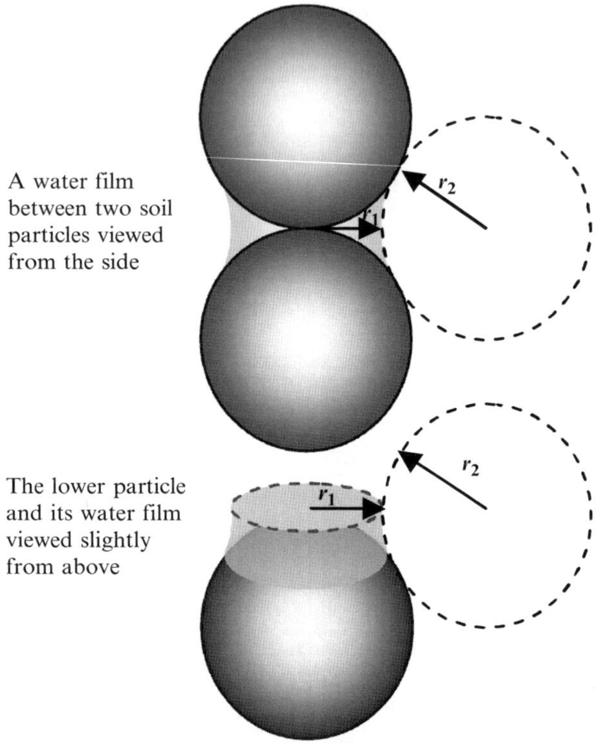

Fig. 4.68 Two radii of curvature (r_1 and r_2) in planes normal to each other in water film between two spherical soil particles

on the interface as shown in Fig. 4.68 (Adamson and Gast 1997; Hillel 1998). The radius of curvature r is taken as positive, if the radius lies within water (such as r_1 in Fig. 4.68). In words, these equations state that at equilibrium, the soil water-air interfaces are characterised by a uniform curvature that may be produced by innumerous combinations of radii of curvature.

According to the Young-Laplace equation, no pressure difference exists across the flat surface of free water. In unsaturated soils, however, the water molecules generally adhere to the solid particles. Owing to high surface tension of water, this produces concavely curved water-air interfaces with contact angles <90° with the solid phase, which decreases the pressure in soil water below that in soil air. In such a case, soil water is under tension (underpressure, suction), the magnitude of which gives the strength of water retention in the capillary pores of soil (unit Pa or J m^{-3}). The radius of curvature of water-air interface (r) is determined by the pore size and the contact angle (α) of water with the solid phase. If the contact angle is taken as zero, as is usual for mineral soil particles, the radius of curvature represents the actual pore size.

The phenomenon of capillarity, outlined above, explains why the largest pores empty first in a drying soil, while the smaller pores remain filled with water (Fig. 4.69), because the curvature of water film and the resulting pressure difference increase (the radius of curvature decreases) with decreasing pore size. Small-sized

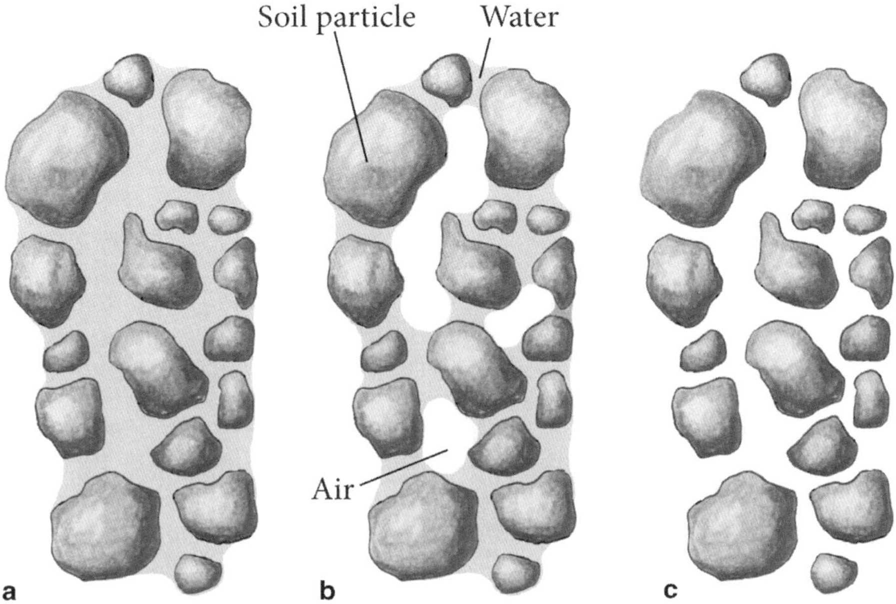

Fig. 4.69 Water and air filling the pore space of soil at three different degrees of water saturation. (**a**) water-saturated soil (%WFPS = 100), (**b**) moist soil with some large air-filled pores (0 < %WFPS < 100) and (**c**) air-filled dry soil (%WFPS = 0). On drying, water drains out of first from the largest pores first and last from the smallest pores (%WFPS = percentage of water-filled pore space)

Fig. 4.70 Soil water retention curves at various depths of mineral soil horizons in a podsol soil at the SMEAR II station

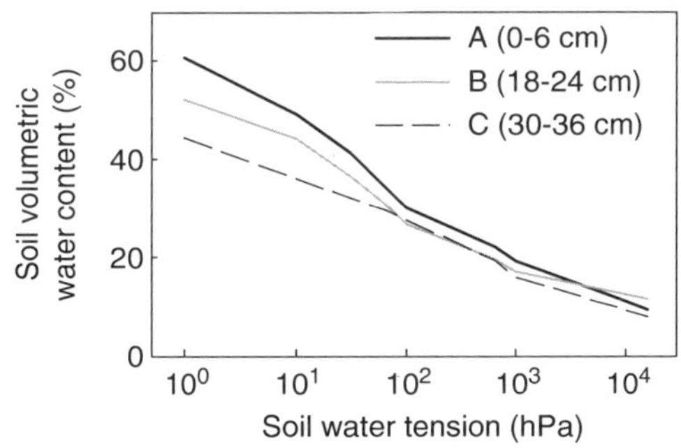

humus is thus important in increasing the retention of water in coarse-textured soils. Capillarity is also important, as the pressure difference between water and air phase drives water flux from lower to higher tension and causes capillary rise from the free ground water table upwards into drier soil until the capillary forces are balanced by the gravity (the weight of water column).

Soil water retention curve (or soil water characteristics, Fig. 4.70) depicts explicitly the water content at a given suction (matric water tension). In reality, the relationship is hysteretic with a drying curve showing larger water contents than a wetting curve at a given suction, but in practice, drying curves are mostly used for hydrological modelling. Teepe et al. (2003) and Puhlmann and von Wilpert (2012) presented pedotransfer functions for the estimation of water retention of forest soils by using such basic soil properties as texture, organic matter content and bulk density. Soil water retention properties determine directly the amount and availability of water for plants and microbes in soil. In addition, they determine indirectly many other soil properties and phenomena that depend strongly on water content, such as air-filled porosity, mechanical strength and the transport of water, solutes, gases and heat.

The hydraulic conductivity of soil (soil water conductivity) decreases very rapidly on drying, as the diameter of water-filled pores decreases and the flow paths of water are blocked by air-filled pores (Fig. 4.69). With further drying, at a soil-specific water content, soil water may occur only as thin layers in disconnected locations between soil particles (Fig. 4.68) allowing no flow of liquid water. Soil water conductivity is also dependent on the viscosity and density of soil water that are functions of chemical composition and temperature, but their role is minor compared with the degree of water saturation.

In contrast to the small pores mostly filled with water, larger soil pores drain out quickly and are air-filled for most of the time. On wetting of soil, the smallest pores are filled with water first, and the largest pores may remain air-filled even in nearly water-saturated soil. Coarse-textured soils of boreal forest are thus usually well aerated. Soil air is in essence atmospheric air nearly saturated with water vapour and continuously modified by consumption and production of gases in soil, such as respiration of plant roots and microbes. Soil solid and liquid phases determine in a complementary manner the amount, shape and distribution of air-filled pores, as well as the related gas transport properties of soil (the gaseous diffusion coefficient and permeability of soil).

To sum up, soil structure is thus an important determinant of soil functions, as it ultimately determines the retention (Fig. 4.70) and flux rates of water, nutrients and gases supporting the processes that change material from one form to another in the soil. In particular, soil water-filled pores are the main habitats for biological activity including such temperature-dependent processes as cellular growth, respiration (O_2 consumption, CO_2 production), water uptake, nutrient uptake and chemical reactions. Soil structure determines the extent to which the transport rates of various chemical substances and heat can support the demands by biological activity in soil.

The above theory on soil water retention can be improved by taking into account of the effects of solutes and non-zero contact angles ($\alpha > 0$) in real soils. Soil solution contains always small amounts of organic and inorganic dissolved species, and especially natural organic compounds containing both hydrophilic and hydrophobic chemical groups do accumulate at the soil water-air interface. However, they typically decrease the surface tension of soil solution only slightly below that in pure water and thus have minor influence on water retention in soil. In contrast, the contact angle may vary from close to zero in wet soils to hydrophobic values ($\alpha > 90°$) on humic coatings on dry mineral soil particles. If the contact angle is positive, the actual pore size is smaller than the radius of curvature of water-air interface. For instance, in a cylindrical tube with a hemispherical water-air interface, the equivalent radius R characterising the pore size is given by $R = r \cos \alpha$. In addition, both the magnitude of surface tension and contact angle depend on temperature. Bachmann and van der Ploeg (2002) and Bachmann et al. (2007) discussed the recent developments in interfacial tension and temperature effects on soil water retention.

Soil Organisms

Besides plant roots, numerous organisms live in the soil. The primary function of soil organisms in the soil ecosystem is the processing and mixing of soil organic matter (Killham 1994). They can be classified according to size into microflora (archae, bacteria, fungi), micro- and mesofauna (e.g. protozoa, rotifers, tardigrades) and macrofauna (e.g. nematodes, earthworms, enchytraeid worms, arthropods) (see, e.g. Blume et al. 2010b). The faunal trophic levels are discussed in detail in Sect. 4.5. The microfauna (less than 100 μm in dimension), also called protozoa, consists of flagellates, ciliates, amoebae and sporozoa (Paul and Clark 1989; Killham 1994). Protozoa live mainly by grazing soil microbes, bacteria and fungi. The most important members of larger soil fauna, meso- (from 100 μm to 1–2 mm) and macrofauna (greater than 1–2 mm) are the earthworms, enchytraeid worms, nematodes, arthropods (e.g. mites and springtails), molluscs and mammals. Macrofauna can move quite freely in soil pore space, and the largest ones can even burrow their own channels through the soil. On the contrary, the smallest organisms cannot move around in the soil and are limited to the small water-filled pores and films covering the soil particles to prevent dehydration by drying and predation by soil animals. Even the growth of plant roots and fungal hyphae, with some degree of freedom, is mainly restricted to soil volumes with available water.

The largest amounts of soil organisms live in the immediate vicinity of roots (rhizosphere) and in soil surface layers containing ample decomposable organic compounds. Some bacteria and fungi live even within plant roots, such as the symbiotic nitrogen-fixing bacteria and mycorrhiza. Moreover, a part of the soil microbiota belongs to the phylum Archaea. It was earlier thought that Archaea can only grow under very extreme conditions (in high salinity or in high temperature or in anaerobic conditions), but it has been shown that Archaea exist also in great numbers in acid boreal forest soils (Jurgens et al. 1997).

Soil microbes are composed of the same organic compound groups as plant litter, except for lignin, tannins, cutins or suberins. The data compiled by Blume et al. (2010a) show the microbes contain up to ten times more protein (50–60% and 12–52% in bacterial and fungal dry matter, respectively) compared with a typical plant litter. In addition, while the bacterial cell walls made of peptidoglycan (murein) are chemically different from the fungal cell walls made of chitin (polymerised glucosamine) and cellulose, they do both contain much nitrogen. Due to the high content proteins and other nitrogenous compounds, the C/N ratios of microbes are only about 5–8 for bacteria and 10–15 for fungi, whereas those for typical plant litter are an order of magnitude higher.

Soil Horizons and Distribution

The soil covering the Earth's surface is not similar everywhere; its composition, structure and properties vary in space. Such differences appear on a global scale and in different vegetation zones as well as within landscapes and individual soil profiles. The differences originate partly from the transport and sedimentation of parent material and partly from the gradual processes of local soil formation. As a consequence, dominant soil texture differs in different geographical areas, and the surface layers of a well-developed soil profile show distinct horizons differing in colour, chemical properties or structure.

Boreal forest soils are mostly coarse textured for several reasons. In the boreal zone, much of the soil parent material originates from the Holocene ice age. During and after deglaciation, much of the fine soil material was transported to more southern latitudes by melting waters, whereas the coarser material remained in place (see, e.g. Salminen 2005). Within the boreal zone, the finer soil fractions have been deposited at the bottom of lakes and seas, whereas the supra-aquatic soils at higher elevations are coarse-textured glacial till. As the land level rose after deglaciation, the lake and sea sediments partly mixed with coarser coastal sediments and became eventually dry land. Later, the fine-textured soils were preferred for agricultural land use rather than forestry due to their fertility.

Soil organic matteraccumulates near the soil surface. This may even form an organic surface horizon (litter or humus layer) above the mineral soil. Deeper in the soil, humus covers larger mineral soil particles and forms complexes with clay particles.

In humid climates such as in the boreal zone, coarse-textured forest soils typically have a podsolic profile (Mazhitova 2006). It is characterised by an acidic litter and humus layer composed of partly decomposed organic matter (O horizon) underlain by a dark-coloured mineral soil horizon (A horizon), a greyish eluviation horizon (albic E horizon) and a reddish dark brown illuviation horizon (spodic B horizon) enriched with humus or sesquioxides (Al and Fe oxides) (Fig. 4.71). The bottom of the profile is practically unchanged parent material. Podsols are identified based on the occurrence of spodic materials and spodic horizons (Buol et al. 1997). The development of different soil horizons varies in response to soil and climatic conditions. In most extreme cases, either the A horizon or more seldom the E horizon may be lacking entirely (Buol et al. 1997). Podsol soils are generally considered the predominant class of boreal mineral soils without permafrost.

Most of the soil organic matter in a podsolic soil profile is located in the humus layer (Figs. 4.72, 4.73 and 4.74). The amount and chemical composition of soil organic matter change during the decomposition of needles, wood and litter at different depths in the soil profile. In general, the contents of hemicellulose and cellulose decrease, while lignin and/or humus as well as proteins seem to become enriched during the decomposition. Such results are in good qualitative agreement with results from decomposition studies carried out using long litterbag incubations and slightly different analytical procedures compared with ours (Berg et al. 1991; Berg 2000; McTiernan et al. 2003; Berg and Dise 2004). In addition, the proportion

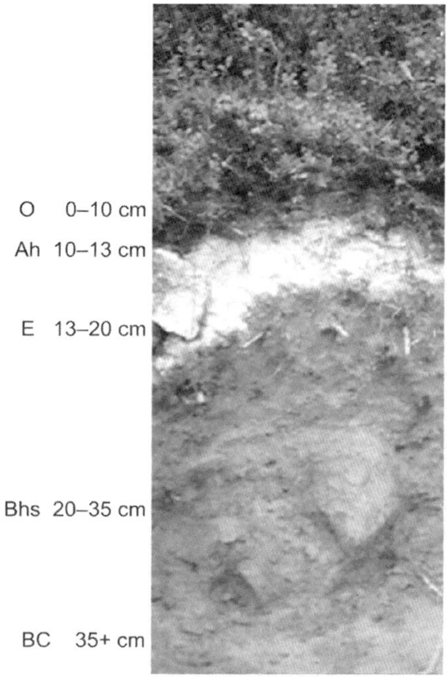

Fig. 4.71 Horizons of a podsolised forest soil (Photo: Markku Yli-Halla)

of protein N to total N seems to increase with increasing soil depth (Fig. 4.74). The same observation has been confirmed under different forest types at several sites in Hyytiälä, Finland. Nevertheless, the extent to which this applies more generally to Podsols and other boreal forest soils remains to be shown by future research.

4.4 Processes in Soil

Asko Simojoki, Mari Pihlatie, Jukka Kurola, Jukka Pumpanen, Mika Kähkönen, Pertti Hari, and Mirja Salkinoja-Salonen

General Considerations About Processes in Soil

The solutes in soil solution originate from such soil processes as weathering of soil minerals, decomposition and mineralisation of soil organic matter, as well as wet

A. Simojoki (✉)
Department of Food and Environmental Sciences, University of Helsinki, P.O. Box 27, 00014 University of Helsinki, Helsinki, Finland
e-mail: asko.simojoki@helsinki.fi

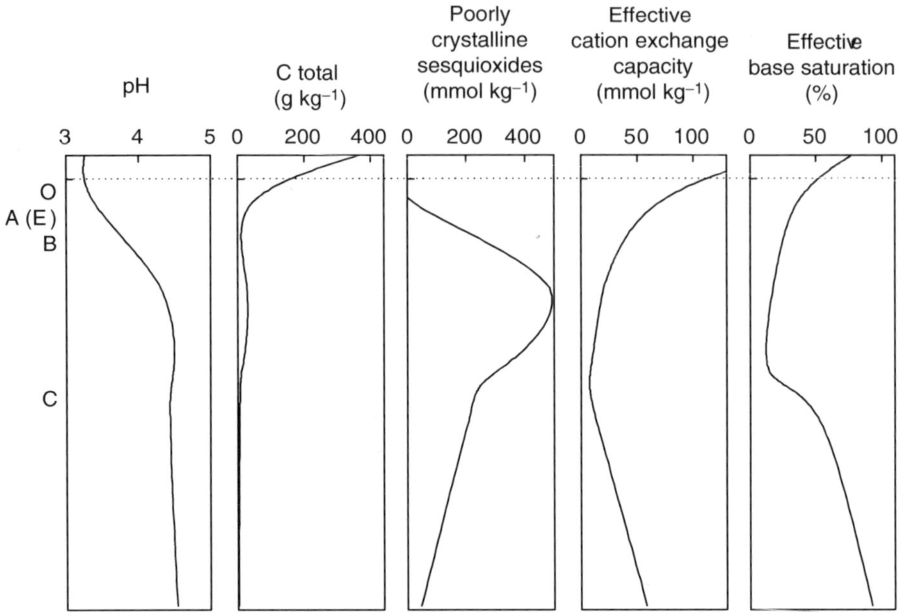

Fig. 4.72 Schematic presentation of typical chemical properties of a podsolic soil profile. Soil surface is illustrated with a *dash line*. Poorly crystalline sesquioxides = Fe and Al oxides extracted by acid ammonium oxalate. Effective cation exchange capacity = the cation exchange capacity at soil pH. Effective base saturation = the sum of charges of exchangeable Ca^{2+}, Mg^{2+}, K^+ and Na^+ cations divided by the effective cation exchange capacity

and dry deposition from the atmosphere. The processes and deposition generate spatial concentration and tension differences, and these differences eventually give rise to fluxes within forest soils, between forest soil and vegetation and between the soil and its surrounding.

The two ecological key components of soil solution are the plant nutrients and the various organic compounds that feed the microbes. Plants take up the nutrients from soil solution mostly as mineral ions (see Sect. 4.2.5). The decomposition of dead tissues and other organic exudates of roots and microbes recycles these elements back into ion form and make them again available for plant uptake and chemical processes in soil solution. The decomposition of organic matter is a major input into soil of such ions as NH_4^+, $H_2PO_4^-$, Ca^{2+}, K^+, Mg^{2+} and Na^+. Thus, the rate of decomposition of soil organic matter, together with the reactions of ion, determines the amount of ions available for plant and microbial growth in soil.

Most ions in the soil solution have originated from the weathering of soil. However, soil nitrogen originates ultimately from the deposition or biological nitrogen fixation from the atmosphere, as nitrogen-containing minerals are scarce in soil. In a like manner, soil carbon originates from the carbon dioxide photosynthesised from the atmosphere. Root growth distributes organic carbon and nitrogen within the soil profile and feeds the soil microbes with organic substrates. Although

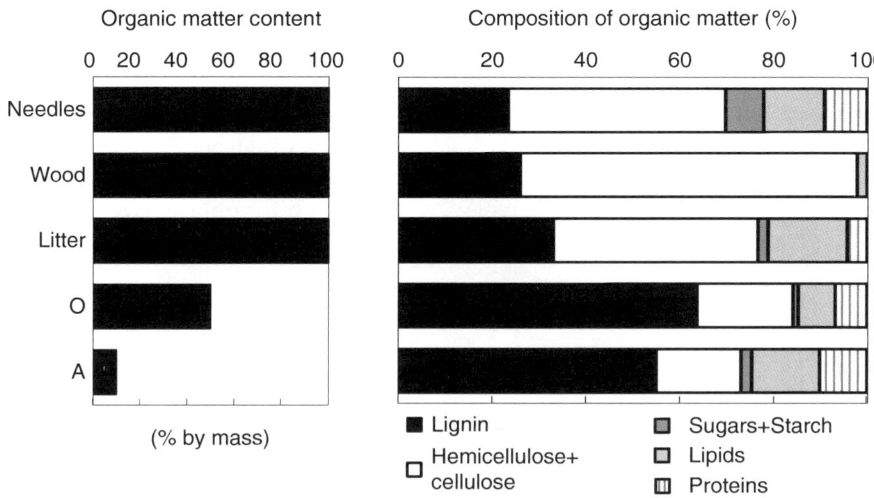

Fig. 4.73 The content and composition of organic matter in plant (green needles, wood, litter) and soil samples (soil horizons O and A) taken from a 40-year-old pine forest growing at SMEAR II in Hyytiälä, in 2002

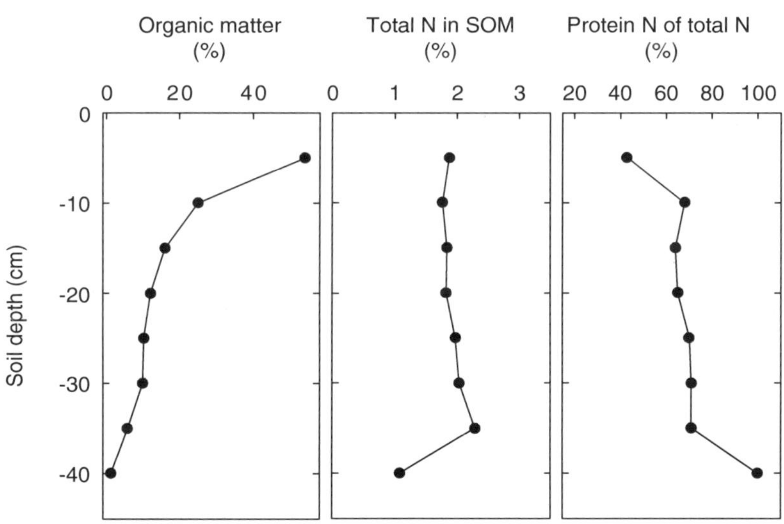

Fig. 4.74 The gravimetric percentages of soil organic matter (*SOM*) content, the content of total N in SOM and the proportion of protein N (6M HCl-hydrolysable N) to total N at 5-cm intervals in a podsolic soil profile under pine forest near SMEAR II in Hyytiälä (Drawn after Rajasekar 2007 unpublished)

the combined biological activities of roots and microbes increase the chemical weathering of primary rock minerals in soil, the rate in boreal forest soils remains low compared with those of nutrient cycling in the plant and soil and may produce appreciable changes only during centuries and millennia of podsol formation. In contrast, the biological activities generally do have a major impact on soil properties. The chemical form and solubility of ions depend both on elemental and soil properties that vary considerably within a podsolic soil profile typical of boreal forests (Fig. 4.72).

Ion Exchange and Retention

Most nutrients in the soil water exist as ions, that is, an atom or groups of atoms, which are either positively charged cations such as ammonium ($NH_4{}^+$), potassium (K^+), calcium (Ca^{2+}) or negatively charged anions such as nitrate (NO_3^-), nitrite (NO_2^-), chloride (Cl^-) and phosphates ($H_2PO_4^-$, HPO_4^-). Soil solution is in a dynamic equilibrium with the contacting particles' surfaces. The extent and mechanisms of ion retention vary greatly depending on the properties of ion and soil. The amounts of cations at the particle surfaces are generally two or three magnitudes higher than those in soil solution. The same is true for anions with a specific adsorption mechanism (e.g. phosphates), whereas other anions (e.g. nitrate, chloride) are in practice present only in soil solution. Any adsorption on the particle surfaces slows down the movements of ions compared with the free, non-adsorbed species.

Soil particles usually have a slightly negative charge that attracts and holds positively charged ions by electrostatic forces on the particle surfaces. They thus form an electrical double layer, where the negative charge is exactly balanced by the positive charges of cations. The distribution of ions in the soil solution surrounding a negative surface is described by the diffuse double layer theory and other electric double layer theories (Singh and Uehara 1999).

The cations adsorbed on the surfaces can be exchanged into the soil solution by other cations. The exchange occurs on an equivalent charge basis: monovalent cations (with one positive charge) occupy one negatively charged site, and divalent cations (with two positive charges) occupy two sites. The ease of exchange is determined by the charge density of exchange surfaces as well as the valence and size of cations. The higher the charge density of exchange surfaces is, the stronger are the cations adsorbed. Multivalent and small cations are adsorbed more strongly than monovalent and large cations, respectively.

The amount of net negative charge per unit mass of soil is termed the cation exchange capacity. It expresses the sum of positive charges of cations adsorbed on soil particle surfaces. The cation exchange capacity increases with increasing clay and organic matter contents. Clay minerals have a permanent negative charge, whereas organic matter has a variable pH-dependent charge. Soil solution is well buffered against pH changes by the reactions of weak acid groups of soil organic matter and some inorganic compounds (such as the hydrated oxides of Al and Fe or sesquioxides). Acidification, such as caused slowly by, for example, podsolisation and acid deposition, decreases the variable charge and the effective cation exchange capacity of soil by increasing the protonation of weak acid groups. *Base saturation* is the proportion of cation exchange sites occupied by the so-called basic cations (Na^+, K^+, Mg^{2+}, Ca^{2+}) rather than acidic cations (H^+, Al^{3+}). Any decrease in the base saturation indicates acidification of soil, even if pH were not affected (see Fig. 4.72). The cations in the soil solution and those adsorbed on soil exchange surfaces are both considered easily available to plants. Cation exchange capacity and base saturation thus describe the availability of plant nutrients in the soil.

Soil particle surfaces generally do not retain anions by electrostatic forces, as the anion exchange capacity of soil is usually negligible. However, the hydroxyl groups on the clay mineral edges and oxide surfaces retain anions by specific mechanisms, such as the ligand exchange of phosphate ions. Neutral molecules are adsorbed on the particle surfaces by weak molecular and hydrophobic interactions. In addition, some weakly soluble molecules may precipitate and become part of the solid phase depending on the ion composition of soil solution.

The reactions in soil solution are important for soil ecology as they impact the ionic concentrations of plant-available nutrients and the activities of exoenzymes in soil solution. In this context, the acidity of soil solution has a master role, as it affects both the solubility and chemical form of elements (e.g. at pH below 5, most inorganic phosphates are $H_2PO_4^-$ rather than $H_2PO_4^{2-}$ ions), the conformation and cofactors of soil enzymes and the pH-dependent properties of soil particle surfaces. Apart from acidity, the chemical form of some elements (e.g. iron and manganese) depends also on the oxidation status of soil. At oxygen deficiency, the oxidised forms of such elements are converted by chemical/biochemical reduction into soluble ions (e.g. Fe^{2+}, Mn^{2+}). Except for some high ground water situations, such incidences in boreal forests are temporary and rare, as podsolic soils are generally coarse textured and readily permeable to water.

In boreal forests, the podsol soils show a range of different reactions and speciation of ions in the different soil horizons. The uppermost organic and mineral soil layers (O and A horizons) are characterised by a large cation exchange capacity due to the high organic matter content and a very low pH due to the organic acids released by the slow microbial decomposition of coniferous litter. The decomposition is further impeded by the cold climate and the lack of nutrients in a soil with a quartz- and silicate-rich parent material and humid climate favouring leaching of nutrients. In these conditions, iron and aluminium oxides are converted into soluble organic complexes and inorganic forms that may consequently be taken up by plants or leached deeper into the soil, and many micronutrient cations behave in a like manner (copper, zinc, manganese). The free Al^{3+} ions in the soil solution exchange cations of lower valence efficiently out of the exchange sites and enhance their leaching. The formation and leaching of organic aluminium complexes moderates the harmful effects of free Al^{3+} ions on plants in the highly acidic O and A horizons. Concurrently, the amounts of plant nutrient cations at the uppermost horizons remain rather high due to the constant mineralisation from the decomposing litter.

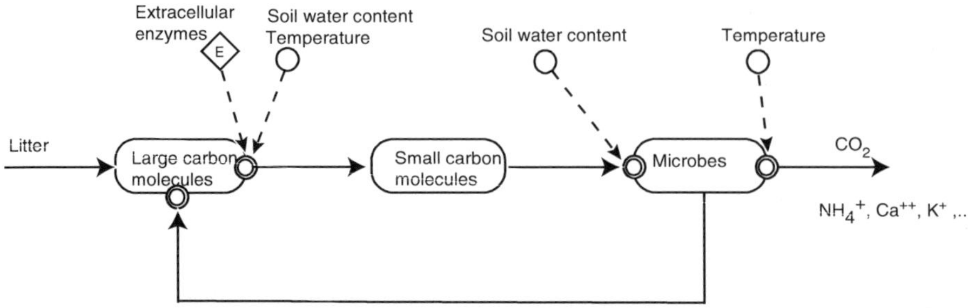

Fig. 4.75 Microbial decomposition of large carbon molecules with extracellular enzymes (The symbols are introduced in Fig. 2.3)

In the less acidic B horizon, the aluminium and iron are again precipitated as oxides due to increased pH and the microbial breakdown of organic complexes. The accumulation of poorly crystalline sesquioxides in the B horizon provides efficient adsorption of anions with a ligand exchange mechanism, such as phosphates, whereas the other anions, such as nitrate and chloride, leach easily down and out of the soil.

Decomposition of Soil Organic Matter

Plant cells have limited lifetime, and they die and fall on the ground as litter (Sect. 6.3.2.4). Microbes start to decompose the macromolecules in the litter and utilise them as a source of energy and raw material. Litter input and rain generate a typical structure in the top layer of the soil (Sect. 7.4). The decomposition has a key role in releasing the nutrients in the litter for reuse by vegetation (Sect. 6.3.2.8) and in the formation of soil structure.

The decomposition of organic macromolecules in soil proceeds in two main chemical steps (Fig. 4.75): (1) the breakdown of macromolecules to small molecules – such as the enzymatic hydrolysis of biopolymers to their monomers – and (2) the subsequent uptake and oxidation of the monomers by soil organisms (see reviews by Sinsabaugh et al. 2002; Tate III 2002; Schimel and Bennett 2004). The small molecules are converted by microbial catabolic reactions partly to inorganic forms, such as carbon dioxide, water and ammonium ions, while the rest is resynthesised by the growth of microbial cells. The release of inorganic species from organic compounds to soil is called mineralisation. The organic decomposition products too large to be taken up by microbes may contribute to chemical reactions producing the decay-resistant organic matter called humus.

Soil microbes deal with the structural complexity of decomposable compounds by producing complex enzyme systems and by growing as communities that produce a large variety of different enzymes (Warren 1996; Sinsabaugh et al. 2002).

This requires adequate amounts of nitrogen available for the microbes for constructing the enzymes. The enzymatic character of soil organic matter decomposition has additionally several other consequences. First, the reaction kinetics is affected not only by the availability of substrates and products but additionally by the activity of enzymes. The activity of enzymes are sensitive to soil properties and conditions such as soil temperate, moisture content and acidity, which all have direct impact on the functioning of enzymes in soil. Secondly, the enzymatic reactions are highly compound specific: Each compound and type of bond requires a specific enzyme of its own (Sinsabaugh et al. 2002). For example, the hydrolysis of starch, cellulose and hemicellulose in soil are catalysed by glucosidases, the hydrolysis of proteins by peptidases, and those of lipids and chitin by esterases and chitinases, respectively. Thirdly, the physical structure of soil and decomposing organic residues adds a spatial component to the factors determining the reaction kinetics. They affect the soil volume available for the reactions, the binding of enzymes with soil and the microscale mass transfer of reacting species (Tate III 2002).

The ease of decomposition of biological tissues is largely determined by their chemical composition (as reviewed by Derenne and Largeau 2001; Kögel-Knabner 2002; Blume et al. 2010a). For example, the cell contents surrounded by the plasma membrane are rapidly decomposed by the numerous active enzymes within the cell. This occurs in part already during the senescence of living cells.

The lipids found in biological membranes and bark are relatively short-chained molecules not bonded chemically to each other. They have both hydrophilic and hydrophobic parts. As a consequence, their decomposition proceeds relatively rapidly in soil. In contrast, the macromolecules in cell walls and extracellular enzymes in the soil are too large and insoluble to enter microbial cells. They must first be hydrolysed to their monomers by the enzymes excreted by microbes. The monomers can then enter the microbial cell and be metabolised for energy needs and cell construction.

Cellulose, a major component of plant cell walls, decomposes slowly in soil, notwithstanding its simple chemical structure of straight chains of 1–4 β-linked glucose units. This is because the linear cellulose chains pack close together and form a crystal-like structure stabilised by hydrogen bonding between the molecules. This gives cellulose mechanical strength and effectively prevents the action of hydrolysing enzymes. Chitin, the 1–4 β-linked polymer of *N*-acetyl-glucosamine units in fungal cell walls and arthropod exoskeletons, resembles cellulose in structure but is mechanically even stronger. In the other hand, hemicellulose and other cell wall carbohydrates composed of different sugar units have more branched and open structure, which facilitates more rapid hydrolysis by the enzymes compared with that of cellulose.

Lignin, another major component of plant cell wall, is a high-molecular-weight substance composed three-dimensionally of phenyl propane units, which makes it particularly recalcitrant to microbial decomposition. Actually, three main types of lignin differing in the methoxylation of monolignol monomers provide 'fingerprints' of lignin origin as gymnosperms, dicotyledonous angiosperms and monocotyledonous plants have slightly different share of each (Boerjan et al. 2003; Harris and DeBolt 2010; Weng and Chapple 2010). White-rot and brown-rot fungi belong to the few organisms that can decompose lignin, but even they cannot utilise lignin as the only carbon or energy source for growth. Lignin is decomposed by an extracellular, co-metabolic radical mechanism that partly breaks the aromatic rings and side chains. This direct oxidation releases small molecules (e.g. organic acids) that may be taken up and metabolised by the microbes, whereas other parts of lignin only change in structure. The chemically modified lignin may react with other soil compounds, such as amino acids and polyphenols, to form humus in a process of humification.

To assess the microbial-induced decomposition of soil organic matter at SMEAR II measuring station, hydrolytic activities of ten enzymes related to the decomposition of organic matter were measured with fluorometric microtiter plate method described by Wittmann et al. (2004). Seven of them are related directly to C and N cycles (beta-cellobiosidase, alpha- and beta-glucosidases, *N*-acetyl-glucosaminidase, beta-xylosidase, asetate esterase and aminopeptidase). For the measurements, 5–7 soil cores (to 25 cm depth) were taken nine times at 2–3 month interval during 2 years at SMEAR II (Fig. 4.76). The activities of enzymes catalysing the decomposition of soil organic matter generally decline with soil depth during warm seasons. In contrast, the activities during cold seasons the activities are lower and less variable with soil depth in comparison with warm season.

Nitrogen Processes in Soil

Most of the nitrogen-containing organic matter in the soil is in non-soluble complex forms that are too large to pass through microbial membranes. Therefore, microbes in the soil produce extracellular enzymes such as proteinases and peptidases that can break down proteins into smaller, water-soluble compounds such as amino acids (see Fig. 4.77). These amino acids are further broken down into mineral forms of nitrogen such as ammonium ($NH_4{}^+$) or nitrate (NO_3^-) or used directly by microbes as they can pass through the microbial membranes via membrane pumps (see Fig. 4.77).

The rate of protein decomposition in the soil is driven by soil temperature and the availability of water, as well as the input of new organic matter and the activity of extracellular enzymes in the soil. In a nitrogen-limited forest soil, mineral forms of nitrogen and their precursors, amino acids, are readily assimilated by microbes and plants and utilised in the build-up of new cellular material.

The release of inorganic forms of nitrogen from the bacterial cells into the soil determines the rate of nitrogen cycling in the soil. Microbes release ammonium (NH_4^+) into the soil as soon as the need for the microbial cell growth is fulfilled. Once released into the soil solution, ammonium ions can be attached on to the negatively charged clay particles or taken up by plants or microbes. Microbial uptake of inorganic nitrogen from the soil solution hence reduces the availability of

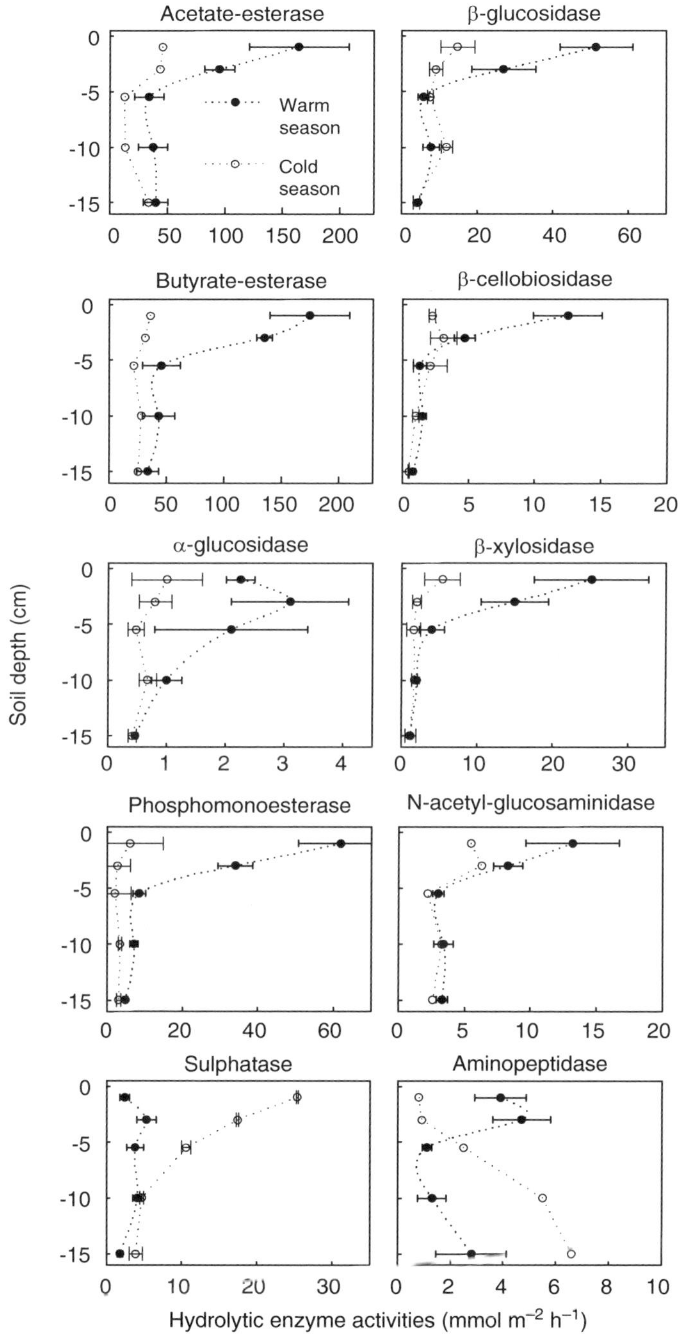

Fig. 4.76 Profiles of hydrolytic activities of enzymes involved in the decomposition of organic matter in soil during warm and cold seasons at SMEAR II. Measured with fluorometric microtiter plate method described by Wittmann et al. (2004)

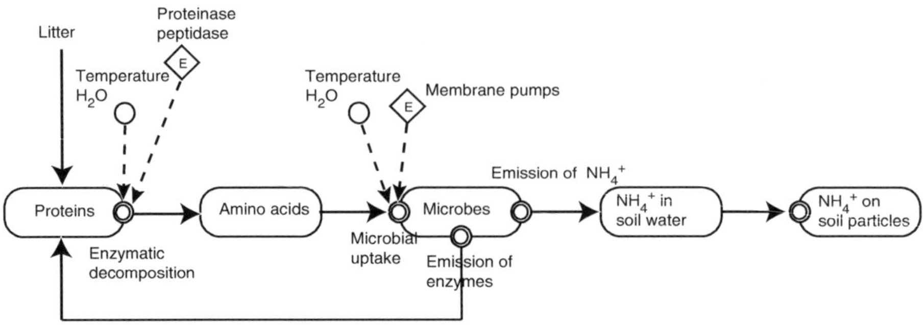

Fig. 4.77 Decomposition chain of organic matter into mineral forms of nitrogen in the soil (The symbols are introduced in Fig. 2.3)

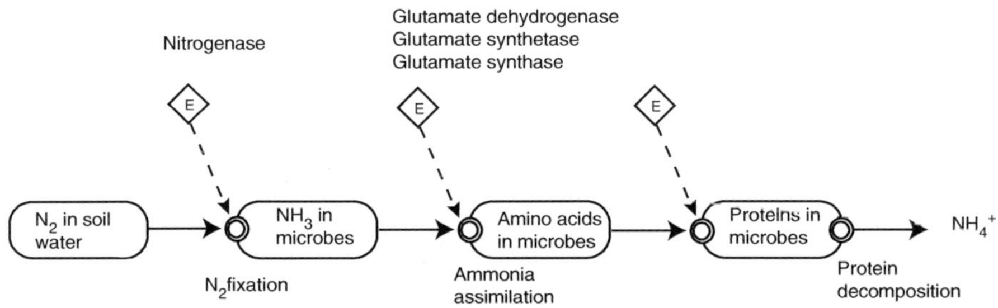

Fig. 4.78 Biological nitrogen fixation and protein synthesis

nitrogen to the growing plants. This process is called immobilisation, and it usually takes place in nitrogen-limited ecosystems where the dissolved organic carbon is not sufficient to meet microbial N need.

Biological nitrogen fixation is a microbial process in which the molecular nitrogen (N_2) from the atmosphere is converted into ammonia (NH_3) and incorporated into amino acids for further protein synthesis (see Fig. 4.78). The reaction chain from the molecular nitrogen present in the soil air space into ammonia (NH_3) and finally into amino acids involves the enzymes nitrogenase and glutamate dehydrogenase (see Fig. 4.78).

The microbes in symbiotic association with plants form nodules in the roots of the parent plant and hence receive carbohydrates from the plant to fulfil their energy needs and in return supply the plant with ammonium ions. In agricultural systems, legumes are well known for their association with bacteria of the genus *Rhizobium*, whereas in natural ecosystems like in boreal forests, alder trees (*Alnus*) are associated with the actinomycete *Frankia.* N fixation by alder trees in the understory of boreal pine forests can increase the N availability and hence bring significant amounts of atmospheric N into the forest (Vogel and Gower 1998; Sanborn et al. 2002). In boreal forests, N fixation by symbiotic associations of a

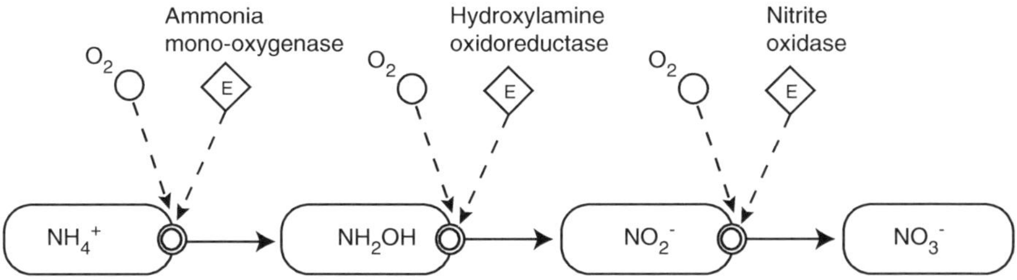

Fig. 4.79 Nitrifications of ammonium ions

cyanobacterium (*Nostoc* sp.) and the mosses *Pleurozium schreberi* or *Sphagnum capillifolium* may also bring N into the forest floor (DeLuca et al. 2002; Zackrisson et al. 2004; Markham 2009). However, the rate of this moss-associated N fixation seems very variable and generally smaller than that of the alder associated. In addition to the symbiotic or associated N fixation, free-living microbes in the soil are able to fix atmospheric N, for instance, during the decomposition of plant litter (Vitousek et al. 2002). Due to the requirements of anaerobicity, this heterotrophic N fixation is more common in wetlands and marine or freshwater ecosystems (Vitousek et al. 2002).

Nitrification is a process in which ammonium (NH_4^+), ammonia (NH_3) or organic N compounds are oxidised to nitrite (NO_2^-) and nitrate (NO_3^-) (Groffman 1991; Bremner 1997; Wrage et al. 2001) (see Fig. 4.79). A range of soil microorganisms are capable of biological nitrification, but the autotrophic ammonia and nitrite oxidising bacteria are one of the most important groups (Killham 1990; Bothe et al. 2000). Bacteria of the genus *Nitrosomonas* and *Nitrosospira* carry out the initial oxidation of NH_4^+ to NO_2^-, and those of the genus *Nitrobacter* oxidise NO_2^- further to NO_3^-. These bacteria obtain their energy for growth from the oxidation of ammonia or nitrite and assimilate carbon in the form of carbon dioxide (CO_2). In contrast, the heterotrophic microbial oxidation of ammonia is not linked to cellular growth (Killham 1990; De Boer and Kowalchuk 2001).

While the general biogeochemistry of bacterial ammonia oxidising is well understood, there are several recent findings suggesting a potential capacity for ammonia oxidation by *Archaea* in soil (Francis et al. 2007). The ammonia-oxidising archaeota seem to be globally distributed and abundant in many terrestrial environments, including forest, agricultural, grassland and alpine soils (Treusch et al. 2005; Leininger et al. 2006; Nicol and Schleper 2006; He et al. 2007). Still, the significance of *Archaea* to the biogeochemical transformations of nitrogen in the soil remains unclear in boreal forest soils.

In autotrophic nitrification, the stepwise oxidation of ammonium (NH_4^+) to nitrate (NO_3^-) is catalysed by several enzymes (see Fig. 4.79). The initial ammonium oxidation to hydroxylamine (NH_2OH) is catalysed by ammonia mono-oxygenase, the oxidation of NH_2OH to NO_2^- by hydroxylamine oxidoreductase and the oxidation of NO_2^- to NO_3^- by nitrite oxidoreductase (McCarty 1999; Wrage et al.

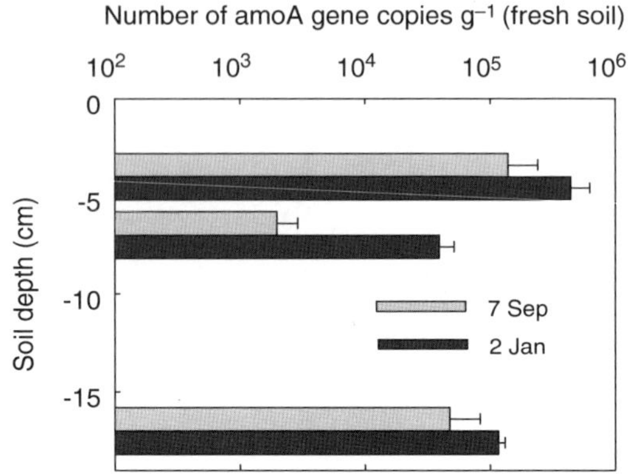

Fig. 4.80 The copy numbers of A subunit of bacterial ammonia mono-oxygenase gene during warm (September) and cold season (January) at SMEAR II soil measured in 2007. *Bars* show standard error of the mean

2001). The oxidation of ammonia is usually the rate-limiting step for nitrification as nitrite is rarely found to accumulate in soils (Kowalchuk and Stephen 2001). In nitrification, availability of oxygen is essential, since each step of the oxidation reaction requires O_2. Gaseous nitrogen oxides (NO and N_2O) can be formed through the chemical decomposition of NH_2OH or NO_2^- (Poth and Focht 1985).

The genes encoding the key enzyme in nitrification, ammonia mono-oxygenase, can be used for determining the number of soil ammonia-oxidising microorganisms in combination with quantitative real-time polymerase chain reaction (PCR) (Francis et al. 2007). In the soil at the SMEAR II station in southern Finland, the number of ammonia-oxidising bacteria was highest in the topsoil humus layer (Fig. 4.80). The number of ammonia mono-oxygenase genes in the soil decreased order of magnitude from the humus to the elluvial layer, but increased again deeper in the soil. This trend was similar in the soils sampled in winter (January) than soils from the early autumn (September).

Factors controlling nitrification in soil are the availability of CO_2, NH_4^+ ions, O_2, moisture and pH (Simek 2000). Carbon dioxide is always present in biologically active soils due to microbial and root respiration, whereas the same factors may lead to localised depletion of soil molecular oxygen. The concentrations of CO_2 and O_2 vary depending on the rates of respiration activity and soil aeration, which is affected by the balance between soil air-filled porosity and soil moisture (Simek 2000; Simojoki 2001). In well-aerated soils, the availability of NH_4^+ is the limiting factor in autotrophic nitrification. Environments with very small soil nitrogen content often have low nitrification activity and consequently low N_2O emissions (Martikainen 1984, 1985; Priha et al. 1999; Priha and Smolander 1999). It has been shown that high atmospheric load of ammonium stimulates nitrification in boreal (Martikainen et al. 1993) and temperate forest soils (van Breemen et al. 1987; Tietema et al. 1992, 1993). Especially, additions of urea and liming increase nitrification activities in acidic forest soils (Martikainen 1984). This effect is partly due to the increase in available mineral nitrogen but also due to the increased soil pH, which favours nitrifying microorganisms (Priha and Smolander 1999; Paavolainen et al. 2000).

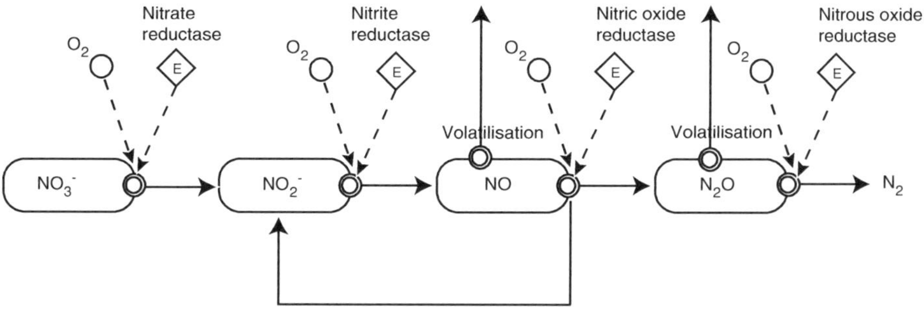

Fig. 4.81 Chain of denitrification steps

Recent findings on the ability of microbial communities to oxidise ammonium anaerobically have indicated that the global nitrogen cycling may be much more diverse than previously considered (Francis et al. 2007). For example, a group of planctomycete-like bacteria can oxidise ammonia with nitrite in strictly anoxic conditions in a process called *anaerobic ammonium oxidation* (anammox) (Schmidt et al. 2002; Strous and Jetten 2004; den Camp et al. 2006). The anammox bacteria have been found in many marine and freshwater ecosystems (Lam et al. 2007; Schmid et al. 2007; Woebken et al. 2007); however, their contribution to ammonium oxidation in terrestrial ecosystems remains largely unknown.

Denitrification is defined as the respiratory bacterial reduction of nitrate (NO_3^-) or nitrite (NO_2^-) to nitrogen oxides or molecular nitrogen (Knowles 1982; Bremner 1997; Einsle and Kroneck 2004) (see Fig. 4.81). Nitrogen monoxide (NO) and nitrous oxide (N_2O) are obligatory intermediates of the reduction process. This chain of the biological reactions is catalysed by the enzymes nitrate reductase, nitrite reductase, nitric oxide reductase and nitrous oxide reductase (Einsle and Kroneck 2004) (see Fig. 4.81). Denitrifying microorganisms are often facultative aerobic bacteria that are able to reduce nitrogen oxides when O_2 becomes limiting. Denitrifying microorganisms derive energy from organic substrates, and hence, denitrification is limited by the amount of readily decomposable organic matter in the soil.

Factors controlling the rate and the end products of denitrification are O_2 concentration, the availability of readily decomposable organic matter and the amount of NO_3^- and NO_2^- (Knowles 1982). The proportion of N_2O as the end product of denitrification increases at low soil pH, as the N_2O reductase is inhibited at low pH (Knowles 1982), whereas, in fully saturated or anaerobic soils, the dominant end product of denitrification is N_2. In general, an increase in soil water content and temperature or addition of fertiliser N, plant residues or animal manure increases denitrification activity and consequently the gaseous losses of NO, N_2O and N_2 from soils (e.g. Davidson 1991, 1993; Kaiser et al. 1998; Baggs et al. 2003; Schindlbacher et al. 2004). In anaerobic soils or anaerobic microsites, N_2O can act as an alternative electron acceptor to O_2 in denitrification. This leads to the net uptake of atmospheric N_2O and the formation of N_2, a process that has recently been intensively studied (see Chapuis-Lardy et al. 2007; Conen and Neftel 2007; Frasier et al. 2010; Vieten et al. 2010).

Denitrification and nitrification are the most important N turnover processes which lead to losses of nitrogen from the soil system. The gaseous emissions of NO and N_2O link the N turnover processes to the atmospheric chemistry and global climate change, as NO participates in the photochemical reactions of the troposphere, and N_2O is a very strong greenhouse gas. Hence, changes in soil conditions such as N availability or temperature will influence the production rates of NO and N_2O and therefore directly influence the soil emissions.

4.5 Forest Animals

Kari Heliövaara and Pertti Hari

The boreal zone is characterised by cold winters and extensive coniferous forests with considerable biological heterogeneity at landscape level. Many mammals and birds require wide home ranges, while some insects and bryophytes may flourish even in small patches. Insects and other arthropods play important roles in the structure and functioning of forest ecosystems, for example, as pollinators, herbivores, predators or decomposers. Forest invertebrates have also diverse biotic interactions with fungi and microbes in soil and decaying wood.

Forest fires create the basic spatial mosaic of forest patches in different successional stages, though small-scale dynamics caused by storms, insects, pathogenic fungi, etc. may predominate locally. The release of nutrients from organic compounds is accomplished by a large number of soil animals and microbes. The detrimental effect of high population densities of certain insect species can cause widespread forest destruction. Well-known examples include the Douglas-fir tussock moth (*Hemerocampa pseudotsugata*), the gypsy moth (*Lymantria dispar*), the eastern and westerns spruce budworms (*Choristoneura* spp.) and southern and western bark beetles (*Dendroctonus* spp.) in North America and spruce bark beetle (*Ips typographus*) and pine sawflies (*Neodiprion sertifer*, *Diprion pini*) in Europe.

K. Heliövaara (✉)
Department of Forest Sciences, University of Helsinki, P.O. Box 27,
00014 University of Helsinki, Helsinki, Finland
e-mail: kari.heliovaara@helsinki.fi

Herbivores

Insects are the major herbivores in many natural ecosystems. Although coniferous trees dominate the boreal zone, deciduous trees still support more herbivores. For example, the numbers of Finnish Macrolepidoptera associated with the Scots pine (*Pinus sylvestris*) and the Norway spruce (*Picea abies*) are only 18 and 19, respectively, while the numbers on deciduous trees are much higher, usually more than 100 species per tree genus (see Vaisanen 1996). The effects of ants extend over most trophic levels. Ants affect forest soils, nutrient cycles, plant dispersal, growth of trees and distribution and abundance of other invertebrates.

Species diversity of boreal forests is related to both the site type and the successional factors. The species richness of understorey vegetation in boreal forests is distinctly associated with site quality: Herb-rich sites have more species than mesic or xeric sites. The species richness of the understorey vegetation usually increases during succession until stands are 10–30 years old and decreases thereafter (Tonteri 1994). It is likely that the high species richness of understorey is reflected in the higher number of herbivorous insects and other associated invertebrates. In addition to insects, there are herbivorous vertebrates such as moose, microtine voles and forest grouses (Fig. 4.82). These herbivores influence ecosystems over different trophic, organisational and spatial scales. Much of a moose's diet (30 kg day^{-1}) comes from terrestrial vegetation, mainly consisting fresh shoots from willow and birch, and pine especially in winter.

Forest canopies support a complex assemblage of herbivorous insects and other arthropods. Almost 30,000 specimens of arthropods, belonging to 512 species, were collected by fogging the canopy of 24 pine trees in Norway. Of these arthropods, the most species rich taxon was Diptera (210 species), followed by Coleoptera (76 species) and Araneae (49 species) (Thunes et al. 2004). In general, old-growth canopies support the greatest diversity of canopy arthropods. It is possible that these canopies provide the greatest and most consistent diversity of foliar, branch and epiphytic habitats and moderate microclimatic conditions. Canopies of younger stands with lower diversity of habitats usually support lower numbers of arthropods (Schowalter 1995).

Foliage consumption during non-outbreak periods has been measured at least as 5–15% of leaf area production in temperate forests, and nominal folivory typically accounts for less than 10% of annual foliage of standing crops (for a review, see Schowalter et al. 1986). In some habitats in Australia, even 80% of the plant material may be ingested by insects. However, much of the food eaten by insect herbivores is rapidly returned to the decomposers through faeces. Trees tolerate low levels of folivory with little or no growth loss, perhaps because only low-value foliage is ordinarily consumed at low herbivore population densities. However, estimates of defoliation in the canopies of tall, mature trees are especially difficult to obtain, as the crown cannot be accessed easily for sample collection. A further difficulty in linking growth and foliar responses of conifers to defoliation is that long-term data on annual fluctuations in defoliator populations are rarely available for study (see Armour et al. 2003).

Fig. 4.82 A grazing male capercaillie (*Tetrao urogallus*), weight 4 kg, feeds almost entirely on Scots pine needles throughout the winter, some 200 g day^{-1} (**a**). Sugars in the needle form the main nutrition of the bird. Other parts of the needles pass through the intestine of the bird (**b**) (Photos: Teuvo Hietajärvi)

Wood Decomposers

Primaeval forests which have large quantities of decaying wood are inhabited by a high number of saproxylic beetle species. Much of the decomposition necessary for mineralisation and recirculation of nutrients is performed by fungi. Mycorrhizas enable trees to take up water and nutrients more efficiently than the roots themselves, and the fungus obtains carbohydrates from the tree. It is possible that grazing by fungivorous invertebrates on mycorrhiza may influence the tree roots more than do the organisms that eat the roots.

Decaying wood is a patchy and unevenly distributed habitat with a limited duration. Dead wood represents an important habitat or substrate for numerous vertebrates, invertebrates, vascular plants, fungi, bryophytes and lichens. Insects inhabiting dead wood fall into a number of larval trophic guilds, including xylophages (wood feeders), fungivores (fungal feeders), saprophages (scavengers and detrital feeders), predators and parasitoids (Vanderwel et al. 2006).

Ambrosia beetles culture symbiotic fungi in their galleries in wood and feed on the fungi. They possess a special organ, which is called mycangium (pl. mycangia), to convey the fungi to new locations. Mycangia have been discovered in several bark beetle species. Recently Tanahashi et al. (2010) report the discovery of the microbe-storage organ in stag beetles (Coleoptera: Lucanidae), which develop in decayed wood. They found that the mycangium, which was located in the abdomen of the beetles, is present in all adult females of 22 lucanid species examined.

Most wood-feeding insects need an association with microbes to utilise wood as food. Wood inhabiting insects have difficulty utilising wood as a food resource because wood is composed mostly of cellulose, lignin and hemicelluloses that together comprise about 90% of the total volume, and these polymers are difficult to digest. Passalid beetles (Coleoptera: Passalidae) and longicorn beetles (Coleoptera: Cerambycidae) are associated with xylose-fermenting yeasts that may help in the digestion of wood hemicelluloses (Suh et al. 2006).

Cellulose is widely distributed in all higher plants, for example, trees. It is the most abundant compound in plant cell walls, contributing to about 20–50% of dry weight in the primary cell walls. In forest ecosystems, decomposition of cellulose, caused by fungi and bacteria, may take tens of years. Since biopolymers such as cellulose are complicated chemical compounds, they cannot pass through the pumps in cell membranes as such. All invertebrates feeding on cellulose need specific enzymes to break down the structure of these molecules as sugars for nutrition. It is generally known that termites and wood-feeding cockroaches have protozoa or bacteria in their digestive organs which produce cellulolytic enzymes. However, recent studies show that several insects from Dictyoptera, Orthoptera and Coleoptera can produce their own endogenous cellulases (i.e. endo-β-1,4-glucanase) in the midgut or the salivary glands (Watanabe and Tokuda 2010). While cellulases from microorganisms are important for cellulose degradation in termites and some Coleoptera larvae, there are numerous examples in lower termites and other insect groups in which endogenous insect cellulases are sufficient for effective cellulose degradation and survival on plant biomass (for a review, see Oppert et al. 2010).

The digestive enzymes in insects are adapted to the diet on which they feed. Polyphagous insects such as cockroaches (Dictyoptera) and phytophagous grasshoppers (Orthoptera) and beetles (Coleoptera) secrete several enzymes including protease, lipase, amylase, invertase and maltase. In general, insects have a high protein content (up to 80%), but plants are composed predominantly of carbohydrates. Protein nitrogen of plant tissue is a critical component for phytophagous insects. Consequently, caterpillars of phytophagous butterflies and moths (Lepidoptera) must digest large amounts of plant material to achieve high growth rates. Digestive proteases catalyse the release of peptides and amino acids from dietary protein. These enzymes are found most abundantly in the midgut region of the insect digestive tract (e.g. Jongsma and Bolter 1997). Several classes of proteases are found in insect guts (for review, see Terra and Ferreira 1994). These include cysteine and aspartate proteinases, metalloproteases and aminopeptidases (Barrett

1986). In addition, there are some plant-produced proteins that are highly resistant to insect proteases. These proteins can interfere with insect digestive processes and participate in the plant antiherbivory defences (see Pauchet et al. 2008).

Soil Fauna

Boreal forests appear to be richer in species below the surface of the soil than above it, thus clearly contrasting with tropical forests. However, much of the ecological research that has focused above ground has ignored those associations occurring below ground. At the same time, only little attention has been paid to ecological research of below-ground biota and its associations with above-ground organisms. The above-ground compartment is responsible for the carbon input in an ecosystem, whereas the below-ground compartment is responsible for most of the decomposition (carbon loss). These two compartments are tightly dependent upon each other because of the role of plants as the source of carbon for soil biota and because the soil biota in turn release nutrients bound up into simpler forms that are more readily taken up by the plants (Wardle 1999). Most mineralisation of nutrients is directly governed by the basal consumer trophic level of the soil food web (bacteria and fungi). However, their activity is stimulated by soil biota at higher trophic levels (e.g. Protozoa, nematodes, mites, springtails, millipedes and earthworms).

One square metre of forest soil harbours millions of invertebrates (Fig. 4.83). The dominant soil animal groups in terms of biomass are enchytraeids and earthworms, but variation between habitats both in number of individuals and biomass can be considerable (Persson et al. 1980; Huhta et al. 1986). In addition to the groups shown in Fig. 4.83, the forest floor also supports several microscopic organisms such decaying fungi (80 g d.w. m^{-2}) and bacteria (20 g d.w. m^{-2}). The biomass of soil fauna in boreal forests consists of more fungivores and less saprovores and herbivores (root feeders) than in nemoral forests, grasslands or arable land (see Huhta et al. 1998). This can be explained by the fact that conifer litter is less rich in nutrients or less palatable for animal consumption than leaf, grass and herb litter. Consequently, conifer litter decomposition can be more dependent on the activity of saprophytic fungi than are other litter types. In acid boreal forests, the functional group known as 'ecosystem engineers' among the soil fauna is predominantly represented by ants, whereas deep-burrowing earthworms are almost lacking (Huhta et al. 1998). Approximately 90% of the soil faunal biomass can be found in the top 10 cm of raw humus forest soil (Persson et al. 1980). The total biomass of soil invertebrates in Finnish coniferous forests varies from 1.3 to 2.2 g d.w. m^{-2} (Huhta et al. 1986), that is, 13–22 kg d.w. per hectare.

Several studies have shown that macroscopic soil invertebrates, often less than 1 mm in length, usually increase the decomposition and mineralisation rate of

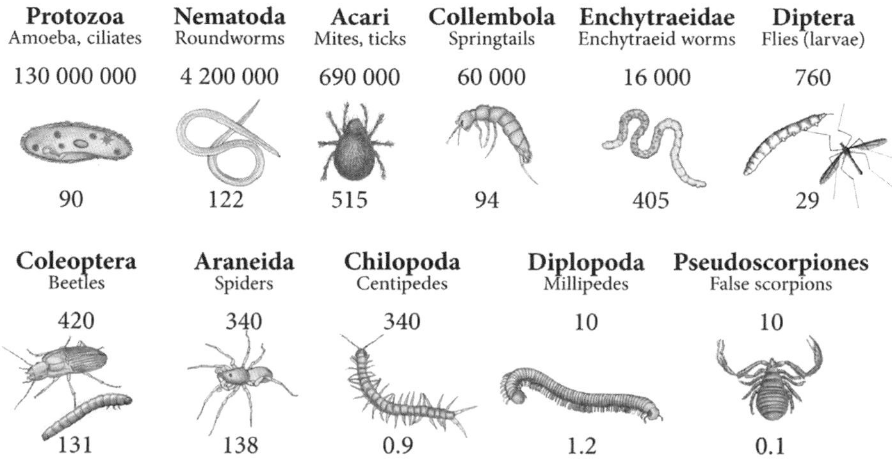

Fig. 4.83 Dominating animal groups in forest soil. Approximate mean numbers (m^{-2}, *above drawings*) and biomass (mg d.w. m^{-2}, *below drawings*) of the most abundant soil invertebrate taxa in boreal coniferous forests. Diptera is mainly represented by larvae in soil (Data from Persson et al. 1980; Huhta et al. 1986; Setälä unpublished)

dead organic matter (Setälä and Huhta 1991; Huhta et al. 1998). Consequently, soil decomposer animals have a strong control on soil structure and soil processes, such as decomposition and nutrient cycling. Environmental conditions, for example, soil temperature and moisture, modify the function of this fauna. However, Haimi et al. (2005) in terms of climate change showed that increasing temperature had only minor effects and elevated air CO_2 concentration had no effects on the numbers of most of the decomposer fauna.

Fungivorous and microbe-detritivorous soil invertebrates may indirectly stimulate the release of plant-available nutrients, reducing nutrient limitation for the seedlings and favouring production of above-ground tissues. When soil fauna is removed from heterotrophic decomposition systems, the activity of soil microbes may decrease dramatically (Setälä et al. 1998). To determine how decomposer food-web architecture might determine plant productivity has been experimentally tested in microcosms (Laakso and Setälä 1999). In their experiment, Laakso and Setälä (1999) established microcosms consisting of boreal forest humus and litter as well as microbial trophic levels, and various functional groups of soil fauna (i.e. fungivores, microbe-detritivores and top predators) added in different combinations. Their results show that the addition of the fungivore and microbe-detritivore trophic groups, either singly or in combination, stimulated growth of planted silver birch (*Betula pendula*) seedlings. This indicates the importance of the consumption of microflora by soil animals in indirectly determining plant productivity. There were also important effects of these two faunal groups on plant root to shoot ratios.

Forest Pests and Decomposition

Conifer foliage is sometimes attacked by a wide range of herbivorous insects that cause considerable amount of damage. Forest insect epidemics are often initiated in mature, slow-growing forests, but during epidemics, it is usually the least vigorous trees of all age classes that are attacked. Outbreaks are highly individual phenomena controlled, for example, by the number of natural enemies and environmental conditions. Effects of insect defoliation on pines and other coniferous trees have been investigated for a long time (e.g. Liebhold et al. 2000). Even though folivorous insects on pine and conifers usually do not kill the plant, their feeding damage might reduce the competitiveness and fitness of the plant. Diprionid sawflies (Hymenoptera, Diprionidae) are among the most serious defoliators of Scots pine in Europe. The European pine sawfly *Neodiprion sertifer* is an early season defoliator causing serious economic losses during outbreaks. During the outbreaks of *Diprion pini,* foliage loss may reach close to 100% of foliage production. In worst cases, after the damage massive amounts of decaying wood are formed in the forest, but also in certain virgin forests, standing dead trees and fallen logs in various stages of decomposition may account for 40% of the total volume of timber present.

Fungi and insects play a predominant interactive role in recycling wood in forest ecosystems. Forest-dwelling insects affect the fungal community of dead wood by carrying various fungi to their target. To enhance access of air-disseminated fungal propagules to wood, some insects, such as bark beetles and especially ambrosia beetles, make holes through the outer bark into the phloem and xylem. Through their tunnelling and chewing activities, insects can affect wood moisture and gaseous composition which will significantly enhance wood decomposition (Müller et al. 2002).

Primary production, as an example of an ecosystem level property, is largely influenced by the activities of various organisms which break down organic material and recycle nutrients bound in dead biomass. The feeding activities of the soil invertebrates have important indirect effects on availability of nutrients in forest soil. Soil animals usually influence decomposition rates and nutrient recycling mainly through their effects on the activity of the microflora, such as grazing on bacteria and fungi, fragmentation of the litter and transport of bacteria and fungal propagules (Edsberg 2000). The role of soil invertebrates should be measured as CO_2 production. However, the direct contribution of the soil fauna to the catabolism of dead organic matter in temperate soils seems to be relatively low (Edsberg 2000). Literature review by Persson (1989) reported that animals in coniferous forest soils contribute only 1–5% of the total heterotrophic respiration. The energy flux through herbivores, decomposers and soil invertebrates is poorly documented, and more research is needed.

References

Aalto T, Juurola E (2001) Parametrization of a biochemical CO_2 exchange model for birch (*Betula pendula* Roth.). Boreal Environ Res 6:53–64

Aalto T, Hari P, Vesala T (2002) Comparison of an optimal regulation model and biochemical model in explaining CO_2 exchange in field conditions. Silva Fenn 36:615–623

Acioli-Santos B, Vieira HEE, Lima CEP, Maia LC (2011) The molecular ectomycorrhizal fungus essence in association: a review of differentially expressed fungal genes during symbiosis formation. In: Rai M, Varma A (eds) Diversity and biotechnology of ectomycorrhizae, vol 25, Soil biology. Springer, Berlin

Adams WW III, Barker DH (1998) Seasonal changes in xanthophyll cycle-dependent energy dissipation in *Yucca glauca* Nuttall. Plant Cell Environ 21:501–511

Adamson AW, Gast AP (1997) Physical chemistry of surfaces, 6th edn. Wiley, New York

Agerer R (2006) Fungal relationships and structural identity of their ectomycorrhizae. Mycol Prog 5:67–107

Ainsworth EA, Rogers A (2007) The response of photosynthesis and stomatal conductance to rising [CO_2]: mechanisms and environmental interactions. Plant Cell Environ 30:258–270

Aitken SN, Hannerz M (2001) Genecology and gene resource management strategies for conifer cold hardiness. In: Bigras FJ, Columbo SJ (eds) Conifer cold hardiness. Kluwer, Dordrecht

Allen JF (2003) Cyclic, pseudocyclic and noncyclic photophosphorylation: new links in the chain. Trends Plant Sci 8:15–19

Allen JF, Forsberg J (2001) Molecular recognition in thylakoid structure and function. Trends in Plant Sci 6:317–326

Amthor JS (1994) Plant respiratory responses to the environment and their effects on the carbon balance. In: Wilkinson RE (ed) Plant–environment interactions. Marcel Dekker, New York

Angeles G, Bond B, Boyer JS et al (2004) Letters. The cohesion-tension theory. New Phytol 163:447–449

Angers DA, Caron J (1998) Plant-induced changes in soil structure: processes and feedbacks. Biogeochemistry 42:55–72

Arena C, Vitale L, Virzo De Santo A (2008) Paraheliotropism in *Robinia pseudoacacia* L.: an efficient strategy to optimise photosynthetic performance under natural environmental conditions. Plant Biol 10:194–201

Armour H, Straw N, Day K (2003) Interactions between growth, herbivory and long-term foliar dynamics of Scots pine. Trees 17:70–80

Atkin OK, Tjoelker MG (2003) Thermal acclimation and the dynamic response of plant respiration to temperature. Trends Plant Sci 8:343–351

Bachmann J, van der Ploeg RR (2002) A review on recent developments in soil water retention theory: interfacial tension and temperature effects. J Plant Nutr Soil Sci 165:468–478

Bachmann J, Deurer M, Arye G (2007) Modeling water movement in heterogeneous water-repellent soil: 1. Development of a contact angle-dependent water-retention model. Vadose Zone J 6:436–445

Baggs EM, Cadisch G, Stevenson M, Pihlatie M, Regar A, Cook H (2003) Nitrous oxide emissions resulting from interactions between cultivation technique, residue quality and fertiliser application. Plant Soil 254:361–370

Baldock JA, Nelson PN (2000) Soil organic matter. In: Sumner ME (ed) Handbook of soil science. CRC Press, Boca Raton

Ball JT, Woodrow IE, Berry JA (1987) A model predicting stomatal conductance and its contribution to the control of photosynthesis under different environmental conditions. In: Biggins I (ed) Progress in photosynthesis research, vol 4. Martinus-Nijhoff, Dordrecht

Bannan MV (1941) Variability in wood structure in roots of native Ontario conifers. Bull Torrey Bot Club 68:73–194

Barber J, Andersson B (1992) Too much of a good thing: light can be bad for photosynthesis. Trends Biochem Sci 12:61–66

Barrett AJ (1986) An introduction to the proteinases. In: Barrett AJ, Salvesen G (eds) Proteinase inhibitors. Elsevier, Amsterdam

Barton CVM, North PRJ (2001) Remote sensing of canopy light use efficiency using the photochemical reflectance index – model and sensitivity analysis. Remote Sens Environ 78:264–273

Beaudet M, Messier C (1998) Growth and morphological responses of yellow birch, sugar maple, and beech seedlings growing under a natural light gradient. Can J For Res 28:1007–1015

Bendall DS (2006) Photosynthesis: light reactions. Encyclopedia of life sciences. Wiley, New York

Bengough AG, Bransby MF, Hans J, McKenna SJ, Roberts TJ, Valentine TA (2006) Root responses to soil physical conditions; growth dynamics from field to cell. J Exp Bot 57:437–447

Berg B (2000) Litter decomposition and organic matter turnover in northern forest soils. For Ecol Manage 133:13–22

Berg B, Dise N (2004) Calculating the long-term stable nitrogen sink in northern European forests. Acta Oecol 26:15–21

Berg B, Booltink H, Breymeyer A, Ewertson A, Gallardo A, Holm B, Johansson M-B, Koivuoja S, Meentemeyer V, Nyman P, Olofsson J, Petterson A-S, Reurslag A, Staaf H, Staaf I, Uba L (1991) Data on needle litter decomposition and soil climate as well as site characteristics for some coniferous forest sites. Part 2. Decomposition data, 2nd edn. Swedish University of Agricultural Sciences, rep 42

Bergh J, McMurtrie RE, Linder S (1998) Climatic factors controlling the productivity of Norway spruce: a model based analysis. For Ecol Manage 110:127–139

Bernacchi CJ, Pimentel C, Long SP (2003) In vivo temperature response functions of parameters required to model RuBP-limited photosynthesis. Plant Cell Environ 26:1419–1430

Berninger F, Nikinmaa E (1994) Geographical variation in the foliage mass – wood cross-sectional area ratios in young Scots pine stands. Can J For Res 24:2263–2268

Betsche T (1983) Aminotransfer from alanine and glutamate to glycine and serine during photorespiration in oat leaves. Plant Physiol 71:961–965

Beyer L (1996) The chemical composition of soil organic matter in classical humic compound fractions and in bulk samples – a review. Z Pflanz Bodenkunde 159:527–539

Beyer L, Vogt B, Kobbemann C (1996) A simple wet chemical extraction procedure to characterize soil organic matter (SOM): 2. Reproducibility and verification. Commun Soil Sci Plan 27:2229–2241

Bilger W, Bjorkman O (1990) Role of the xanthophyll cycle in photoprotection elucidated by measurements of light-induced absorbance changes, fluorescence and photosynthesis in leaves of *Hedera canariensis*. Photosynth Res 25:173–185

Björkman O (1981a) Responses to different quantum flux densities. In: Lange OL, Nobel PS, Osmond CB, Ziegler H (eds) Encyclopedia of plant physiology, vol 12A. Springer, Heidelberg

Björkman O (1981b) The response of photosynthesis to temperature. In: Grace J, Ford ED, Jarvis PG (eds) Plants and their atmospheric environment. The 21st symposium of the British Ecological Society, Edinburgh. Blackwell, Oxford

Blume H-P, Brümmer GW, Horn R, Kandeler E, Kögel-Knabner I, Kretzschmar R, Stahr K, Wilke B-M (2010a) Organische Bodensubstanz. In: Scheffer/Schachtschabel. Lehrbuch der Bodenkunde, 16. Aufl. Spektrum, Heidelberg

Blume H-P, Brümmer GW, Horn R, Kandeler E, Kögel-Knabner I, Kretzschmar R, Stahr K, Wilke B-M (2010b) Bodenorganismen und ihr Lebensraum. In: Scheffer/Schachtschabel. Lehrbuch der Bodenkunde, 16. Aufl. Spektrum, Heidelberg

Boerjan W, Ralph J, Baucher M (2003) Lignin biosynthesis. Annu Rev Plant Biol 54:519–546

Bothe H, Jost G, Schloter M, Ward BB, Witzel K-P (2000) Molecular analysis of ammonia oxidation and denitrification in natural environments. FEMS Microbiol Rev 24:673–690

Bowes G (1991) Growth at elevated CO_2. Photosynthetic responses mediated through Rubisco. Plant Cell Environ 14:795–806

Boyer JS, Silk WK, Watt M (2010) Path of water for root growth. Funct Plant Biol 37:1105–1116

Bremner JM (1997) Sources of nitrous oxide in soils. Nutr Cycl Agroecosyst 49:7–16

Britto DT, Kronzucker HJ (2002) NH4+ toxicity in higher plants: a critical review. J Plant Physiol 159:567–584

Britto DT, Glass ADM, Kronzucker HJ, Siddiqi MY (2001) Cytosolic concentrations and trans-membrane fluxes of NH_4^+/NH_3. An evaluation of recent proposals. Plant Phys 125:523–526

Britto DT, Kronzucker HJ (2002) NH4+ toxicity in higher plants: a critical review. J Plant Physiol 159:567–584

Brooks A, Farquhar GD (1985) Effect of temperature on the CO_2-O_2 specificity of ribulose-1,5-bisphosphate carboxylase/oxygenase and the rate of respiration in the light: estimates from gas exchange measurements on spinach. Planta 165:397–406

Brownlee C (2001) The long and the short of stomatal density signals. Trends Plant Sci 6:441–442

Buchanan BB, Gruissem W, Jones RL (eds) (2000) Biochemistry and molecular biology of plants. American Society for Plant Physiology, Rockville

Buckley TN (2008) The role of stomatal acclimation in modelling tree adaptation to high CO_2. J Exp Bot 59:1951–1961

Bunce JA (2000) Acclimation to temperature of the response of photosynthesis to increased carbon dioxide concentration in *Taraxacum officinale*. Photosynth Res 64:89–94

Buol SW, Hole FD, McCracken RJ, Southard RJ (1997) Soil genesis and classification, 4th edn. Iowa State University Press, Ames

Burch-Smith TM, Stonebloom S, Xu M, Zambryski PC (2011) Plasmodesmata during development: re-examination of the importance of primary, secondary, and branched plasmodesmata structure versus function. Protoplasma 248:61–74

Burdon J (2001) Are the traditional concepts of the structures of humic substances realistic? Soil Sci 166:752–769

Butler WL, Kitajima M (1975) Fluorescence quenching in photosystem II of chloroplasts. Biochim Biophys Acta 376:116–125

Cairney JWG (2011) Ectomycorrhizal fungi: the symbiotic route to the root for phosphorus in forest soils. Plant Soil 344:51–71

Campbell BD, Grime JP, Mackey JML (1991) A trade-off between scale and precision in resource foraging. Oecologia 87:532–538

Canham CD (1985) Suppression and release during canopy recruitment in *Acer saccharum*. Bull Torrey Bot Club 112:134–145

Caraglio Y, Barthélémy D (1997) Revue critique des termes relatifs à la croissance et à la ramification des tiges des végétaux vasculaires. In: Bouchon J, de Reffye P, Barthélémy D (eds) Modélisation et simulation de l'architecture des végétaux. INRA Édit, Versailles

Casper BB, Schenk HJ, Jackson RB (2003) Defining a plant's belowground zone of influence. Ecology 84:2313–2321

Ceulemans R, Mousseau M (1994) Effects of elevated atmospheric CO_2 on woody plants. New Phytol 127:425–446

Chalot M, Javelle A, Blaudez D, Lambilliote R, Cooke R, Sentenac H, Wipf D, Botton B (2002) An update on nutrient transport processes in ectomycorrhizas. Plant Soil 244:165–175

Chalot M, Blaudez D, Brun A (2006) Ammonia: a candidate for nitrogen transfer at the mycorrhizal interface. Trends Plant Sci 11:263–266

Chaparro-Suarez IG, Meixner FX, Kesselmeie J (2011) Nitrogen dioxide (NO_2) uptake by vegetation controlled by atmospheric concentrations and plant stomatal aperture. Atmos Environ 45:5742–5750

Chapuis-Lardy L, Wrage N, Metay A, Chotte J-L, Bernoux M (2007) Soils, a sink for N_2O? A review. Glob Change Biol 13:1–17

Chen HH, Li PH (1978) Interactions of low temperature, water stress, and short days in the induction of stem frost hardiness in red osier dogwood. Plant Physiol 62:833–835

Chitnis PR (2001) Photosystem I: Function and Physiology. Annu Rev Plant Physiol Plant Mol Biol 52:593–626

Christensen-Dalsgaard KK, Ennos AR, Fournier M (2008) Are radial changes in vascular anatomy mechanically induced or an ageing process? Evidence from observations on buttressed tree root systems. Trees 22:543–550

Chuine I, Cour P, Rousseau DD (1998) Fitting models predicting dates of flowering of temperate-zone trees using simulated annealing. Plant Cell Environ 21:455–466
Chuine I, Cour P, Rousseau DD (1999) Selecting models to predict the timing of flowering of temperate trees: implications for tree phenology modelling. Plant Cell Environ 22:1–13
Chuine I, Kramer K, Hänninen H (2003) Plant development models. In: Schwartz MD (ed) Phenology: an integrative environmental science. Kluwer, Dordrecht
Clapham D, Ekberg I, Little CHA, Savolainen O (2001) Molecular biology of conifer frost tolerance and potential applications to tree breeding. In: Bigras FJ, Columbo SJ (eds) Conifer cold hardiness. Kluwer, Dordrecht
Collatz GJ, Ball JT, Givet C, Berry JA (1991) Physiological and environmental regulation of stomatal conductance, photosynthesis and transpiration: a model that includes a laminar boundary layer. Agric For Meteorol 54:107–136
Collatz GJ, Ribas-Carbo M, Berry JA (1992) Coupled photosynthesis-stomatal conductance model for leaves of C4 plants. Aust J Plant Physiol 19:519–538
Conen F, Neftel A (2007) Do increasingly depleted $\delta^{15}N$ values of atmospheric N_2O indicate a decline in soil N_2O reduction? Biogeochemistry-US 82:321–326
Corey AT, Logsdon SD (2005) Limitations of the chemical potential. Soil Sci Soc Am J 69:976–982
Cowan IR (1977) Stomatal behaviour and environment. Adv Bot Res 4:117–228
Cowan IR, Farquhar GD (1977) Stomatal function in relation to leaf metabolism and environment. In: Jennings DH (ed) Integration of activity in the higher plant. Cambridge University Press, Cambridge
Coyne MS (1999) Soil microbiology: an exploratory approach. Delmar Publishers, Albany
Dahl E, Mork E (1959) Om sambandet mellom temperatur, ånding og vekst hos gran, *Picea abies* (L.) Karst. Medd Norske Skogforsøksvesen 53:83–93
Davidson EA (1991) Fluxes of nitrous oxide and nitric oxide from terrestrial ecosystems. In: Rogers JE, Whitman WB (eds) Microbial production and consumption of greenhouse gases: methane, nitrogen oxides, and halomethanes. American Society for Microbiology, Washington, DC
Davidson EA (1993) Soil water content and the ratio of nitrous oxide to nitric oxide emitted from soil. In: Oremland, R.S. (ed.) Biogeochemistry of global change radiatively active trace gases. Chapman & Hall, New York, USA, p. 369–386
De Boer W, Kowalchuk GA (2001) Nitrification in acid soils: micro-organisms and mechanisms. Review. Soil Biol Biochem 33:853–866
DeLuca TH, Zackrisson O, Nilsson M-C, Sellstedt A (2002) Quantifying nitrogen-fixation in feather moss carpets of boreal forests. Nature 419:917–920
Demidchik V, Maathuis FJM (2007) Physiological roles of nonselective cation channels in plants: from salt stress to signalling and development. New Phytol 175:387–404
Demmig-Adams B, Adams WW III (2006) Photoprotection in an ecological context: the remarkable complexity of thermal energy dissipation. New Phytol 172:11–21
Demmig-Adams B, Ebbert V, Mellman DL, Mueh KE, Schaffer L, Funk C, Zarter CR, Adamska I, Jansson S, Adams WW III (2006) Modulation of PsbS and flexible vs sustained energy dissipation by light environment in different species. Physiol Plantarum 127:670–680
den Camp HJMO, Kartal B, Guven D, van Niftrik LAMP, Haaijer SCM, van der Star WRL, van de Pas-Schoonen KT, Cabezas A, Ying Z, Schmid MC, Kuypers MMM, van de Vossenberg J, Harhangi HR, Picioreanu C, van Loosdrecht MCM, Kuenen JG, Strous M, Jetten MSM (2006) Global impact and application of the anaerobic ammonium-oxidizing (anammox) bacteria. Biochem Soc Trans 34:17–178
Derenne S, Largeau C (2001) A review of some important families of refractory macromolecules: composition, origin and fate in soils and sediments. Soil Sci 166:833–847
Dibb JE, Arsenault M, Peterson MC, Honrath RE (2002) Fast nitrogen oxide photochemistry in Summit, Greenland snow. Atmos Environ 36:2501–2511
Drake BG, Gonzàlez-Meler MA, Long SP (1997) More efficient plants: a consequence of rising atmospheric CO_2? Annu Rev Plant Phys 48:609–639

Eamus D, Jarvis P (1989) The direct effects of increase in the global atmospheric CO_2 concentration on natural and commercial temperate trees and forests. Adv Ecol Res 19:1–55
Edsberg E (2000) The quantitative influence of enchytraeids (Oligochaeta) and microarthropods on decomposition of coniferous raw humus in microcosms. Pedobiologia 44:132–147
Edwards GE, Baker NR (1993) Can CO_2 assimilation in maize leaves be predicted accurately from chlorophyll fluorescence analysis? Photosynth Res 37:89–102
Ehleringer J (1981) Leaf absorptances of Mohave and Sonoran desert plants. Oecologia 49:366–370
Ehlers K, Kollmann R (2001) Primary and secondary plasmodesmata: structure, origin, and functioning. Protoplasma 216:1–30
Einsle O, Kroneck PMH (2004) Structural basis of denitrification. Biol Chem 385:875–883
Endo T, Shikanai T, Takabayashi A, Asada K, Sato F (1999) The role of chloroplastic NAD(P)H dehydrogenase in photoprotection. FEBS Lett 457:5–8
Ensminger I, Sveshnikov D, Campbell DA, Funk C, Jansson S, Lloyd J, Shibistova O, Öquist G (2004) Intermittent low temperatures constrain spring recovery of photosynthesis in boreal Scots pine forests. Glob Change Biol 10:995–1008
Enstone DE, Peterson CA, Ma F (2003) Root endodermis and exodermis: structure, function, and responses to the environment. J Plant Growth Regul 21:335–351
Evert RF (2006) Esau's plant anatomy: meristems, cells, and tissues of the plant body: their structure, function, and development, 3rd edn. Wiley, Holboken
Fall R (2003) Abundant oxygenates in the atmosphere: a biochemical perspective. Chem Rev 103:4941–4951
Farley RA, Fitter AH (1999) The responses of seven co-occurring woodland herbaceous perennials to localized nutrient-rich patches. J Ecol 87:849–859
Farquhar GD, von Caemmerer S, Berry JA (1980) A biochemical model of photosynthetic CO_2 assimilation in leaves of C3 species. Planta 149:78–90
Fayle DCF (1968) Radial growth in tree roots; distribution, timing, anatomy. Technical report 9, University of Toronto
Fernandez MD, Pieters A, Donoso C, Tezara W, Azkue M, Herrera C, Rengifo E, Herrera A (1998) Effects of a natural source of very high CO_2 concentration on the leaf gas exchange, xylem water potential and stomatal characteristics of plants of *Spatiphylum cannifolium* and *Bauhinia multinervia*. New Phytol 138:689–697
Filella I, Porcar-Castell A, Munné-Bosch S, Bäck J, Garbulsky M, Peñuelas J (2009) PRI assessment of long-term changes in carotenoids/chlorophyll ratio and short-term changes in de-epoxidation state of the xanthophyll cycle. Int J Remote Sens 30:4443–4455
Finér L, Messier C, De Grandpré L (1997) Fine-root dynamics in mixed boreal conifer – broad-leafed forest stands at different successional stages after fire. Can J For Res 27:304–314
Fitter AH (1991) The ecological significance of root system architecture: an economic approach. In: Atkinson D (ed) Plant root growth. An ecological perspective. Blackwell, Oxford
Flexas J, Briantais J-M, Cerovic Z, Medrano H, Moya I (2000) Steady-state and maximum chlorophyll fluorescence responses to water stress in grapevine leaves: a new remote sensing system. Remote Sens Environ 73:283–297
Francis CA, Beman JM, Kuypers MMM (2007) New processes and players in the nitrogen cycle: the microbial ecology of anaerobic and archaeal ammonia oxidation. Minireview. ISME J 1:19–27
Frasier R, Ullah S, Moore TR (2010) Nitrous oxide consumption potentials of well-drained forest soils in southern Quebec, Canada. Geomicrobiol J 27:53–60
Fry SC (2001) Plant cell walls. In: Encyclopedia of life sciences. Wiley Online Library, Chichester. www.els.net
Fuchigami LH, Weiser CJ, Kobayashi K, Timmis R, Gusta LV (1982) A degree growth stage (°GS) model and cold acclimation in temperate woody plants. In: Li PH, Sakai A (eds) Plant cold hardiness and freezing stress. Mechanisms and crop implications, vol 2. Academic, New York

Gaastra P (1959) Photosynthesis of crop plants as influenced by light, carbon dioxide, temperature and stomatal diffusion resistance. Mededelingen van de Landbouwhogeschool te Wageningen 59:1–68

Gadd GM (2007) Geomycology: biogeochemical transformations of rocks, minerals, metals and radionuclides by fungi, bioweathering and bioremediation. Mycol Res 111:3–49

Gamon JA, Peñuelas J, Field CB (1992) A narrow-waveband spectra lindex that tracks diurnal changes in photosynthetic efficiency. Remote Sens Environ 41:35–44

Gamon JA, Serrano L, Surfus JS (1997) The photochemical reflectance index: an optical indicator of photosynthetic radiation use efficiency across species, functional types, and nutrient levels. Oecologia 112:492–501

Garbulsky MF, Peñuelas J, Papale D, Filella I (2008) Remote estimation of carbon dioxide uptake by a Mediterranean forest. Glob Change Biol 14:2860–2867

Garrett PW, Zahner R (1973) Fascicle density and needle growth responses of red pine to water supply over two seasons. Ecology 54:1328–1334

Geßler A, Rienks M, Rennenberg H (2002) Stomatal uptake and cuticular adsorption contribute to dry deposition of NH_3 and NO_2 to needles of adult spruce (*Picea abies*) trees. New Phytol 156:179–194

Geiger D (2011) Plant sucrose transporters from a biophysical point of view. Mol Plant 4:395–406

Genty B, Briantais J-M, Baker NR (1989) The relationship between the quantum yield of photosynthetic electron transport and quenching of chlorophyll fluorescence. Biochim Biophys Acta 990:87–92

Ghashghaie J, Cornic G (1994) Effect of temperature on partitioning of photosynthetic electron flow between CO_2 assimilation and O_2 reduction and the CO_2/O_2 specificity of Rubisco. J Appl Physiol 143:643–650

Gielen B, Jach ME, Ceulemans R (2000) Effects of season, needle age and elevated atmospheric CO_2 on chlorophyll fluorescence parameters and needle nitrogen concentration in Scots pine (*Pinus sylvestris*). Photosynthetica 38:13–21

Givnish TJ (1985) On the use of optimality arguments. In: Givnish TJ (ed) On the economy of plant form and function. Cambridge University Press, Cambridge

Givnish TJ (1995) Plant stems: biomechanical adaptation for energy capture and influence on species distributions. In: Gartner BL (ed) Plant stems: physiology and functional morphology. Academic, San Diego

Glass ADM, Britto DT, Kaiser BN, Kinghorn JR, Kronzucker HJ, Kumar A, Okamoto M, Rawat S, Siddiqi MY, Unkles SE, Vidmar JJ (2002) The regulation of nitrate and ammonium transport systems in plants. J Exp Bot 53:855–864

Govindarajulu M, Pfeffer PE, Jin H, Abubaker J, Douds DD, Allen JW, Bückling H, Lammers PJ, Shachar-Hill Y (2005) Nitrogen transfer in the arbuscular mycorrhizal symbiosis. Nature 435:819–823

Greacen EL, Oh JS (1972) Physics of root growth. Nat New Biol 235:24–25

Greer DH (1983) Temperature regulation of the development of frost hardiness in *Pinus radiate* D. Don. Aust J Plant Physiol 10:539–547

Gregory PJ (2006) Roots, rhizosphere and soil: the route to a better understanding of soil science? Eur J Soil Sci 57:2–12

Grime JP (2002) Plant strategies, vegetation processes, and ecosystem properties, 2nd edn. Wiley, Chichester

Groffman P (1991) Ecology of nitrification and denitrification in soil evaluated at scales relevant to atmospheric chemistry. In: Rogers JE, Whitman WB (eds) Microbial production and consumption of greenhouse gases: methane, nitrogen oxides, and halomethanes. American Society for Microbiology, Washington, DC

Gross KL, Pregitzer KS, Burton AJ (1995) Spatial variation in nitrogen availability in three successional plant communities. J Ecol 83:357–367

Gunderson CA, Wullschleger SD (1994) Photosynthetic acclimation in trees to rising atmospheric CO_2: a broader perspective. Photosynth Res 39:369–388

Häkkinen R (1999) Analysis of bud-development theories based on long-term phenological and air temperature time series: application to *Betula* sp. leaves. Finnish Forest Research Institute, Research paper 754
Häkkinen R, Linkosalo T, Hari P (1995) Methods for combining phenological time series: application to bud burst in birch (*Betula pendula*) in Central Finland for the period 1896–1955. Tree Physiol 15:721–726
Häkkinen R, Linkosalo T, Hari P (1998) Effects of dormancy and environmental factors on timing of bud burst in *Betula pendula*. Tree Physiol 18:707–712
Hänninen H (1995) Effects of climatic change on trees from cool and temperate regions: an ecophysiological approach to modelling of bud burst phenology. Can J Bot 73:183–199
Hänninen H (2006) Climate warming and the risk of frost damage to boreal forest trees: identification of critical ecophysiological traits. Tree Physiol 26:889–898
Hänninen H, Kramer K (2007) A framework for modelling the annual cycle of trees in boreal and temperate regions. Silva Fenn 41:167–205
Haimi J, Laamanen J, Penttinen R, Räty M, Koponen S, Kellomäki S, Niemelä P (2005) Impacts of elevated CO_2 and temperature on the soil fauna of boreal forests. Appl Soil Ecol 30:104–112
Hallé F, Oldeman RAA, Tomlinson PB (1978) Tropical trees and forests: an architectural analysis. Springer, Berlin
Hänninen H (1990) Modelling bud dormancy release in trees from cool and temperate regions. Acta For Fenn 213:1–47
Hari P (1972) Physiological stage of development in biological models of growth and maturation. Ann Bot Fenn 9:107–115
Hari P, Häkkinen R (1991) The utilization of old phenological time series of budburst to compare models describing annual cycles of plants. Tree Physiol 8:281–287
Hari P, Mäkelä A (2003) Annual pattern of photosynthesis of Scots pine in the boreal zone. Tree Physiol 23:145–155
Hari P, Leikola M, Räsänen P (1970) A dynamic model of the daily high increment of plants. Ann Bot Fenn 7:375–378
Hari P, Heikinheimo P, Mäkelä A, Kaipiainen L, Korpilahti E, Salmela J (1986) Trees as a water transport system. Silva Fenn 20:205–210
Hari P, Raivonen M, Vesala T, Munger JW, Pilegaard K, Kulmala M (2003) Ultraviolet light and leaf emission of NO_x. Nature 422:134
Harju AM, Venäläinen M, Anttonen S, Viitanen H, Kainulainen P, Saranpää P, Vapaavuori E (2003) Chemical factors affecting the brown-rot decay resistance of Scots pine heartwood. Trees 17:263–268
Harper JL, Jones M, Sackville-Hamilton NR (1991) The evolution of roots and the problems of analysing their behaviour. In: Atkinson D (ed) Plant root growth. An ecological perspective. Blackwell, Oxford
Harris D, DeBolt S (2010) Synthesis, regulation and utilization of lignocellulosic biomass. Plant Biotechnol J 8:244–262
He J-Z, Shen J-P, Zhang L-M, Zhu Y-G, Zheng Y-M, Xu M-G, Di H (2007) Quantitative analyses of the abundance and composition of ammonia-oxidizing bacteria and ammonia-oxidizing archaea of a Chinese upland red soil under long-term fertilization practices. Environ Microbiol 9:2364–2374
Helmisaari H (1992) Nutrient retranslocation within the foliage of *Pinus sylvestris*. Tree Physiol 10:45–58
Hendrickson L, Furbank RT, Chow WS (2004) A simple alternative approach to assessing the fate of absorbed light energy using chlorophyll fluorescence. Photosynth Res 82:73–81
Hikosaka K, Hirose T (1998) Leaf and canopy photosynthesis of C_3 plants at elevated CO_2 in relation to optimal partitioning of nitrogen among photosynthetic components: theoretical prediction. Ecol Model 106:247–259
Hilker T, Coops NC, Wulder MA, Black TA, Guy RD (2008a) The use of remote sensing in light use efficiency based models of gross primary production: a review of current status and future requirements. Sci Total Environ 404:411–423

Hilker T, Coops NC, Hall FG, Black TA, Wulder MA, Nesic Z, Krishnan P (2008b) Separating physiologically and directionally induced changes in PRI using BRDF models. Remote Sens Environ 112:2777–2788
Hill CA (1971) Vegetation: a sink for atmospheric pollutants. J Air Pollut Control Assoc 21:341–346
Hillel D (1998) Introduction to environmental soil physics. Academic, San Diego
Hinsinger P, Gobran GR, Gregory PJ, Wenzel WW (2005) Rhizosphere geometry and heterogeneity arising from root mediated physical and chemical processes. New Phytol 168:293–303
Hinsinger P, Bengough AG, Vetterlein D, Young IM (2009) Rhizosphere: biophysics, biogeochemistry and ecological relevance. Plant Soil 321:117–152
Hishi T (2007) Heterogeneity of individual root within the fine root architecture: causal links between physiological and ecosystem functions. J For Res 12:126–133
Hodge A (2004) The plastic plant: root responses to heterogeneous supplies of nutrients. New Phytol 162:9–24
Hoefnagel MHN, Atkin OK, Wiskich JT (1998) Interdependence between chloroplasts and mitochondria in the light and the dark. BBA-Bioenergetics 1366:235–255
Hölttä T, Vesala T, Sevanto S, Perämäki M, Nikinmaa E (2006) Modeling xylem and phloem water flows in trees according to cohesion theory and Münch hypothesis. Trees Struct Funct 20:67–78
Hölttä T, Mencuccini M, Nikinmaa E (2009) Linking phloem function to structure: analysis with a coupled xylem–phloem transport model. J Theor Biol 259:325–337
Holmes MG, Keiller DR (2002) Effects of pubescence and waxes on the reflectance of leaves in the ultraviolet and photosynthetic wavebands: a comparison of a range of species. Plant Cell Environ 25:85–93
Horn HS (1971) The adaptive geometry of trees. Princeton University Press, Princeton
Horton P, Ruban A (2005) Molecular design of the photosystem II light-harvesting antenna: photosynthesis and photoprotection. J Exp Bot 56:365–373
Hose E, Clarkson DT, Steudle E, Schreiber L, Hartung W (2001) The exodermis: a variable apoplastic barrier. J Exp Bot 52:2245–2264
Houghton JT, Ding Y, Griggs DJ, Noguer M, van der Linden PJ, Xiaosu D (eds) (2001) Climate change 2001. The scientific basis. Contribution of working group I to the third assessment report of the Intergovernmental Panel on Climate Change (IPCC). Cambridge University Press, Cambridge
Hsiao TC, Xu L-K (2000) Sensitivity of growth of roots versus leaves to water stress: biophysical analysis and relation to water transport. J Exp Bot 51:1595–1616
Huhta V, Hyvönen R, Kaasalainen P, Koskenniemi A, Muona J, Mäkelä I, Sulander M, Vilkamaa P (1986) Soil fauna of Finnish coniferous forests. Ann Zool Fenn 23:345–360
Huhta V, Persson T, Setälä H (1998) Functional implications of soil fauna diversity in boreal forests. Appl Soil Ecol 10:277–288
Huner NPA, Öquist G (2003) Photostasis in plants, green algae and cyanobacteria: the role of light harvesting antenna complexes. In: Green BR, Parson WW (eds) Light-harvesting antennas in photosynthesis. Kluwer, Dordrecht
Ilomäki S, Mäkelä A, Nikinmaa E (2003) Crown rise due to competition drives biomass allocation in silver birch (*Betula pendula* L.). Can J For Res 33:2395–2404
Ilvesniemi H, Liu C (2001) Biomass distribution in a young Scots pine stand. Boreal Environment Research 6:3–8
Isayenkov S, Isner JC, Maathuis FJM (2010) Vacuolar ion channels: roles in plant nutrition and signalling. FEBS Lett 584:1982–1988
Jach ME, Ceulemans R (2000) Effects of season, needle age and elevated atmospheric CO_2 on photosynthesis in Scots pine (*Pinus sylvestris*). Tree Physiol 20:145–157
Jackson RB, Caldwell MM (1993) Geostatistical patterns of soil heterogeneity around individual perennial plants. J Ecol 81:683–692
Jarvis AJ, Mansfield TA, Davies WJ (1999) Stomatal behaviour, photosynthesis and transpiration under rising CO_2. Plant Cell Environ 22:639–648

Johnson DA, Richards RA, Turner NC (1983) Yield, water and related characteristics in dryland environments, and its relations, gas exchange, and surface reflectance or near-isogenic wheat lines differing in glaucousness. Crop Sci 23:318–325

Jones DL, Healey JR, Willett VB, Farrar JF, Hodge A (2005) Dissolved organic nitrogen uptake by plants – an important N uptake pathway? Soil Biol Biochem 37:413–423

Jongsma M, Bolter C (1997) The adaptation of insects to plant protease inhibitors. J Insect Physiol 43:885–895

Jordan DB, Ogren WL (1984) The CO_2/O_2 specificity of ribulose 1,5-bisphosphate carboxylase/oxygenase. Dependence on ribulose bisphosphate concentration, pH and temperature. Planta 161:308–313

Jurgens G, Lindstrom K, Saano A (1997) Novel group within the kingdom Crenarchaeota from boreal forest soil. Appl Environ Microbiol 63:803–805

Juurola E (2003) Biochemical acclimation patterns of *Betula pendula* and *Pinus sylvestris* seedling to elevated carbon dioxide concentration. Tree Physiol 23:85–95

Juurola E (2005) Photosynthesis, CO_2 and temperature – an approach to analyse the constraints to acclimation of trees to increasing CO_2 concentration. Dissertation, University of Helsinki

Kaiser E-A, Kohrs K, Kücke M, Schnug E, Heinemeyer O, Munch JC (1998) Nitrous oxide release from arable soil: importance of N-fertilization, crops and season. Soil Biol Biochem 30:1553–1563

Kanninen M, Hari P, Kellomäki S (1982) A dynamic model for above ground growth of dry matter production in a forest community. J Appl Ecol 19:465–476

Keeling CD, Whorf TP (2005) Atmospheric CO_2 records from sites in the SIO air sampling network. In: Trends: a compendium of data on global change. Carbon Dioxide Information Analysis Center, Oak Ridge National Laboratory, Oak Ridge

Kellomäki S, Oker-Blom P (1983) Canopy structure and light climate in a young Scots Pine stand. Silva Fenn 17:1–21

Kesselmeier J, Staudt M (1999) Biogenic Volatile Organic Compounds (VOC): an overview on emission, physiology and ecology. J Atmos Chem 33:23–88

Killham K (1990) Nitrification in coniferous forest soils. Plant Soil 128:31–44

Killham K (1994) Soil ecology. Cambridge University Press, Cambridge

Kimball BA, Idso SB, Johnson S, Rillig MC (2007) Seventeen years of carbon dioxide enrichment of sour orange trees: final results. Glob Change Biol 13:2171–2183

Knowles R (1982) Denitrification. Microbiol Rev 46:43–70

Kögel-Knabner I (2000) Analytical approaches for characterizing soil organic matter. Org Geochem 31:609–625

Kögel-Knabner I (2002) The macromolecular organic composition of plant and microbial residues as inputs to soil organic matter. Soil Biol Biochem 34:139–162

Körner C (2006) Plant CO_2 responses: an issue of definition, time and resource supply. New Phytol 172:393–411

Kolari P, Lappalainen HK, Hänninen H, Hari P (2007) Relationship between temperature and the seasonal course of photosynthesis in Scots pine at northern timberline and in southern boreal zone. Tellus 59B:542–552

Kollist H, Jossier M, Laanemets K, Thomine S (2011) Anion channels in plant cells. FEBS J 278:4277–4292

Koski V, Sievänen R (1985) Timing of growth cessation in relation to the variations in the growing season. In: Tigerstedt PMA, Puttonen P, Koski V (eds) Crop physiology of forest trees. Helsinki University Press, Helsinki

Kowalchuk GA, Stephen JR (2001) Ammonia-oxidizing bacteria: a model for molecular microbial ecology. Annu Rev Microbiol 55:485–529

Kozaki A, Takeba G (1996) Photorespiration protects C3 plants from photooxidation. Nature 384:557–560

Kramer K (1994a) Selecting a model to predict the onset of growth of *Fagus sylvatica*. J Appl Ecol 31:172–181

Kramer K (1994b) A modelling analysis of the effects of climatic warming on the probability of spring frost damage to tree species in The Netherlands and Germany. Plant Cell Environ 17:367–377
Kramer DM, Avenson TJ, Edwards GE (2004a) Dynamic flexibility in the light reactions of photosynthesis governed by both electron and proton transfer reactions. Trends Plant Sci 9:349–357
Kramer DM, Johnson G, Kiirats O, Edwards GE (2004b) New fluorescence parameters for the determination of Q_A redox state and excitation energy fluxes. Photosynth Res 79:209–218
Krause GH, Weis E (1991) Chlorophyll fluorescence and photosynthesis: the basics. Annu Rev Plant Physiol 42:313–349
Kronzucker HJ, Siddiqi MY, Glass ADM (1996) Kinetics of NH_4^+ influx in spruce. Plant Physiol 110:773–779
Kühn C, Grof CPL (2010) Sucrose transporters of higher plants. Curr Opin Plant Biol 13:288–298
Küppers M (1989) Ecological significance of above ground patterns in woody plants; a question of cost-benefit relationships. Tree 4:375–379
Kuuluvainen T (2002) Natural variability of forests as a reference for restoring and managing biological diversity in boreal Fennoscandia. Silva Fenn 36:97–125
Kuuluvainen T, Pukkala T (1987) Effect of crown shape and tree distribution on the spatial distribution of shade. Agric For Meteorol 40:215–231
Kuuluvainen T, Pukkala T (1989) Simulation of within-tree and between-tree shading of direct radiation in a forest canopy: effect of crown shape and sun elevation. Ecol Model 49:89–100
Laakso J, Setälä H (1999) Sensitivity of primary production to changes in the architecture of belowground food webs. Oikos 87:57–64
Lam P, Jensen MM, Lavik G, McGinnis DF, Müller B, Schubert CJ, Amann R, Thamdrup B, Kuypers MMM (2007) Linking crenarchaeal and bacterial nitrification to anammox in the Black Sea. Proc Natl Acad Sci USA 104:7104–7109
Landsberg JJ (1974) Apple fruit bud development and growth; analysis and an empirical model. Ann Bot (Lond) 38:1013–1023
Lawlor DW, Delgado E, Habash DZ, Driscoll SP, Mitchell VJ, Mitchell RAC, Parry MAJ (1995) Photosynthetic acclimation of winter wheat to elevated CO_2 and temperature. In: Mathis P (ed) Photosynthesis: from light to biosphere, vol 5. Kluwer, Dordrecht
Lee KE, Foster RC (1991) Soil fauna and soil structure. Aust J Soil Res 29:745–775
Lehto T, Zwiazek JJ (2011) Ectomycorrhizas and water relations of trees: a review. Mycorrhiza 21:71–90
Leininger S, Urich T, Schloter M, Schwark L, Qi J, Nicol GW, Prosser JI, Schuster SC, Schleper C (2006) Archaea predominate among ammonia-oxidizing prokaryotes in soils. Nature 442:806–809
Leinonen I (1996) A simulation model for the annual frost hardiness and freeze damage of Scots pine. Ann Bot (Lond) 78:687–693
Lerdau MT, Munger JW, Jacob DJ (2000) The NO_2 flux conundrum. Science 298:2291–2293
Leuning R (1995) A critical appraisal of a combined stomatal-photosynthesis model for C3 plants. Plant Cell Environ 18:339–355
Leverenz JW, Öquist G (1987) Quantum yields of photosynthesis at temperatures between –2°C and 35°C in cold-tolerant C3 plant (*Pinus sylvestris*) during the course of one year. Plant Cell Environ 10:287–295
Lew RR (2011) How does a hypha grow? The biophysics of pressurized growth in fungi. Nat Rev Microbiol 9:509–518
Lewis JD, Tissue DT, Strain BD (1996) Seasonal response of photosynthesis to elevated CO_2 in loblolly pine (*Pinus taeda* L.) over two growing seasons. Glob Change Biol 2:103–114
Liberloo M, Tulva I, Raïm O, Kull O, Ceulemans R (2007) Photosynthetic stimulation under long-term CO_2 enrichment and fertilization is sustained across a closed *Populus* canopy profile (EUROFASE). New Phytol 173:537–549
Liebhold A, Elkinton J, Williams D, Muzika RM (2000) What causes outbreaks of the gypsy moth in North America? Popul Ecol 42:257–266

Liesche J, Martens HJ, Schulz A (2011) Symplasmic transport and phloem loading in gymnosperm leaves. Protoplasma 248:181–190
Lin J, Jach ME, Ceulemans R (2001) Stomatal density and needle anatomy of Scots pine (*Pinus sylvestris*) are affected by elevated CO_2. New Phytol 150:665–674
Linder S, Troeng E (1980) Photosynthesis and transpiration of 20-year-old Scots pine. In Persson T (ed) Structure and functioning of northern coniferous forests – an ecosystem study. Ecol Bull 32:165–181. Stockholm
Linkosalo T, Lappalainen HK, Hari P (2008) A comparison of phenological models of leaf bud burst and flowering of boreal trees using independent observations. Tree Physiol 28:1873–1882
Lloyd J, Shibistova O, Zolotoukhine D, Kolle O, Arneth A, Wirth C, Styles JM, Tchebakova NM, Schulze ED (2002) Seasonal and annual variations in the photosynthetic productivity and carbon balance of a central Siberian pine forest. Tellus 54B:590–610
Long SP (1991) Modification of the response of photosynthetic productivity to rising temperature by atmospheric CO_2 concentration: has its importance been underestimated? Plant Cell Environ 14:729–739
Longeutaud F, Mothe F, Leban J-M, Mäkelä A (2006) *Picea abies* sapwood width: variations within and between trees. Scand J For Res 21:41–53
Loque D, von Wirén N (2004) Regulatory levels for the transport of ammonium in plant roots. J Exp Bot 55:1293–1305
Lundmark T, Bergh J, Strand M, Koppel A (1998) Seasonal variation of maximum photochemical efficiency in boreal Norway spruce stands. Trees 13:63–67
Luo Y, Sims DA, Griffin KL (1998) Nonlinearity of photosynthetic responses to growth in rising atmospheric CO_2: an experimental and modelling study. Glob Change Biol 4:173–183
Luomala E-M, Laitinen K, Sutinen S, Kellomäki S, Vapaavuori E (2005) Stomatal density, anatomy and nutrient concentrations of Scots pine needles are affected by elevated CO_2 and temperature. Plant Cell Environ 28:733–749
Luu D-T, Maurel C (2005) Aquaporins in a challenging environment: molecular gears for adjusting plant water status. Plant Cell Environ 28:85–96
Lynch J (1995) Root architecture and plant productivity. Plant Physiol 109:7–13
MacCurdy E (ed) (2002) The notebooks of Leonardo DaVinci, definitive edition in one volume. Konecky & Konecky, Old Saybrook
Mäkelä A, Hari P, Berninger F, Hänninen H, Nikinmaa E (2004) Acclimation of photosynthetic capacity in Scots pine to the annual cycle of temperature. Tree Physiol 24:369–376
Manninen S, Huttunen S, Perämäki P (1997) Needle S fractions and S to N ratios as indices of SO_2. Water Air Soil Pollut 95:277–298
Markham JH (2009) Variation in moss-associated nitrogen fixation in boreal forest stands. Oecologia 161:353–359
Martikainen PJ (1984) Nitrification in two coniferous forest soils after different fertilization treatments. Soil Biol Biochem 16:577–582
Martikainen PJ (1985) Nitrous oxide emission associated with autotrophic ammonium oxidation in acid coniferous forest soil. Appl Environ Microbiol 50:1519–1525
Martikainen PJ, Lehtonen M, Lång K, De Boer W, Ferm A (1993) Nitrification and nitrous oxide production potentials in aerobic soil samples from the soil profile of a Finnish coniferous site receiving high ammonium deposition. FEMS Microbiol Ecol 13:113–122
Martin F, Kohler A, Duplessis S (2007) Living in harmony in the wood underground: ectomycorrhizal genomics. Curr Opin Plant Biol 10:204–210
Martino E, Perotto S (2010) Mineral transformations by mycorrhizal fungi. Geomicrobiol J 27:609–623
Maurel C (1997) Aquaporins and water permeability of plant membranes. Annu Rev Plant Physiol 48:399–429
Mazhitova G (2006) Soils of the boreal forest. In: Encyclopedia of soil science. Taylor & Francis, New York
McCarty GW (1999) Modes of action of nitrification inhibitors. Biol Fertil Soils 29:1–9

McTiernan KB, Coûteaux M-M, Berg B, Berg MP, de Anta RC, Gallardo A, Kratz W, Piussi P, Remacle J, De Santo AV (2003) Changes in chemical composition of *Pinus sylvestris* needle litter during decomposition along a European coniferous forest climatic transect. Soil Biol Biochem 35:801–812

Mencuccini M, Grace J, Fiovaranti M (1997) Biomechanical and hydraulical determinants of tree structure in Scots pine: anatomical characteristics. Tree Physiol 17:105–113

Mengel K, Kirkby EA (2001) Principles of plant nutrition, 5th edn. Kluwer, Dordrecht

Meroni M, Busetto L, Colombo R, Guanter L, Moreno J, Verhoef W (2010) Performance of spectral fitting methods for vegetation fluorescence quantification. Remote Sens Environ 114:363–374

Miller AJ, Cramer MD (2004) Root nitrogen acquisition and assimilation. Plant Soil 274:1–36

Morgan PB, Bollero GA, Nelson RL, Dohleman FG, Long SP (2005) Smaller than predicted increase in aboveground net primary production and yield of field-grown soybean under fully open-air [CO_2] elevation. Glob Change Biol 11:1856–1865

Morrell JJ, Gartner BL (1998) Wood as a material. In: Bruce A, Palfreyman JW (eds) Forest products biotechnology. Taylor & Francis, London

Moya I, Camenen L, Evain S, Goulas Y, Cerovic ZG, Latouche G, Flexas J, Ounis A (2004) A new instrument for passive remote sensing 1. Measurements of sunlight-induced chlorophyll fluorescence. Remote Sens Environ 91:186–197

Müller P, Li X-P, Niyogi KK (2001) Non-photochemical quenching. A response to excess light energy. Plant Physiol 125:1558–1566

Müller M, Varama M, Heinonen J, Hallaksela A-M (2002) Influence of insects on the diversity of fungi in decaying spruce wood in managed and natural forests. For Ecol Manage 166:165–181

Müller T, Avolio M, Olivi M, Benjdia M, Rikirsch E, Kasaras A, Fitz M, Chalot M, Wipf D (2007) Nitrogen transport in the ectomycorrhiza association: The Hebeloma cylindrosporum-Pinus pinaster model. Phytochem 68:41–51

Mulroy TW (1979) Spectral properties of heavily glaucous and non-glaucous leaves of a succulent rosette-plant. Oecologia 38:349–357

Näsholm T, Kielland K, Ganeteg U (2009) Uptake of organic nitrogen by plants. New Phytol 182:31–48

Nehls U, Mikolajewski S, Magel E, Hampp R (2001) Carbohydrate metabolism in ectomycorrhizas: gene expression, monosaccharide transport and metabolic control. New Phytol 150:533–541

Nehls U, Göhringer F, Wittulsky S, Dietz S (2010) Fungal carbohydrate support in the ectomycorrhizal symbiosis: a review. Plant Biol 12:292–301

Nelson PN, Baldock JA (2005) Estimating the molecular composition of a diverse range of natural organic materials from solid-state 13C NMR and elemental analysis. Biogeochemistry 72:1–34

Nelson DW, Sommers LE (1996) Total carbon, organic carbon, and organic matter. In: Bigham JM (ed) Methods of soil analysis. Part 3. Chemical methods, vol 5, SSSA book series. Soil Science Society of America, American Society of Agronomy, Madison

Nichol CJ, Lloyd J, Shibistova O, Arneth A, Roser C, Knohl A, Matsubara S, Grace J (2002) Remote sensing of photosynthetic-light-use efficiency of a Siberian boreal forest. Tellus 54B:677–687

Nicol GW, Schleper C (2006) Ammonia-oxidising Crenarchaeota: important players in the nitrogen cycle? Trends Microbiol 14:207–212

Niyogi KK (1999) Photoprotection revisited: genetic and molecular approaches. Annu Rev Plant Physiol 50:333–359

Nobel PS (1991) Physicochemical and environmental plant physiology. Academic, San Diego

Noe SM, Giersch C (2004) A simple dynamic model of photosynthesis in oak leaves: coupling leaf conductance and photosynthetic carbon fixation by a variable intracellular CO_2 pool. Funct Plant Biol 31:1195–1204

Oades JM (1993) The role of biology in the formation, stabilization and degradation of soil structure. Geoderma 56:377–400

Oaks A (1994) Efficiency of nitrogen utilization in C3 and C4 cereals. Plant Physiol 106:407–414

Odum HT (1983) Systems Ecology: An Introduction. John Wiley, New York
Oker-Blom P, Smolander H (1988) The ratio of shoot silhouette area to total needle area in Scots pine. For Sci 34:894–906
Olascoaga et al. (manuscript in preparation)
Öquist G, Huner NPA (2003) Photosynthesis of overwintering evergreen plants. Annu Rev Plant Biol 54:329–355
Oppert C, Klingeman WE, Willis JD, Oppert B, Jurat-Fuentes JL (2010) Prospecting for cellulolytic activity in insect digestive fluids. Comp Biochem Phys B 155:145–154
Paavolainen L, Fox M, Smolander A (2000) Nitrification and denitrification in forest soil subjected to sprinkling infiltration. Soil Biol Biochem 32:669–678
Parrent JL, Vilgalys R (2009) Expression of genes involved in symbiotic carbon and nitrogen transport in *Pinus taeda* mycorrhizal roots exposed to CO_2 enrichment and nitrogen fertilization. Mycorrhiza 19:469–479
Parson WW, Nagarajan V (2003) Optical spectroscopy in photosynthetic antennas. In: Green BR, Parson WW (eds) Light-harvesting antennas in photosynthesis. Kluwer, Dordrecht
Patrick JW (1997) Phloem unloading: sieve element unloading and post-sieve element transport. Annu Rev Plant Phys 48:191–222
Pauchet Y, Muck A, Svatos A, Heckel DG, Preiss S (2008) Mapping the larval midgut lumen proteome of *Helicoverpa armigera*, a generalist herbivorous insect. J Proteotome Res 7:1629–1639
Paul EA, Clark FE (1989) Soil microbiology and biochemistry. Academic, San Diego
Pelkonen P (1980) The uptake of carbon dioxide in Scots pine during spring. Flora 169:386–397
Pelkonen P, Hari P (1980) The dependence of the springtime recovery of CO_2 uptake in Scots pine on temperature and internal factors. Flora 169:398–404
Peñuelas J, Filella I (1998) Visible and near-infrared reflectance techniques for diagnosing plant physiological status. Trends Plant Sci 3:151–156
Pensa M, Sellin A (2002) Needle longevity of Scots pine in relation to foliar nitrogen content, specific leaf area, and shoot growth in different forest types. Can J For Res 32:1225–1231
Persson T (1989) Role of soil animals in C and N mineralisation. Plant Soil 115:241–245
Persson T, Bååth E, Clarholm M, Lundkvist H, Söderström BE, Sohlenius B (1980) Trophic structure, biomass dynamics and carbon metabolism of soil organisms in a Scots pine forest. Ecol Bull 32:419–459
Persson J, Gardeström P, Näsholm T (2006) Uptake, metabolism and distribution of organic and inorganic nitrogen sources by *Pinus sylvestris*. J Exp Bot 57:2651–2659
Peterson CA, Enstone DE, Taylor JH (1999) Pine root structure and its potential significance for root functions. Plant Soil 217:205–213
Peterson MG, Dietterich HR, Lachenbruch B (2007) Do Douglas-fir branches and roots have juvenile wood? Wood Fiber Sci 39:651–660
Pettersson R, McDonald AJS, Stadenberg I (1993) Response of small birch plants (*Betula pendula* Roth.) to elevated CO_2 and nitrogen supply. Plant Cell Environ 16:1115–1121
Piccolo A (2001) The supramolecular structure of humic substances. Soil Sci 166:810–832
Pichersky E, Gershenzon J (2002) The formation and function of plant volatiles: perfumes for pollinator attraction and defense. Curr Opin Plant Biol 5:237–243.
Pietarinen I, Kanninen M, Hari P, Kellomäki S (1982) A simulation model for daily growth of shoots, needles and stem diameter in Scots pine trees. For Sci 28:573–581
Pinelli P, Loreto F (2003) (CO_2)–C12 emission from different metabolic pathways measured in illuminated and darkened C3 and C4 leaves at low, atmospheric and elevated CO_2 concentration. J Exp Bot 54:1761–1769
Plassard C, Dell B (2010) Phosphorus nutrition of mycorrhizal trees. Tree Physiol 30:1129–1139
Plotkin JB (2011) The lives of proteins. Science 331:683
Porcar-Castell A (2011) A high-resolution portrait of the annual dynamics of photochemical and non-photochemical quenching in needles of *Pinus sylvestris*. Physiol Plantarum 143:139–153
Porcar-Castell A, Juurola E, Ensminger I, Berninger F, Hari P, Nikinmaa E (2008a) Seasonal acclimation of photosystem II in *Pinus sylvestris*. II. Studying the effect of light environment through the rate constants of sustained thermal dissipation and photochemistry. Tree Physiol 28:1483–1491

Porcar-Castell A, Juurola E, Nikinmaa E, Berninger F, Ensminger I, Hari P (2008b) Seasonal acclimation of photosystem II in *Pinus sylvestris*. I. Estimating the rate constants of sustained thermal energy dissipation and photochemistry. Tree Physiol 28:1475–1482

Porcar-Castell A, Juurola E, Ensminger I, Berninger F, Hari P, Nikinmaa E (2008c) Seasonal acclimation of photosystem II in *Pinus sylvestris*. II. Using the rate constants of sustained thermal energy dissipation and photochemistry to study the effect of the light environment. Tree Physiol 28:1483–1491

Porcar-Castell A, García-Plazaola JI, Nichol C, Kolari P, Olascoaga B, Kuusinen N, Fernández-Marín B, Pulkkinen M, Juurola E, Nikinmaa E (2012). Physiology of the seasonal relationship between photochemical reflectance index and photosynthetic light use efficiency. Oecologia 170:313–323

Poth M, Focht DD (1985) ^{15}N kinetic analysis of N_2O production by *Nitrosomonas europaea*: an examination of nitrifier denitrification. Appl Environ Microbiol 49:1134–1141

Priha O, Smolander A (1999) Nitrogen transformations in soil under *Pinus sylvestris, Picea abies* and *Betula pendula* at two forest sites. Soil Biol Biochem 31:965–977

Priha O, Grayston SJ, Pennanen T, Smolander A (1999) Microbial activity related to C and N cycling and microbial community structure in the rhizospheres of *Pinus sylvestris*, *Picea abies* and *Betula pendula* seedlings in an organic and mineral soil. FEMS Microbiol Ecol 30:187–199

Pritchard SG, Peterson CM, Prior SA, Rogers HH (1997) Elevated atmospheric CO_2 differentially affects needle chloroplast ultrastructure and phloem anatomy in *Pinus palustris*: interaction with soil resource availability. Plant Cell Environ 20:461–471

Pritchard SG, Rogers HH, Prior SA, Peterson CM (1999) Elevated CO_2 and plant structure: a review. Glob Change Biol 5:807–837

Pritsch K, Garbaye J (2011) Enzyme secretion by ECM fungi and exploitation of mineral nutrients from soil organic matter. Ann For Sci 68:25–32

Prusinkiewicz P, Lindenmayer A (1990) The algorithmic beauty of plants. Springer, Berlin

Puhlmann H, von Wilpert K (2012) Pedotransfer functions for water retention and unsaturated hydraulic conductivity of forest soils. J Plant Nutr Soil Sci 175:221–235

Raghothama KG, Karthikeyan AS (2005) Phosphate acquisition. Plant Soil 274:37–49

Rahman AF, Cordova VD, Gamon JA et al (2004) Potential of MODIS ocean bands for estimating CO_2 flux from terrestrial vegetation: a novel approach. Geophys Res Lett 31:1–4

Raivonen M, Keronen P, Vesala T, Kulmala M, Hari P (2003) Measuring shoot-level NO_x flux in field conditions: the role of blank chambers. Boreal Environ Res 8:445–455

Raivonen M, Bonn B, Sanz MJ, Vesala T, Kulmala M, Hari P (2006) UV-induced NO_y emissions from Scots pine: could they originate from photolysis of deposited HNO_3? Atmos Environ 40:6201–6213

Rajasekar R (2007) Quantification of protein nitrogen in boreal forest soils. Pro gradu, University of Helsinki

Rautio P, Huttunen S, Lamppu J (1998) Element concentrations in Scots pine needles on radial transects across a subarctic area. Water Air Soil Poll 102:389–405

Rennie EA, Turgeon R (2009) A comprehensive picture of phloem loading strategies. Proc Natl Acad Sci USA 106:14162–14167. www.pnas.org_cgi_doi_10.1073_pnas.0902279106

Repo T, Pelkonen P (1986) Temperature step response of hardening in *Pinus sylvestris* seedlings. Scand J For Res 1:271–284

Repo T, Leinonen I, Wang K-Y, Hänninen H (2006) Relation between photosynthetic capacity and cold hardiness in Scots pine. Physiol Plantarum 126:224–231

Rey A, Jarvis PG (1998) Long-term photosynthetic acclimation to increased atmospheric CO_2 concentration in young birch trees (*Betula pendula*). Tree Physiol 18:441–450

Richardson EA, Seeley SD, Walker DR (1974) A model for estimating the completion of rest for 'Redhaven' and 'Elberta' peach trees. HortScience 9:331–332

Riikonen J, Holopainen T, Oksanen E, Vapaavuori E (2005) Leaf photosynthetic characteristics of silver birch during three years of exposure to elevated concentrations of CO_2 and O_3 in the field. Tree Physiol 25:549–560

Ritchie RJ (2006) Estimation of cytoplasmic nitrate and its electrochemical potential in barley roots using $^{13}NO_3^-$ and compartmental analysis. New Phytol 171:643–655
Roberntz P, Stockfors J (1998) Effects of elevated CO_2 concentration and nutrition on net photosynthesis, stomatal conductance and needle respiration of field-grown Norway spruce trees. Tree Physiol 18:233–241
Rondón A, Granat L (1994) Studies on the dry deposition of NO_2 to coniferous species at low NO_2 concentrations. Tellus 46B:339–352
Rondón A, Johansson C, Granat L (1993) Dry deposition of nitrogen dioxide and ozone to coniferous forests. J Geophys Res 98:5159–5172
Room PM, Maillette L, Hanan JS (1994) Module and metamer dynamics and virtual plants. Adv Ecol Res 25:105–157
Sakai A, Larcher W (1987) Frost survival of plants. Responses and adaptation to freezing stress. Springer, Berlin
Salminen R (ed) (2005) Geochemical atlas of Europe. Part 1: Background information, methodology and maps. Geological Survey of Finland, Espoo
Sanborn P, Preston C, Brockley R (2002) N_2-fixation by Sitka alder in a young lodgepole pine stand in central interior British Columbia, Canada. For Ecol Manage 167:223–231
Sarvas R (1972) Investigations on the annual cycle of development of forest trees. Active period. Commun Inst For Fenn 76:1–110
Sarvas R (1974) Investigations on the annual cycle of development of forest trees. II. Autumn dormancy and winter dormancy. Commun Inst For Fenn 84:1–101
Satoh S (2006) Organic substances in xylem sap delivered to above-ground organs by the roots. J Plant Res 119:179–187
Sauer N (2007) Molecular physiology of higher plant sucrose transporters. FEBS Lett 581:2309–2317
Scarascia-Mugnozza G, De Angelis P, Matteucci G, Valentini R (1996) Long-term exposure to elevated [CO_2] in a natural *Quercus ilex L.* community: net photosynthesis and photochemical efficiency of PSII at different levels of water stress. Plant Cell Environ 19:43–654
Schimel JP, Bennett J (2004) Nitrogen mineralization: challenges of a changing paradigm. Ecology 85:591–602
Schindlbacher A, Zechmeister-Bolternstern S, Butterbach-Bahl K (2004) Effects of soil moisture and temperature on NO, NO_2, and N_2O emissions from European forest ecosystems. J Geophys Res 109:D17302
Schmid MC, Risgaard-Petersen N, van de Vossenberg J, Kuypers MMM, Lavik G, Petersen J, Hulth S, Thamdrup B, Canfield D, Dalsgaard T, Rysgaard S, Sejr MK, Strous M, Op den Camp HJM, Jetten MSM (2007) Anaerobic ammonium-oxidizing bacteria in marine environments: widespread occurrence but low diversity. Environ Microbiol 9:2364–2374
Schmidt I, Sliekers O, Schmid M, Cirpus I, Strous M, Bock E, Kuenen JG, Jetten MSM (2002) Aerobic and anaerobic ammonia oxidizing bacteria competitors or natural partners? – minireview. FEMS Microbiol Ecol 39:175–181
Schnitzer M (2000) A lifetime perspective on the chemistry of soil organic matter. Adv Agron 68:1–58
Schowalter TD (1995) Canopy arthropod communities in relation to forest age and alternative harvest practices in western Oregon. For Ecol Manage 78:115–125
Schowalter TD, Hargrove WW, Crossley DA Jr (1986) Herbivory in forested ecosystems. Annu Rev Entomol 31:177–196
Schreiber U, Schliwa U, Bilger W (1986) Continuous recording of photochemical and non-photochemical chlorophyll fluorescence quenching with a new type of modulation fluorometer. Photosynth Res 10:51–62
Schüßler A, Martin H, Cohen D, Fitz M, Wipf D (2006) Characterization of a carbohydrate transporter from symbiotic glomeromycotan fungi. Nature 444:933–936
Schulz B (2011) Functional classification of plant plasma membrane transporters. In: Murphy AS, Peer W, Schulz B (eds) The plant plasma membrane, vol 19, 1st edn, Plant cell monographs. Springer, Berlin

Scurlock JMO, Asner GP, Gower ST (2001) Worldwide historical estimates and bibliography of Leaf Area Index, 1932–2000. ORNL Technical Memorandum TM-2001/268, Oak Ridge National Laboratory, Oak Ridge

Segschneider H-J, Wildt J, Förstel H (1995) Uptake of $^{15}NO_2$ by sunflower (*Helianthus annuus*) during exposures in light and darkness: quantities, relationship to stomatal aperture and incorporation into different nitrogen pools within the plant. New Phytol 131:109–119

Seinfeld JH, Pandis SN (1998) Atmospheric chemistry and physics. Wiley, New York

Setälä H, Huhta V (1991) Soil fauna increase *Betula pendula* growth: laboratory experiments with coniferous forest floor. Ecology 72:665–671

Setälä H, Laakso J, Mikola J, Huhta V (1998) Functional diversity of decomposer organisms in relation to primary production. Appl Soil Ecol 9:25–31

Sharkey T (1988) Estimating the rate of photorespiration in leaves. Physiol Plantarum 73:147–152

Shinozaki K, Yoda K, Hozumi K, Kira T (1964a) A quantitative analysis of plant form – the Pipe Model theory. I. Basic analyses. Jpn J Ecol 14:97–105

Shinozaki K, Yoda K, Hozumi K, Kira T (1964b) A quantitative analysis of plant form – the Pipe Model theory. II. Further evidence of the theory and its application in forest ecology. Jpn J Ecol 14:133–139

Sievänen R, Lindner M, Mäkelä A, Lash P (2000) Volume growth and survival graphs: a method for evaluating process-based models. Tree Physiol 20:357–365

Silveira ML, Comerford NB, Reddy KR, Cooper WT, El-Rifai H (2008) Characterization of soil organic carbon pools by acid hydrolysis. Geoderma 144:405–414

Simek M (2000) Nitrification in soil – terminology and methodology. Rost Vyroba 46:385–395

Simojoki A (2001) Oxygen supply to plant roots in cultivated mineral soils. Dissertation, University of Helsinki

Simojoki A, Garcia H, Pihlatie M, Pumpanen J, Kurola J, Salkinoja-Salonen M, Hari P (2008) Environmental factors in soil. In: Hari P, Kulmala L (eds) Boreal forest and climate change, vol 34, Advances in global change research. Springer, Dordrecht

Singh U, Uehara G (1999) Electrochemistry of the double layer: principles and applications to soils. In: Sparks DL (ed) Soil physical chemistry, 2nd edn. CRC Press, Boca Raton

Sinsabaugh RL, Carreiro MM, Alvarez S (2002) Enzyme and microbiological dynamics of litter composition. In: Burns RG, Dick RP (eds) Enzymes in the environment. Marcel Dekker, New York

Sjöström E (1993) Wood chemistry. Fundamentals and applications, 2nd edn. Academic, San Diego

Smith SE, Read DJ (1997) Mycorrhizal symbiosis. Academic, San Diego

Smith SE, Read DJ (2008) Mycorrhizal symbiosis, 3rd edn. Academic, New York

Sokolova TA (2011) The role of soil biota in the weathering of minerals: a review of literature. Eurasian Soil Sci+ 44:56–72

Sparks JP, Monson RK, Sparks KL, Lerdau M (2001) Leaf uptake of nitrogen dioxide (NO_2) in a tropical wet forest: implications to tropospheric chemistry. Oecologia 127:214–221

Springer CJ, Thomas RB (2006) Photosynthetic responses of forest understory tree species to long-term exposure to elevated carbon dioxide concentration at the Duke Forest FACE experiment. Tree Physiol 27:25–32

Stafsform JP (1995) Developmental potential of shoot buds. In: Gartner BL (ed) Plant stems: physiology and functional morphology. Academic, San Diego

Steinberg G (2007) Hyphal growth: a tale of motors, lipids, and the Spitzenkörper. Eukaryot Cell 6:351–360

Stemmler K, Ammann M, Donders C, Kleffmann J, George C (2006) Photosensitized reduction of nitrogen dioxide on humic acid as a source of nitrous acid. Nature 440:195–198

Steudle E (2001) The cohesion-tension mechanism and the acquisition of water by plant roots. Annu Rev Plant Physiol 52:847–875

Steudle E, Petersen CA (1998) How does water get through roots? J Exp Bot 49:775–788

Stitt M (1991) Rising CO_2 levels and their potential significance for carbon flow in photosynthetic cells. Plant Cell Environ 14:741–762

Stitt M, Schulze D (1994) Does Rubisco control the rate of photosynthesis and plant growth? An exercise in molecular ecophysiology. Plant Cell Environ 17:465–487

Strous M, Jetten MSM (2004) Anaerobic oxidation of methane and ammonium. Annu Rev Microbiol 58:99–117

Suh SO, McHugh JV, Pollock DD, Blackwell M (2006) The beetle gut: a hyperdiverse source of novel yeasts. Mycol Res 109:261–265

Sutinen M-L, Arora R, Wisniewski M, Ashworth E, Strimbeck R, Palta J (2001) Mechanisms of frost survival and freeze-damage in nature. In: Bigras FJ, Colombo SJ (eds) Conifer cold hardiness. Kluwer, Dordrecht

Sutinen S, Partanen J, Viherä-Aarnio A, Häkkinen R (2009) Anatomy and morphology in developing vegetative buds on detached Norway spruce branches in controlled conditions before bud burst. Tree Physiol 29:1457–1465

Sutinen S, Partanen J, Viherä-Aarnio A, Häkkinen R (2012) Development and growth of primordial shoots in Norway spruce buds before visible bud burst in relation to time and temperature in the field. Tree Physiol 32(8):987–997

Sutton R, Sposito G (2005) Molecular structure in soil humic substances: the new view. Environ Sci Technol 39:9009–9015

Taiz L, Zeiger E (2006) Plant physiology, 4th edn. Sinauer, Sunderland

Takahashi S, Badger RB (2010) Photoprotection in plants: a new light on photosystem II damage. Trends Plant Sci 16:53–60

Tanahashi M, Kubota K, Matsushita N, Togashi K (2010) Discovery of mycangia and the associated xylose-fermenting yeasts in stag beetles (Coleoptera: Lucanidae). Naturwissenschaften 97:311–317

Tate RL III (2002) Microbiology and enzymology of carbon and nitrogen cycling. In: Burns RG, Dick RP (eds) Enzymes in the environment. Marcel Dekker, New York

Taylor JH, Peterson CA (2005) Ectomycorrhizal impacts on nutrient uptake pathways in woody roots. New For 30:203–214

Taylor LL, Leake JR, Quirk J, Hardy K, Banwart SA, Beerling DJ (2009) Biological weathering and the long term carbon cycle: integrating mycorrhizal evolution and function into the current paradigm. Geobiology 7:171–191

Teepe R, Dilling H, Beese F (2003) Estimating water retention curves of forest soils from soil texture and bulk density. J Plant Nutr Soil Sci 166:111–119

Teklemariam TA, Sparks JP (2006) Leaf fluxes of NO and NO_2 in four herbaceous plant species: the role of ascorbic acid. Atmos Environ 40:2235–2244

Terashima I, Hikosaka K (1995) Comparative ecophysiology of leaf and canopy photosynthesis. Plant Cell Environ 18:1111–1128

Terra WR, Ferreira C (1994) Insect digestive enzymes: properties, compartmentalization and function. Comp Biochem Physiol 109:1–62

Thoene B, Rennenberg H, Weber P (1996) Absorption of atmospheric NO_2 by spruce (*Picea abies*) trees. II. Parameterization of NO_2 fluxes by controlled dynamic chamber experiments. New Phytol 134:257–266

Thunes KH, Skartveit J, Gjerde I et al. (44 authors) (2004) The arthropod community of Scots pine (*Pinus sylvestris* L.) canopies in Norway. Entomol Fenn 15:65–90

Tietema A, De Boer W, Riemer L, Verstraten JM (1992) Nitrate production in nitrogen-saturated acid forest soils: vertical distribution and characteristics. Soil Biol Biochem 16:577–582

Tietema A, Riemer L, Verstraten JM, van der Maas MP, van Wijk AJ, van Voorhuyzen I (1993) Nitrogen cycling in acid forest soils subject to increased atmospheric nitrogen input. For Ecol Manage 57:29–44

Tisdall JM, Oades JM (1982) Organic matter and water-stable aggregates in soils. J Soil Sci 33:141–163

Tjoelker MG, Oleksyn J, Reich PB (1998) Seedlings of five boreal tree species differ in acclimation of net photosynthesis to elevated CO_2 and temperature. Tree Physiol 18:715–726

Tognetti R, Minnocci A, Peñuelas J, Raschi A, Jones MB (2000) Comparative field water relations of three Mediterranean shrub species co-occurring at a natural CO_2 vent. J Exp Bot 51:1131–1146
Tonteri T (1994) Species richness of boreal understorey forest vegetation in relation to site type and successional factors. Ann Zool Fenn 31:53–60
Treusch AH, Leininger S, Kletzin A, Schuster SC, Klenk H-P, Schleper C (2005) Novel genes for nitrite reductase and Amo-related proteins indicates a role of uncultivated mesophilic Crenarchaeota in nitrogen cycling. Environ Microbiol 7:1985–1995
Tucker CJ (1979) Red and photographic infrared linear combinations for monitoring vegetation. Remote Sens Environ 8:127–150
Turgeon R (2010) The role of phloem loading reconsidered. Plant Physiol 152:1817–1823
Turgeon R, Wolf S (2009) Phloem transport: cellular pathways and molecular trafficking. Annu Rev Plant Biol 60:207–221
Turpin DH, Bruce D (1990) Regulation of photosynthetic light-harvesting by nitrogen assimilation in the green alga *Selenastrum minutum*. FEBS Lett 263:99–103
Turunen M, Huttunen S (1996) Scots pine needle surfaces on radial transects across the north boreal area of Finnish Lapland and the Kola Peninsula of Russia. Environ Pollut 93:175–194
Twigg MM, House E, Thomas R, Whitehead J, Phillips GJ, Famulari D, Fowler D, Gallagher MW, Cape JN, Sutton MA, Nemitz E (2011) Surface/atmosphere exchange and chemical interactions of reactive nitrogen compounds above a manured grassland. Agric For Meteorol 151:1488–1503
Tyree MT, Zimmermann MH (2002) Xylem structure and the ascent of sap, 2nd edn. Springer, New York
Vaisanen R (1996) Boreal forest ecosystems. IUFRO-95 papers and abstracts, IUFRO XX Word Congress, Tampere
van Breemen N, Mulder J, van Grinsven JJM (1987) Impacts of acid atmospheric deposition on woodland soils in the Netherlands. II. Nitrogen transformations. Soil Sci Soc Am J 51:1634–1640
van Schöll L, Kuyper TW, Smits MM, Landeweert R, Hoffland E, van Breemen N (2009) Rock-eating mycorrhizas: their role in plant nutrition and biogeochemical cycles. Plant Soil (2008) 303:35–47
Vanderwel MC, Malcolm JR, Smith SM, Islam N (2006) Insect community composition and trophic guild structure in decaying logs from eastern Canadian pine-dominated forests. For Ecol Manage 225:190–199
Vanninen P, Mäkelä A (2005) Carbon budget for Scots pine trees: effects of size, competition and site fertility on growth allocation and production. Tree Physiol 25:17–30
Vieten B, Conen F, Neftel A, Alewell C (2010) Respiration of nitrous oxide in suboxic soil. Eur J Soil Sci 60:332–337
Viil J, Pärnik T (1995) The rate constant for the reaction of CO_2 with enzymebound ribulose 1,5-bisphosphate in vivo. J Exp Bot 46:1301–1307
Vitousck PM, Cassman K, Cleveland C, Crews T, Field CB, Grimm NB, Howarth RW, Marino R, Martinelli L, Rastetter EB, Sprent JI (2002) Towards an ecological understanding of biological nitrogen fixation. Biogeochemistry 57(58):1–45
Vogel JG, Gower ST (1998) Carbon and nitrogen dynamics of boreal jack pine stands with and without a green alder understory. Ecosystems 1:386–400
Vogg G, Heim R, Hansen J, Schäfer C, Beck E (1998) Frost hardening and photosynthetic performance in Scots pine (*Pinus sylvestris* L.) needles. I. Seasonal changes in the photosynthetic apparatus and its function. Planta 204:193–200
Vuokko R, Kellomäki S, Hari P (1977) The inherent growth rhythm and its effect on the daily height increment of plants. Oikos 29:137–142
Wada M, Kagawa T, Sato Y (2003) Chloroplast movement. Annu Rev Plant Biol 54:455–468
Walker J, Sharpe PJH, Penridge LK, Wu H (1989) Ecological field theory: the concept and field tests. Vegetatio 83:81–95
Wardle DA (1999) How soil food webs make plants grow. Trends Ecol Evol 14:418–420

Warren RAJ (1996) Microbial hydrolysis of polysaccharides. Annu Rev Microbiol 50:183–212
Warren CR, Adams MA (2004) Evergreen trees do not maximize instantaneous photosynthesis. Trends Plant Sci 9:270–274
Warren CR, Dreyer E, Adams MA (2003) Photosynthesis-Rubisco relationships in foliage of *Pinus sylvestris* in response to nitrogen supply and the proposed role of Rubisco and amino acids as nitrogen stores. Trees 17:359–366
Watanabe H, Tokuda G (2010) Cellulolytic systems in insects. Annu Rev Entomol 55:609–632
Weiser CJ (1970) Cold resistance and injury in woody plants. Science 169:1269–1278
Weng J-K, Chapple C (2010) The origin and evolution of lignin biosynthesis. New Phytol 187:273–285
West GB, Brown JH, Enquist BJ (1997) A general model for the origin of allometric scaling laws in biology. Science 276:122–126
Wildt J, Kley D, Rockel A, Rockel P, Segschneider HJ (1997) Emission of NO from several higher plant species. J Geophys Res 102:5919–5927
Wilkins D, Van Oosten J-J, Besford RT (1994) Effects of elevated CO_2 on growth and chloroplast proteins in *Prunus avium*. Tree Physiol 14:769–779
Wingler A, Lea PJ, Quick WP, Leegood RC (2000) Photorespiration: metabolic pathways and their role in stress protection. Philos Trans Roy Soc B Biol Sci 355:1517–1529
Wittmann C, Kähkönen MA, Ilvesniemi H, Kurola J, Salkinoja-Salonen MS (2004) Areal activities and stratification of hydrolytic enzymes involved in the biochemical cycles of carbon, nitrogen, sulphur and phosphorus in podsolized boreal forest soils. Soil Biol Biochem 36:425–433
Woebken D, Fuchs BM, Kuypers MMM, Amann R (2007) Potential interactions of particle-associated anammox bacteria with bacterial and archaeal partners in the Namibian upwelling system. Appl Environ Microbiol 73:4648–4657
Woodrow IE (1994) Optimal acclimation of the photosynthetic system under enhanced CO_2. Photosynth Res 39:401–412
Woodrow IE, Berry JA (1988) Enzymatic regulation of photosynthetic CO_2 fixation in C_3 plants. Annu Rev Plant Phys 39:533–594
Woodward FI (1987) Stomatal numbers are sensitive to increases in CO_2 from pre-industrial levels. Nature 327:617–618
Woodward FI, Kelly CK (1995) The influence of CO_2 concentration on stomatal density. New Phytol 131:311–327
Wrage N, Velthof GL, van Beusichem ML, Oenema O (2001) Role of nitrifier denitrification in the production of nitrous oxide. Soil Biol Biochem 33:1723–1732
Wu T (2011) Can ectomycorrhizal fungi circumvent the nitrogen mineralization for plant nutrition in temperate forest ecosystems? Soil Biol Biochem 43:1109–1117
Wynn JG (2003) Towards a physically based model of CO_2-induced stomatal frequency model. New Phytol 157:391–398
Xu JM, Cheng HH, Koskinen WC, Molina JAE (1997) Characterization of potentially bioreactive soil organic carbon and nitrogen by acid hydrolysis. Nutr Cycl Agroecosyst 49:267–271
Yamasaki H, Sakihama Y (2000) Simultaneous production of nitric oxide and peroxynitrite by plant nitrate reductase: in vitro evidence for the NR-dependent formation of active nitrogen species. FEBS Lett 468:89–92
Yu F, Berg VS (1994) Control of paraheliotropism in two *Phaseolus* species. Plant Physiol 106:1567–1573
Zackrisson OT, DeLuca H, Nilsson MC, Sellstedt A, Berglund LM (2004) Nitrogen fixation increases with successional age in boreal forests. Ecology 85:3327–3334
Zeide B (1987) Analysis of the 3/2 power law of self-thinning. For Sci 33:517–537
Zhou X, He Y, Huang G, Thornberry TD, Carroll MA, Bertman SB (2002) Photochemical production of HONO on glass sample manifold wall surface. Geophys Res Lett 29:1681

Chapter 5
Fluxes of Carbon, Water and Nutrients

Teemu Hölttä, Pertti Hari, Kari Heliövaara, Eero Nikinmaa, Jukka Pumpanen, Timo Vesala, Pasi Kolari, Samuli Launiainen, Üllar Rannik, Liisa Kulmala, Kourosh Kabiri Koupaei, Minna Pulkkinen, Mari Pihlatie, Janne F.J. Korhonen, Asko Simojoki, Antti-Jussi Kieloaho, Jaana Bäck, and Markku Kulmala

Contents

T. Hölttä (✉)
Department of Forest Sciences, University of Helsinki, P.O. Box 27, 00014 University of Helsinki, Helsinki, Finland
e-mail: teemu.holtta@helsinki.fi

P. Hari et al. (eds.), *Physical and Physiological Forest Ecology*, DOI 10.1007/978-94-007-5603-8_5,

Abstract The metabolic and physical processes result in concentration, pressure and temperature differences that generate fluxes within ecosystems and between ecosystems and their surroundings. We apply the same approach combining metabolic and physical processes with transport in the analysis of very different phenomena from water and sugar transport within trees to the fluxes of carbon and nutrients in the forest ecosystem.

Keywords Concentration • Pressure or temperature difference • Fluxes of CO_2 • Water and heat • Annual cycle • Photosynthesis • Respiration of microbes and vegetation

5.1 Water and Sugar Transport in Trees

Teemu Hölttä, Pertti Hari, Kari Heliövaara, and Eero Nikinmaa

5.1.1 Introduction

Trees as Transporting Structures

Tree stems and crowns are magnificent entities that keep on intriguing human imagination (Milner 2010). However, already centuries ago, Leonardo da Vinci linked in his notes the tree architecture with water movement (Leonardo's notes, e.g. Horn 2000). Since then, several other authors have associated the tree architecture and structure with transport processes (e.g. Huber 1935; Zimmermann 1983; West et al. 1999).

The stem and crown structure result from the action of two meristematic tissues. The primary meristem at the end of tree axis creates new shoots, and secondary meristem on the surface of the axis elements makes new conducting tissue in the xylem and the phloem resulting in radial growth of the woody axis. The crown development can be viewed as birth, growth and death of elementary structural units. In botanical literature, units such as a "metamer" and "growth unit" (Room et al.

T. Hölttä (✉)
Department of Forest Sciences, University of Helsinki, P.O. Box 27, 00014 University of Helsinki, Helsinki, Finland
e-mail: teemu.holtta@helsinki.fi

1994; Caraglio and Barthélémy 1997) have been identified. A metamer is defined as an internode with axillary bud(s) and leaf (leaves) in its upper end but without any shoots resulting from growth of the axillary buds (Caraglio and Barthélémy 1997). A growth unit, initially called a "unit of extension" by Hallé et al. (1978), is the part of the shoot resulting from uninterrupted extension growth. Room et al. (1994) describe it as "extension of the contents of a previously dormant apical bud followed by growth of neoformed leaves (if any) and formation of a new, dormant, apical bud". An axis is "a sequence of growth units in the same general direction from one (monopodial) or more (sympodial) meristems" (Room et al. 1994). From the point of view of transport of material between the distal ends of the axes, the properties of the axis that results from the action of secondary meristem are most interesting.

When a cross section of any woody axis with secondary growth is studied, we can distinguish the xylem core surrounded by secondary cambium, the phloem tissue and outer bark tissue with possible cork cambium and cork tissue making the surface boundary with air. In the centre of the tissue, there are remnants of the tissue formed by the primary meristem during the shoot's extension growth. In thicker segments, part of the xylem tissue may have become non-conducting heartwood with very low water content and possibly distinguishable chemical composition (Saranpää 1990). The outer layer of xylem is water-conducting sapwood. Xylem consists of dead cells with an inelastic cell wall and perforation, forming a conducting tube, while the phloem is living material. Often clearly distinguishable annual growth rings can be seen in the xylem tissue consisting of early wood and latewood. The early wood is normally thicker consisting of conducting xylem elements (vessels or tracheids) that have large conducting cavities relative to cell walls, while in the latewood, the cell walls can be comparable or larger than the conducting cavity. Also, the latewood cell size is smaller.

The actual conducting tissue consists of rather small tracheids in conifers and much larger vessels in broadleaved trees. The xylem tissue in the latter group is much more versatile as there are also fibres and tracheids, but their role is less important in water transport. The cambial cells and the newest bark tissue consist of living cells, and there are radial intrusions of living cells also into xylem tissue that are called rays. Rays may consist of entirely living cells, or they may also have dead conducting cells. In conifers, they are often associated with resin ducts. In the phloem tissue, there are conducting sieve elements (in broadleaved) or sieve cells (in conifers) with associated living companion cells that make together the conducting phloem tissue. Closer to stem surface, there are dead bark cells and possibly cork cells as mentioned above.

The size of the total conducting area and the cross-sectional area of the cavities of the conducting cells is important from the transport point of view. However, also the connectivity between the conducting cells in the axial direction is important. The cells and vessels are connected to each other through various pits that have porous pit membranes. In broadleaved trees, several vessel elements may be connected to each other without end walls making long continuous vessels, while in conifers, the individual tracheids are linked to others through pits. In the middle of the pit cavity, there is the pit membrane that can have central impermeable thickening called torus particularly in some conifers. The small pores of pit membranes can cause considerable proportion of the total water transport resistance, but they also have an important role in confining possible air bubbles forming in the xylem tissue when water is under large tension.

In the dynamic development of tree crowns and stems, important criteria for tree survival are an access to favourable light conditions and a provision of sufficient transport capacity that supplies the leaves with sufficient nutrient and water resources. Trees could be viewed as modular organisms in their developmental pattern. The reiteration of simple structures repeating over time has been identified as the process responsible for tree crown development (e.g. Hallé et al. 1978). Many physiological studies have revealed that branches are autonomous with respect to their carbon requirement (e.g. Dickson and Isebrands 1991). However, branches are not independent on the rest of the tree in their development as they rely on the rest of the tree structure in their supply of other necessary substances. The properties of the transporting tissue, resulting from the action of secondary meristem, have a central role in providing the connection between the distal parts of the tree.

The concept of an integrated physiological unit (IPU, Watson and Casper 1984) has been introduced to describe and used in modelling the tree crown development (Nikinmaa 1992). It stresses the importance of a balanced supply and transport of both above- and belowground resources for undisturbed tree functioning. The linkage between foliar development and wood axis thickening leads naturally to direct implications for crown development. Foliage at different heights (i.e. different distances from the roots) has different carbon consumption requirements (Stevens and Perkins 1992) but also different roles in crown dynamics. Young, first-order shoots in the upper crown are unlikely to be immediately shaded, and they play an important role in supporting tree structure for many years. On the other hand, shoots in the lower crown, even if at first in a strong light environment, become shaded and less able to support other shoots (Nikinmaa et al. 2003).

Tree crown development leads often to distinct hydraulic architecture (Zimmermann 1983). The (thicker) main axes are often hydraulically preferred. They may have manyfold hydraulic conductivity per unit leaf area compared to more suppressed side branches. This preference is reflected in higher transpiration rates in given conditions, higher supply of nutrients with the transpiration stream and thus higher carbon uptake rate and growth. The linkage between the hydraulic preference and growth leads to positive feedback: vigorous branches grow more, while less vigorous branches are gradually shed (Goulet et al. 2000; Nikinmaa et al. 2003). The linkage is particularly obvious during high water demand. In those parts of the crown that have low leaf-specific water conductivity (e.g. side branches, Zimmermann 1983), the shoot water tension would increase the most, which may result in stomatal closure to avoid cavitation (Williams et al. 1996). The closing of stomata in one part of the crown would facilitate transpiration in the other crown parts (Whitehead et al. 1996).

No one can deny that old stems and tree crowns are visually impressive. However, they also reveal a very interesting success story of trees life if we can read it. The intimate linking between the metabolism, transport and structure stresses the role of stem and crown structure in long-term acclimation of trees to their growing site (Nikinmaa 1992). In this chapter, we deal with the mechanisms of the long-distance transport in trees and how they are connected to structure of the transporting tissue and the metabolism of the living cells.

General Transport Mechanisms

Water needs to be transported to the leaves as water is lost in transpiration as the leaves would otherwise desiccate. If water transport rate decreases below the transpiration rate, water tension in the leaf increases and the tree is eventually forced to close the stomata to avoid excess tension, which also causes a decrease in the CO_2 assimilation rate. Efficient water transport from soil to the leaves is therefore important for tree performance. Tree carbon assimilation and photosynthetic production are largely dependent on the supply of water to the leaves (Tyree and Sperry 1988; Bond and Kavanagh 1997). As a tree grows in height, the transport distance from soil to leaves increases and the supply of water to the leaves grows increasingly limiting for tree performance and growth (Koch et al. 2004). At the same time, the construction and maintenance of the water transport system, the xylem, typically requires a larger proportion of the carbon assimilated by the tree (Mäkelä 1988).

Long-distance transport of sugar in the phloem is a much less studied subject than water flow in xylem. Sugars need to be transported from the leaves to sugar sinks, or sugar accumulation in the leaves would force the stomata to close and/or photosynthesis to be downregulated. The general transport mechanism for the xylem and phloem is the same; in both cases, we are dealing with mass flow driven by a pressure gradient, but the pressure gradient for xylem transport relies on water tension, while the phloem transport has an osmotically driven water pressure. The pressure gradient required for transport is in both cases dependent on the hydraulic conductance and on the rate of transport.

5.1.2 Upward Flow of Water in Xylem

5.1.2.1 Cell Scale

Water in trees moves predominantly in the xylem tissue from soil to atmosphere. The elementary unit of transport is a tracheid cell in conifers and vessel in broadleaved species. Tracheids are some tens of micrometres wide and some hundreds of micrometres long. They are thick walled but dead and without protoplasts when functional. Tracheids are connected to neighbouring tracheids through holes known as pits, which are very small, only some micrometres wide. Pits are more numerous

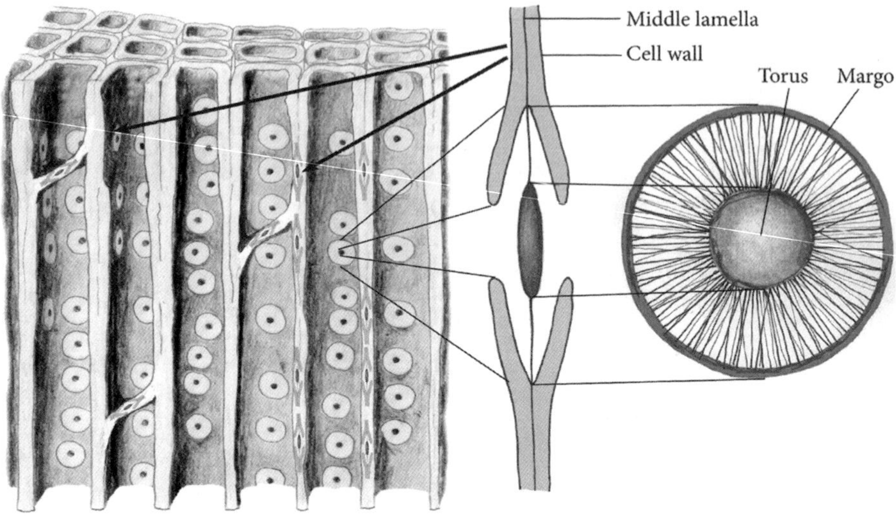

Fig. 5.1 Schematic illustration of the structure of xylem in a conifer tree and of pit structure with torus (*black centre*) and margo (*surrounding netlike structure*). One tracheid is typically a few millimetres long and a few tens of micrometres in diameter

at both ends of tracheids, but there are some of them along the whole length of tracheid wall (Kärkkäinen 2007). The pit opening is narrowest at the surface of the tracheid cavity, and it widens towards the centre of the cell wall. In the middle of the pit between the two adjacent tracheids, there is the pit membrane that often has impermeable thickening, the torus, at the centre and a netlike margo surrounding it (Fig. 5.1). The structure is such that it allows the movement of the torus against the pit walls, which completely blocks the pit if sufficient pressure difference develops between adjacent tracheids (Hacke et al. 2004). The tracheid cell walls are very strong as a consequence of the secondary cell wall formation, and the pits can withstand pressure difference of several MPa (Hacke et al. 2001).

Broadleaved vessels are much wider and longer than tracheids. One vessel consists of several vessel elements where the cell walls of adjacent vessel elements are either partially or completely removed. The pit membranes of broadleaved trees do not have the torus.

Water loss from the liquid surfaces in transpiration and the cohesion between water molecules cause water to be under tension in the xylem of all trees. Transpiration causes a net evaporation of water molecules from water surfaces on mesophyll cells and increases the concavity of the surface of the water film between cells (Fig. 5.2) or into pores in cell wall. Surface tension is relatively high on water, 0.073 N m^{-1}, owing to strongly polar structure of water molecules; see Fig. 5.3. Surface tension results from asymmetric forces between water molecules at the surface, tends to minimise the area of the surface and pulls the water molecules towards the surface. This results in pressure drop in water below the curved surface. The tension difference across the conducting cell drives the water flow.

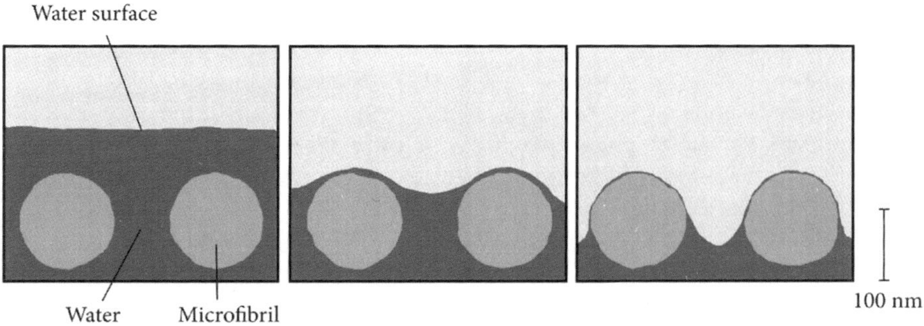

Fig. 5.2 Schematic illustration of decreasing amount of water in a leaf from *left* to *right*. The surface tension generates water tension due to increasing water surface area

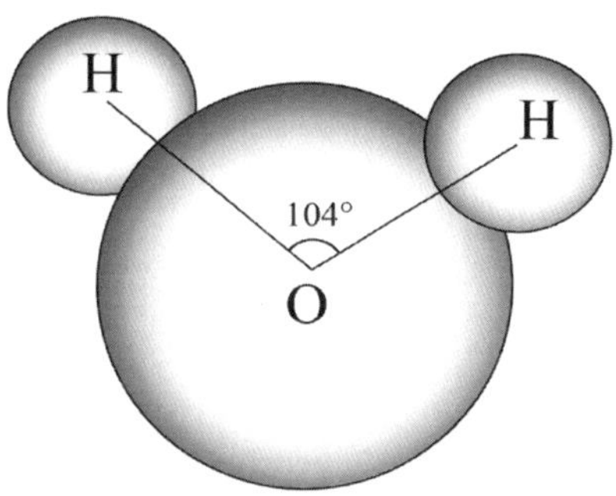

Fig. 5.3 A schematic illustration of the composition of a water molecule

The speed of flow (J, $m^3\ s^{-1}$) depends on the pressure gradient ($\Delta P/\Delta x$, Pa m^{-1}), viscosity of the liquid (μ, $Pa^{-1}\ s^{-1}$) and fourth power of the radius (r, m) of the transporting capillary expressed with well-known Hagen–Poiseuille equation (Zimmermann 1983, Eq. 5.1):

$$J = \frac{\pi r^4}{8\mu}\frac{\Delta P}{\Delta x}. \tag{5.1}$$

The dependency of flow on the cross-sectional area makes the width of the transporting pathway from cell cavity to pits and pit membrane pores critical. The pits and pit membranes constitute about 50% of the total resistance to the flow and the size of the cell cavities about the same.

The large resistance caused by the pits and pit membranes seems harmful to trees. They are, however, important in preventing the spreading of gas bubbles in xylem. This is because the existence of tension in the xylem means that trees rely on metastable water for their water and nutrient supply, which is very rare elsewhere in nature and also in engineering applications. A metastable state is a condition of a phase in which a disturbance will cause the appearance of a new phase and the system will fall to a lower energy level (Debenedetti 1996). A liquid is metastable when its pressure is lower than its saturation vapour pressure. In these circumstances, the vapour phases would give a lower energy state, but before a phase change can occur, an energy barrier due to surface energy between the two phases has to be overcome by nucleation.

Nucleation can either be homogeneous where the energy barrier to nucleation is overcome by random thermodynamical fluctuations in the bulk liquid itself or heterogeneous where impurities suspended in the liquid or irregularities in the walls in contact with liquid can catalyse a phase change by lowering the surface energy barrier to nucleation. According to the classical nucleation theory and experimental evidence, the pressure of pure water can be decreased down to approximately –150 MPa at ambient temperatures before homogeneous cavitation occurs. In a tree, classical heterogeneous nucleation is also likely to play a minor role in cavitation induction as it would require the existence of very hydrophobic surfaces (contact angle between solid and gas $>160°$) which are unlikely to be found. "Nonclassical" nucleation is likely to be in more important role based on both theoretical considerations and experimental manipulations (e.g. Tyree and Sperry 1988; Cochard et al. 2009).

The xylem always contains air-filled conduits, and in practice, gas emboli are thought to spread from one xylem conduit to another. The cell structure has an important role in avoiding the spreading of gas bubbles as it only takes place if the capillary seal formed in the pores of interconduit pit membranes fails in a process termed air seeding. The pore sizes in the pit membranes separating air and water-filled conduits are known to be a few nanometres in radius, and they also stretch under mechanical stress (Choat et al. 2004), so capillary seal failure according to Laplace's law will induce cavitation at tensions typically experienced in plants.

After a bubble is nucleated into a xylem conduit, it grows very fast to fill the entire conduit with vapour and subsequently with air. This is called embolism. Embolised vessels become hydraulically isolated from water-conducting vessels and cease conducting water (Zimmermann 1983; Holbrook and Zwieniecki 1999). Water in the transpiration stream has to find alternate pathways around the embolised vessels. This is possible since the conducting cells are always connected to more than one neighbouring cells. However, this introduces "bottlenecks" in transport tissue compared with non-embolised situation, and the hydraulic conductivity of the xylem is reduced.

Another consequence of cavitation which has received less attention is that cavitation should be associated with a release of tension in the transpiration stream. As a xylem conduit cavitates, it becomes filled with water vapour or gas, and nearly all of the liquid water inside the xylem conduit is freed to the transpiration stream. As xylem conduits are quite inelastic and liquid water itself is practically incompressible, even a small amount of extra water released from an embolising conduit will bring about a relatively large increase in the water pressure of the surrounding xylem tissue. Cavitation will therefore affect the water status of a plant positively in the short term due to the capacitive effect.

Embolism formation is a common occurrence in the life of trees, and embolised cells can be refilled again with water in many species. The exact mechanisms for embolism refilling are not yet known (Zwieniecki and Holbrook 2009). Refilling has also been found to occur in some species during transpiration, while the xylem is under considerable water tension (e.g. Tyree et al. 1999; Melcher et al. 2001). What makes the refilling process problematic is that water must be forced into the gas-filled conduit, where pressure is close to atmospheric, from the surrounding tissue, where the water is under considerable tension, that is, it has negative pressure. Theoretical calculations by Vesala et al. (2003) showed that embolised cells may refill even in such conditions if they are connected to living cells and semipermeability of the cell membrane would change. As the refilling process is near to completion, it is necessary that all gas voids in the refilling vessel have to be dissolved before hydraulic conductivity is re-established with the adjacent conduits (Holbrook and Zwieniecki 1999). Otherwise, these residual gas pockets would seed new embolisms when hydraulic conductivity is established to the adjacent water-filled conduits.

It has been hypothesised that radial water flux from the phloem to xylem can induce embolism refilling (Milburn 1996). Model calculations by Hölttä et al. (2006a) reveal that this hypothesis can induce refilling when xylem water tension is high, provided that the radial hydraulic conductance is low enough and radial water flux is high enough. Salleo et al. (2004) have also clearly demonstrated that embolism refilling occurs in Laurel tree (*Laurus nobilis*) only when phloem integrity is maintained.

5.1.2.2 Tissue Scale

Xylem transport pathway consists of vessels/tracheids and pits adjoining the vessels/tracheids. Xylem is only part of the total transport pathway, as the soil and root tips and the mesophyll tissue, intercellular airspace and the boundary layer also contribute to transport resistance. As mentioned previously, the water-transporting sapwood consists of conducting elements that have dimensional differences depending whether they are formed at early or late phases of growth period. The cell walls of the latewood are thicker than those of the early wood, and also their pit openings are narrower. There are also specialised rays that run in radial direction. They may consist of living cells only but there are also radial ray tracheids. The proportion of rays of the wood volume is relatively low in many species. The inner part of the stem of coniferous trees is non-water-conducting heartwood.

5.1.2.3 Tree Scale

Transport Pathway

Radial water exchange between the xylem and phloem is a requirement for phloem transport according to the Münch hypothesis (Taiz and Zeiger 2002). Phloem sap needs to be diluted in osmotic concentration from the plant top to bottom, and constant influx of water along the axial pathway is required for this (Hölttä et al. 2009). At the sugar sinks, water is freed and directed towards the xylem drawn by water tension.

The xylem tissue runs as a continuous pathway from roots in the soil to leaves in the crown. At entering and leaving, the water flow is under active control by the living tissue. However, even if the xylem tissue is dead, its properties will greatly influence the water transport. As previous chapter demonstrated depending on the structure of the xylem elements, the permeability of the tissue can vary considerably. At tree level, another dimension enters the picture as the size of the transpiring surface relative to pathway conductivity influences the water transport and the water tensions under which the tree is operating. Apart from the tissue properties also, the tissue size influences the pathway conductivity.

Several theories exist which describe the structure of the xylem. The pipe model theory states that there is a linear relationship between leaf and sapwood area (Shinozaki et al. 1964). Experimental evidence supports the pipe model theory fairly well (Waring et al. 1982; Hari et al. 1986; Berninger et al. 2005). The pipe model was originally formulated without an explicit interpretation of tree hydraulic properties (Shinozaki et al. 1964). However, it has been later interpreted to signify that conduit size remains constant and conduit number increases in relation to leaf area as a tree grows (West et al. 1999; McCulloh et al. 2003). It is true that this type of hydraulic architecture will lead to an unsustainable decrease in xylem water transport capacity per leaf area with increasing transport distance (West et al. 1999). West et al. (1999) formulated from general biological scaling laws that xylem conduits should increase in size from the apex downwards according to a power law to prevent the loss of water transport capacity due to increased transport length. Anatomical measurements of the axial distribution of conduit size support this theory fairly well (e.g. Anfodillo et al. 2006; Weitz et al. 2006; Mencuccini et al. 2007; Petit et al. 2008). However, many argue that the theory is too general and inaccurate to be applicable to real trees (e.g. Mäkelä and Valentine 2006).

Other theories for optimal xylem structure include the maximisation of hydraulic conductance per units of carbon invested in its construction, which leads to Murray's law (McCulloh et al. 2003), and the hypothesis that plants should maximise their net carbon gain, that is, the amount of carbon that a plant can assimilate with a given xylem structure minus the construction costs of the system (Mencuccini et al. 2007).

Sapwood vs. Heartwood

Another important determinant for the transport is the functional longevity of sapwood. This is particularly important from the point of view of annual investment needed for maintenance of sufficient transport capacity relative to leaf productivity. Modelling analyses have shown that sapwood turnover to heartwood has a very significant influence on tree development when the carbon balance is considered, if we assume that leaf area requires certain quantifiable amount of sapwood in stem (Mäkelä 1988; Nikinmaa 1992).

Different tree species are very different as regards to the proportion of sapwood area in stem. Ring-porous trees that have very wide vessels often exhibit very narrow sapwood, while in diffuse porous trees, there is not always clear heartwood, but even the inner parts of the stem may be still conducting. In conifers, the heartwood is practically always present in larger stem segments.

Numbers of theories have been suggested for the mechanisms leading to heartwood formation. Ageing and consequent loss of vitality, toxic effect of metabolic products and growth-regulating substances have been proposed as the cellular level mechanisms of sapwood senescence. The outcome of these mechanisms is more or less that the sapwood rings have a maximum age that has also empirically observed (Saranpää 1990). Another line of reasoning is that since the sapwood is supporting foliage (conducting water) (Kaipiainen and Hari 1985), its senescence must be related to death of foliage, but no conclusive general theory has yet been put forwards.

There is also a big difference between the area of functional sapwood and actual heartwood. The sapwood–heartwood boundary can be distinguished from moisture differences, but often also different chemicals are excreted into the heartwood, producing a colour difference (e.g. Saranpää 1990). Nikinmaa et al. (1996) showed that there was a connection between the number of living branch whorls and number of tree rings transporting actively water. It has been shown that the transport velocity of sap in xylem varies considerably in radial direction (Čermák and Nadezhdina 1998). The velocity changes in radial direction reflect the leaf area distribution by height. It has often been also observed that the sapwood area increases from crown base downwards Long and Smith 1988). With increasing tree size, the contribution of stored water in the stem in daily transpiration becomes more important (Meinzer et al. 2001; Phillips et al. 2003; Koch et al. 2004). Increasing sapwood area below the crown could be a similar response to hydraulic limitation (see below, Ryan and Yoder 1997) as the structural scaling of conduits suggested by West et al. (1999) and Perämäki et al. (2005), or it could also be means of increasing the stem water storing capacity.

5.1.2.4 Whole Tree Mechanisms

Hydraulic Limitation

The capacity of the xylem to deliver water for transpiration is perhaps the main factor in limiting the maximum tree height (Koch et al. 2004). Raising water against gravity up to over 100 m presents a challenge, but the viscous pressure losses which are dependent on transport length and transport rate constitute even a larger problem, as a majority of the pressure losses occur due to viscosity and not height.

Why the pressure loss and compensation of gravitational forces limit the productivity is linked to the cavitation process discussed above. The pressure gradient increase that would compensate viscous pressure losses and would allow maintenance of the same transpiration rates would require such a large tension development in the xylem that large-scale cavitation would eventually follow as trees often operate at the brink of catastrophic xylem malfunctioning (Tyree and Sperry 1988). To avoid such an event, trees need to reduce transpiration, but this simultaneously limits assimilation. With increasing tree size, there is thus a tendency of larger carbohydrate demand in order to build up xylem relative to leaf area (Mäkelä 1988) and decreasing leaf productivity which will lead to suggested hydraulic limitation of tree size (Ryan and Yoder 1997).

Domec et al. (2008) found that xylem conductance decreased almost to zero at the tops of the very tall trees. As discussed in the connection with the cavitation and embolism processes, one of the mechanisms that allow trees to survive even very low negative pressures is to reduce the size of the pit pores between adjacent conducting vessels as this will effectively hinder spreading of air bubbles. However, it will also decrease the wood conductivity.

It has also been suggested that the turgor pressure reduction due to gravity is limiting the size to which the conduits can reach at the top of very tall trees (Woodruff et al. 2004). High gravitational pull of water means that it is under high tension even when there is no transpiration. This will mean that in order to grow, there have to be very high solute concentrations of sugars that would create sufficient turgor pressure to expand cell walls. This together with reduced assimilation makes cell growth more difficult with increasing tree size.

Water Storage and Embolism

Cavitation causes detrimental effects on the water transport capacity of the xylem, but cavitation could also play an important role as a water release mechanism. Studies of whole tree water use have suggested the cavitation plays a role in the water storage capacity of trees (e.g. Schulze et al. 1985; Meinzer et al. 2001; Čermák et al. 2007). The capacitive effect of cavitation has been measured to be much larger than other water storage compartments at physiologically important, high xylem water tensions (Tyree and Yang 1990). Various measurements have also demonstrated that conifer stem water content can vary seasonally by many tens of percents (Waring and Running 1978; Waring et al. 1979; Nikinmaa et al. 1996).

Also, modelling shows that during drought and/or for large trees where the transpiration rate decreases in comparison to xylem total volume, the capacitive effect of cavitation grows in importance in comparison to the water transport capacity effect (Hölttä et al. 2009). Upon cavitation, xylem tension is relieved and leaf water potential remains higher than without cavitation. However, cavitation can only provide the leaves with water for a limited time period, after which the embolised xylem conduits require refilling.

The hydraulic structure of trees is such that the conduit size and hydraulic conductivity increases from tree apex towards the tree base (West et al. 1999) which presumably means that also pit pores are able to prevent the spreading of embolisms there under lower water tensions (Domec et al. 2008). From the water storage point of view, this would efficiently distribute also the cavitation over the whole volume of the stem avoiding formation of large bottlenecks in water conductivity along the stem axis.

Perämäki et al. (2005) pointed out that there seems to exist capacitive discharge–recharge of water storage along the whole stem that could also be linked to radial water movement, as well as stores in more elastic bark tissue, which does not delay the fast propagation of tension changes in the sapwood that variation in transpiration rate creates. They modelled rapid tension propagation throughout the stem and demonstrated the phenomenon in real plants by measurements of xylem diameter variation in different heights. Now, increased tension inside the sapwood also causes radial tension propagation and water flow from the living tissues external to the xylem, that is, phloem and bark (Zweifel et al. 2001). Because radial permeability is several orders of magnitude smaller than axial permeability (Zwieniecki et al. 2001), and because the elastic modulus of the phloem is smaller than that of the xylem (Nobel 1991), the radial tension propagation is slower. This explains why the diameter change of the whole stem has been reported to lag behind the xylem diameter change (Sevanto et al. 2002). When transpiration ceases, the flow from soil fills up the sapwood quite rapidly, but the slower recharge of the phloem and bark water reserves maintains a gradual flow and slow tension relaxation during the night. This produces the "typical tail" in diurnal sap flow and diameter change curves and explains the discrepancy between modelled and measured tension relaxation reported by Perämäki et al. (2001).

Soil–Atmosphere Boundaries

In terrestrial environments, vegetation is an important "valve" controlling the water flow from soil to the atmosphere. As mentioned, the driving force for transpiration is the dryness of air that drives the water molecules to evaporate from the water surfaces. Unlike open soil surfaces, vegetation can regulate the evaporation. Changes in the size of the stomatal openings allow rapid control of water loss, while variation of the leaf area influences the interception of radiation that provides the energy for evaporation. These two factors determine largely the demand for water in given environment, while the total conductivity of water transport pathway from soil to roots and from there to leaves via xylem in roots, stem and branches in trees determines the water supply to leaves.

As mentioned in the context of structural scaling, there has to be a balance between the water demand and supply on the long run, which imposes strong requirements for the plant and tree structure. However, that is also the case between the atmospheric demand, soil supply and vegetation structure. This feature is highlighted in the global vegetation classification systems, such as that of Köppen (Kottek et al. 2006). When there is an ample supply of water in the soil, leaf area and tree conductivity determine the maximum rate of transpiration and water loss from the soil. However, as the soil dries out, the water conductivity in soil rapidly decreases and becomes the bottleneck for transpiration (Duursma et al. 2007). In such conditions, variations in plant properties have little influence on transpiration apart from the root density and extension. High root density and mycorrhizal symbiosis decrease the average transport distance of water in the soil matrix decreasing the relative importance of soil transport. On the other hand, higher rooting depth allows water harvesting from thicker soil layer and bring larger water resources available for plant use. The role of the root system in providing transport channels in drying soil is demonstrated in the so-called water lifts (Dawson 1993). It has been shown that some deeply rooted trees may leak water from roots to drying surface soil layers.

Evaporative demand, level and frequency of precipitation and soil matrix properties determine the level of soil water content. In conditions when it is frequently low, large leaf area or high water conductivity will not give any added value for plants, while deep root systems do. In such conditions, soil profile often contains sufficient oxygen that also allows growth of deep-rooted plants. In addition, vegetation above ground is open, while a large proportion of the plant biomass is below ground. In moist conditions, the opposite is true, and plants that are able to maintain high transpiration rate are also able to capture much carbon and gain competitive advantage.

The behaviour of transpiration of terrestrial vegetation has a strong influence on the climate. The latent heat of evaporation is an important component of the global heat balance, and the circulation of water directly influences climate (Bonan 2008). Seneviratne et al. (2006) demonstrated that vegetation behaviour may actually enhance heatwaves, such as experienced in Europe 2003. Present-day climate models do not yet fully incorporate the impact of structural acclimation and adaptation of vegetation to transpiration. However, depending on the vegetation structure and hence its water supply capacity from soil, the transpiration dynamics may be very different in given pattern of evaporative demand and precipitation.

Carbon Cost of Water Transport

Previous chapters have shown that plants seem to follow optimal water usage strategies, that is, transpiration rate depends not only on the atmospheric demand and soil-tree supply capacity but also on the rate of carbon intake that photosynthesis may maintain. This can also be understood within the larger scheme on carbon allocation within trees. The carbon cost of water transport is the thick lignified secondary cell walls of xylem conduits that plants, and trees in particular, need to build in order to withstand the large pressure variations that occur during transpiration (Hacke et al. 2001; Mencuccini et al. 2007). Conduit wall thickness has been shown to scale with lumen radius raised to the second power. This conclusion is reached both on the basis of a theoretical treatment of the mechanical stresses caused by water tension and also on a wide set of empirical data analysis (Hacke et al. 2001). As this is combined with the structural scaling requirements and obvious needs to balance between water uptake and transpiring surfaces, it is clear that under given pattern of conditions, there is a trade-off between carbon cost of water transport and the potential carbon gain associated with transpiration. Increasing the structural investment will allow higher transpiration rate and potentially larger carbon uptake rate, but there is necessarily a threshold level when higher marginal investment to water uptake and transport will actually give lower carbon gain. On short timescale, such behaviour is possible, but if sustained, it will necessarily lead to tree death.

Quantitative Treatment of Water Transport in Woody Structures

For the quantitative treatment of water transport, the cell-level features can be considered at tissue level, and the xylem becomes porous material with position-specific properties. We can now consider the emergent water transport properties of a wood element consisting of all the sapwood at the height in consideration. The wood elements form a chain, and tension pulls water upwards in the chain (Fig. 5.4).

Specified Basic Concepts and Ideas

Basic idea WT1: The water flow between two sequential wood elements is proportional to the pressure difference between the elements. Viscous pressure losses along the transport pathway that depend on cell structure can be simplified with conductivity of porous material.
The wood reacts to the tension.

Basic idea WT2: Wood elements are elastic material, and they react reversibly to changes in tension.

The basic ideas WT1 and WT2 can be formulated as dynamic model of the amount of water in a wood element. Let $g_{H_2O}^{\text{in}\,i}(t_j)$ denote the water flow into the

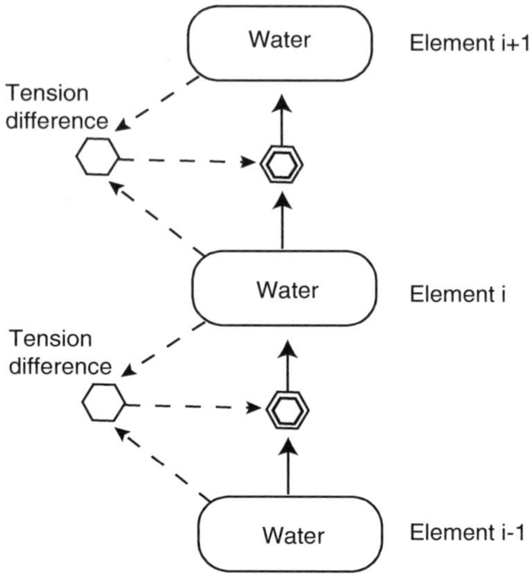

Fig. 5.4 Water tension difference pulls water through a stem element. Symbols explained in Fig. 2.3

wood element i at the moment t_j and $\overset{\text{out}\,i-1}{g_{H_2O}}(t_j)$ outflow correspondingly. The wood elements form a chain, and this is why

$$\overset{\text{in}\,i}{g_{H_2O}}(t_j) = \overset{\text{out}\,i-1}{g_{H_2O}}(t_j). \tag{5.2}$$

Let further $P^i(t_j)$ denote the water tension in the element i at the moment t_j. According to the basic idea TW1, the tension difference generates the flow between elements; thus,

$$\overset{\text{in}\,i}{g_{H_2O}}(t_j) = -k(P^{i-1}(t_j) - P^i(t_j) + \rho g)A, \tag{5.3}$$

where k is the area-specific hydraulic conductance and A is the sapwood area of the sapwood element. Let r denote the radius of the stem and r_{hw} the radius of heartwood. Then,

$$A = \pi(r^2 - {r_{hw}}^2). \tag{5.4}$$

The water flows change the amounts of water in the element i, $m^i_{H_2O}$, and we obtain from the conservation of mass

$$m^i_{H_2O}(t_j) = m^i_{H_2O}(t_{j-1}) + (\overset{\text{in}\,i}{g_{H_2O}}(t_{j-1}) - \overset{\text{out}\,i+1}{g_{H_2O}}(t_{j-1}))(t_j - t_{j-1}). \tag{5.5}$$

The water amount in the volume changes if the in and outflows in an element differ from each other. This changes the tension in the element, and according to basic idea WT2, the elastic element reacts to the change in the tension. The change in diameter, Δr, is proportional to the product of change in water tension and the radial elastic modula of the xylem E_r:

$$\Delta r = E_r(r - r_{hw})(P(t_j) - P(t_{j-1})). \tag{5.6}$$

The water uptake, transport and transpiration system expand from soil to the top of the tree. The transport within this system is driven by water tension differences as well as the water uptake from the soil. Transpiration is driven by environmental factors and controlled by stomatal action.

As explained before, the pulling force of water transport initiates through transpiration at water air interface in the foliage. The surface of the leaf mesophyll surface is rather uneven. There is roughness at the scale of 10 nm caused by the microfibrils running in the mesophyll cell walls. When leaf water supply relative to transpiration is abundant, then the surface tension makes the surface of the water film on the mesophyll cell wall even and it is not following the roughness of the underlying cell wall. When the water film on the mesophyll cell wall becomes thinner, it starts to follow the underlying roughness (Fig. 5.2). Surface tension then generates tension in the water on the mesophyll cell wall. During high transpiration, the needle tension is approximately 2 MPa and consequently the thickness of the water film on the mesophyll wall is only approximately 10 nm.

In the fine roots, a cylinder of living cells surrounds the conductive tissue. The sugar content is high in the living fine root cells next to the soil particles. They are covered with a water layer that has a low solute concentration. The large solute concentration difference between fine root cells and soil water generates osmotic water flow from soil solution into fine root cells. Transpiration from leaves generates water tension that propagates into the tracheids of the central vascular tissue in the fine root cells. The high sugar concentration in the endodermis cells pulls water from the soil through apoplastic pathway, and a tension difference between xylem and endodermis cell moves water into xylem (Sect. 4.1.2). The water transport system is visualised in Fig. 5.5.

Two basic ideas cover the generation of the water tension into the tracheids next to mesophyll cells.

Basic idea WT3: Surface tension generates tension into the water on mesophyll surface.

Basic idea WT4: Water tension on mesophyll surface pulls water from tracheids into the water film on the mesophyll cells.

Basic idea WT5: The thickness of the water film on the mesophyll cells is rather stable.

Thickness variations of the water film at air water interphase are negligible and do not need to be considered in the water volume calculations.

The water uptake by fine roots provides additional material for the transport system.

Basic idea WT6: The tension and sugar concentration difference between soil and xylem water determine the water flux from the soil into the water transport system.

The basic ideas WT3–WT5 result that the water flow out of the stem volume element next to the needles equals with transpiration. Assume that there are n woody

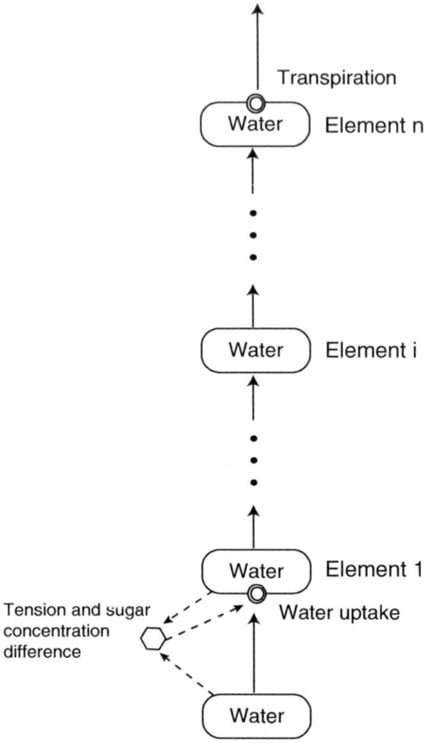

Fig. 5.5 Water uptake and flow through a tree

elements in the water path way. Let t_r denote transpiration rate from the needles. The water balance equation for the top volume element n is

$$m^n_{H_2O}(t_j) = m^n_{H_2O}(t_{j-1}) + (g^{\text{in}\,n}_{H_2O}(t_{j-1}) - t_r(t_{j-1}))(t_j - t_{j-1}). \tag{5.7}$$

The basic idea WT6 allows the characterisation of the influx into the root volume element next to the fine roots. Let u_{H_2O} denote the water uptake rate and P_S water tension in the soil. The tension and sugar concentration difference between soil and xylem water determines the water uptake according to the basic idea WT6. We omit for simplicity, however, the sugar concentration difference in the water uptake and get

$$u_{H_2O}(t_j) = k_{\text{soil}}(P_1(t_{j-1}) - P_S(t_{j-1})). \tag{5.8}$$

Parameter k_{soil} value is fairly constant, except in the case of drought when it is strongly dependent on soil water tension. The same approach, that is, mass balance, can also be used to the lower end of the woody water pipe from soil to the needles. Then, the mass balance equation for the first element in the stem is

$$m^1_{H_2O}(t_j) = m^1_{H_2O}(t_{j-1}) + (u_{H_2O}(t_{j-1}) - g^{\text{out}\,1}_{H_2O}(t_{j-1}))(t_j - t_{j-1}). \tag{5.9}$$

Equations 5.2, 5.3, 5.4, 5.5, 5.6, 5.7, 5.8 and 5.9 characterise the water transport in the woody structures in a tree.

Field Measurements

Small but still easily detectable changes in tree stem and branch diameter reflect variations in plant water status (e.g. Klepper et al. 1971; Simmoneau et al. 1993; Zweifel et al. 2001) as volume of living cells (Nobel 1991) and xylem conduits (Irvine and Grace 1997) vary according to pressures they are subjected to. Also cambial growth can also be clearly distinguished in the stem diameter (Génard et al. 2001; Steppe et al. 2006). The changes in cell and xylem conduit pressure arise mainly due to changes in the transpiration rate which determines the water tension in the xylem. Water tension in the bark tissue follows the water tension variation of the xylem as tension tends to equilibrate within the tree. After the removal of the outer bark, phloem is the main extra-xylem tissue which contributes to the whole tree stem diameter variation.

Changes in xylem diameter d_x reflect changes in xylem water pressure P_x according to Hooke's law (Irvine and Grace 1997; Perämäki et al. 2001):

$$\frac{d_x(t_j) - d_x(t_{j-1})}{t_j - t_{j-1}} = \frac{d_{x,0}}{E_{r,x}}(P_x(t_j) - P_x(t_{j-1})), \tag{5.10}$$

where $d_{x,0}$ is the xylem diameter at a zero tension and $E_{r,x}$ is the radial elastic modulus of the xylem tissue. The same can be written for the inner bark diameter d_b:

$$\frac{d_b(t_j) - d_b(t_{j-1})}{t_j - t_{j-1}} = \frac{d_{x,0}}{E_{r,b}}(P_b(t_j) - P_b(t_{j-1})), \tag{5.11}$$

where P_b is now the inner bark pressure, $d_{b,0}$ is the inner bark diameter at a reference pressure and $E_{r,b}$ is the radial elastic modulus of the inner bark tissue.

Simultaneous xylem and stem diameter measurements have been conducted for 10 years at the Helsinki University research station in Hyytiälä, southern Finland, on now 47-year-old, approximately 15 m tall, Scots pine (*Pinus sylvestris*) trees (DBH approximately 15 cm) at heights of 1.5 and 10 m. The diameter variation measurement system consists of two pen-like diameter variation sensors (LVDT; Solartron AX/5.0/S, Solartron Inc., West Sussex, UK) attached next to each other to a rectangular metal frame mounted around the stem. We have also measured the temperature of the stem (1 cm depth) and the metal frame using copper–constantan thermocouples and corrected the data for the effects of thermal expansion of wood and the frame (see Sevanto et al. 2005). The data have been collected with frequency min^{-1}. Stem diameter was 15 cm (at 1.5 m) and 5 cm (at 10 m) at the measuring heights, and the inner bark thickness was approximately 2 mm at both heights. The inner bark diameter change was simply calculated to be the whole stem diameter change (excluding the bark) minus the xylem diameter change.

Figure 5.6 shows variation of xylem diameter at two different heights during 1 day, 3 July 2007. Xylem diameter follows transpiration with a time lag of the order of few minutes. This fast pressure propagation in xylem allows the different parts of the tree sense changes in water status in other parts of the tree almost immediately,

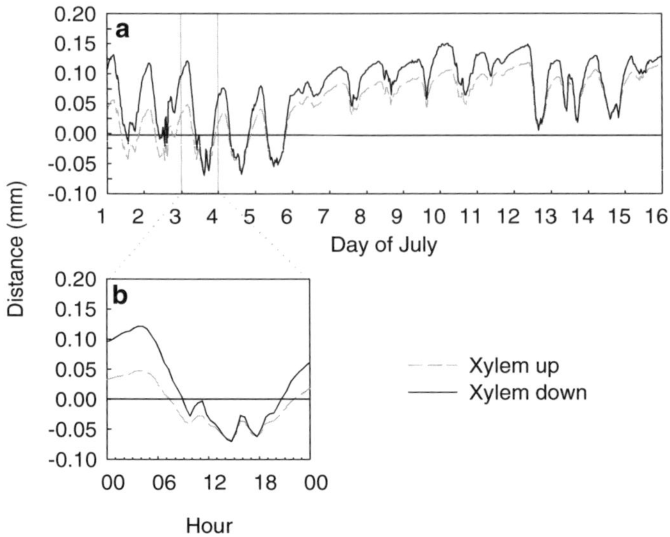

Fig. 5.6 Distance of xylem surface from a reference point at two different heights during 1–16 July 2007 (**a**) and a detailed view on the daily pattern during 3 July (**b**)

while the transport time of individual water molecules, ions and hormones is of the order of hours or days. The time lag in pressure and diameter grows (not shown) with increasing tree size and with decreasing xylem conductance and elasticity modulus of the tissue.

Test of the Theory with Field Measurements

Estimation of Parameter Values. For calculating xylem tension from diameter change $E_{r,x}$, the radial elastic modulus of the xylem tissue was 750 MPa (Perämäki et al. 2001), and $E_{r,b}$, the radial elastic modulus of the inner bark tissue, was estimated to be 20 MPa.

Prediction of Characteristic Features of Water Transport. Figure 5.7 shows measured and modelled xylem diameter change for 9 days in August 2007. The transpiration rate, fed as input to the model, was taken from Eddy Covariance measurements (see Sect. 5.4.3 in this book). The model predicts the pattern of diameter changes rather well; however, the simulated diameter increases too fast, especially at low tensions. Direct comparison between simulated and observed diameter change is not possible as about 30% of the transpiration measured with Eddy Covariance originates from ground vegetation. However, also Perämäki et al. (2001, 2005) paid attention to slower than predicted diameter increase in the evenings as discussed earlier. This delay probably reflects the storage effect of the living stem tissue.

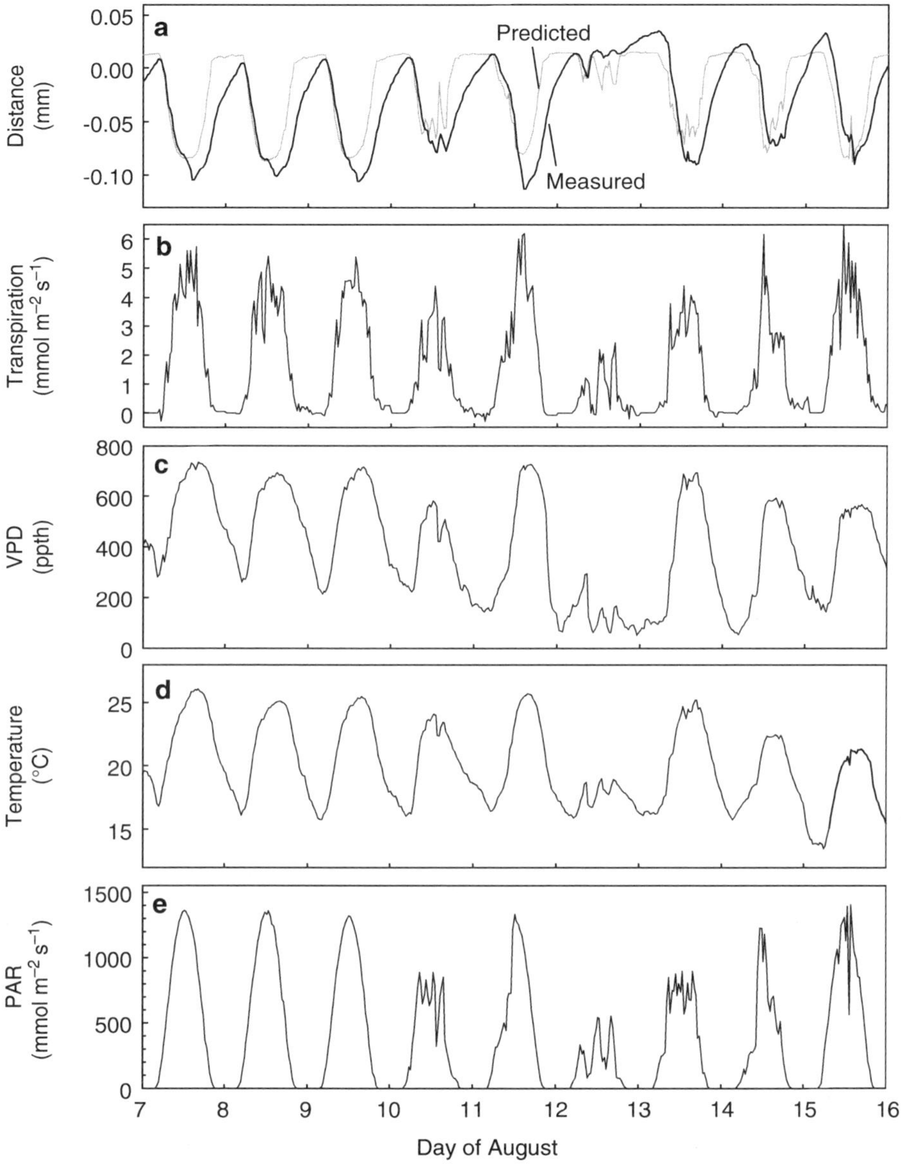

Fig. 5.7 Measured and predicted distance of xylem surface from a reference point (**a**), transpiration (**b**), water pressure deficit VPD (**c**), temperature (**d**) and photosynthetically active radiation PAR (**e**) at SMEAR II, southern Finland, during 7–16 August 2007

There are seldom cloudy, moist and non-rainy days at SMEAR II; during these days, the diameter changes are rather small, and the measured value fluctuates rather randomly around stable mean value. The model predicts very similar behaviour

(Fig. 5.8). During rainy days, especially after a dry spell, there are some additional phenomena involved and the diameter increases in a quite irregular manner. The transpiration pattern is fairly regular during most of the summer days, but there are days when heavy clouds interrupt the travelling of small clouds across the sky. During these days, the transpiration is suddenly dropped to low values and the diameter responses to the change in transpiration (Fig. 5.8). The predicted and observed diameters show similar behaviour.

Adequacy of the Model Structure. It is typical that measured xylem diameter lags somewhat behind the modelled xylem diameter, particularly at nigh upon refilling of the stem. This is probably due to water storage/withdrawal from storage, which is not included in the model, except for the case of elastic deformation of tracheids, that is, the water storage, for example, within capillaries at the end of tracheids and fibres (Brough et al. 1986; Tyree and Zimmermann 2002), or the water released by cavitation are not included in the model. The residual figures of the diameter change demonstrate these phenomena (Fig. 5.9). The residual between the measured and predicted diameter increases when transpiration decreases (xylem diameter increases) and vice versa for all of the cases. This highlights the effect of water storage seen in the xylem diameter change which is lacking from the model.

Explaining Power of the Theory. Water tension induced by transpiration explains the observed xylem diameter changes. The discrepancy at low tensions indicates possible problems in the model structure, and it is possible that some minor phenomenon is missing in the model structure. The problem could be caused by, for example, non-linearity between tension and xylem diameter at low tensions or by charging and retrieving water from water stores other than those caused by elastic deformation of tracheids (e.g. "capillary storage", see, e.g. Tyree and Zimmermann (2002), or by embolism and refilling of embolisms, see Hölttä et al. (2002)). The response to rains requires additional understanding of the behaviour of water tension in the soil and water uptake of trees. However, one also needs to point out that some of the non-explained variation may stem because the transpiration was derived from Eddy Covariance measurements that correspond to whole ecosystem evapotranspiration and do not represent directly tree transpiration.

The modelled diameter explained 0.61% of the variance of the measured values. Higher proportions could possibly be achieved if xylem conduit properties and leaf area were distributed more realistically in the model structure (now it was extremely simple) and if parameter values were fitted to the data (instead of chosen to represent values for a typical Scots pine tree)

Comparison with Other Approaches. Xylem water flow and water tension in a single tree have been modelled in many previous studies. However, typically models are based on the use of electrical circuit analogy and not on the underlying physical principles as with the model presented here. Another problem with analogy approaches is that their results are not directly comparable with field measurements. Another, and a more widely used method, for measuring xylem water tension is the

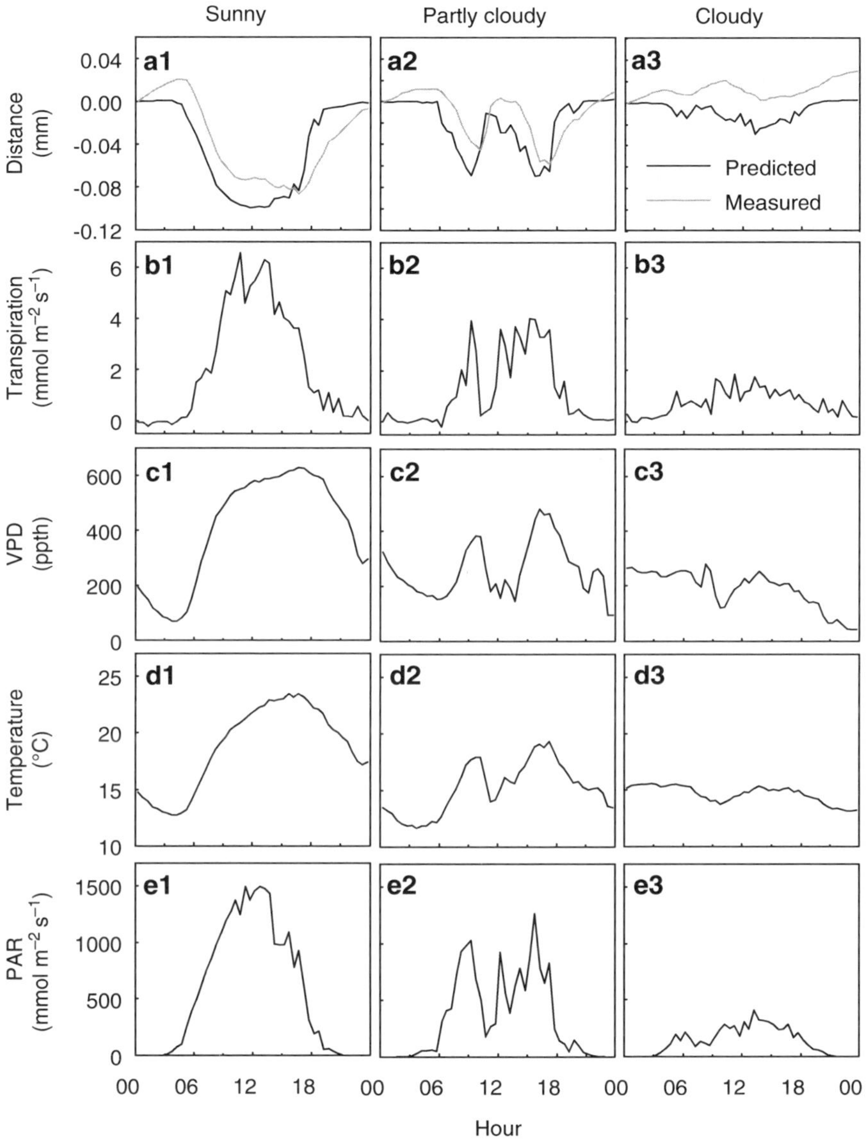

Fig. 5.8 Measured and predicted distance of xylem surface from a reference point (**a1–a3**), transpiration (**b1–b3**), water pressure deficit VPD (**c1–c3**), air temperature (**d1–d3**) and photosynthetically active radiation PAR (**e1–e3**) at SMEAR II, southern Finland, during sunny, partly cloudy and cloudy and moist days in July 2007

xylem pressure chamber (Scholander et al. 1965), and other widely used methods for measuring water flow in stem are heat-based methods such as the Granier sap flow sensors (Granier 1987) and cuvette measurements.

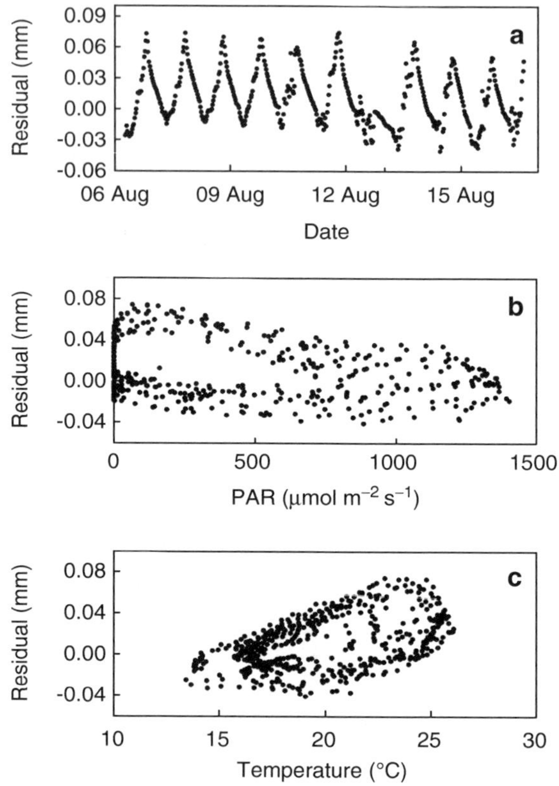

Fig. 5.9 Residuals between measured and predicted distances from a reference point at SMEAR II, southern Finland, during 7–16 August 2007 as a function of time (**a**), photosynthetically active radiation PAR (**b**) and temperature (**c**)

Annual Cycle in Stem Water Flow

The annual cycle in metabolism is strong, particularly in boreal zone where trees are inactive and dormant in the winter. Trees recover during spring, and the activity peaks in summer and declines to dormancy during autumn. The strong annual cycle in the environment generates also an annual cycle in the water flow in the stems. The strong annual metabolic cycle raises the question of annual changes in the water transport properties of stems.

The amplitude of the daily diameter changes reflects the need for water transport during a day, and there should be a linear relationship between the daily transpiration and daily amplitude of diameter changes. Figure 5.10a shows a rather clear linear relationship explaining 64% of the variance of the daily diameter change amplitude.

If the annual changes in wood reduce the water transport properties in the stem, then the residuals of the linear regression should show systematic behaviour in the spring and in the autumn. Figure 5.10 indicates no systematic behaviour in the residuals. Thus, evidently the water transport properties of stem wood are quite stable, and we can consider the annual cycle in stem wood negligible. Factors which would influence water transport properties, that is, xylem hydraulic conductance, are

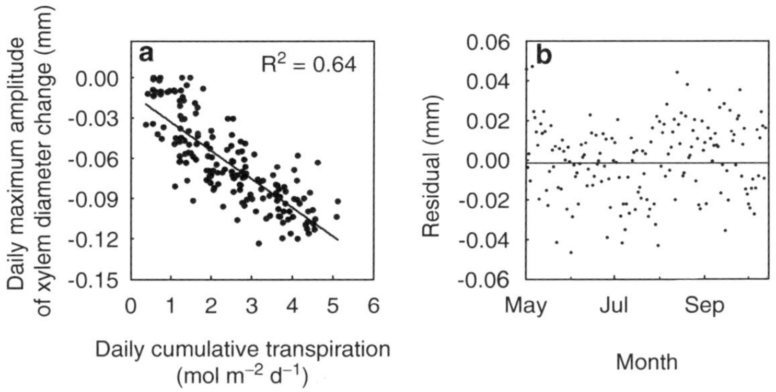

Fig. 5.10 Comparison between daily maximum amplitude of xylem diameter change and daily cumulative transpiration (**a**) and residual of the linear regression in the panel 'a' as a function of date (**b**) at SMEAR II, southern Finland, during the summer 2007. Each point represents one calendar day between 1 May and 15 October

changes in xylem sap viscosity mediated by changes in temperature (e.g. Thompson 2006), changes in the ratio of embolised conduits (e.g. Hölttä et al. 2009) and changes in pit membrane structure due to, for example, changes in xylem sap ionic concentration (e.g. Zwieniecki et al. 2001).

5.1.3 Assimilate Transport in Phloem

5.1.3.1 Cell Scale

Transport of assimilation products from mature leaves to the stem, roots, immature leaves and other organs occurs via phloem. Sugars are transported in the phloem along a continuous pathway of sieve space made out of highly specialised, living phloem sieve cells (gymnosperms) or sieve elements (angiosperms). The phloem functional conduit occurs via the cell-to-cell connections at their perforated end walls or through pore-like openings in the conifers, where these pores are rather evenly distributed, in contrast to perforated end walls of sieve tubes in angiosperms. The sieve elements are living cells although they have lost their nucleus. However, there are a number of cell organelles left such as plastids and mitochondria in a thin cytoplasmic layer anchored in the plasma membrane, and the sieve elements are thus capable of individual metabolism, for example, sealing damaged sieve elements from the vascular connection and thus preventing leakages from the phloem. In gymnosperm sieve cells, there are also remnants of endoplasmic reticulum (ER) that is lacking in angiosperms. One or several mature sieve cells are closely connected to large Strasburger cells in conifers, while sieve elements are connected to companion cells in angiosperms.

Both companion cells and Strasburger cells have numerous mitochondria and ribosomes indicative of high metabolic activity. These pairs of cells are connected to each other through a number of plasmodesmata, and these companion cells have an important role in controlling the loading and unloading of sugars into the sieve cells. This is why the phloem transport tissue normally consists of sieve element–companion cell (SE–CC) complexes. The plasmodesmata between the companion cells and the surrounding tissue (transfusion parenchyma) are far fewer in plants that are active loaders, while in many trees, the SE–CC complex is well connected to mesophyll with plasmodesmata (Turgeon 2010). In gymnosperms, there is also transfusion tissue and endodermis-like bundle sheath that is lacking in angiosperms. In the transfusion tissue, there are both dead tracheids and living parenchyma cells (Liesche et al. 2010). The former has been suggested to be involved in the conifer drought resistance and winter hardiness (see review by Liesche et al. 2010), while the latter provides symplastic continuum from bundle sheath to Strasburger cells.

5.1.3.2 Transport Element Mechanisms

Loading–Unloading Turgor

The driving force and mechanisms for phloem transport have been under dispute. The pressure-flow model, as proposed by Ernst Münch in the 1930s, is now quite commonly agreed to explain phloem transport and has also gained experimental support (Knoblauch and van Bel 1998). According to the pressure-flow model, the active loading of solutes onto mature leaves, "sources", lowers the osmotic pressure of the semipermeable sieve elements and draws in water from the surrounding tissue. As the cell walls are rigid, the turgor pressure of the cells increases and creates a positive pressure that pushes water bulk flow. There are no membranes between the sieve elements, and water is free to flow along the pressure gradient. An opposite reaction takes place at the sites where the sugars are unloaded from SE–CC complexes. The unloading increases the osmotic pressure, and water leaves phloem which decreases the turgor pressure and enhances the turgor pressure gradient from sources to sinks (Fig. 5.11).

The loading at the source is predominantly active in herbaceous plants, while in trees, it may take place also passively down sugar concentration gradient through symplastic connections from mesophyll to phloem (Turgeon 2010). The active loading takes place either through symplastic connections between the surrounding cells and the companion cells, or it may be apoplastic meaning that the sugars are taken in to the companion cells from the spaces between the cells. The active symplastic loading does not require active transport as such, but it is based on polymer trapping where sugars diffusing into companion cells are transformed into other forms such as raffinose that cannot diffuse back to mesophyll and are transported in axial phloem (Turgeon 2010). Apoplastic loading involves active transport across the membranes to companion cells.

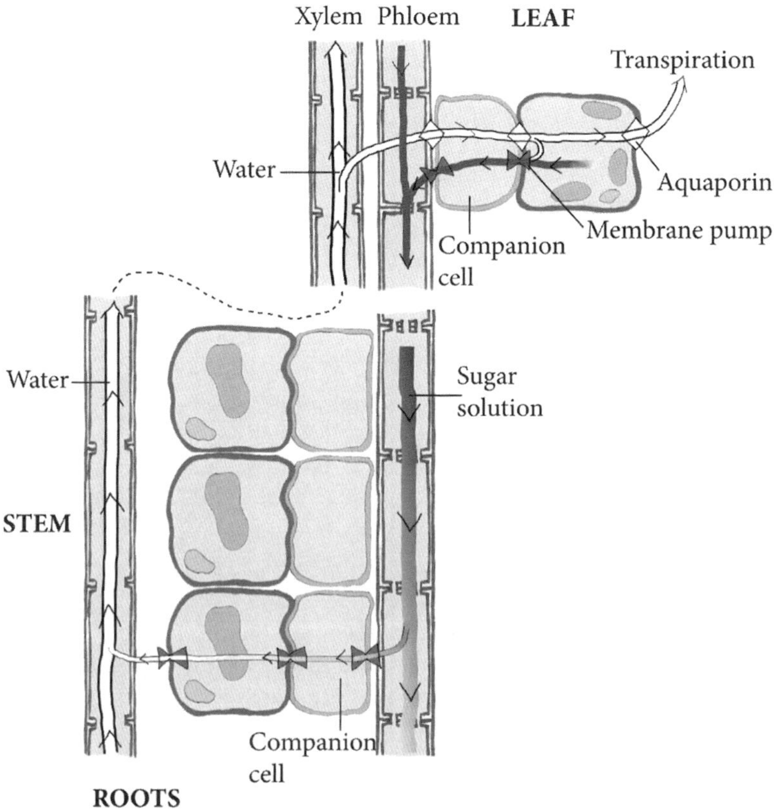

Fig. 5.11 Xylem and phloem transport at the whole tree level. Water tension pulls water in the xylem to leaves and pressure difference transports sugars in phloem from leaves to roots

Apoplastic loading is able to accumulate sugars (mostly oligosaccharides, typically sucrose predominates) into transporting veins from surrounding tissue, while in particular, the passive loading takes place down the concentration gradient from assimilating tissue to phloem. Accordingly, the companion cell structure between the different loaders is different. While there are only very few plasmodesmata in the companion cells of apoplastic loaders, they may be numerous in passive loaders (Turgeon 2010). Turgeon (2010) suggested that these differences are not to do with the plants ability to transport sugars over distances as many of the passive loaders are angiosperm and presumably also gymnosperm (Liesche et al. 2010) trees, but they are more to do with benefit of growth or avoidance downregulation of photosynthesis as leaf concentration of sugars can be maintained at lower levels (Turgeon 2010).

Viscosity

There is an absolute upper limit to the phloem sugar transport set by the increasing viscosity of phloem sap with increasing sugar concentration as sap viscosity, and phloem conductance, is an exponentially rising function of sugar concentration. The viscosity rise could be alleviated, yet not entirely avoided, by a contribution of other osmolytes, such as potassium, that do not increase viscosity even in high concentrations while contributing to the osmotic pull of water to the phloem. Potassium has been measured in the phloem at concentrations which are comparable to sucrose (Lang 1983; Thompson and Zwieniecki 2005). However, viscosity remains an important issue for sugar transport as the leaves need to be able to move sugars away from the sources. Too low a transport rate will cause sugar accumulation in the photosynthesising leaves which leads to downregulation of photosynthetic capacity and production losses (Paul and Foyer 2001). This is likely in particular in passive-loading trees.

Unloading

The unloading of sugars from sieve elements takes place passively in symplastic pathway from high concentration in sieve elements to low concentration in surrounding tissue. It has been suggested that the transport is driven by diffusion, but also mass flow needs to be considered, in particular, when the sieve element turgor is high (Patrick 1997). However, there is also some evidence on facilitated transport to the sink tissue that would help to maintain the axial carbohydrate transport (Patrick 1997). In addition to symplastic unloading, there is also apoplastic unloading or leakage of sugars from sieve elements to apoplast (McQueen et al. 2005). Such apoplastic leakage and retrieval have an important role in maintaining phloem transport during short-term imbalances between source and sink processes. Minchin and Thorpe (1984) showed that considerable amounts of sugars may be found in stem apoplast. McQueen et al. (2005) showed that sugars are also retrieved from there. The leakage of sugars from sieve elements to stem apoplast may also be important for water transport in xylem. Hölttä et al. (2006b) showed that it is feasible that elevated turgor pressure in phloem tissue could be responsible for refilling of embolised xylem tissue under certain conditions. If there was associated leakage of sugars into xylem tissue, it would facilitate refilling (Salleo et al. 2004).

Wound Blocking

In phloem, the sap is under positive turgor pressure in contrast to the negative water pressure in xylem. In this way, the transport system of assimilates is comparable to that of humans as any puncture in the transport system may result into leakage and loss of pressure driving the transport. No wonder the phloem tissue is often under thick bark layer that protects it. As such, phloem is attractive for different feeders due to its high content of sugars but also the numerous other substances that there are in the phloem sap, including amino acids (Turgeon 2010).

In case there is a puncture in phloem, the p-proteins in sieve elements undergo coagulation blocking the punctured wall. The p-proteins are attached to the walls of the intact sieve elements, but upon wounding, they are released and transported with mass flow to the punctured site and to sieve pores effectively blocking the flow.

5.1.3.3 Tissue Scale: Phloem Transport Mechanisms

Pressure Propagation in Phloem

The loading of sugars to the phloem near the leaves reduces the osmotic potential and draws in water from the surrounding tissue and raises the water pressure, that is, turgor pressure. Similarly, the unloading of sugars elsewhere lowers water pressure. According to the pressure-flow theory, this axial water pressure gradient in the phloem sieve cells is the driving force for phloem translocation.

Pressure propagates in the phloem slower in comparison to the xylem as the phloem tissue is more elastic. In addition to pressure propagation, also phloem sap sugar concentration can propagate much faster in comparison to individual sugar molecules. Information on local changes in water pressure and sugar concentration may therefore be transmitted faster than individual molecules in the form of pressure and concentration waves, provided that the ratio of solution osmotic concentration to turgor pressure differential is large (Thompson 2006). A meta-study by Mencuccini and Hölttä (2010) provided empirical evidence that variation in leaf photosynthetic rate is propagated faster to the roots and measured as soil respiration than what tracing of individual sugar molecules transported through the phloem would indicate. This indicates the significance of pressure and concentration waves in the phloem propagating information at a higher velocity than that of bulk phloem transport.

Diameter Variation

Most of the change in whole stem diameter (excluding bark) is induced by changes in the phloem tissue. If changes in xylem water tension would be the sole factor driving diameter variation of the inner bark, then its diameter, discarding cambial growth, should simply be expected to follow changes in xylem diameter in a predictable way depending on the hydraulic coupling between the xylem and the inner bark and the elastic properties of the tissues. The inner bark tends towards water potential equilibrium with the xylem by exchanging water with the xylem when in disequilibrium. The changes in whole stem diameter follow the xylem water tension and the general water status of the tree quite closely (e.g. Klepper et al. 1971), but xylem water potential alone cannot completely explain the stem diameter dynamics. Phloem sugar dynamics are therefore also believed to influence the inner bark diameter variation (Sevanto et al. 2002, 2003).

Phloem Relays

A solute relay mechanism has been suggested. According to the formulation by Lang (1979), the phloem could consist of various sieve tubes connected in series. Translocation would occur according to the Münch hypothesis within a single sieve tube, while solutes would be metabolically transferred from one sieve tube to the next. However, the presence of relays remains hypothetical at the moment as there has been no substantial experimental evidence in favour or against the existence of solute relays, although continuous leakage and reloading of solutes has been observed along the phloem translocation pathway (Minchin and Thorpe 1987; McQueen et al. 2005). Lang (1979) estimated, based on the 1:1 stoichiometry of moles of ATP used per mole of sucrose transported, that approximately 2% of sugars would be used up in each relay. Theoretically, this metabolic cost could be compared with the carbon cost of building more and wider continuous phloem sieve tubes without relays to achieve identical sugar translocation rates. The marginal return of adding more and more relays on the number of required conduits tended to decline rapidly, and with more than just a few relays, the benefits became very small.

5.2 Water Dynamics in Forest Soil

Jukka Pumpanen

5.2.1 Small-Scale Water Fluxes in Forest

Typically, the soil matrix consists of mineral and organic particles and space between them. The soil porosity in the mineral soil matrix ranges between 40 and 50% and in organic soil up to 90%. This pore space is filled with water or air depending on the soil water status. Water retention curves show the relationship between soil water tension and volumetric water content of the soil (see, e.g. Hillel

J. Pumpanen (✉)
Department of Forest Sciences, University of Helsinki, P.O. Box 27, 00014 University of Helsinki, Helsinki, Finland
e-mail: jukka.pumpanen@helsinki.fi

1982, Fig. 4.69). The water fluxes within a forest ecosystem and between forest ecosystem and its surroundings are large.

The annual precipitation falling on forest ecosystems varies greatly between geographic locations from a few hundreds of millimetres to several metres. The amount of water entering the soil depends on the canopy structure and density. In typical boreal forest such as in Hyytiälä SMEAR II, the canopy interception is significant. It captures even as much as one-third of the annual precipitation before it enters the forest floor. The rest, throughfall, is falling on the soil surface either in snow or water and is available for plant uptake or soil processes.

Trees transpire usually about half of the rainfall entering the soil, and the other half is transported away either as surface runoff or as seepage flow depending on the depth of the soil, soil type and slope. In soils with poor hydraulic conductivity, for example, soils having high clay or peat content, the seepage flow is slow which can result in significant runoff during intensive rain events. Major surface flow also takes place when soil is already saturated with water or frozen and prevents the downward moving water flux. The proportion of surface flow is highest during snowmelt in spring when the soil is still frozen from the surface. Surface flow also takes on places where the solid bedrock prevents the vertical water flux, for example, on steep slopes or ridges.

5.2.2 A Case Study at SMEAR II, Southern Finland

The seepage flux is a major problem in studying the water balance of a forest ecosystem, since it is nearly impossible to measure. The soils in Finland and Scandinavia are formed by glaciers and the waters of the melting glaciers during the last ice age 115,000–10,000 years ago. The soil was transported by the ice from its original place resulting in glacial tills in places where the ice was melted and rocks, stones and soil particles of all size classes were left on the site. The area covered by ice obtained its mineral soil from the melting ice, and it forms a shallow layer on the bedrock. When we planned SMEAR II measuring station (see Chap. 9), we detected tiny watersheds on the top of a hill near the Forestry Field Station of Helsinki University at Hyytiälä. These mini watersheds enabled the measuring of seepage flux since the bedrock is solid granite, and there is a layer of silt having high clay content on the bedrock. We can say that the watersheds are natural lysimeters.[1] We constructed dams and instrumentation to measure the water flow on the bedrock out of the ecosystem. In addition, we measure rainfall and water storage in the soil to get a complete picture of the water balance of the ecosystem.

---

[1]A *lysimeter* is a measuring device which can be used to measure the amount of actual evapotranspiration which is released by plants, usually crops or trees. By recording the amount of precipitation that an area receives and the amount lost through the soil, the amount of water lost to evapotranspiration can be calculated.

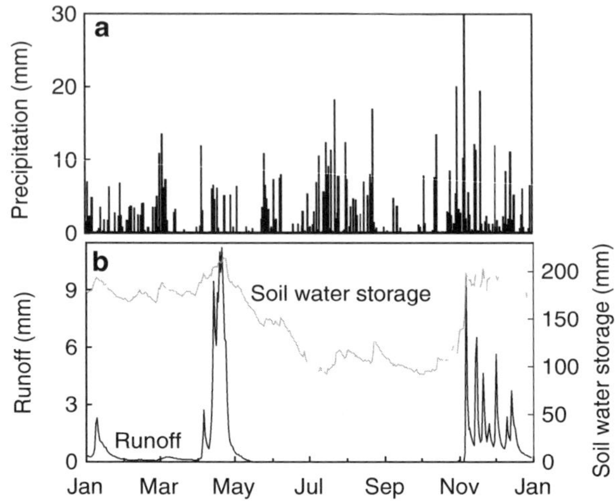

Fig. 5.12 Precipitation (**a**) and runoff and soil water storage (**b**) during a typical year 2001 at SMEAR II station, southern Finland

Figure 5.12 presents soil water storage, precipitation and runoff measured at SMEAR II in 2001. There is a typical seasonal pattern in soil water storage being highest during the snowmelt in spring decreasing towards the end of the summer. The maximum soil water storage at SMEAR II is about 200 mm. The average soil thickness at the site is 540 mm, so about half of the soil volume can be filled with water at maximum.

Water storage decreases during the summer by about 50% to only 100 mm at SMEAR II. During average years, lack of water does not limit photosynthesis and transpiration, but during extremely dry seasons, the water storage can go down to 50 mm. This corresponds to volumetric water content of 0.10 $m^3\ m^{-3}$ in the surface soil and 0.15 $m^3\ m^{-3}$ in the deeper soil layers. It also corresponds to soil water tension of –2,000 kPa in the surface soil and –700 kPa in the deep soil. When the autumn rains start, soil water storage recovers quite quickly to 150–170 mm. The water storage has recovered from summer droughts on average years by September and October.

The average annual runoff from the SMEAR II catchment is about 32% of the precipitation which was about 700 mm in the years between 1998 and 2006. Average long-term (30-year annual average between 1971 and 2000) precipitation value based on the daily recordings of Finnish Meteorological Institute at Hyytiälä Forestry Field Station is 713 mm (Drebs et al. 2002). The runoff takes place after the spring thaw in April and in most years during October and November when the soil is saturated with water.

Average annual *evapotranspiration,* that is, the sum of evaporation and plant transpiration at SMEAR II, has been determined with two methods: with Eddy Covariance and as a difference between precipitation and the sum of surface and seepage flows. The latter method is called as the *difference method* from herein. These two methods give rather different values the Eddy Covariance method giving 295 mm and the difference method 496 mm. The difference method gives surprisingly similar values to those presented by Vakkilainen (1986), 400–500 mm year^{-1} in central and south-western Finland. According to Vakkilainen (1986), the annual average evapotranspiration in Finland varies between 100 mm in northern Lapland and 500 mm in Central Finland.

The reason for the difference between these two methods is still unclear. Both methods have their benefits and drawbacks. The difference method is basically very robust due to its simplicity and well-established measurement methodology. However, the measuring accuracy of precipitation, surface and seepage flow has technical challenges. For example, precipitation measurements are very sensitive to wind and area-specific wind correction functions. Recently, a new type of infrared-based precipitation sensors have been introduced (e.g. Vaisala FD12P, Vaisala Oyj, Finland), which can overcome these problems.

Runoff is basically easy to determine from two watertight weirs held with concrete on the bedrock at SMEAR II. The weirs have been equipped with flow metres that record flow readings automatically every minute. In addition, the automated readings have been calibrated with manual flow readings and annual calibrations of the flow metres in the laboratory. But despite of the basically robust methodology, there are two things which may result in uncertainty in the runoff results: (1) the surface area of the catchment and (2) possible leakages through the bedrock. The surface area of the catchment remains the most uncertain quantity when determining the water balance of the site. However, the difference of 201 mm in the evapotranspiration values of the two methods would require that the surface area of the two catchments should be 45% smaller than the measured surface area. An error of this magnitude in the determination of the surface area is impossible.

The Eddy Covariance method has also its own problems although it starts to be rather well-established method for determining evapotranspiration fluxes. Wilson et al. (2001) have conducted an analysis of a large number of eddy sites and showed that the energy balance cannot be closed by Eddy Covariance measurements possibly because the latent heat exchange, that is, evapotranspiration, may be underestimated. Thus, a bias of tens of percents can be possible for water vapour exchange. Also Baldocchi et al. (1997) have shown Eddy Covariance giving underestimation for water fluxes in boreal coniferous forests.

Another major issue when comparing the evapotranspiration determined by the Eddy Covariance and the difference method is the fact that the footprint area of the Eddy Covariance tower is larger than our catchments which were only about 1,200 m^2 in size. The theoretical footprint area of the eddy tower is about 20,000 m^2. Although the forest area around SMEAR II station is quite homogeneous, it still contains some microsite variation that affects the hydrological properties and water flow at the site.

The tree canopy retains a significant proportion of the precipitation, about 33% on an annual basis at the Scots pine stand at SMEAR II. In the summertime, the *canopy interception* is even higher, about 37%. The canopy interception depends on the intensity of the rain the interception being higher at low rain intensities. Some part of the precipitation is also directed to the soil in stem flow, but this proportion is only negligible, about 8 mm, which is about 1.1% of the annual precipitation.

Snow is also an important component of the water balance in boreal zone. In southern Finland, the air temperature is below zero about 120–130 days per year, and consequently, the precipitation during those days is coming as snow. At SMEAR II, the annual amount of precipitation as snow is on average 190 mm of the total 713-mm precipitation. Most of the snow remains on the soil throughout the winter, and therefore, the water storage in snow is between 100 and 150 mm in March and early April. According to Solantie (1981), the average maximum water storage in snow at SMEAR II area was 120 mm and ranges in Finland from 80 mm in south-western Finland to 200 mm in eastern and northern Lapland between years 1961 and 1975. When the snow thaws, usually during a very short-time period in late April, this water storage is released to the soil resulting in runoff flow, evidently sublimation plays a minor role.

5.3 Carbon Dioxide in Soil

Jukka Pumpanen

When the energy captured in photosynthesis is utilised in the soil for metabolism of roots, microbes and soil fauna, CO_2 is released into the soil air resulting in carbon dioxide flux from soil to the atmosphere. Being about 50–70% of the annual gross primary productivity of the trees and ground vegetation, the CO_2 efflux from soil plays an important role in the carbon budget of forest ecosystems.

5.3.1 Structure and Metabolism of Forest Soil

Forest soils are generally layered: litter and organic macromolecules dominate the top layers, and the mineral material dominates deeper layers in the soil. Litter input on the surface and water transport within soil generate the layered structure (Sect. 4.3). Several different soil types have been described each of them having characteristic features. We concentrate here on podzolic soil, because it is the most common in northern forests where rainfall exceeds evapotranspiration resulting in seepage flow deeper into the soil. As a result of this downward water flux typical layered structure with distinct soil horizons, organic layer containing litter and humus on top of the soil and mineral soil A-, B- and C-horizons has been formed since the soils were released from the sea after the last glacial period. Metabolism of roots and microbes requires energy, the necessary ATP is produced in respiration and simultaneously CO_2 is released into soil water. The mitochondria converting

J. Pumpanen (✉)
Department of Forest Sciences, University of Helsinki, P.O. Box 27,
00014 University of Helsinki, Helsinki, Finland
e-mail: jukka.pumpanen@helsinki.fi

ADP to ATP utilise small carbon molecules such as carbohydrates produced in photosynthesis. Microbes obtain the energy from root exudates from roots or from small carbon molecules originating from decomposition of soil macromolecules by extracellular enzymes.

Respiration and decomposition of macromolecules are enzymatic reactions, and temperature affects them, as all other enzymatic reactions. Thus, the CO_2 production in the soil depends on temperature. When soil becomes dry, the water layer on soil particles gets thinner and the decomposition of soil macromolecules by extracellular enzymes slows down. The availability of small carbon molecules is reduced, and microbial respiration responses to the decrease in the energy source. Photosynthesis reacts also to drought, because the availability of organic substrates produced as root exudates is reduced (Ruehr et al. 2009). Thus, drought slows down respiration in the soil and also the CO_2 efflux from soil.

Root and microbial respiration generate CO_2 into the soil layers resulting in increased carbon dioxide concentration. The CO_2 concentrations can be rather high, even one order of magnitude higher than in the atmosphere. The concentration differences generate diffusion between soil layers, from higher to lower concentrations resulting in CO_2 efflux from the soil into the atmosphere (Fig. 5.13).

5.3.2 Basic Concepts and Ideas

A layered structure is present by definition in podzolic soils: organic material dominates the top layers and mineral material the lower ones. A special concept is needed to characterise the respiration in each layer. The respiration rate in each layer per unit area is defined as the CO_2 formed in the layer during short-time interval divided by the mass in the layer of unit area and by the length of the time interval.

The differences between different layers in the forest soil are large, and the layers are rather homogeneous. This gives rise to the first basic idea.

Basic idea SR1: Each layer has its characteristic respiration rate.

Respiration is an enzymatic process, and consequently, it responds to temperature.

Basic idea SR2: Respiration in each layer depends on the temperature in the layer.

Extracellular enzymes decompose macromolecules and large organic compounds. The resulting small carbon molecules are available to microbes. The action of extracellular enzymes slows down when the soil gets dry and the water film on soil particles becomes thinner. Thus, during a drought, the availability of small carbon molecules for microbial metabolism is reduced. Drought also affects photosynthesis by reducing sugars available for root respiration and for root exudates.

Basic idea SR3: Drought slows down respiration rate in the soil.

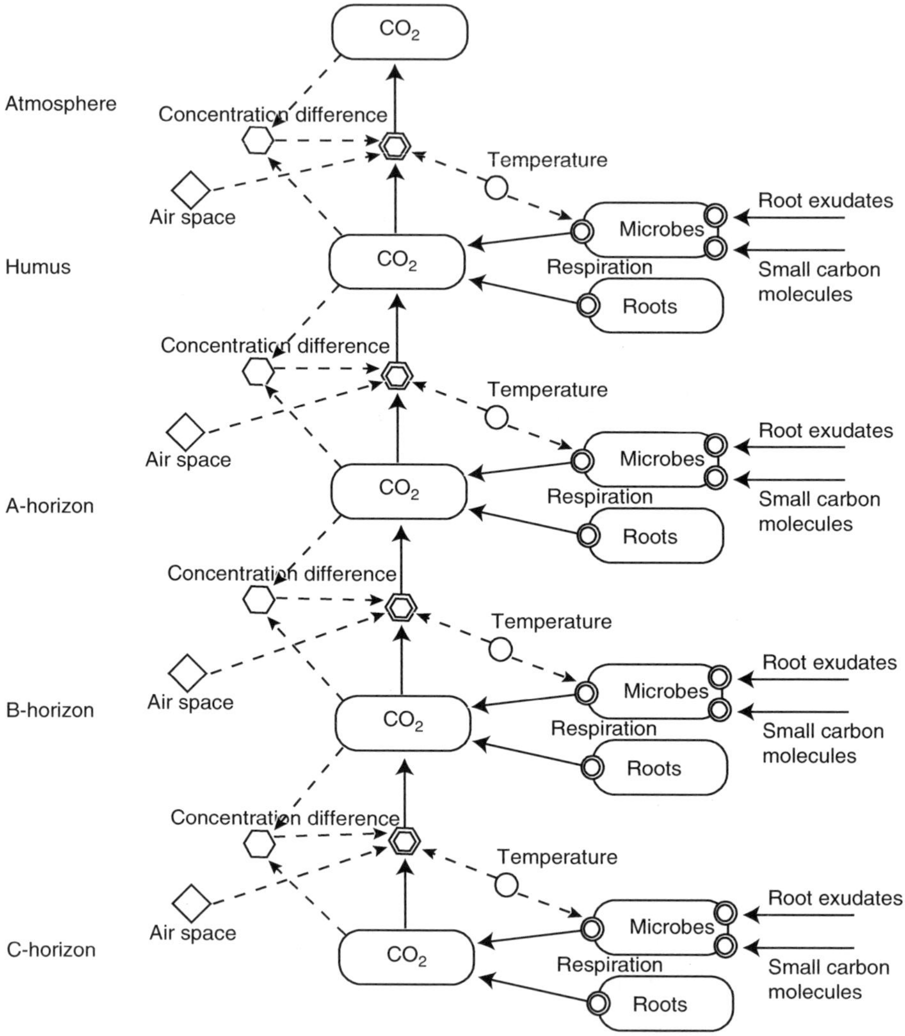

Fig. 5.13 Schematic presentation of the processes involved in soil CO_2 efflux. The *circles* represent environmental factors, *squares* structures, *hexagons* physical properties, *double hexagons* transport and *double circles* processes (Fig. 2.1)

Respiration requires oxygen, and this is why excess water inhibits respiration because then the transport cannot provide enough oxygen for respiration.

Basic idea SR4: Very wet soil slows down respiration rate.

The strong local CO_2 emissions, caused by microbial and root respiration, generate concentration differences in CO_2 in the soil.

Basic idea SR5: Diffusion transports CO_2 in forest soils generating CO_2 efflux from the soil.

5.3.3 Theoretical Model

The respiration in soil (r) is formed by microbial and root respiration. Temperature and moisture both affect microbial activity in the soil, and it is rather difficult to separate the effect of soil moisture and temperature explicitly. We can give more concrete form for the basic ideas SR1–SR4 as follows:

$$r(T, \theta) = f_T(T) f_\theta(\theta_v). \tag{5.12}$$

In the equation, $f_T(T)$ is the dependence of respiration of microbes on temperature, and $f_\theta(\theta)$ describes the dependence of microbial respiration on soil water content. The same kind of multiplicative approach has previously been used in several studies (e.g. Schlentner and van Cleve 1985; Davidson et al. 1998; Fang and Moncrieff 1999; Moncrieff and Fang 1999).

The root and microbial respirations are temperature dependent since they are enzymatic processes (Basic idea SR2). There are many possible expressions to relate the dependence of respiration in soil on temperature. Here, we use an exponential function for the temperature response of respiration $f_T{}^{\mathrm{i}}(T)$ in the layer i, an approach used by Boone et al. (1998), Buchmann (2000) and Widén and Majdi (2001):

$$f_T{}^{\mathrm{i}}(T) = \alpha_{\mathrm{i}} e^{\beta\, T} l \rho_{\mathrm{i}}, \tag{5.13}$$

where T is the temperature (°C), α_{i} are layer-specific parameters and the parameter β is not layer specific. We use similar temperature responses for both microbial respiration and root respiration within each soil horizon. This is justified by studies showing similar temperature responses for these components of soil respiration.

Respiration depends also on soil moisture and merely on the availability of substrates and oxygen for the microbes (Basic idea SR3 and SR4). Numerous studies have shown the relationship between soil moisture and microbial activity (Greaves and Carter 1920; Linn and Doran 1984; Davidson et al. 1998). Skopp et al. (1990) presented an equation taking into account both the effects of drought and anoxic conditions in wet soils approaching water saturation:

$$f_\theta(\theta_{\mathrm{i}}) = \min\left\{a\theta_{\mathrm{i}}{}^d, b(E_{\mathrm{i}} - \theta_{\mathrm{i}})^g, 1\right\}, \tag{5.14}$$

where θ_{i} is the volumetric water content ($m^3\ m^{-3}$) in the layer i and E_{i} is the total porosity ($m^3\ m^{-3}$) in the layer i. Parameters a, b, d and g are empirical constants that are fixed for a given soil (Skopp et al. 1990). At low water contents, water availability limits respiration activity in soil. This aerobic microbial activity increases with soil water content until a point is reached where water starts to restrict the diffusion and availability of oxygen. According to Linn and Doran (1984), many studies involving a wide range of soil types indicate that a soil water content equivalent to 60% of a soil's water-filled pore space achieves maximum aerobic microbial activity. In the study of Doran et al. (1988), the maximum aerobic microbial respiration for 16 soils of varying texture occurred at volumetric water content equal to 0.55–0.61 times the value of total porosity.

CO_2 concentration in the soil airspace between soil particles is often an order of magnitude higher than in the atmosphere (Fernandez and Kosian 1987; Suarez and Šimůnek 1993) resulting in a large-concentration gradient between the soil and the atmosphere. The primary mechanism for transporting CO_2 from the soil to the atmosphere is molecular diffusion (Freijer and Leffelaar 1996). According to Fick's first law, the gas flux is dependent on the concentration gradient and the diffusivity of the soil. Thus, the CO_2 flux in the soil is usually upwards, resulting in a CO_2 efflux out of the soil. The diffusion rate is dependent on the total porosity of subsequent soil layers, soil water content, the distance and the concentration gradient between the layers.

A layered structure is very characteristic for podzolic soils. This is why we have treated the soil as a structure consisting of distinctive layers and formulated our flux equations in discrete form. As an example, we present here the equations for the flux between O and A layers. The diffusion transports CO_2 in the soil between the layers according the basic idea SR5; thus, the flux between layers depends on the concentration difference. The soil layers are specified and denoted with capital letters referring to the horizons O, A, B and C. Other horizons can be obtained by changing the indexes referring to respective layers. The CO_2 flux between an A- and an O-horizon is

$$g_{\mathrm{CO_2}}^{\mathrm{AO}} = -D_{\mathrm{AO}} \frac{C_{\mathrm{CO_2}}^{\mathrm{O}} - C_{\mathrm{CO_2}}^{\mathrm{A}}}{(l_{\mathrm{O}} + l_{\mathrm{A}})/2}, \tag{5.15}$$

where $g_{\mathrm{CO_2}}^{\mathrm{AO}}$ is the flux (g CO_2 m^{-2} s^{-1}); D_{AO} is the average diffusion coefficient of CO_2 in the O- and A-horizons (m^2 s^{-1}); and $C_{\mathrm{CO_2}}^{\mathrm{O}}$, $C_{\mathrm{CO_2}}^{\mathrm{A}}$, l_{O} and l_{A} are the CO_2 concentration (g CO_2 m^{-3}) and thickness (m) of the O- and A-horizons, respectively.

The diffusion coefficient of CO_2 (D) in a soil layer is a fraction of the diffusion coefficient of CO_2 in air D_{o} (m^2 s^{-1}) since water in the soil hinders diffusion (Troeh et al. 1982):

$$\frac{D}{D_{\mathrm{o}}} = \left(\frac{E_{\mathrm{g}} - u}{1 - u}\right)^{h}, \tag{5.16}$$

where E_{g} is the air-filled porosity of soil (m^3 m^{-3}) and u and h are empirical parameters obtained from the literature (Glinski and Stepniewski 1985). D was determined separately for each layer.

When the fluxes in each layer are combined with conservation of CO_2 (Basic idea 12, Chap. 2), we obtain difference equations for the CO_2 concentration in each layer, and finally, the combination of the layer models result into model that describes the concentrations in the layers, the fluxes between the layers and the efflux from the soil.

5.3.4 *Field Measurements*

Continuous measurements of soil CO_2 efflux were made using automated chambers described in detail by Hari et al. (1999) and by Pumpanen et al. (2001). In the system, compensation air with known CO_2 concentration was introduced into the chamber at 3 dm^{-3} min^{-1} flow rate, and equal amount of air was pumped from the chamber to the CO_2 analyser. The compensation air was taken from above the tree canopy and pumped through a 0.05-m^3 steel container to eliminate possible fluctuations in CO_2 concentrations in the compensation air. The flow rates of the compensation air and the sample air were regulated by two separate pumps and mass flow controllers (5850E, Brooks Instrument, Veenendaal, Netherlands). Air in the chamber was continuously mixed by a small fan, which was installed in the middle of the chamber 10 cm above the soil surface.

The chambers were closed during the measurement periods automatically for 70 s once an hour. During the measurement period, the CO_2 concentration was recorded every 5 s with infrared CO_2 analyser ("URAS 4", Mannesmann, Hartmann & Braun, Frankfurt am Main, Germany). The same analyser was used for measuring the compensation air CO_2 concentration immediately before and after each 70-s measurement period. The calculation of CO_2 efflux was based on the mass balance equation (see Chap. 2).

We also measured CO_2 effluxes manually with manual portable system from plastic collars on randomly selected spots at the measurement site to cover the spatial variation in the CO_2 efflux at the site. Later, the readings from the automated chambers were scaled over the whole area using the data from the manual chambers.

The spatial variation in the soil CO_2 efflux was high. The coefficient of variation ranged between 0.18 and 0.45, the variation being largest during high efflux. The spatial variation in the soil CO_2 concentration was also high, the coefficient of variation, CV, ranging from 0.37 in the humus layer to 0.51 in the B-horizon. Seasonal pattern in the spatial variation was observed only in the humus layer, the variation being at its lowest in December–February (CV 0.1) and highest in April–May (CV 0.7). The great spatial variation in the soil CO_2 efflux is a major problem in the determination of the absolute level of the flux.

5.3.5 *Test with Field Measurements*

Estimation of Parameter Values

We wanted to minimise the number of the parameters estimated from field measurements, so instead we utilised values obtained from literature. The values of the parameters a, b, d and g in the dependence of respiration on soil water content (Eq. 5.14) were obtained from the research by Skopp et al. (1990).

Table 5.1 Parameters for respiration and transport functions

Soil horizon	α	β	u	h	ρ (Mg m^{-3})	l (m)	E_o (m^3 m^{-3})
O	7.613	0.1167	0	1.1	0.50	0.050	0.70
A	0.430	0.1026	0	1.4	1.68	0.054	0.61
B	0.237	0.1026	0	1.4	1.47	0.174	0.58
C	0.0800	0.1026	0	1.4	1.63	0.543	0.50

We determined the values for the parameters α and β in the dependence of respiration on temperature (Eq. 5.13) for the O-horizon using laboratory incubations carried out for humus samples collected from the measurement site in July 1998. The α- and β-parameters for the humus layer were obtained by fitting the respiration values (μg CO_2 g^{-1} h^{-1}) measured in the laboratory at four different temperatures ranging from 2 to 17°C. The respiration rates per unit of mass were converted to unit area by multiplying the mass-based values with bulk density and thickness of the soil layer. The incubations were carried out without roots, but in the model simulations, we assumed that the root and rhizosphere respiration were of the same magnitude than the microbial respiration. Thus, the respiration values obtained from the incubations were multiplied by two.

For the A-, B- and C-horizons, we used a constant value of 0.1026 for the β-parameter assuming a Q_{10}-value of 2.8 based on the studies of Boone et al. (1998) and Kähkönen et al. (2001), and the α-parameter values were scaled based on incubations carried out for mineral soil at constant +7°C temperature. The conversion from mass-based units to area based was done in the similar way as in the humus layer and assuming that the root and rhizosphere respiration was of the same magnitude than the microbial respiration of the incubations.

The value for the total porosity of the soil (E_o) was obtained from soil water retention curves determined separately for each soil layer (Mecke and Ilvesniemi 1999).

We measured the thickness of the soil and the soil density at SMEAR II. The parameter values, soil thickness and density values are summarised in Table 5.1.

Prediction of Characteristic Features of Soil CO_2 Efflux

There are four characteristic features in the behaviour of soil CO_2 efflux and in the CO_2 concentrations in the soil. The annual pattern in the soil CO_2 efflux is very clear: we observed a low efflux in early spring, an increase during spring and early summer, a high efflux during summer and finally a declining trend in late autumn. Our theory was able to predict the annual behaviour of the CO_2 efflux (Fig. 5.14).

The annual pattern in the soil CO_2 efflux is strong, high in summer and low in winter (Fig. 5.14). The predicted CO_2 efflux has very similar pattern; however, the prediction is too high in winter.

There is clear day-to-day variation in the efflux. Our model predicts also this day-to-day variation (Fig. 5.15). The daily pattern in the efflux is also clear, especially on sunny days. This daily pattern is also clear in the prediction of our model (Fig. 5.15).

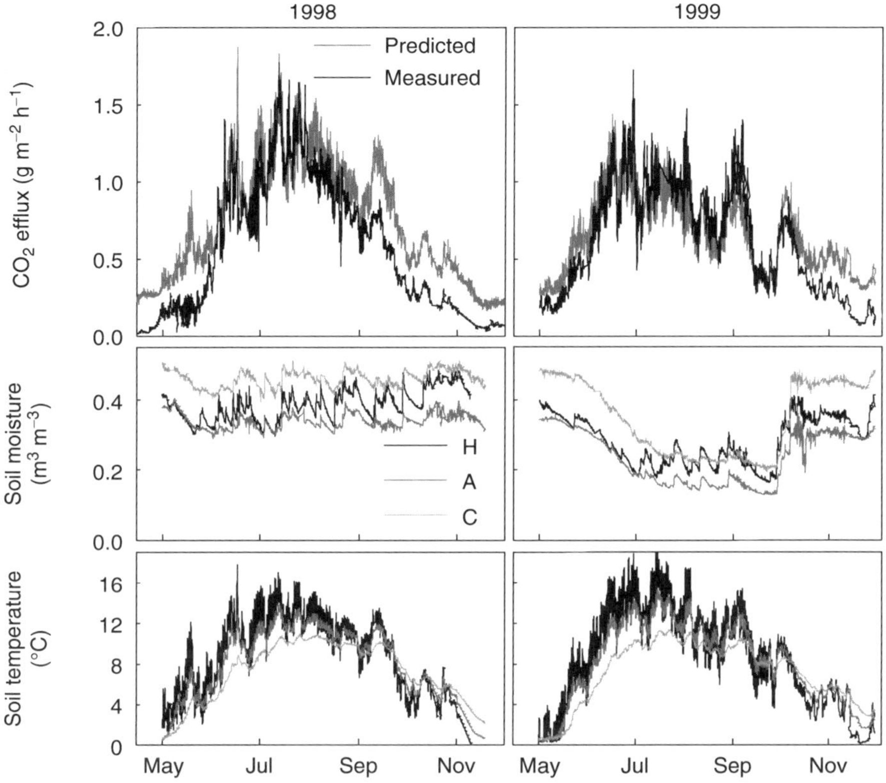

Fig. 5.14 The annual pattern in the soil CO_2 efflux is very clear (*top panel*), soil moisture at various depths in the soil (*middle panel*) and temperatures at various depths (*bottom panel*). The measurements are from SMEAR II in southern Finland in 1998 and 1999

Adequacy of the Model Structure

The carbon in the CO_2 efflux can originate from decomposition of soil organic matter, from root respiration or from microbial utilisation of root exudates. We have implicitly assumed that the CO_2 flux from decomposition of root exudates depends on temperature as the flux from decomposition of soil organic matter. This is evidently not justified since there is strong annual cycle in the carbon metabolism of trees and the emission of root exudates is high in the summer and low in the winter and early spring. Our implicit assumption may introduce unjustified model structure and consequently also bias in the seasonal cycle of the predicted CO_2 effluxes.

Recent studies have shown that plants act as carbohydrate sources, by emitting root exudates which sustain mycorrhizal fungal symbionts, but also other groups of specialised microorganisms (Högberg and Read 2006; Pumpanen et al. 2009; Heinonsalo et al. 2010). The flux of carbon belowground stimulates the decomposition of soil organic matter and nitrogen uptake of trees (Drake et al. 2011;

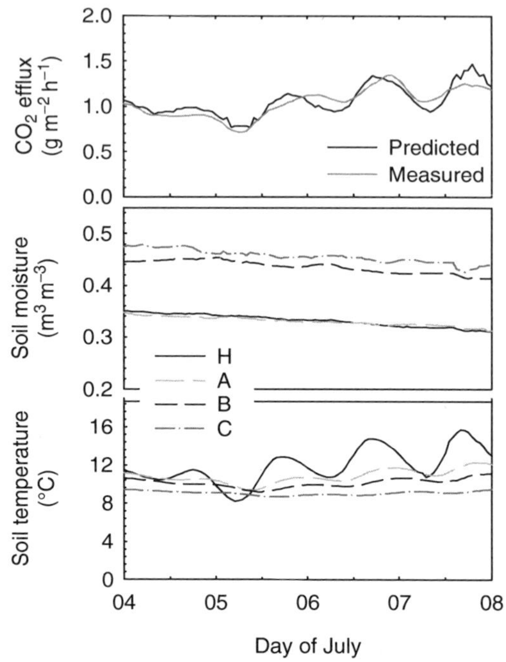

Fig. 5.15 The CO_2 efflux in southern Finland at SMEAR II during 4–8 July 1998. There is a clear day-to-day variation in the efflux. The daily pattern in the efflux is also clear, especially on sunny days

Phillips et al. 2011). It has been suggested that root exudates and other easily decomposable carbon could enhance the decomposition of old soil organic matter (Fontaine et al. 2007; Kuzyakov 2010). The ability of the model to predict the soil CO_2 effluxes in the spring and in the autumn could probably be improved by introducing the seasonal cycle of root exudation into the model. However, this would require knowledge on the seasonal pattern of the belowground carbon allocation pattern of the trees, for example, sugar transport in the phloem and seasonality of fine root production and mycelia.

CO_2 can also move between the soil layers as it dissolves in water (Šimůnek and Suarez 1993). Also, mass flow of CO_2 by convection caused by wind or atmospheric pressure fluctuations may affect the gas movement in soil especially in deep soils. However, the contribution of convection to the transport of CO_2 in shallow soils such as at SMEAR II is small. Diffusion is the dominating mechanism in the CO_2 transport within soil. Other mechanisms of gas movement than diffusion have been shown to account for less than 10% of the CO_2 lost from the upper soil and even less for the deeper unsaturated zone (Wood and Petraitis 1984). Thus, the lack of other transport mechanisms than diffusion may bias our model structure.

The residuals between measurements and predictions show evident systematic behaviour (Fig. 5.16). There is clear relationship between residuals and soil temperature in late summer and autumn. Also, the prediction gives too high efflux in the winter. The annual pattern of root exudates explains evidently the systematic features in the residuals.

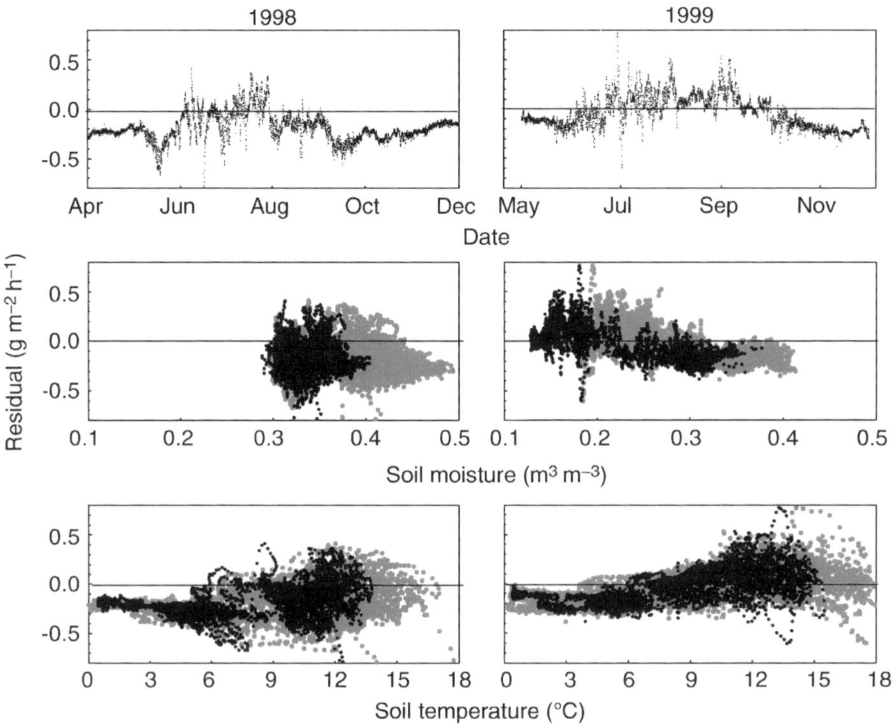

Fig. 5.16 The residuals of measured and predicted CO_2 efflux as function of time, soil moisture and soil temperature. The *black dots* indicate the soil temperature and moisture in humus and the *grey dots* in A-horizon

Explaining Power of the Theory

The strong annual cycle in temperature explains the annual pattern in soil CO_2 efflux. The winter soil temperature of about 0°C reduces the microbial and root metabolism resulting in very low CO_2 production and efflux. In summer, the high soil temperatures of about 15–20°C accelerate metabolism, and also, the efflux is higher. The soil temperature explains also variation between days and the daily pattern of efflux. The inefficient transport by diffusion generates the high concentrations of CO_2 in the deep layers of soil.

Because the diffusion of CO_2 in air is about 10,000 times faster than in water, the annual variation in weather conditions and consequent soil water content had a substantial effect on CO_2 concentration in the soil profile. In 1998, the soil in the young forest was close to field capacity, and the CO_2 concentrations in deeper soil were nearly twice as high compared to dry conditions in the following summer. High soil water content also affected the CO_2 production in deeper soil layers resulting in a very low respiratory activity. But still, due to slow diffusion, even a minor respiration was enough to maintain high CO_2 concentration. Contrary to this, during the drought, the respiration of the C-horizon was higher and exceeded that of the A-horizon, but still the CO_2 concentrations in deeper soil layers decreased because of faster diffusion in increased air-filled pore space. Soil water content may also have affected the soil CO_2 concentration by limiting microbial activity A- and B-horizons during the extreme drought.

Test runs with the model showed that soil air CO_2 concentration was even more sensitive to soil water content than the CO_2 efflux. If the effect of water was not included in the model, predicted CO_2 concentrations were 3–16 times higher than what was actually measured and predicted when the water content was taken into account. Similar results were also presented by Šimůnek and Suarez (1993). Their sensitivity analysis with different water contents showed that, in wet soil, CO_2 concentrations were as much as 15 times higher than in dry soil even if the lack of oxygen was limiting respiration in wet soil. According to Šimůnek and Suarez (1993) and Magnusson (1995), soil volumetric water content and ground-water level are dominating factors controlling soil air CO_2 concentration. The soil and air temperatures were normally of secondary importance.

Comparison with Other Approaches

There are also other models for predicting soil CO_2 effluxes from concentration gradient than presented here (e.g. Tang et al. 2003; Jassal et al. 2005; Turcu et al. 2005; DeSutter et al. 2008) or for estimating the depth of CO_2 production in the vertical soil profile (Šimůnek and Suarez 1993). All models are based on similar principle using either Fick's 1st or 2nd law. However, there are differences between the models in how the diffusivity in the soil is calculated. The two most common approaches for calculating it in the soil are models presented by Millington and Quirk (1961) and Moldrup et al. (2000a, b). Both of the methods take the changes in soil water content into account through soil air-filled porosity.

Another approach used for diffusion calculations is based on simultaneous concentration and flux measurements of ^{222}Rn isotope (Koehler et al. 2010). ^{222}Rn is produced in the soil in radioactive decay of ^{226}Ra isotope, which is distributed quite uniformly in the soil. By measuring simultaneously the production of ^{222}Rn in the soil and its profile and flux, it is possibly to estimate the changes in diffusion rates of other gases as well. The disadvantage of this method is, however, that separate ^{222}Rn analysers are needed in addition to the target gases (e.g. CO_2) to be measured.

5.4 CO_2, Water Vapour and Energy Fluxes Between Forest Ecosystem and Atmosphere

Pertti Hari, Timo Vesala, Pasi Kolari, Samuli Launiainen, Üllar Rannik, Liisa Kulmala, Kourosh Kabiri Koupaei, and Eero Nikinmaa

The CO_2, water vapour and energy fluxes convey the principal interactions between forest ecosystems and the atmosphere. Thus, they are focal aspects in forest ecology. Prevailing environmental factors effect on these fluxes, thus, they vary during days and years. The Eddy Covariance method started a new era in the research of the fluxes between forest ecosystems and the atmosphere. H.T. Odum would have liked to measure energy and material fluxes, but at that time, the 1960s, the instrumentation was not available. Thus, the Eddy Covariance method is a fulfilment of his dreams. The interpretation of the measured results is rather problematic since several processes and transport phenomena contribute to the fluxes making the relationship between metabolic processes and the measurements rather complicated. When these fluxes are accumulated over time, we get the amounts of material and energy lost or gained by the ecosystem.

5.4.1 Structural, Metabolic and Physical Background of the Fluxes

Hierarchical structure is characteristic for forest ecosystems: cells form individuals, and individuals form populations and further communities. The capture of carbon dioxide and loss of water take place in leaves that form the canopy. Stems transport water and, as living organisms trees, need energy, that is, ATP for metabolism, and they emit CO_2. Roots are necessary for water and nutrient uptake. Forest soil is a combination of mineral and organic material, living and respiring microbes and roots.

Photosynthesis is the engine of the life in forest ecosystems. It converts solar radiation energy into chemical form as sugars to be utilised either in maintenance of the living cells or for growth. Photosynthesis consumes atmospheric CO_2, and it generates carbon dioxide flux from the atmosphere into ecosystems.

Living cells need energy, and it is obtained from sugars when they are burned in respiration to produce ATP. Without the energy provided by ATP, life is impossible and all living components including leaves, sapwood, roots and microbes burn small

P. Hari (✉)
Department of Forest Sciences, University of Helsinki, P.O. Box 27, 00014 University of Helsinki, Helsinki, Finland
e-mail: pertti.hari@helsinki.fi

carbon molecules to CO_2 to obtain ATP for the energy needed in living cells. Thus, the fulfilment of energy needs of life generates CO_2 efflux from vegetation and microbes into the atmosphere.

The pathway of CO_2 from the atmosphere to the interior of the leaf goes via stomatal pores into the water on the surface of mesophyll cells and then to the chloroplasts where the CO_2 is finally fixed, maintaining the gradient of concentration so that diffusion continues. In similar way, diffusion transports water vapour out of the leaf via stomatal pores, and thus, photosynthesis and transpiration are strongly coupled to each other: photosynthesis is impossible without transpiration. The driving force, that is, the concentration difference between stomatal cavity and the atmosphere, is often for water vapour about 10 $g(H_2O)\ m^{-3}$ and for CO_2 about 0.2 $g(CO_2)\ m^{-3}$. Thus, the transpiration is often nearly two orders of magnitude larger than photosynthesis. After rainfall and dew formation, evaporation from various surfaces in the forest is significant.

Photosynthesis, respiration and transpiration consume and produce CO_2 and water vapour into the airspace in ecosystems resulting in decrease or increase in the concentrations of carbon dioxide or water vapour. Air movements, especially its vertical component, smear the concentration differences. If the mixing of the air by wind is strong enough, then the concentration differences inside forest canopies and also temporally are rather small around some stable mean value.

Solar radiation is the ultimate source of energy for life. When light quanta are travelling through the atmosphere, they scatter, especially from aerosol particles, giving rise to diffuse radiation that comes everywhere from the sky. The radiation coming directly from the solar disc is called direct radiation. In clear sky conditions at high solar, elevation up to 80% is direct radiation, and at low solar elevation nearly, all radiation is diffuse. During cloudy weather, all radiation is diffuse.

When light quanta hit leaves or other structures in forest ecosystems, usually its energy is converted to heat, and only small proportion is utilised in photosynthesis or reflected away.

5.4.2 Processes Generating CO_2 and Water Vapour Fluxes Between Ecosystems and the Atmosphere

Several processes in vegetation and soil contribute to the CO_2 and water vapour flux between a forest ecosystem and its surrounding. We treat the processes separately, because each of them has its characteristic aspects, and thereafter join the fluxes together. We pay more attention to the photosynthesis and transpiration of trees and treat the remaining processes with rather simple descriptive approach.

5.4.2.1 Photosynthesis and Respiration of Trees

Basic Concepts and Ideas

Absorption of light quanta in the canopy generates strong spatial and temporal variation in light. This is why we have to define a new concept that is able to cope with the varying nature of light inside the canopy. The *flux of photosynthetic active radiation at point x and moment t* is defined as the number of visible light quanta falling on a small horizontal plane element around the point x during small time interval divided by the product of the area of the plane element and the length of the time interval. The plane element has to be so small and the time interval so short that there is no variation in the flux of light quanta within the plane element during the time interval.

The spatial and temporal variation in light affects strongly on processes taking place in the canopy, especially in photosynthesis. Thus, we have to define a new set of concepts dealing with space, time, structure and processes, and it needs to be coherent with the definition of photosynthetic radiation. *Space and time element* is, by definition, so small volume and so short-time interval that the environmental factors, especially photosynthetic active radiation, are stable within the volume element during the time interval.

The structure of living organisms is characterised with mass densities that are analogous to the concept used in physics to describe nonhomogeneous masses. The *density of living mass*, especially of leaves, is defined as the living mass in a small volume element divided by the volume of the element.

We define the *specific photosynthetic rate at a point and moment* as the amount of CO_2 consumed in a small volume element around the point and short-time interval around the moment divided with the leaf mass in the volume element and the length of the interval. We used rather similar definition already in Chap. 4.

We define analogously the *specific respiration rate* at a point and moment as the amount of CO_2 produced in metabolism to produce ATP by living mass in a small volume element around the point and short-time interval around the moment divided with the amount of living mass in the volume element and the length of the interval.

The *specific transpiration rate* at a point and moment is defined as the amount of transpired water in a small volume element around the point and short-time interval around the moment divided with the amount of living mass in the volume element and the length of the interval.

The photosynthetic and respiratory processes are extremely variable in space and time due to the variation in environmental factors. We solve this problem of variation by dividing the space into such small space elements and time into such short intervals that we can omit variation within the space element and time interval.

Basic idea EF1: The fluxes between forest ecosystem and the atmosphere are generated by metabolism and physical phenomena in small volume elements during short-time intervals.

The environmental factors, especially photosynthetically active radiation, affect the metabolism of living organisms in the volume elements.

Basic idea EF2: Absorption of light quanta in the canopy reduces the photosynthetically active radiation. Temperature, CO_2 and water vapour concentration are stable within the canopy.

The results obtained in the Chap. 4 can be utilised to the fluxes generated by metabolism between volume elements and the atmosphere.

Basic idea EF3: We can apply the theories of photosynthesis, transpiration and respiration to obtain the CO_2 and water vapour fluxes between each volume element and the atmosphere.

Derivation of the Theoretical Model

The basic ideas EF1–EF3 are very general statements, and they must be specified to be able to use quantitative approach and to test the ideas with measurements. We have to combine several processes to obtain all relevant phenomena into our theoretical model. The theory and the mathematical formulations of the different processes were originally documented and analysed in more detail in Stenberg (1996), Mäkelä et al. (2006) and Kolari et al. (2009).

Description of the Structure. Trees of different species, size or leaf area density are represented as size classes. In each size class, the trees can have different canopy shape and dimensions. The trees are assumed to be randomly distributed (Poisson distribution) in the stand to make calculation of the shading by the neighbouring crowns more simple. Tree crowns are described as ellipsoids or cones filled with shoots randomly distributed within the crown volume and the individual crowns consist of a homogeneous medium. The shoot orientation is spherical, that is, the shoots are randomly pointing in all directions, and there is equal amount of leaf surface facing all possible directions. The crowns are further divided into cubical volume units that have the same leaf area density.

Extinction of Light in the Canopy. Spatial distribution or arrangement of biomass affects the environmental factors inside vegetation. The leaves, branches and trunks of trees absorb radiation, and its intensity rapidly declines when descending into the foliage. The microenvironment around each phytoelement is affected by phytoelements in the same tree and by the neighbouring trees.

We determine the flux of photosynthetic active radiation at a given point in the canopy as a function of the distance travelled by the beam inside the absorbing medium. The light environment calculation is based on the canopy light model of Stenberg (1996). In the model, canopy foliage mass is distributed evenly inside the individual crowns, that is, there is no explicitly defined structure inside the crowns. Incident radiation is divided into direct and diffuse components. Direct radiation arrives at a given elevation angle, while diffuse radiation is assumed to enter equally from all directions of the sky. Attenuation of the radiation components is determined separately. The canopy is further divided into Sun shoots illuminated by both direct beam and diffuse radiation and shade shoots that are only receiving diffuse light. The calculation of light environment and photosynthesis is performed for volume elements in the crown.

Description of Photosynthesis and Respiration. The analysis of the annual cycle of photosynthesis and transpiration resulted in theoretical model presented in the Boxes 4.1 and 4.2. We use them as theoretical models for photosynthesis and transpiration in each volume element. We omit the annual cycle in respiration and use the same exponential relationship between respiration and temperature.

We divide the volume of the canopy into N small volume elements ΔV_i, $i = 1, 2, .., N$. Let x_i denote the centre point of the volume ΔV_i. We obtain the photosynthetically active radiation $I(x_i, t)$ at the point x_i at moment t from the consideration of light extinction in the canopy. Let $A(\Delta V_i, t)$ denote the photosynthetic rate in the volume element ΔV_i at moment t. We obtain it utilising the definition of specific photosynthetic rate p:

$$A(\Delta V_i, t) = p\, m(\Delta V_i), \tag{5.17}$$

where $m(\Delta V_i)$ is the needle mass in the volume ΔV_i.

Let $g^{C}_{H_2O}$ denote the transpiration rate from the canopy. We obtain the canopy photosynthesis by summing volume elements together and Eqs. 4.23, 4.26, 4.27 and 4.29:

$$g_C(t) = \sum_{i=1}^{N} A(\Delta V_i, t) = \sum_{i=1}^{N} p(I(x_i, t))\, m(\Delta V_i). \tag{5.18}$$

Diffusion of water vapour molecules takes place simultaneously with that of CO_2, and photosynthesis is connected with transpiration. Also, the water vapour flux generated by transpiration varies greatly in space and time since the water vapour concentration difference between stomatal cavity and ambient air and the stomatal action changes within canopy and during day. This problem of variability in the canopy was solved in the case of photosynthesis, and we can apply the same approach also to transpiration.

Let $g^{C}_{H_2O}$ denote the transpiration rate from the canopy. We obtain the canopy transpiration by summing the transpiration rates in volume elements as in the case of photosynthesis:

$$g^{C}_{H_2O}(t) = \sum_{i=1}^{N} g_{H_2O}(x_i, t)\, m(\Delta V_i). \tag{5.19}$$

We obtain the transpiration rate in the point x_i, $g_{H_2O}(x_i, t)$ from the Eqs. 4.30, 4.31, 4.32 and 4.34.

5.4.2.2 Respiration of Stems

The tracheids in coniferous wood are nonliving, but the rays are living, and they need energy obtained from ATP. In addition, the phloem and cambium are active and living. The conversion of ADP to ATP results in release of CO_2 in the xylem, phloem and cambium and consequently into CO_2 flux from the stem. The area-specific stem CO_2 efflux is defined, analogously to specific needle respiration rate. We define the area-specific stem efflux as the amount of released CO_2 from a small stem-area element during a short-time interval divided with product of the area of the area element and the length of the time interval.

Respiration in stem is an enzymatic biochemical process.

Basic idea SR1: Area-specific stem CO_2 efflux depends exponentially on temperature.

Let g_s denote the stem CO_2 efflux:

$$g_s = \alpha e^{bT_b}, \tag{5.20}$$

where T_b is bole temperature (Kolari et al. 2009). The bole temperature $T_b(t_i)$ at moment of time t_i is estimated as a delayed response to air temperature T_a with a time constant τ of 4 h:

$$T_b(t_i) = T_b(t_{i-1}) + (t_i - t_{i-1})\frac{T_a(t_i) - T_b(t_{i-1})}{\tau}. \tag{5.21}$$

This is relatively good approximation to describe the actual dynamics of stem CO_2 efflux (Höltta and Kolari 2009).

5.4.2.3 Photosynthesis and Respiration of Ground Vegetation

Ground vegetation forms an additional leaf layer in forest ecosystems. In forest ecosystems having closed canopy, the role of ground vegetation is clearly smaller than that of trees, but if the canopy is so scarce that light can rather freely penetrate down to the ground vegetation, then the mosses, herbs, grasses and dwarf shrubs may have an important role in generating the CO_2 flux between the ecosystem and its surroundings.

There are usually several species in the ground vegetation, and it forms an additional layer of leaves under forest canopy.

Basic idea GV1: The solar radiation above the canopy and the shading by the canopy determines photosynthetically active radiation falling on ground vegetation.

The photosynthesis of ground vegetation is based on the same functional substances as in trees, but there is, however, variation between the species in the timing and in the level of photosynthesis. The biochemical regulation system controls the activities of the functional substances.

Basic idea GV2: The species-specific state of the photosynthetic functional substances and photosynthetically active radiation determine the specific photosynthetic rates.

The momentary photosynthesis of ground vegetation, p_G (gC m^{-2} s^{-1}), can be predicted by species-specific light response curves (Kulmala et al. 2008). The photosynthesis usually increases with light but saturates to species-specific level usually at relative low light intensities due the shade-adopted nature of most ground vegetation species existing below closed canopies. The annual cycle in the photosynthesis of ground vegetation is rather strong. We denote with p_i the specific photosynthetic rate of the species i and the maximum level with $p_{max}^{i}(t)$ (gC g^{-1} s^{-1}). The relationship between species-specific photosynthetic rate and light intensity (photon flux, I) is

$$p_i(t) = \frac{p_i^{max}(t)I(t)}{b_i + I(t)}, \tag{5.22}$$

where b_i introduces the species-specific non-linearity in the light response.

We obtain the photosynthesis of all ground vegetation species as the sum over the species:

$$p_G(t) = \sum_i m_i p_i(t), \tag{5.23}$$

where m_i (g^{-1} m^{-2}) is the species-specific biomass of leaves.

The action of the biochemical regulation system changes the photosynthetic functional substances. This action is reflected into the maximum level of photosynthesis that changes during the season, that is, $p_{max}^{i}(t)$. The changes in the maximum level of photosynthesis are temperature driven, in a similar way than in Scots pine (Eq. 4.29). Feather mosses are ectohydric plants without root systems drying out without external moisture supply. Therefore, the $p_{max}^{i}(t)$ values of mosses are affected by recent precipitation (Kulmala et al. 2011).

We consider evapotranspiration, $g_{H_2O}^{G}$, from the forest floor as combination of transpiration of forest floor vegetation as well as evaporation from the ground. The driving factor of evaporation and transpiration is the solar energy available on the forest floor. We use the so-called Priestley–Taylor equation (Priestley and Taylor 1972) to simulate evapotranspiration (mol m^{-2} s^{-1}) from ground vegetation and ground:

$$g_{H_2O}^{G} = \alpha \frac{f_S(T)}{f_S(T) + \gamma} \frac{I_n}{\rho c_v}, \tag{5.24}$$

where I_n is available radiation on the forest floor (W m^{-2}), $f_S(T)$ the slope of the saturation vapour concentration–temperature relationship (mol m^{-3} K^{-1}) at temperature T, γ psychrometric constant (mol m^{-3} K^{-1}), ρ the molar density of water (mol m^{-3}) and c_v the latent heat of vaporisation (J mol^{-1} K). Ideally, net radiation should be used as not all radiation entering the forest floor is absorbed, but for simplicity, we use global radiation transmitted through the tree canopy. Parameter α (dimensionless) was determined from H_2O flux measurements with soil chambers.

5.4.2.4 Respiration of Microbes and Roots in Soil

Microbes break down the macromolecules, as cellulose, lignin and proteins, with extracellular enzymes. Resulting small molecules penetrate the microbial cell membrane, and they are the source of raw material and energy for the microbial metabolism. In addition, trees emit large amounts of sugars from roots. These root exudates are also an important source of energy for life in the soil.

The conversion of chemical energy to ATP in respiration releases CO_2 into the soil. The layered structure is characteristic for forest soil, and also, the fine root and microbial respiration decreases strongly when going deeper in the soil. We omit, however, the diffusion in the soil and assume that the soil efflux depends on temperature as respiration does.

To a first approximation, all enzymatic reactions depend exponentially on temperature over the range of temperatures normally occurring in the boreal environment; thus, also the relationship between microbial respiration and temperature is exponential. Let g_S denote the soil CO_2 efflux per unit area and T_s soil temperature. Then,

$$g_S = r_o {Q_{10}}^{T_S/10}, \tag{5.25}$$

where Q_{10} and r_o are parameters.

5.4.3 CO_2 and Water Vapour Fluxes Between Forest Ecosystem and the Atmosphere

Processes in vegetation and soil generate CO_2 and water vapour concentration differences between the air near the surfaces and further away resulting in fluxes. Each of these fluxes can be derived from the properties of the process generating the flux. We obtain the theoretical model of the CO_2 and water vapour fluxes between forest ecosystem and the atmosphere by adding together all the fluxes generated by the processes. Let g_E denote the CO_2 flux between the forest ecosystem and the atmosphere. It is

$$g_E(t) = g_C(t) + g_S(t) + g_G(t) + g_s(t). \tag{5.26}$$

We obtain the water vapour flux between the forest ecosystem and the atmosphere in analogous way. Let $g_{H_2O}^{E}$ denote the transpiration rate of the ecosystem. Then,

$$g_{H_2O}^{E}(t) = g_{H_2O}^{C}(t) + g_{H_2O}^{G}(t). \tag{5.27}$$

5.4.3.1 Measurement of Fluxes of CO_2, Water Vapour and Heat

Our Eddy Covariance measuring system of CO_2 flux has been in operation since 1996, and it is still in operation providing 30-min averages of carbon dioxide and water vapour flux between the stand and the atmosphere day and night, summer and winter. We measure simultaneously also temperature and water vapour and CO_2 concentration (see Chap. 9).

Eddy Covariance fails to detect the actual ecosystem exchange under stable atmospheric stratification when there is little turbulent vertical movement of air, especially at night. In order to determine cumulative exchange of matter end energy, missing flux records are filled and measurements that are expected to be biased are rejected and replaced with calculated values that are based on regularities in the accepted fluxes. This procedure is called gap filling. The most common gap-filling procedure for CO_2 exchange is based on employing simple empirical models for ecosystem respiration and photosynthesis and using the accepted flux data to estimate the values of the model parameters (review of methods in Falge et al. 2001). The missing or rejected fluxes are then calculated as the combination of modelled photosynthesis and respiration. Gap filling of the CO_2 flux data shown here was documented in more detail in Kolari et al. (2009). Gap filling of the heat fluxes was based on simple regressions of flux on net radiation.

5.4.3.2 Test of the Theory with Field Measurements

Estimation of the Parameter Values

We utilise the chamber measurements for determination of the parameter values to obtain a more severe test. The PhenPhoto model, tested with the data obtained at SMEAR I about 600 km north from SMEAR II, predicted also successfully the CO_2 measured at SMEAR II. There is, however, a small difference: the relationship between the state of the photosynthetic functional substances and the efficiency of the photosynthetic functional substances seems to saturate at midsummer. This is introduced into the PhenPhoto with a slight modification as follows:

$$E_P(S_P) = \begin{cases} 0, & \text{if } S_P < S_0 \\ \dfrac{E_P^{max} S_P}{S_P^{max} - S_o}, & \text{if } S_0 < S_P < S_P^{max}, \\ E_P^{max}, & \text{if } S_P > S_P^{max} \end{cases} \tag{5.28}$$

where S_0 is the state of photosynthetic functional substances above which the functional substances are active, P_P^{max} is the state of the photosynthetic functional substances at which the response saturates and E_P^{max} is the maximal efficiency of the photosynthetic functional substances.

The parameter values for photosynthesis and the annual cycle models were based on mean values of several shoots and years derived from gas exchange measurements on Scots pine shoots at SMEAR II (Mäkelä et al. 2006; Kolari et al. 2007). The studied shoots were located at the top of the canopy. We utilised results (e.g. Kolari et al. 2006) for the parameter values for photosynthesis of ground vegetation.

The estimation resulted in the following parameter values: E_P^{max}=0.00075 m s^{-1}, $\gamma = 828$ μmol m^{-2} s^{-1}, $\lambda = gCO_2/gH_2O$, $g_{max} = 1$ mm s^{-1}, $\tau = 200$ h and $E_P^{max} = 16°C$.

Characteristic Features in CO_2 Fluxes

The CO_2 flux between forest ecosystem and the atmosphere in summer has a clear daily pattern. The ecosystem emits carbon dioxide in the night and consumes it during daytime. The theoretical model is able to predict this pattern; in addition, it is also able to simulate the rather small fluctuations in the CO_2 flux during daytime (Fig. 5.17).

Most of the summers are rather moist at SMEAR II. There are, however, infrequent dry summers when the water tension in the soil limits the availability of water and the stomata close partially during afternoons reducing photosynthesis. The summer of 2006 in August was dry, and we could see reduction in the CO_2 flux between the forest ecosystem and the atmosphere (Fig. 5.18). Our theory and the model based on it predicted similar decline in the CO_2 flux.

There are clear differences between the days. The clouds are travelling across the sky during normal summer days causing large variation in the photosynthetically active radiation. The CO_2 flux between ecosystem and the atmosphere varies strongly during these days. Our theoretical model is able to predict the variation in the CO_2 flux (Fig. 5.19a).

During exceptionally sunny days having clear skies without reduction in photosynthetically active radiation by clouds, the decline in CO_2 flux in the afternoon is evident. Our model is able to predict this daily pattern (Fig. 5.19a1). During very cloudy and often also rainy days, the CO_2 flux is clearly reduced and it varies rather strongly. Our model is able to predict the CO_2 exchange also during cloudy days (Fig. 5.19a2). The days of intermittent cloudiness dominate the weather during summer days. Then, the strong variation in light generates variation also in the measured and predicted CO_2 flux between the ecosystem and the atmosphere (Fig. 5.19a3).

The daily CO_2 fluxes between the forest ecosystem and the atmosphere have a very clear annual cycle, with a source of carbon dioxide in the winter and strong sink in the summer (Fig. 5.18). Our theoretical model is able predict this annual pattern.

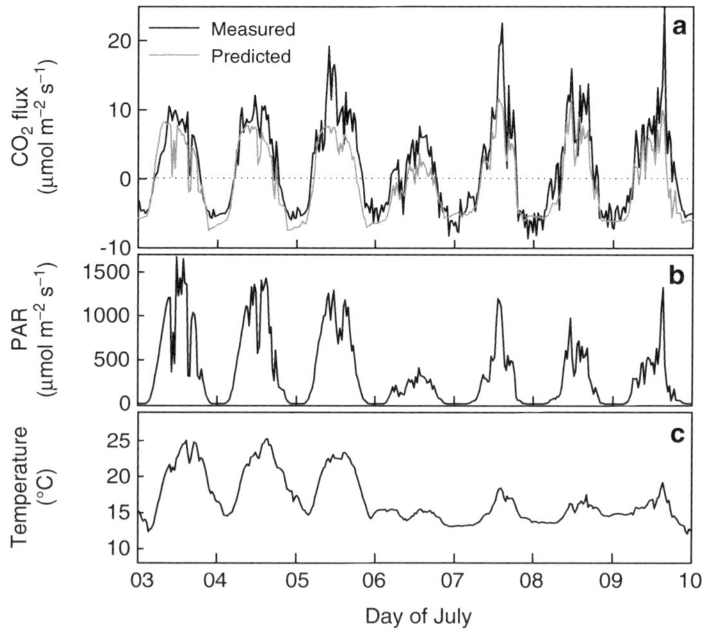

Fig. 5.17 Measured and predicted half-hourly CO_2 flux between forest ecosystem and the atmosphere (**a**), photosynthetically active radiation PAR (**b**) and temperature (**c**) at SMEAR II, southern Finland during 3–10 July 2007. We follow the old ecological sign convection of CO_2 flux, that is, flux towards the ecosystem is positive as has been for decades in the analysis of field measurements of CO_2 exchange or photosynthesis

The onset of CO_2 uptake varies rather much between years; during early spring, the uptake of carbon dioxide can start a month earlier than during a cool one. Our model is able to predict also this characteristic feature (Fig. 5.20). There is a rather clear discrepancy between predicted and measured CO_2 flux in late summer: the predicted flux is quite often positive indicating release of carbon dioxide. However, the measured flux is negative; thus, the forest ecosystem is binding CO_2.

Adequacy of the Model Structure

Our theoretical model of the fluxes between forest ecosystem and the atmosphere is based on the analysis of several processes generating fluxes in the ecosystem. The stem respiration is so small that it is unable to introduce any major discrepancy between modelled and measured values. Also, the role of ground vegetation in a closed stand, as the trees around SMEAR II station, is so small that shortcomings in the treatment of ground vegetation are of lesser importance. The canopy photosynthesis and the respiratory efflux from soil dominate the predicted CO_2 flux

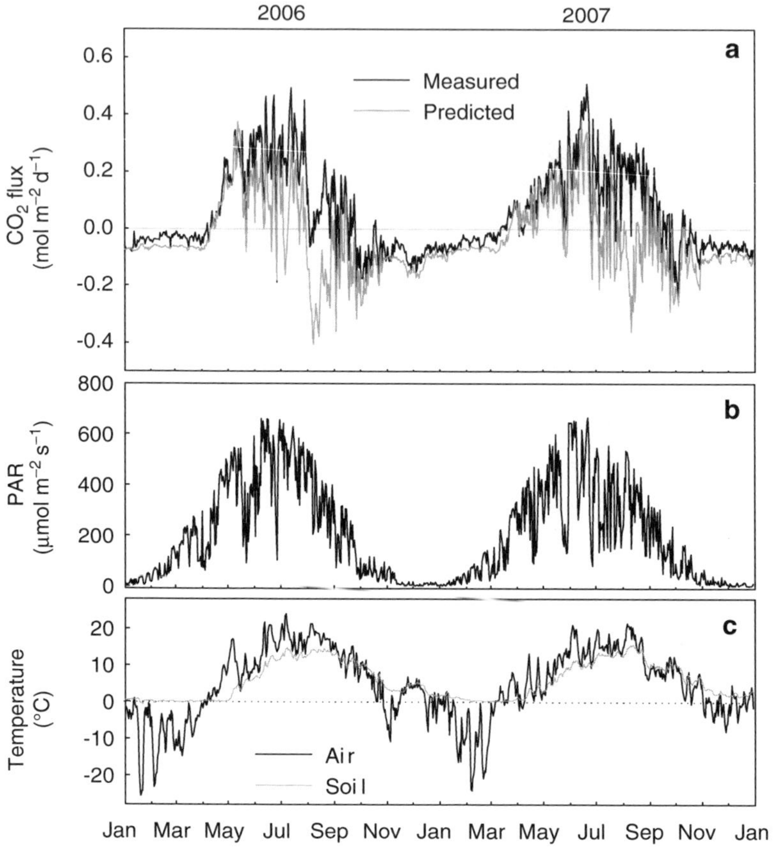

Fig. 5.18 The daily measured and predicted net CO_2 fluxes between the forest ecosystem and the atmosphere (**a**), averages of photosynthetically active radiation PAR (**b**) and air and soil temperature in A-horizon (**c**) at SMEAR II, southern Finland in 2006–2007. We follow the old ecological sign convection of CO_2 flux, that is, flux towards the ecosystem is positive as has been for decades in the analysis of field measurements of CO_2 exchange or photosynthesis

between the forest ecosystem and the atmosphere. These fluxes are of the same magnitude, but they have opposite directions, and consequently, the ecosystem flux is the difference between them. This decreases the signal and increases a little the noise; thus, the noise to signal ratio increases.

The half-hourly residuals between the predicted and observed CO_2 indicate only rather small systematic features in the summer (Fig. 5.21). On the other hand, the residuals of the daily means show clear annual pattern (Fig. 5.22a), and the relationship between soil temperature and the residuals is very clear (Fig. 5.22d). This large discrepancy between measured and predicted fluxes may be connected with issues relating to the soil efflux. We found already in the treatment of soil efflux that evidently the annual cycle of root exudates generates a rather large flux

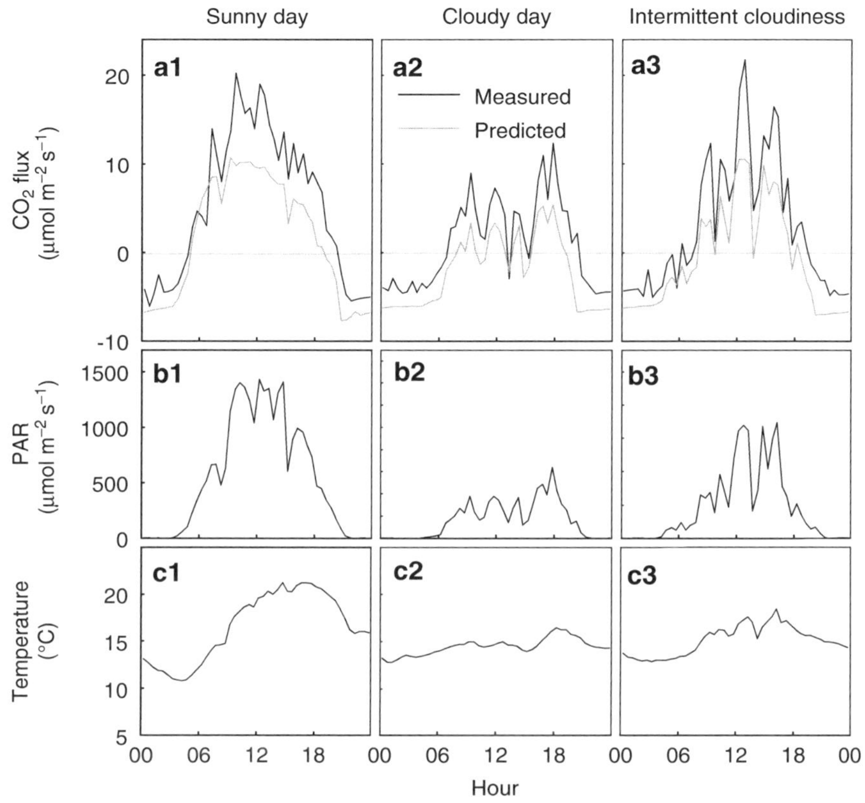

Fig. 5.19 Measured and predicted half-hourly CO_2 flux between forest ecosystem and the atmosphere (**a1–a3**), photosynthetically active radiation PAR (**b1–b3**) and temperature (**c1–c3**) at SMEAR II, southern Finland, during a day with intermittent cloudiness (25 July), a sunny day (21 July) and a cloudy day (24 July) in 2006. We follow the old ecological sign convection of CO_2 flux, that is, flux towards the ecosystem is positive as has been for decades in the analysis of field measurements of CO_2 exchange or photosynthesis

from the soil that is poorly introduced into the analysis of soil efflux. The observed problems in the daily residuals are in agreement with the analysis of soil efflux.

We can separate the photosynthesis of the tree canopy and the ground vegetation from the measured ecosystem CO_2 flux. We determine first ecosystem respiration by applying the regression of night-time fluxes on soil surface temperature to daytime conditions. Thereafter, we obtain the stand photosynthesis by subtracting the respiration from the measured flux. If the measured flux is missing, we calculate photosynthesis with a simple light response function. The prediction of the canopy photosynthesis is rather successful as can be seen in the Fig. 5.23 also during the drought period in August 2006, since the agreement between the observed and predicted canopy photosynthesis is rather good (Fig. 5.23). The residuals of the canopy photosynthesis do not show any clear systematic behaviour.

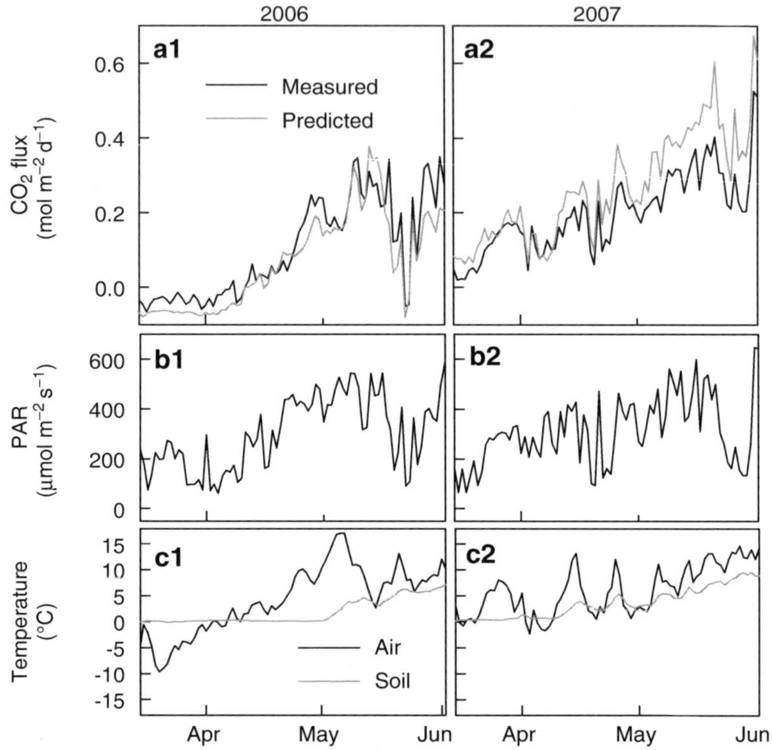

Fig. 5.20 The measured and predicted daily CO_2 fluxes in the spring 2006 (**a1**) and 2007 (**a2**), photosynthetically active radiation PAR (**b1–b2**) and temperature (**c1–c2**) at SMEAR II, southern Finland

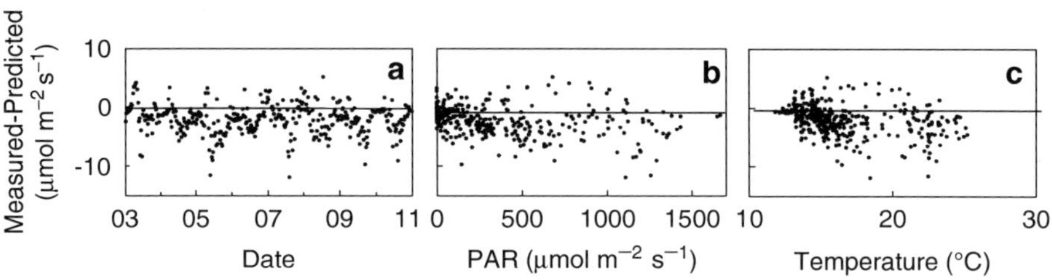

Fig. 5.21 The residuals between measured and predicted CO_2 flux (30 min average) against date (**a**), photosynthetically active radiation PAR (**b**) and air temperature (**c**) at SMEAR II, southern Finland, during 3–10 July 2007

Explaining Power

Microbial and root respiration in soil generates a rather strong efflux from soil into the atmosphere. The soil efflux is rather stable since the variation in temperature, the driving environmental factor of respiration, is rather small, only about 3°C during

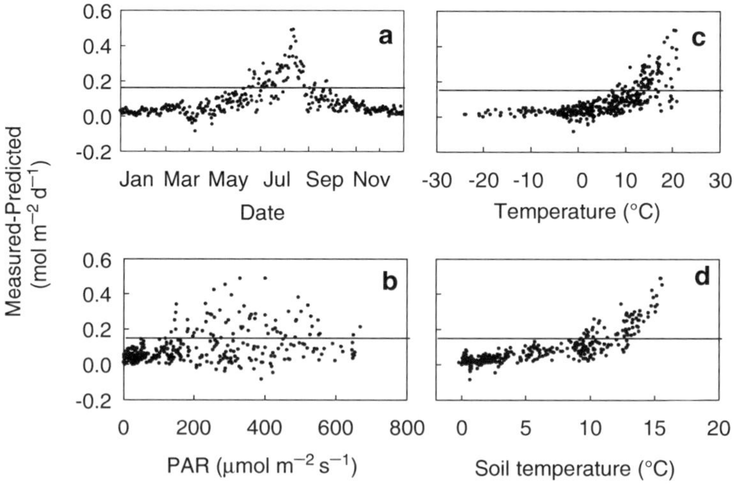

Fig. 5.22 The residuals between measured and predicted daily cumulative CO_2 flux during a day against time of year (**a**), daily average photosynthetically active radiation (**b**), air temperature (**c**) and soil temperature (**d**) at SMEAR II, southern Finland, in 2006

a day (Fig. 5.15). In contrast, the photosynthetically active radiation varies strongly between day and night causing strong photosynthesis in daytime and photosynthesis is missing during nights. The photosynthetic consumption of CO_2 is about twofold when compared with the production of CO_2 in respiration (Fig. 5.17). Our theory explains the daily patterns in ecosystem CO_2 flux with the stable production in respiration, especially in the soil and with intensive daytime and missing night-time photosynthesis.

Photosynthetically active radiation explains also the characteristic features during normal sunny summer day when clouds are travelling across the sky and during cloudy days. The great reduction in light intensity caused by small summer clouds generates corresponding reduction also in photosynthesis and great variation in the CO_2 flux between the ecosystem and the atmosphere. Heavy clouds during rainy days reduce strongly photosynthetically active radiation and also photosynthesis, and the CO_2 flux from the atmosphere into the ecosystem is reduced.

The reduction in the CO_2 flux in the afternoon from the atmosphere to forest ecosystem during exceptional sunny days is also a characteristic feature in the carbon dioxide exchange between forest ecosystems and the atmosphere. Our theory explains this with the action of stomata; evidently, the water transport system is unable to provide sufficiently water during these days, and the stomata close partially.

The annual cycle in the ecosystem CO_2 flux is clear, and the forest is a carbon sink in summer and source in winter. The annual cycle in environmental factors and in the functional substances of vegetation explains this alternation of sink and source behaviour. The air and soil temperatures are very different in winter, the

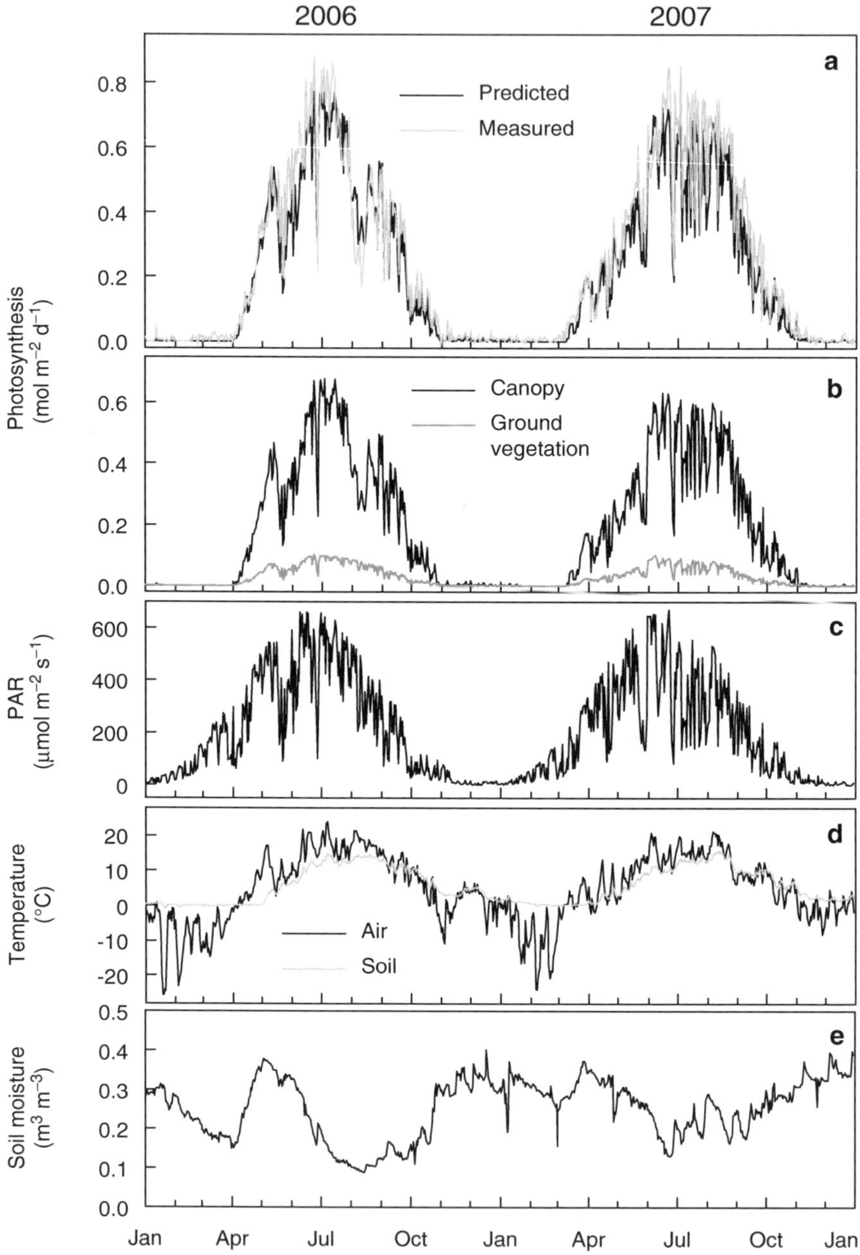

Fig. 5.23 Predicted and measured daily photosynthesis of canopy and ground vegetation (**a**), measured daily photosynthesis of ground vegetation and the canopy (**b**), photosynthetically active radiation PAR (**c**), air and soil temperature (**d**) and soil moisture (**e**) at SMEAR II, southern Finland, in 2006–2007

air temperature varies in a wide range from –35 to +5°C and the soil is stable around zero because snow isolates effectively. However, the CO_2 efflux from soil is considerable also in winter.

The metabolic annual cycle contributes also to the annual pattern of CO_2 flux between the forest ecosystem and the atmosphere. The biochemical regulation system deactivates the photosynthetic functional substances resulting in very low efficiency of the photosynthetic functional substances in late winter and spring. The photosynthetically active radiation is often high, and temperature is about 5°C in early April. However, the forest is still a sink of CO_2 although the prevailing environmental factors are favourable for photosynthesis (Fig. 5.20). This indicates the crucial role of the efficiency of the photosynthetic functional substances. The biochemical regulation system changes the state of the photosynthetic functional substances that is reflected in the efficiency of the photosynthetic functional substances and also into the specific photosynthetic rate. The slow change in the state of photosynthetic functional substances explains the recovery of photosynthesis during spring.

There is great variation in the onset of the CO_2 flux from the atmosphere into the forest ecosystem: in the warm spring of 2007, the carbon fixation started about a month earlier than during the cold spring of 2006 (Fig. 5.20). The onset of photosynthesis was very slow during the early spring and very fast in late spring. The state of the photosynthetic functional substances responds to temperature according to the basic idea EF3 in such a way that warm weather accelerates and cold inactivates photosynthesis. The different behaviour of the state of the functional substances during the springs 2006 and 2007 explains the difference in the onset of carbon fixation.

The proportion of explained variance by our theoretical model is about 85%. The measuring noise of Eddy Covariance method is about 10% (Rannik et al. 2006). Thus, it explains a considerable portion of the residual variation. The inadequate behaviour of the soil component reduces the proportion of explained variance. In addition, the treatment of light evidently contributes to the residuals.

Comparison with Other Theories

Simultaneous measurements of CO_2 fluxes in a forest ecosystem with Eddy Covariance and chambers are rather rare, and consequently, very few research teams can predict the forest ecosystem flux utilising chamber measurements. We do not know any other study that predicts, using only chamber measurements, the ecosystem flux by adding photosynthetic fluxes by trees and ground vegetation and respiration fluxes by trees, ground vegetation and microbes together and by integrating over space.

Kramer et al. (2002) compared how well different models can predict forest gas exchange at flux measurement sites. They concluded that in general the component fluxes (e.g. photosynthesis of trees and soil CO_2 efflux) could be modelled more accurately than net ecosystem exchange. The uncertainties of the component fluxes accumulate in the ecosystem flux (NEE), while the absolute magnitude of NEE is smaller than many of the component fluxes. The proportion of explained variance in canopy photosynthesis (GPP) and ecosystem respiration was in 83–94%, whereas in NEE, it ranged between 29 and 93%.

In the studies of Amthor et al. (2001) and Kramer et al. (2002), models with different underlying assumptions could predict the gas exchange of a forest stand with relatively good accuracy. In some models, soil CO_2 efflux was linked to the amount of roots and soil organic matter and in others to measure CO_2 effluxes; however, both approaches describe the seasonality of soil CO_2 efflux with an exponential temperature response. One notable difference to the theory tested here was that none of the models incorporated the annual cycle of photosynthesis. Most models attribute seasonal variability of photosynthetic rate to instantaneous responses and do not consider the changes in the photosynthetic functional substances. Such models often overestimate photosynthetic rates in spring (Bergh et al. 1998).

There are also different approaches in calculation of light environment within the canopy. This creates systematic and time-independent differences in stand photosynthesis between the models. Eventually, the accuracy of simulations of stand photosynthesis comes down to the fraction of absorbed light that largely explains the variability of long-term photosynthetic productivity in forest stands of the same species but different structure (Medlyn et al. 2003; Mäkelä et al. 2008). Using different leaf photosynthesis models with the same canopy light environment model, however, can result in range of variation of about 10% in stand photosynthesis (Launiainen et al. 2011).

Conclusion

Our theory was able to predict the characteristic features in the behaviour of the half-hourly CO_2 flux. The predicted and measured daily CO_2 fluxes were close to each other in the spring and early summer, but the discrepancy between predicted and measured daily CO_2 fluxes was clear during normal and wet summers. This discrepancy is believed to be caused by the annual cycle in root exudates, since the omitting the annual cycle of root exudates in the theory of soil CO_2 efflux caused systematic behaviour in the residuals. During the dry late summer 2006, the predicted and measured daily CO_2 fluxes were quite close to each other. In addition, the prediction of the photosynthesis of the ecosystem is rather successful during all studied years. Our theory of ecosystem CO_2 fluxes gained corroboration in the tests. However, the failure in the predictions of daily fluxes in late summer and autumn reduces the corroboration.

5.4.3.3 Test of Water Fluxes with Field Data

Estimation of the Values of the Parameters

The parameters in the models of CO_2 and water vapour fluxes are common for both fluxes, and no additional parameter values for water vapour flux are needed. The parameter values are based on the chamber measurements at SMEAR II.

Prediction of Characteristic Features of Water Vapour Flux

The characteristic features in water vapour flux between forest ecosystem and the atmosphere are rather similar with those we observed in the CO_2 flux. The strong daily patterns dominate the water vapour flux during summer time, the flux is large during most of the days and the flux is nearly missing in night-time. Our theory and the theoretical model are able to predict the transpiration pattern in the summer time (Fig. 5.24). However, the predicted water fluxes are clearly smaller than the measured ones. Also, the daily patterns of water vapour flux are clear.

On any normal summer day, when small clouds are travelling across the sky, the water vapour flux begins in the morning, peaks in the afternoon and finally declines towards evening. Our model predicts this clear daily pattern (Fig. 5.25), but the level is clearly too low. During rainy days, the water vapour flux is missing or very small. Our prediction is in reasonable agreement with the measurements (Fig. 5.25). Decline of water vapour flux in the afternoon is evident during sunny and warm summer days. Again, the prediction produces the same pattern (Fig. 5.25).

The early August in the year 2006 was exceptionally dry at SMEAR II, and the water vapour flux was slowly declining during the drought (Fig. 5.26). Our prediction was rather similar with the measured behaviour.

The annual pattern in the water vapour flux is clear, low flux in the early spring, increase during the spring, rather stable in the midsummer and decline in the autumn. The predicted pattern of daily transpirations is rather similar with the measured one, but the level on predicted transpiration is only about 70% from the measured one.

Adequacy of the Model Structure

Evaporation from moist surfaces and from the water film on mesophyll cells dominates the water flux. The stomatal action is the only biological aspect in the water vapour flux from forest ecosystem into the atmosphere. The evaporation is a physical phenomenon, and it depends on the concentration differences between the stomata interior and ambient air. The optimal stomata control of photosynthesis and transpiration is evidently rather adequate description of the stomatal action. Thus, the model structure is evidently well justified.

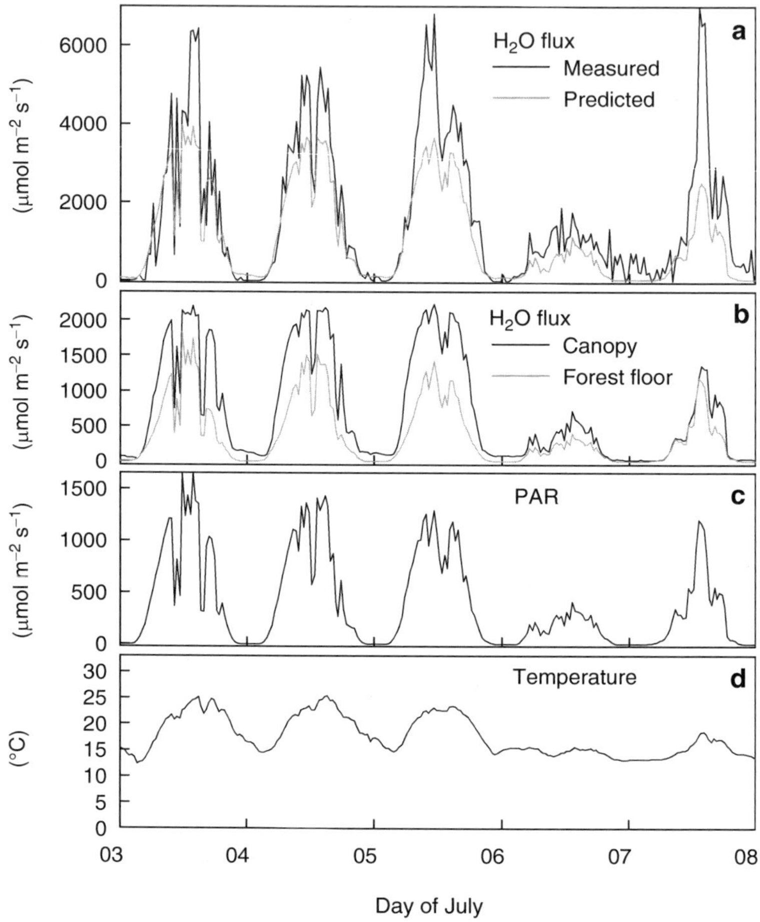

Fig. 5.24 Measured and predicted half-hourly water vapour flux from forest ecosystem to the atmosphere (**a**), measured transpiration by canopy and forest floor (**b**), photosynthetically active radiation PAR (**c**) and temperature (**d**) at SMEAR II, southern Finland, during 3–7 July 2007. We follow the old ecological and meteorological sign convection of water vapour flux, that is, flux towards the ecosystem is positive

The predicted values are clearly too low, and further analysis of the model's behaviour is not justified at the present state of the model performance. We should first find the reason for the predicted values being low. The systematic measuring errors in the chamber measurements or in the approximated leaf temperature are the most probable reasons for these low-predicted water fluxes. In addition, we omitted the acclimation of photosynthesis and of needle structure in the canopy.

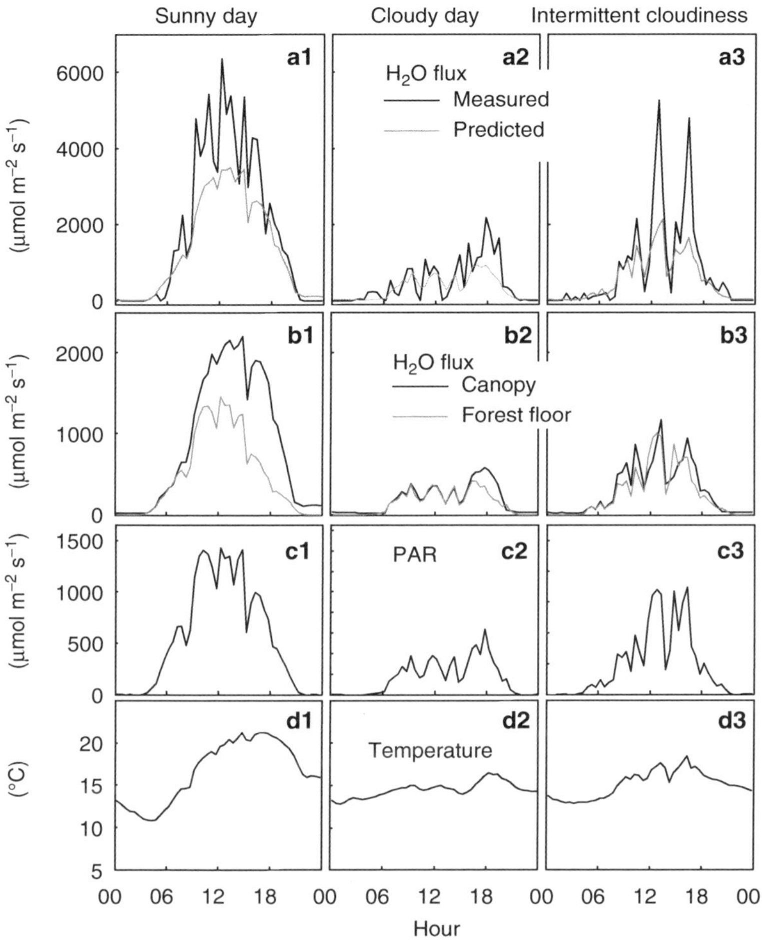

Fig. 5.25 Measured and predicted water vapour flux between the atmosphere and the forest ecosystem (**a1–a3**), predicted evapotranspiration from canopy and the forest floor (**b1–b3**), photosynthetically active radiation PAR (**c1–c3**) and temperature (**d1–d3**) at SMEAR II, southern Finland, during a sunny day (21 July), a cloudy day (24 July) and a day with intermittent cloudiness (25 July) in 2006

Explaining Power of the Theory

Solar radiation warms the surfaces and the air, while thermal radiation streaming back to space cools air giving rise to clear daily pattern in temperature, often 10–20°C higher in the day than in the night. The water vapour saturation deficit (VPD) is the difference of water-saturated air and actual water vapour concentration. The concentration of water-saturated air depends strongly on temperature, and the actual concentration reflects the properties of the prevailing air mass. Thus, the water

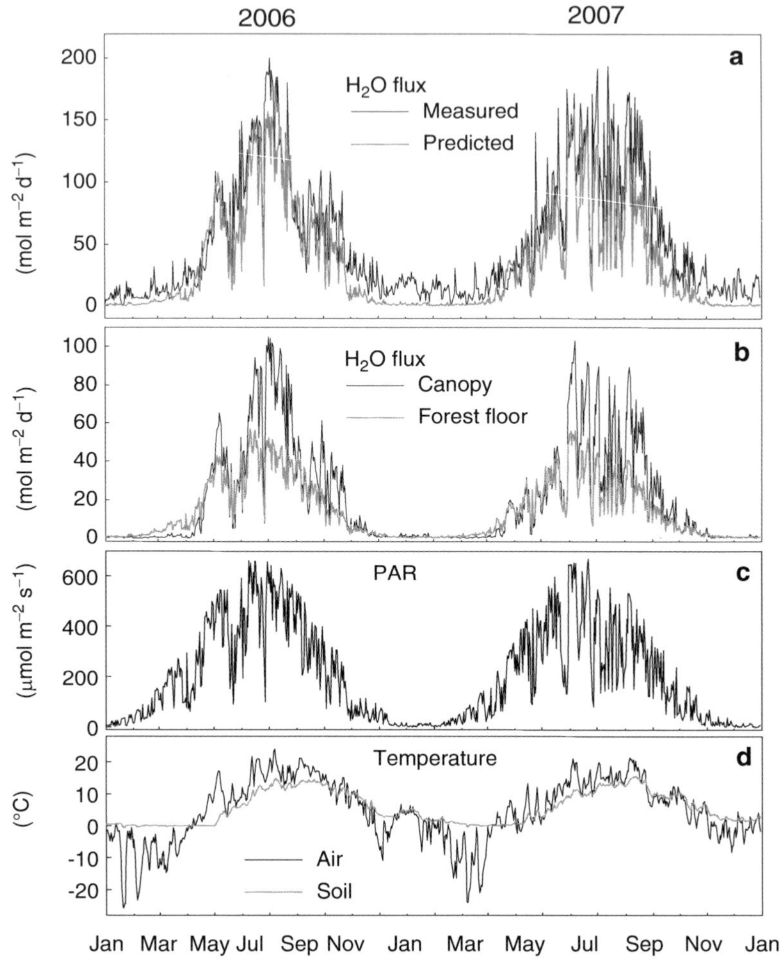

Fig. 5.26 Predicted and measured daily transpiration from the forest ecosystem (**a**), predicted transpiration from forest floor and tree canopy (**b**), photosynthetically active radiation PAR (**c**) and temperature (**d**) at SMEAR II, southern Finland, in 2006

vapour saturation deficit has a clear daily pattern that is reflected into the evaporation and transpiration. In addition, the stomatal action reduces transpiration during sunny days and during drought.

The proportion of explained variance is meaningless in this case, because of the systematic underestimation in the prediction. In the model comparisons done by Amthor et al. (2001) and Kramer et al. (2002), evapotranspiration models performed clearly poorer than models of stand photosynthesis or net CO_2 exchange although transpiration and CO_2 uptake are strongly linked via stomatal control. This indicates that different models of stomatal responses may result in vastly different rates of transpiration when integrated to the stand level. Model comparison done by Launiainen et al. (2011) showed that different model formulations that gave the same leaf-level behaviour at the top of the canopy resulted in discrepancies of as much as 25% in stand transpiration. Also, the lower quality of water flux measurements used in estimating model parameters reduces the prediction power.

Conclusion

Our theory predicted the daily and annual patterns of water vapour flux between forest ecosystem and the atmosphere. The predictions of the daily and annual patterns were rather successful. However, the predictions of absolute values were too low, the discrepancy being about 30%. We have to postpone further conclusions dealing with our theory until this systematic discrepancy is solved.

5.4.4 Energy Fluxes Within and Over Forest

There are several energy fluxes between forests and their surroundings. These fluxes include short-wave and long-wave radiation and turbulent heat fluxes. At the forest soil or leaf surface, there is also a transfer of sensible and latent heat, through a laminar boundary layer surrounding every surface. These fluxes change the heat storage in the forest, that is, the temperature of the forest elements. If the sum of the fluxes is zero, then the temperature does not change. This fact makes the energy balance an important tool in the studies of temperature in forests.

Most input of the energy is from the Sun. The solar energy is short-wave radiation, containing most of its energy at waveband from 0.1 to 5.0 μm (see Chap. 3). The incident solar energy may be reflected and absorbed by the Earth's surface or the atmosphere. The reflectivity of the forest canopy is low, and the reflected radiation is significantly lower than downwelling solar radiation (Fig. 5.27).

5.4.4.1 Long-Wave, Thermal Radiation and Radiation Balance

Earth's surface and atmosphere also emit the long-wave radiation, energy being radiated at (invisible) thermal infrared wavelengths between 4 and 25 μm, with maximum emission occurring at 9.7 μm. Thus, the canopy elements as well as the forest floor emit thermal radiation. The vegetated surface behaves almost as a full radiator[4] with emissivity over 0.9.

---

[4]A thermal radiator capable of providing a spectrum dependent on the temperature alone according to Planck's law is called a full radiator. The law states that the energy emission of the black body is proportional to the fourth power of its absolute temperature.

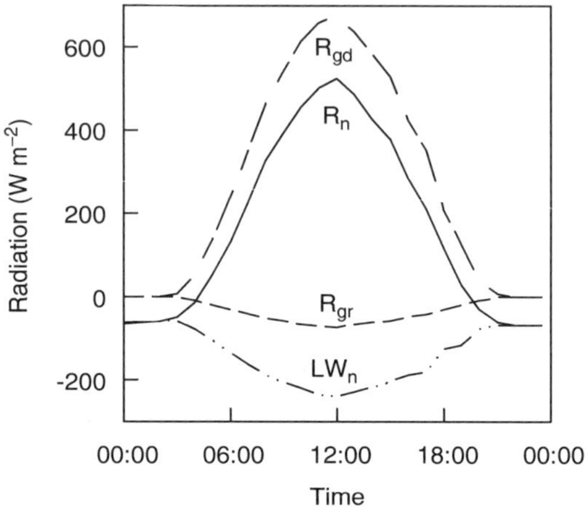

Fig. 5.27 Diurnal patterns of downward global (R_{gd}), reflected global (R_{gr}), net long-wave (LW_n) and net radiation (R_n). Averaged over July 2006 data. The radiation balance component is positive when surface gains energy and negative in case of loss

Terrestrial radiation, that is, long-wave radiation emitted by the Earth's surface, is mostly absorbed by the atmospheric greenhouse gases. These gases in turn emit long-wave radiation, part of which is lost to outer space and part is propagating downwards. The downward thermal radiation of the atmosphere is strongly dependent on cloudiness. The lower surface of the clouds emits approximately as a full radiator, and the total radiation reaching the surface is supplemented by emission from clouds in wavebands that the gaseous emissions lack. Large fraction of the downward flux reaching the Earth surface originates from the lower troposphere of about 1 km. The difference between upward and downward long-wave radiation, the net long-wave radiation, is typically in the order of 100 W m^{-2} at summer night (Fig. 5.27). During daytime, the long-wave radiation loss is a few times higher.

The sum of the downward and upward solar and thermal radiation forms the net radiation, the radiation balance. Net radiation at surface is typically positive in daytime (surface gains energy) and negative at night (loss of energy). Net radiation is sensitive to both short-wave and long-wave components, both being affected by cloudiness. Naturally, the long-wave radiation has significant variability, mainly based on surface temperature (annual cycle, diurnal cycle) and cloudiness. Net radiation above the canopy top is typically somewhat less than net global (short-wave) radiation in daytime, while at night, it is –100 W m^{-2} or less (thermal cooling). Figure 5.27 shows an example of a daily course of net radiation measured at SMEAR II above the forest canopy. This is an average of about 1-month period, thus being an average of clear sky and cloudy conditions at the particular site.

Together with thermal radiation, the solar radiation input generates strong daily patterns in temperature (Fig. 3.2). Clouds reduce solar radiation reaching the forests, and they also reduce the thermal radiation escaping to the space. Therefore, the difference between night and daytime temperatures is smaller on cloudy days than on sunny ones.

5.4.4.2 Energy Balance

The net radiation is used to large extent for heating up canopy elements and the air which is in contact with canopy elements. The absorption and heating occurs differentially with height inside the canopy, most of the radiation being absorbed at the upper part of the canopy. The short-wave radiation absorbed at the surface of each element in the canopy (e.g. needle, branch, part of trunk) is consumed in variety of processes such as evaporation, transpiration, heat conduction and photosynthesis. Part of the absorbed energy goes to heating the surface, part is emitted back to the atmosphere as long-wave radiation or transported as sensible heat, the heat flux driven by temperature differences between the surface and the atmosphere or latent heat away from the surface.

Vaporising liquid water requires energy that is provided by the surroundings, mainly in the form of absorbed solar radiation. Therefore, evaporation is also energy transport driven by H_2O concentration difference between the evaporating surface and the ambient air, called latent heat flux. In forest ecosystems and in vegetation in general, there are two kinds of water sources: evaporation from moist surfaces and evaporation from leaves through stomata, called transpiration, which plants control by closing and opening their stomata. The total flux of water vapour is called evapotranspiration.

The energy fluxes change the energy storage Q and consequently temperatures in the forest canopy and soil. When we combine the fluxes of net radiation R_n, sensible heat H, latent heat of evapotranspiration LE, heat conduction into the surface G and energy used in photosynthesis Q_p with the change in the energy storage using conservation of energy (Chap. 2, Eq. 2.2), we obtain

$$\frac{dQ}{dt} = R_n - H - LE - G - Q_p. \tag{5.29}$$

The net radiation in the equation above is defined positive towards the surface and the other fluxes positive away from the surface (H, LE, G) or positive when consuming energy by photosynthesis (Q_p). The heat storage in the forest ecosystem increases, that is, it becomes warmer, if net radiation is larger than the sum of the other energy fluxes and cools down in the other case. The negative net radiation implies surface cooling, driving the sensible heat flux towards the surface and potentially condensation of water on surfaces. The upward thermal radiation plays an important role during cold and calm nights in winter resulting in formation of ice crystals on surfaces facing the sky (Fig. 5.28).

Clearly, the largest terms of the fluxes are R_n, H and LE. Energy used in photosynthesis accounts only few percentages of net radiation. For a given net radiation R_n, partitioning of energy into sensible and latent heat fluxes depends strongly on the surface characteristics, vegetation functioning and weather conditions. When the canopy is wet, the greater part of R_n is consumed in evaporation of the intercepted rain. In dry conditions, especially when transpiration from leaves

Fig. 5.28 A beautiful phenomenon generated by thermal radiation into space. Birches that grow in the yard of Hyytiälä are in the *left photo* during a normal cloudy winter day (3 March 2012). The *right photo* is of the same birches after cloudless sky conditions when thermal radiation into space has cooled the twigs to be colder than the surrounding air (2 February 2012). Condensation of water vapour has generated ice crystals on the twigs. Photos: Juho Aalto

and soil moisture availability is restricted, a large fraction of R_n goes into heating the leaves which consequently increases temperature difference between them and the air and enhances sensible heat flux.

5.4.4.3 Diurnal Variation of Energy Balance Components

Figure 5.29 shows the main energy fluxes measured during 3 days in a pine forest in June, 2006. The first day was relatively sunny and cloudless, and net radiation (R_n) is symmetric over the day following the course of incoming solar radiation. In daytime, net radiation is positive (the surface gains more energy by radiation than it emits) and negative at night (net loss of energy by radiation). During the afternoon of the second day, R_n varies rapidly because of passing clouds. The net radiation is very sensitive to incoming short-wave radiation, but also, the long-wave (thermal) radiation is affected by the cloudy sky.

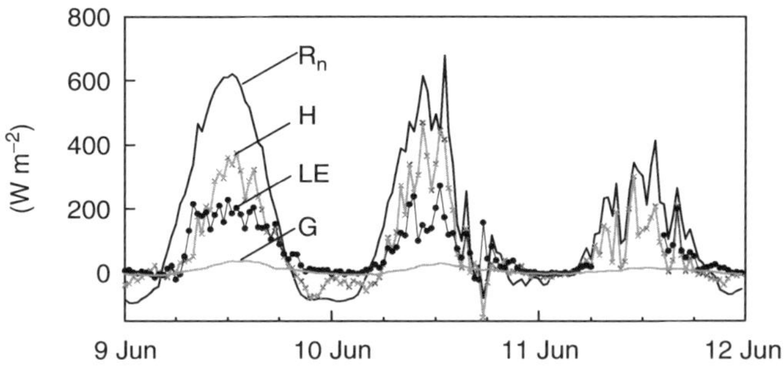

Fig. 5.29 Energy fluxes over 2 days in June 2006 at SMEAR II. The net radiation (R_n), sensible heat flux (H) and latent heat flux (LE) were obtained above forest canopy. G represents conductive heat flux to forest soil. Evapotranspiration (LE) and sensible heat exchange (H) are defined positive when directed towards the atmosphere. G is positive when directed towards soil, and R_n is positive when the surface is gaining energy by radiative components. LE is the total forest evapotranspiration rate multiplied with latent heat of vaporisation

The transport of water vapour and heat energy between a forest and the atmosphere occurs by turbulent motions, and these are the two major fluxes which balance the net radiation excess or deficit at the surface. Figure 5.29 shows how evapotranspiration takes place mainly during daylight hours and remains small during night. During nights with clear skies and low temperature, the latent heat flux can be directed towards the surface which means that condensation of water vapour occurs to surfaces. When the foliage is heated by the absorbed radiation, a temperature gradient between the surface and air develops. This gives rise for sensible heat flux, which peaks during midday/early afternoon. During clear nights, sensible heat flux is often towards the surface (air is warmer than foliage) to balance the net radiation deficit. The heat flux to the soil (G) is typically small below the forest canopy because the forest floor is efficiently shadowed by the crowns. This makes also the diurnal amplitude of soil temperature much smaller than over the areas where no vegetation exists. Note that the rapid variations of LE and H compared to net radiation during the first day in Fig. 5.29 are associated with natural (random) variability of turbulent fluxes as determined by Eddy Covariance micrometeorological flux measurement technique. During the second day, we also observe how LE and H vary in phase with changes in net radiation.

The remaining terms of the energy balance equation are much smaller in the case of forest canopy. This applies to the rate of change of sensible and latent heat storage when integrated over the forest column (the left-hand side of Eq. 5.29 integrated from forest floor up to the forest height). In particular, the energy consumed by photosynthesis Q_p is in the order of 5 W m^{-2} at midday of a summer month, thus being two orders of magnitudes smaller than the net radiation.

The soil heat flux G has a maximum ($\sim$25 W m^{-2}) in the afternoon (Fig. 5.29) lagging the net radiation R_n by about 2 h and remains positive or close to zero throughout the day. The soil at the site is still warming in June, which explains this diurnal behaviour. Over the pine forest at SMEAR II, the sensible and latent heat fluxes above the canopy are approximately equal in the summer. Above the forest floor, the fluxes are significantly reduced compared to those measured above the canopy, corresponding to lower net radiation. The latent heat flux *LE* dominates the turbulent energy exchange at the forest floor, being approximately a quarter of the respective flux above the canopy. It reaches about 45 W m^{-2} around noon in the summer while *H* is approximately 25 W m^{-2} (not shown). The evapotranspiration starts earlier in the morning and lasts longer in the evening than the sensible heat transfer.

5.4.4.4 Annual Cycle of Energy Fluxes

The climate in the boreal zone is characterised by strong seasonal variation with cold, dark winters and rather warm summers with high solar radiation input. Accordingly, the energy fluxes between the forest and the atmosphere have a strong annual cycle as shown in Fig. 5.30. Besides the seasonality of environmental conditions, partitioning of the net radiation is also strongly modulated by the seasonal cycle of vegetation. In March–April, the radiation intensity is already high, but photosynthetic capacity still low because of slowness of dormancy recovery (Sect. 4.2.2.4). Consequently, the photosynthetic demand for atmospheric CO_2 is scanty, and the water uptake and transport may be restricted. In these conditions, trees tend to keep their stomata relatively closed, and a large fraction of available energy is transformed into sensible heat transport to the atmosphere. In spring, 3–6 times more energy is transported as sensible heat than latent heat. Along with the recovery of photosynthesis, also transpiration increases and latent heat flux reaches maximum in July–August, about 2 months after the sensible heat flux (Fig. 5.30). In late summer, slightly more energy is consumed in evapotranspiration than is transported as sensible heat, and the ratio of the sensible heat flux over the latent heat flux, the Bowen ratio (H/LE), is typically 0.8–0.9. During periods when the forest suffers from intensive drought, as in July–August 2006, the transpiration is greatly reduced and sensible heat flux increases accordingly to balance the net radiation excess.

Partitioning of evapotranspiration into its components varies between different types of ecosystems. In absence of vegetation, or in wetlands dominated by mosses, most of the water flux is by evaporation. There are also strong differences between forests: the coniferous species are in general more conservative in their water use and transpire less than deciduous trees. Pine forests in boreal zone have typically a well-developed, extensive overstory layer and a shallow understory layer beneath it, as does the stand at SMEAR II in Hyytiälä. In these ecosystems, the forest floor (understory and soil) is well coupled to the atmosphere and contributes

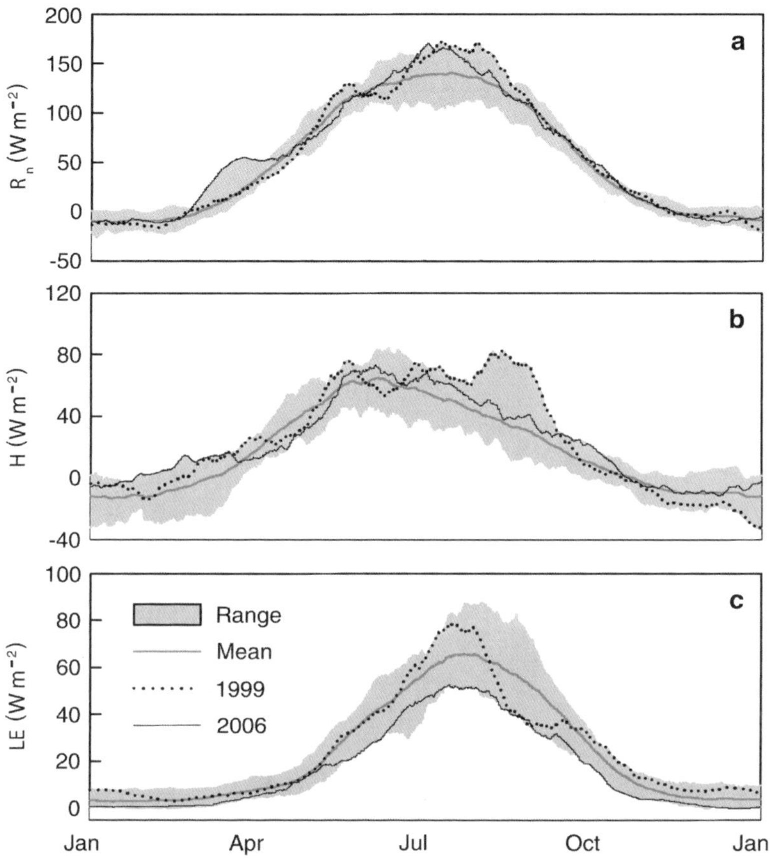

Fig. 5.30 Seasonal cycles of net radiation (R_n), sensible heat flux (H) and latent heat flux (LE, evapotranspiration) at SMEAR II. Shown are 30-day running averages; the *grey area* corresponds to the variability range, *thick black curve* is average and years 1999 and 2006 represent a moist and very dry summer seasons, respectively

significantly to forest scale carbon and water fluxes. In SMEAR II pine forest, the annual forest floor evapotranspiration has varied from 56 to 76 mm, which is between 18 and 25% of the whole-forest values (Launiainen et al. 2005; Launiainen 2010). These fractions are typical for boreal coniferous stands although significantly larger values have been reported in sparser forests (Baldocchi and Vogel 1996; Baldocchi et al. 1997; Kelliher et al. 1998; Constantin et al. 1999). The forest floor contribution peaks in spring and decreases slightly towards the autumn but seems to remain remarkably constant over wide range of conditions including an intense drought period.

5.5 Statistical Analysis of the Daily Photosynthetic Production of Coniferous Canopies Consistent with Process Knowledge

Minna Pulkkinen

The approach that we took in the previous section requires very detailed measurements of photosynthesis, canopy structure and light attenuation in the canopy. There are several sites having Eddy Covariance measurements of CO_2 flux, but the additional measurements, such as at SMEAR II, are rather limited. An alternative approach is to summarise the fast, leaf-element responses at a longer-term, whole-canopy scale and take longer-term average weather data as the model input. Examples of such models are the widely used Forest-BGC and Biome-BGC (Running and Gower 1991; Running and Hunt 1993; Kennedy et al. 2006) with a daily time step and 3-PG (Landsberg and Waring 1997) with a monthly time step. Moving from a fast timescale compatible with the usual shoot-level photosynthesis measurements to a coarser scale requires integration over time and space to relate the available measurements with the summary model characteristics. Because of uncertainties in the methods of summarising, parameter values need usually be calibrated to make model predictions agree with observations (Law et al. 2000; Kramer et al. 2002). With the increasing availability of canopy-level estimates of photosynthesis, GPP, from Eddy Covariance measuring stations, this calibration process has become more feasible than ever.

On the other hand, the availability of Eddy Covariance canopy-level GPP estimates from different regions and biomes also opens up a third approach: developing canopy-level photosynthesis models using statistical model fitting. A few studies have been conducted using hourly (van Dijk et al. 2005), daily (Yuan et al. 2007) and monthly (Maselli et al. 2006) timescales. However, if we want the fitted models to be applicable as generally as possible, we should base the model structure and parameters on our biological understanding of the processes generating the flux. The structures of the summary-type photosynthesis models that have been shown to be consistent with canopy-level measurements (Landsberg and Waring 1997; Thornton et al. 2002) would therefore seem good candidates for empirical model fitting. One such candidate is the widely used light use efficiency (LUE) approach (McMurtrie et al. 1994; Landsberg and Waring 1997). In the following, we summarise a study (Mäkelä et al. 2008) where we applied forest ecological knowledge and the LUE approach for analysis of daily photosynthetic

M. Pulkkinen (✉)
Department of Forest Sciences, University of Helsinki, P.O. Box 27,
00014 University of Helsinki, Helsinki, Finland
e-mail: minna.pulkkinen@iki.fi

production of coniferous canopies in boreal and temperate conditions. In particular, we introduce the annual cycle of photosynthesis into the analysis.

The LUE model assumes that in theoretical optimal conditions, where no environmental factors restrict photosynthesis, canopy GPP depends linearly on absorbed photosynthetically active radiation (APAR). The proportionality constant is referred to as *potential* LUE, and it gives the amount of carbon that the canopy is able to photosynthesise with one unit of APAR within one unit of time. For real (suboptimal) conditions, the potential LUE need be modified by the effects of restricting environmental factors; as a result, we get *actual* LUE, which change in time together with the environmental factors. For the daily canopy GPP, we decided to try the following model:

$$P(t) = \beta I_{\mathrm{APAR}}(t) f_{\mathrm{L}}(t) f_{\mathrm{S}}(t) f_{\mathrm{D}}(t) f_{\mathrm{W}}(t) + \varepsilon(t), \tag{5.30}$$

where $P(t)$ is canopy GPP, (g C m^{-2}) during day t; β (g C (mol APAR)$^{-1}$) is potential LUE; $I_{\mathrm{APAR}}(t)$ is APAR (mol m^{-2}) during day t; $f_{\mathrm{L}}(t)$, $f_{\mathrm{S}}(t)$, $f_{\mathrm{D}}(t)$ and $f_{\mathrm{W}}(t)$ are light, temperature, water vapour pressure deficit (VPD) and soil water content modifier functions assuming values in [0, 1] in day t; and $\varepsilon(t)$ is random error during day t. The actual LUE of the canopy in day t is the product of β, $f_{\mathrm{L}}(t)$, $f_{\mathrm{S}}(t)$, $f_{\mathrm{D}}(t)$ and $f_{\mathrm{W}}(t)$.

Although fairly linear over monthly or annual time periods, GPP has been found to be strongly nonlinear with respect to APAR at the daily scale (Medlyn et al. 2003; Turner et al. 2003). This is because the unavailability of CO_2 for carbon reactions reduces photosynthesis at high photosynthetically active radiations (Sect. 4.2.2). To account for this non-linearity, we defined the light modifier so as to yield the rectangular hyperbola when multiplied with the linear response included in the LUE model:

$$f_{\mathrm{L}}(t) \equiv \frac{1}{\gamma I_{\mathrm{APAR}}(t) + 1}, \tag{5.31}$$

where γ (m^2 mol^{-1}) is an empirical parameter.

In boreal and temperate conifers, the main effect of temperature has been proposed to be through the biochemical regulation system of the annual cycle that changes the state of the photosynthetic functional substances and consequently also the concentrations of the functional substances. This annual cycle effect can be described as a temperature-dependent dynamic delay process (Mäkelä et al. 2004; van Dijk et al. 2005). Accordingly, we modelled the effect of annual cycle on daily GPP using a temperature transformation $S(t)$ similar to the concept of the state of the functional substances of photosynthesis S_{P} (Sect. 4.2.2.3).

First, we calculated the delayed temperature, $X(t)$, from the mean daily ambient temperature, $T(t)$ (°C), as

$$X(t) = X(t-1) + \frac{1}{\tau}\left[T(t) - X(t-1)\right], \tag{5.32}$$

where the parameter τ (d) is the time constant of the delay process and where the initial state is

$$X(1) = T(1). \tag{5.33}$$

To avoid negative values of $P(t)$, we formed a piecewise linear function

$$S(t) = \max\{X(t) - X_0, 0\}, \tag{5.34}$$

with X_0 (°C) as a threshold value of the delayed temperature. The temperature modifier was then defined as

$$f_S(t) \equiv \min\left\{\frac{S(t)}{S_{max}},\ 1\right\}, \tag{5.35}$$

where the empirical parameter S_{max} (°C) determines the value of $S(t)$ at which the temperature modifier attains its saturating level. A saturating function was considered more appropriate than the commonly used parabolic temperature effect (McMurtrie et al. 1994; Landsberg and Waring 1997), as the parabolic function has been observed to cause underestimating bias during the warm season (Thornton et al. 2002).

Stomata react to high water saturation deficit in the air. Following Landsberg and Waring (1997), we thus defined the VPD modifier as a decreasing exponential function:

$$f_D(t) \equiv e^{\kappa D(t)}, \tag{5.36}$$

where $D(t)$ (g H_2O m^{-3}) is VPD in day t and $\kappa < 0$ (m^3 g^{-1} H_2O) is an empirical parameter.

Furthermore, stomata also control water loss when the water pool in the soil is small. Following Landsberg and Waring (1997) again, we modelled the reduction caused by decreasing soil water content as

$$f_W(t) \equiv \left[1 + \left(\frac{1 - W(t)}{\alpha}\right)^{\nu}\right]^{-1}, \tag{5.37}$$

where $W(t)$ is the relative extractable water (REW) in day t and α and ν are empirical parameters. REW was defined as

$$W(t) \equiv \min\left\{\frac{\theta(t) - \theta_{WP}}{\theta_{FC} - \theta_{WP}},\ 1\right\}, \tag{5.38}$$

where $\theta(t)$ is volumetric soil water content (SWC) (m^3 m^{-3}) and θ_{WP} and θ_{FC} are SWC at permanent wilting point and at field capacity, respectively.

The model was fitted to data from the following five European coniferous forest sites with Eddy Covariance measurement towers in them: Sodankylä in Finland (67°22′N, 26°38′E; *Pinus sylvestris*), Hyytiälä in Finland (61°51′N, 24°18′E; *Pinus sylvestris*), Norunda in Sweden (60°05′N, 17°29′E; *Pinus sylvestris*, *Picea abies*), Tharandt in Germany (50°58′N, 13°34′E; *Picea abies*) and Bray in France (44°42′N, 0°46′W; *Pinus pinaster*). There were 2–8 years of daily data in each site, making altogether 18 site-years. The sites were contrasting not only in terms of their geographical locations, which formed a good north–south gradient, but also in terms of leaf area index, stand structure and age. After fitting, the model was tested against data from two coniferous American sites: northern old black spruce (NOBS) stand in northern Manitoba, Canada (55°53′N, 98°29′W; *Picea mariana*), and Metolius in Oregon, USA (44°27′N, 121°33′W; *Pinus ponderosa*). These are not standard sites exactly: in NOBS, the mosaic of vegetation within 500 m from the Eddy Covariance tower consists of about 75% imperfectly to very poorly drained wetlands, whereas Metolius is a dry sandy soil site at a fairly high elevation in a semiarid region subject to summer drought. There were 3 years of daily data in each of the test sites.

Daily *canopy* APAR was taken to be a constant, site-specific fraction of daily total above-canopy PAR; the fraction was estimated either from below-canopy PAR measurements or from the canopy leaf area index using the Lambert–Beer law with a site-specific extinction coefficient. Daily *ecosystem* GPP was estimated from eddy flux measurements in the same way as at SMEAR II (Chap. 9). As we could not separate out the effects of soil and ground vegetation on the ecosystem GPP, we ended up in a slight inconsistency: fitting a canopy GPP model to the data of ecosystem GPP.

The model parameters were first estimated separately for each site and year (site-year-specific models) and for all years in each site (site-specific models). The idea was to study the role of the different modifiers and to investigate the between-years and between-sites variation in the parameter estimates. A set of common parameters for all the sites and years was then sought by fitting the model with the statistically significant modifiers over the whole data set (whole-data model). Finally, a modification of the whole-data model was fitted where the LUE parameter β was allowed to vary between sites, while the other parameters were shared (variable-LUE model).

The soil water modifier improved the model fit significantly only in few site-years, whereas the other modifiers became statistically significant in practically all site-years and sites. Accordingly, we discuss below only the results from the model without the soil water modifier.

The site-year-specific models fitted fairly well in the data; however, the site-specific models in each site performed almost equally well (Figs. 5.31 and 5.33). In other words, allowing between-years variation in the GPP response to APAR, temperature and VPD did not bring much benefit compared to keeping the response constant in each site. Further, the response could be assumed similar even across sites as long as the LUE parameter β was allowed to vary between sites, as the variable-LUE model fitted to the data almost as well as the site-specific models

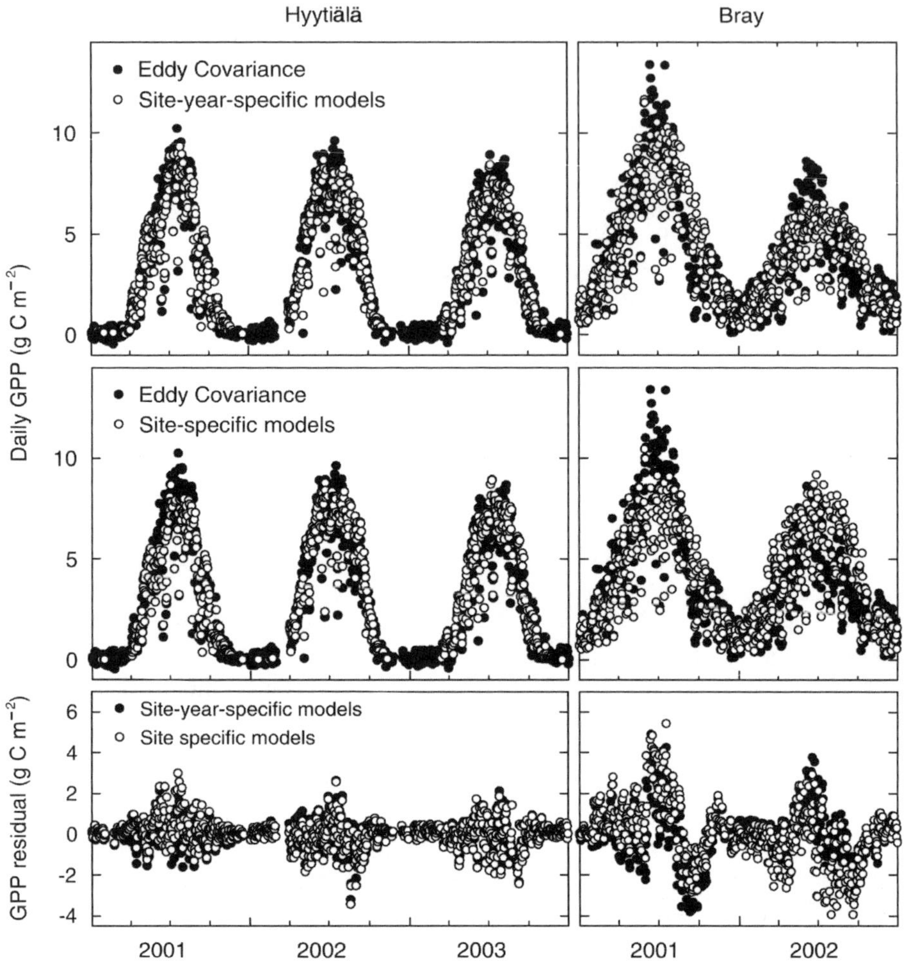

Fig. 5.31 The best (Hyytiälä) and the worst (Bray) fits of the site-year-specific (*top panel*) and site-specific models (*middle panel*) containing light (f_L), temperature (f_S) and VPD (f_D) modifiers and residuals (*bottom panel*)

(Figs. 5.32 and 5.33). This means that the LUE parameter largely governed the level of GPP in each site and that the gain from the site-wise estimation of the other parameters was negligible.

The whole-data and variable-LUE models were assessed against the test data from the two North American sites. For the variable-LUE model, the LUE parameter β was estimated for each test site conditional on the values obtained from the whole European data for the other parameters. The prediction performance of the variable-LUE model in the test sites was comparable to its fit in the estimation sites, whereas the whole-data model did rather badly in both test sites (Fig. 5.33). Interestingly,

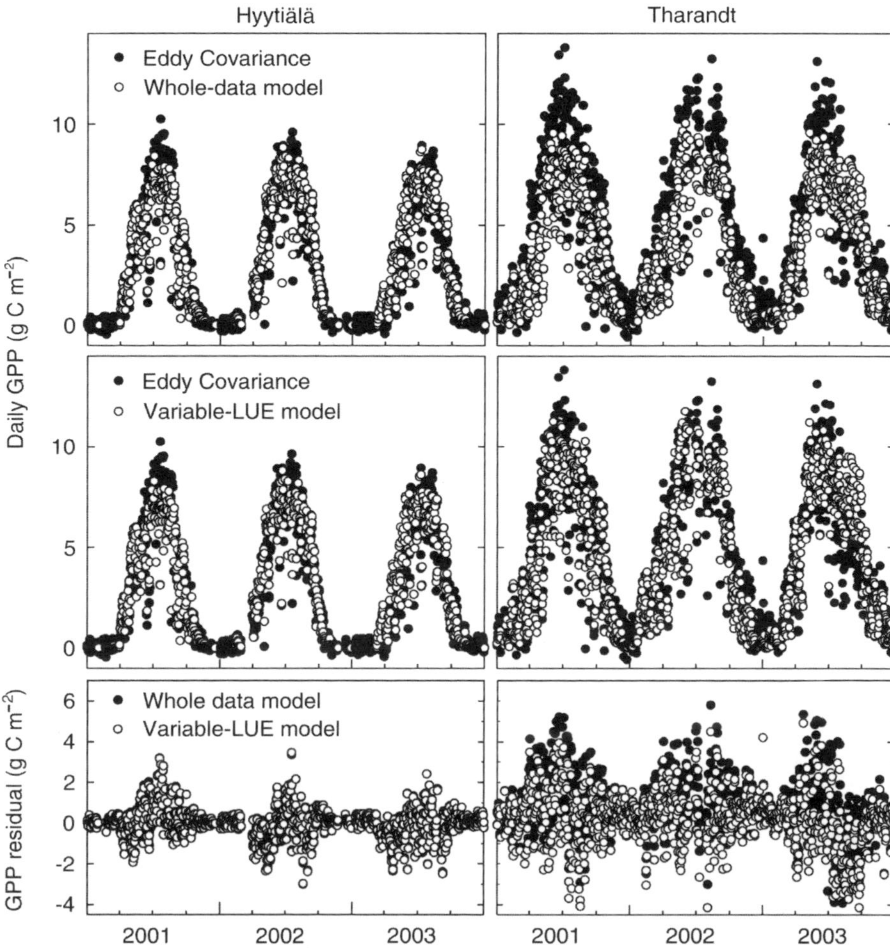

Fig. 5.32 The best (Hyytiälä) and the second worst (Tharandt) fits of the whole-data (*top panel*) and variable-LUE models (*middle panel*) containing light (f_L), temperature (f_S) and VPD (f_D) modifiers and residuals (*bottom panel*)

the temporal pattern of the residuals in the dry Metolius site resembled that of Bray (Fig. 5.31), the only European site in this study where water was likely to be limited.

The main finding above was that while the same response to the environmental driving variables can be assumed across different sites and species, the level of GPP is still site-specific. If we assume that the biochemical mechanism of photosynthetic production is universal and largely independent of species (Landsberg and Waring 1997), this suggests that some site-specific factors affecting the level of light use efficiency have not been included, or have been misrepresented, in the model. An obvious candidate for such a factor is foliar nitrogen needed to construct functional substances and suggested in many studies to be a key determinant of canopy

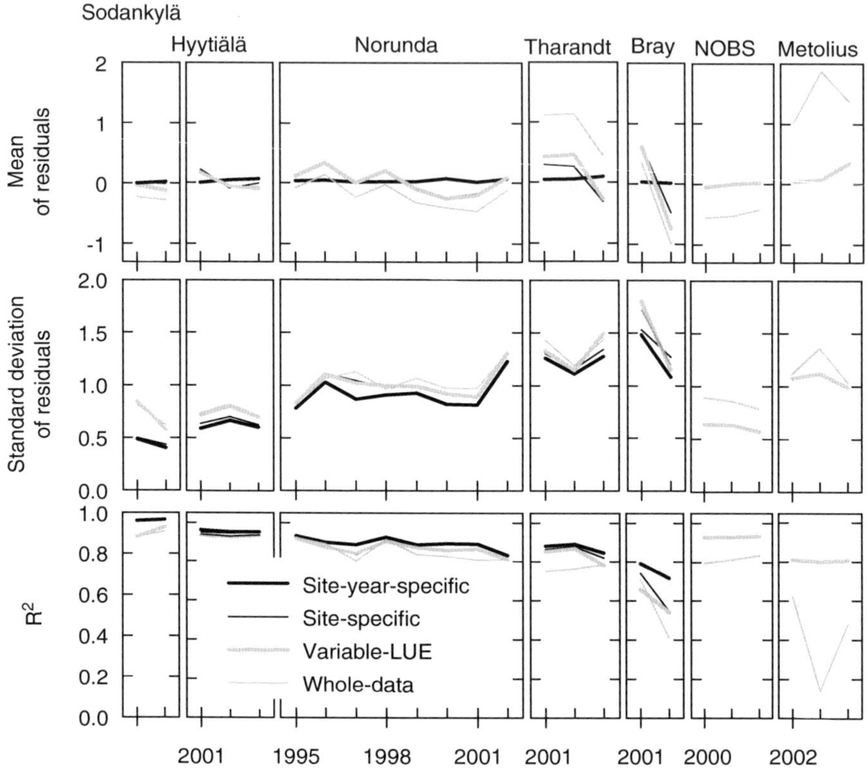

Fig. 5.33 GPP residual (g C m^{-2} day^{-1}) diagnostics for the models at seven sites containing light (f_L), temperature (f_S) and VPD (f_D) modifiers in all the years in the estimation and test sites. R^2 is the unadjusted coefficient of determination

photosynthesis (Ågren 1996; Smith et al. 2002). Indeed, when plotting the variable-LUE model estimates of β against available information on foliar N per leaf mass in the sites, a clear relationship was detected. Other factors causing differences in β could be differences in ground vegetation and uncertainty in estimating the fraction of absorbed PAR in the sites.

The relative importance of the significant modifying factors somewhat varied between the sites. The more northern the site, the more important the role of the state of the photosynthetic functional substances in the annual cycle appeared to be: while the temperature modifier together with linear light explained almost 90% of the variation in daily GPP in Sodankylä and Hyytiälä and around 80% in Norunda, the effect was less pronounced in Tharandt and Bray and came up mainly through an interaction with VPD and light. In all the site-years, the effect of the annual cycle was a better explanatory variable than current temperature. The choice of a saturating light function instead of a linear one and the incorporation of a reducing effect of high VPD values had a smaller but still significant influence, fairly similar in all the sites. It was surprising that soil water did not become an important explanatory factor.

However, this lack of response may be not because of a lack of effect, but because of the sparse data available on soil water. Also, part of the soil water effect was probably embedded in VPD, as drought periods tend to be accompanied by high VPD.

In conclusion, the study showed that the day-to-day variation of GPP over a wide geographical range of temperate and boreal coniferous forests can be rather generally explained by the variation in absorbed PAR, annual cycle of photosynthesis and VPD. The explanation seemed more robust in the more boreal, less drought-limited sites, but no definite improvement was gained in this study by including the volumetric soil water content as an explanatory variable. The absorbed PAR is a function of total PAR and leaf area index, possibly modified to some extent by stand structure (Duursma and Mäkelä 2007). However, after accounting for the three modifying factors, some between-sites variation remained in the potential LUE. For practical applications of the model, the site-specific values of potential LUE can relatively easily be estimated if Eddy Covariance data are available.

To summarise, the findings in the statistical analysis of daily GPP, that is, daily photosynthetic CO_2 fluxes between forests and the atmosphere, are as follows: (1) daily mean absorbed photosynthetically active radiation was the dominant explanatory factor; (2) the annual cycle of photosynthesis was important to take into account, especially in the spring and (3) the stomatal action reduced photosynthesis during high evaporation. These findings provide additional evidence that the previous theory of CO_2 flux covers the most important features in the fluxes between forest ecosystems and the atmosphere.

5.6 Annual Energy, Carbon, Nitrogen, Water and Ion Fluxes and Amounts at SMEAR II

Pertti Hari, Eero Nikinmaa, Pasi Kolari, Mari Pihlatie, Jukka Pumpanen, Janne F.J. Korhonen, Liisa Kulmala, Asko Simojoki, Antti-Jussi Kieloaho, Jaana Bäck, Timo Vesala, and Markku Kulmala

Previous sections have described metabolic processes and transport phenomena generating the interactions, that is, fluxes, between forest ecosystems and the environment. Energy, water and carbon fluxes in a forest ecosystem are large, while those of nitrogen are small. The quantification of the fluxes requires a versatile field station that is able to follow continuously the material and energy fluxes. The planning and construction principle of our field station, SMEAR II in Hyytiälä, southern Finland, was to measure all relevant material and energy fluxes and amounts in the forest ecosystem around the station and to understand the factors

P. Hari (✉)
Department of Forest Sciences, University of Helsinki, P.O. Box 27, 00014 University of Helsinki, Helsinki, Finland
e-mail: pertti.hari@helsinki.fi

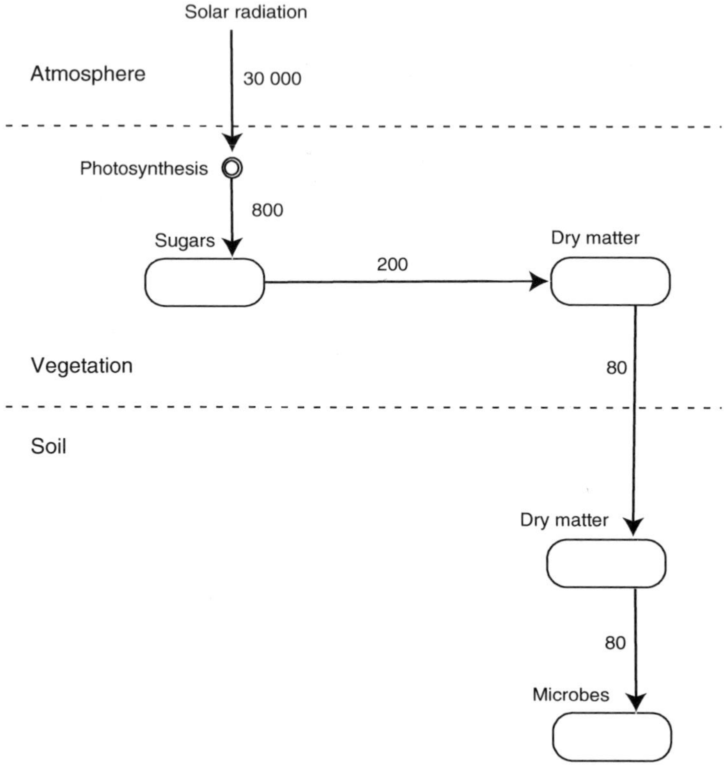

Fig. 5.34 The energy fluxes (MJ ha^{-1} $year^{-1}$) as solar radiation and as carbon compounds in the ecosystem around at SMEAR II

affecting the metabolic processes and physical phenomena generating these fluxes. Using the direct measurements and approaches described previously, we can now summarise the fluxes to the annual mean fluxes obtained in measurements.

5.6.1 Energy Fluxes as Radiation and as Organic Matter

Solar radiation is the source of energy available to the vegetation (Fig. 5.34). Its annual input at SMEAR II is about 30,000 MJ ha^{-1} $year^{-1}$. Photosynthesis converts solar radiation energy into chemical form as sugars: the efficiency of conversion is low, less than 3% at annual level. However, the sugars, formed in photosynthesis, are the source of energy for plant metabolism and the main source of raw material for growth of new plant structures and through their input also source of energy for all other living organisms in the ecosystem. The annual amount of energy bound in photosynthesis is 800 MJ ha^{-1} $year^{-1}$. The biggest usage of sugars is respiration and root exudates and fine root turnover. The formation of new permanent structures consumes clearly less, only 200 MJ ha^{-1} year^{-1}; thus, only less than 1% of solar radiation is bound into new cells of the trees and ground vegetation. The energy flux associated to litter production is rather small, 80 MJ ha^{-1} year^{-1}. If we assume that the soil organic matter does not accumulate, then the flux from soil organic matter to microbes is also 80 MJ ha^{-1} year^{-1}.

5.6.2 Carbon Pools and Fluxes

Carbon enters the forest ecosystem in photosynthesis both in the canopy (37,000 kg CO_2 ha^{-1} year^{-1}) and in the ground vegetation (4,000 kg CO_2 ha^{-1} year^{-1}, Fig. 5.35). Sugars are utilised for the production of ATP and the formation of new structures. Vegetation respiration, formation of ATP, consumes around 11,000 kg CO_2 ha^{-1} year^{-1}. The remaining sugars are used for growth of new leaves, xylem elements for transporting water and fine roots and to root respiration and exudates. Decomposing microbes in the soil live on the litter of needles, wood and fine roots that are altogether 3,200 kg dry matter ha^{-1} year^{-1}. If we assume that the soil organic matter is no longer accumulating, then the microbial respiration based on decomposition is 6,400 kg (CO_2) ha^{-1} year^{-1}.

The lifetime of fine roots is about 1 year, of needles 3 years and of wood about 100 years. The needle litter fall is approximately 3,200 kg dry weight ha^{-1} year^{-1}. The wood litter production during the lifetime of trees is small, and fine root litter is so difficult to measure that we do not have direct measurements of its annual production. The different litter components have specific chemical composition. When they are combined, we get annual litter components: 80 kg proteins ha^{-1} year^{-1}, 260 kg starch ha^{-1} year^{-1}, 540 kg lipids ha^{-1} year^{-1}, lignin 1,050 kg ha^{-1} year^{-1} and 1,500 kg cellulose ha^{-1} year^{-1}.

5.6.2.1 Mean Residence Time of Carbon

The mean residence time in a pool/storage of any substance provides a good understanding of the dynamics of the substance in the pool. It is defined as the time required to accumulate or to deplete the pool with the mean in- or outflow. The mean residence times can be determined only for major routes in the circulation of material in the ecosystem. We can approximate the mean residence time by dividing the pool of substances with the mean in- or outflows. The deaths of structures form the outflow from the vegetation as litter production. The mean residence time of carbon in the vegetation is about 20–30 years.

Microbial decomposition determines the residence times of macromolecules in the soil. The amounts of starch, lipids, cellulosc, lignin and proteins have been measured at SMEAR II, and the annual inputs can be obtained from litter fall and their macromolecule concentrations. These results show that the mean residence time of organic macromolecules is high: for proteins, starches, lipids and cellulose,

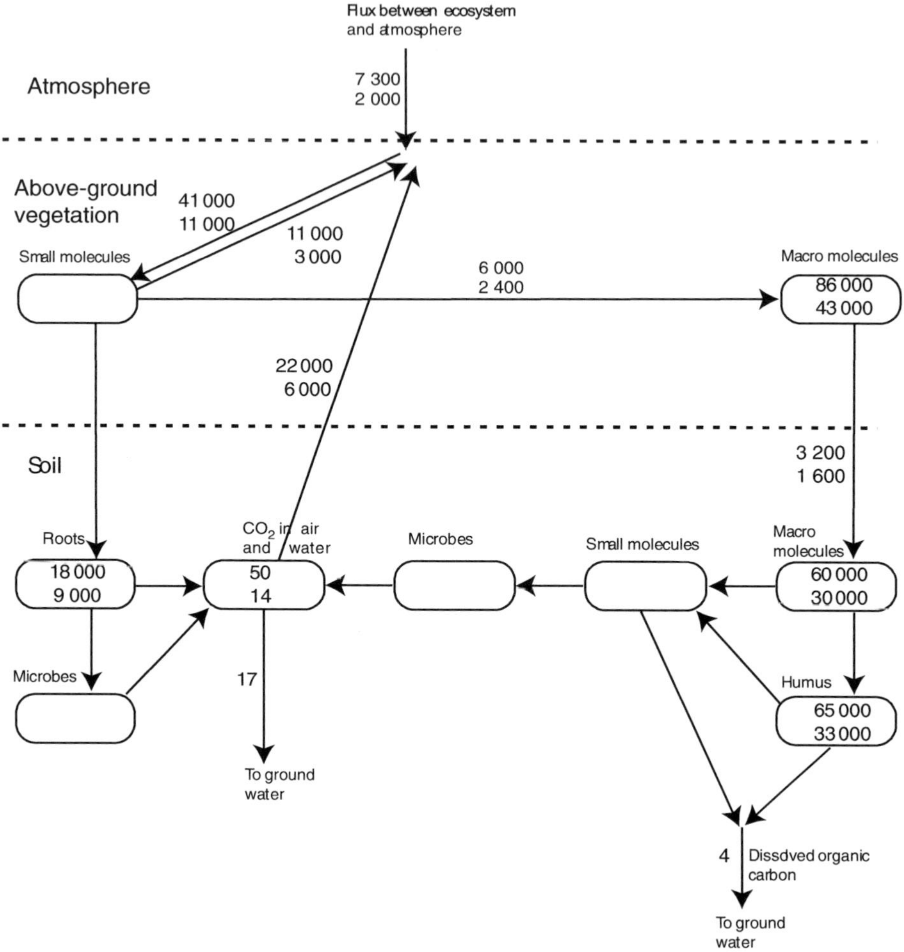

Fig. 5.35 The carbon amounts and fluxes in the ecosystem around SMEAR II. *Upper numbers* indicate the masses (kg dry weight ha^{-1}) and fluxes of the compounds (kg CO_2 ha^{-1} year^{-1} or kg dry weight ha^{-1} year^{-1}). *Lower numbers* indicate the masses (kg C ha^{-1}) and fluxes (kg C ha^{-1} year^{-1}) as carbon

they are 45, 15, 37 and 16 years, respectively. We cannot determine the mean residence time for lignin, since our method to analyse the chemical composition of soil organic matter does not separate lignin and humus. (cf. Simojoki et al. 2008).

Humus compounds dominate the soil organic matter (65,000 kg dry weight ha^{-1} year^{-1}). The lifetime of humus is very long, on the scale of millennia (Liski et al. 2005). If we assume that the mean residence time is 1,000 years and that the amount of humus compounds is in steady state, we get a rough approximation for the annual

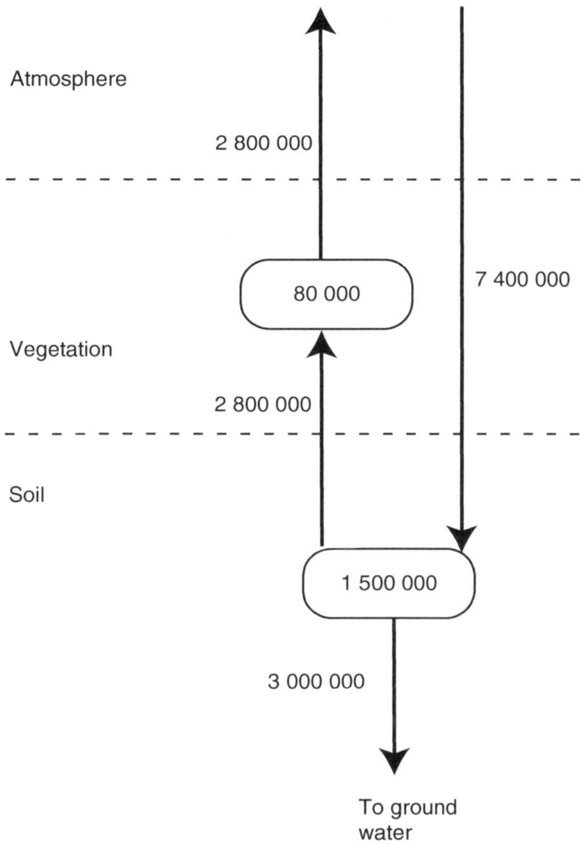

Fig. 5.36 Annual water fluxes (kg (H_2O) ha^{-1} year^{-1}) and amounts (kg (H_2O) ha^{-1}) in the ecosystem around SMEAR II

formation and decomposition of humus (65 kg dry weight ha^{-1} year^{-1}). The CO_2 pool in soil air is small, and the efflux from soil is large resulting in a small mean residence time, only slightly below a day.

5.6.3 Water Pools and Fluxes

All biological processes occur in liquid phase, and, in addition, water is an important transport media at long-distance within trees. These facts make water essentially important for living organisms. On the other hand, photosynthesis and transpiration are strongly coupled, since CO_2 dissolves in water on the mesophyll cell walls and diffusion transports CO_2 and water through stomata pores: CO_2 going in and H_2O coming out. Water input in forest ecosystems is rainfall, and transpiration and runoff are outputs (Fig. 5.36). The rainfall is in boreal forests often around 300–800 mm in a year, at SMEAR II 7,400,000 kg (H_2O) ha^{-1} $year^{-1}$ during the years 1998–2002. Water goes away from the forest ecosystem either as evapotranspiration to the atmosphere or as runoff on soil surface or as seepage in deeper layers in the soil. The solid bedrock near the surface and small water catchments enable measurement of the runoff flux in the soil from our ecosystem. The evapotranspiration measured with the Eddy Covariance system is 2,800,000 kg (H_2O) ha^{-1} $year^{-1}$ and runoff 3,000,000 kg (H_2O) ha^{-1} $year^{-1}$, and seepage water is missing due to the shallow soil on solid bedrock. The difference between out- and inflows at SMEAR II is 1,600,000 kg (H_2O) ha^{-1} $year^{-1}$ indicating measuring problems. The systematic errors in rainfall and evapotranspiration measurements are large, and the present imbalance is disturbing.

Most of the water in the forest ecosystem is either loosely bound on the surfaces of soil particles or in the wood. Both these amounts are large at SMEAR II, on soil particles 1,500,000 kg (H_2O) ha^{-1} and in the wood 80,000 kg (H_2O) ha^{-1}.

5.6.3.1 Mean Residence Time of Water

The water pools in the forest ecosystem at SMEAR II are large, as they are in most forests. Evapotranspiration is minimal in winter due to humid cold weather and stomatal closure. In contrast, the evapotranspiration is intensive in summer because the temperatures are often over 20°C, humidity is low and stomata open. The mean residence time is reasonable only during summer because water fluxes are negligible in winter. The mean annual residence time for water in soil is 60 days and in wood 8 days. If we take into consideration that transpiration is very small in winter, we can say that the mean residence time of water is in the soil about 30 days and in the wood 4 days.

5.6.4 Nitrogen Pools and Fluxes

Molecular nitrogen is not available for plants; it has to be converted into ions, either ammonium or nitrate, before nitrogen can be utilised in metabolism. Nitrogen ions enter the SMEAR II ecosystem either in deposition from the atmosphere or in microbial nitrogen fixation. Thereafter, nitrogen circulates in the system either as proteins, amino acids or ions (Fig. 5.37). Plants take ammonium and nitrate ions from soil. The nitrate is converted to ammonium and used in amino acid synthesis together with ammonium ions taken from soil. The amino acids are utilised in protein synthesis to build the needed functional substances, that is, enzymes, membrane pumps and pigments. When the tissue dies, the proteins are decomposed back to amino acids, which are used for protein synthesis elsewhere in living tissues. However, about 40% of the proteins in the dying tissue are lost as litter

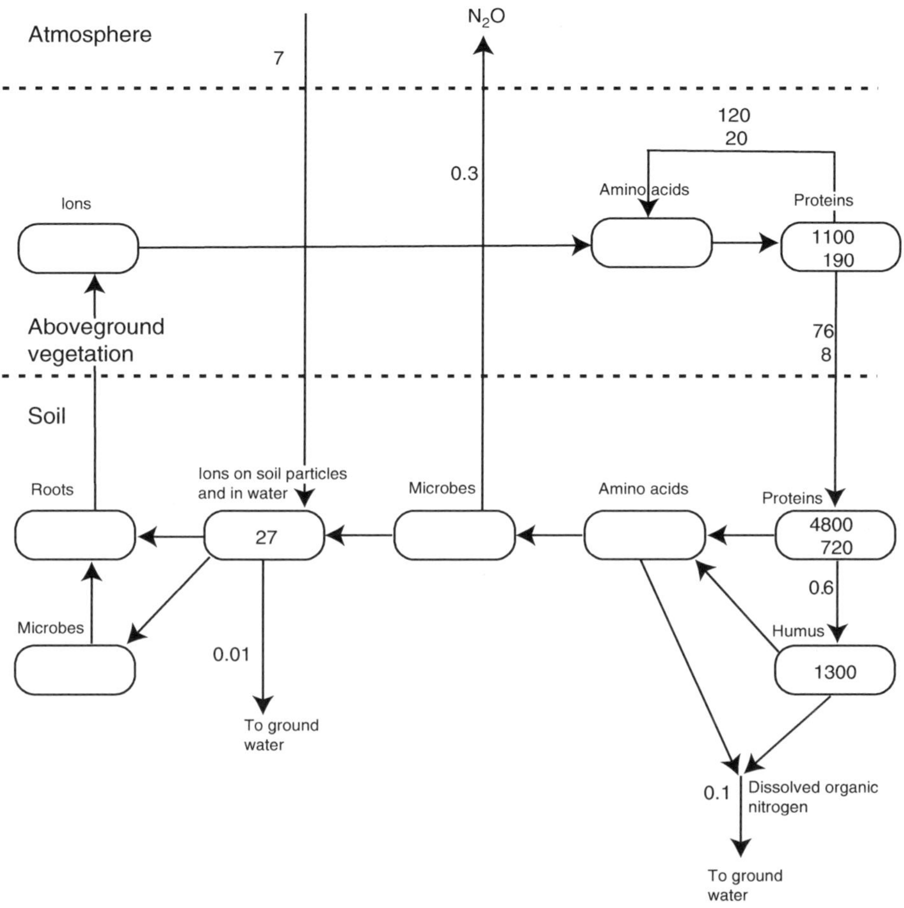

Fig. 5.37 The nitrogen amounts and fluxes in the ecosystem around SMEAR II. *Upper numbers* indicate the masses kg proteins ha^{-1} and fluxes of the compounds and single kg proteins ha^{-1} year^{-1} or dry weight of organic matter kg proteins ha^{-1} year^{-1}. *Lower numbers* indicate the masses kg N ha^{-1} and fluxes kg (N) ha^{-1} year^{-1}

to the soil. Microbes decompose with extracellular enzymes the proteins in the soil into amino acids. A small proportion of nitrogen is consumed in the formation of humus substances. Microbes take up the amino acids to construct proteins or utilise the energy in amino acids and release ammonium ion into soil water.

The internal cycling within the stand is about 120 kg (proteins) ha^{-1} year^{-1}. Fluxes with the surroundings are small when compared with internal cycling. The input is about 4 kg (N) ha^{-1} year^{-1} by wet deposition at SMEAR II. The dry deposition is about 3.5 kg N ha^{-1} year^{-1} (Flechard et al. 2011). The role of nitrogen fixation is unclear; we assume that it can be neglected. The outflows are still smaller; in drainage flow, 0.1 kg (N) ha^{-1} year^{-1} is lost, mainly in form of organic nitrogen.

There are short episodes in autumn and spring when the N_2O flow from the system is above detection limit of chamber measurements, but the flux is so small that it can be regarded as very marginal for N budget (<0.5 kg N ha^{-1} $year^{-1}$). NO_x flux between the needles and atmosphere is clearly measurable when the Sun is shining. The NO_x comes evidently from a UV radiation-driven process on the needle surface. We assume that the nitrogen is not coming from the metabolism of trees and it can be omitted in the nitrogen budget of the ecosystem.

The differences between sizes of nitrogen pools in forest ecosystems are several orders of magnitude, reflecting very active and rather passive processes. The pool of nitrogen as ions in the soil is 27 kg (N) ha^{-1}. The amount of proteins in vegetation is large reflecting the fact that functional substances are proteins. The value is around 600 kg proteins ha^{-1}. The biggest protein nitrogen pool is in the soil, 4,800 kg proteins ha^{-1}. The nitrogen in humus is in a very inactive form, and it does not participate in the normal nitrogen cycle in the forest ecosystem, but its amount is rather large, 1,300 kg (N) ha^{-1}.

5.6.4.1 Mean Residence Time of Nitrogen

The variation in nitrogen pool sizes and in the fluxes is large in the forest ecosystem around SMEAR II, which is reflected also in mean residence times in various components of the system. The pool of nitrogen ions in the soil is small, and the nitrogen uptake by vegetation is large resulting in a short mean residence time, only 3.5 years. Most of the proteins in living tissues have rather short lifetimes, but reuse of the amino acids is efficient, resulting in rather long mean residence time in vegetation, about 8 years. The decomposition of proteins by extracellular enzymes in the soil is a slow process. In addition, the circulation in soil generated by microbes prolongs the stay of nitrogen in the soil. The mean residence time of proteins in the soil at SMEAR II is long, about 60 years. We have no quantitative information of nitrogen dynamics in the large pool of humus substances. However, the residence time of nitrogen in humus is large in the scale of millennia.

The nitrogen pool in the forest ecosystem around SMEAR II is large: when expressed in the units nitrogen per hectare, we get 2,100 kg (N) ha^{-1}. There are short and weak N_2O and N_2 formation periods when the oxygen diffusion is blocked by water but their emissions can be neglected. The leakage as ions is small compared with that of dissolved organic nitrogen (DON). The mean nitrogen residence time in the forest ecosystem is long, according to our measurements about 10,000 years.

Extensive forest fires are an essential part of development of boreal forests. Then, all green components, small woody parts and the top layer of soil burn and release the protein nitrogen as NO_x. Fire frequency varies according to the location, but it can be assumed to be several times in a millennium. If we assume that proteins remaining in the unburned wood and that burned from the soil are equal and if we assume further that there are five fires in a millennia, we get that the outflow is 3 kg (proteins) ha^{-1} $year^{-1}$ and the nitrogen residence time caused by forest fires is about 1,500 years.

5.6.5 Comparison of Fluxes, Pools and Residence Times

The differences of fluxes between water, carbon and nitrogen are large. Water flux from the atmosphere into SMEAR II ecosystem is 7,400,000 kg (H_2O) ha^{-1} $year^{-1}$, and nitrogen flux out of the ecosystem is about 0.2 kg (N) ha^{-1} $year^{-1}$. The carbon flux into the system (41,000 kg CO_2 ha^{-1} $year^{-1}$) is two magnitudes smaller than water flux and five magnitudes bigger than nitrogen flux out of the system.

Although water flux into the ecosystem is two magnitudes larger than the carbon flux, the carbon pool is clearly larger than the pool of water. The water is running fast through the system, and carbon remains for long periods either in wood or in soil organic matter resulting in large pools. There are functional substances in all living cells enabling processes. The nitrogen content in vegetation and macromolecules in soil is around 1%, and thus, nitrogen pool in the ecosystem is about two magnitudes smaller than that of carbon.

The residence time of water in the ecosystem is small, in summer time about 30 days. In contrast, the nitrogen stays in the system for millennia if it has been synthesised into organic form. This difference is again five magnitudes reflecting the very different behaviour of water and nitrogen in the ecosystem, water is flowing through and nitrogen is accumulating. The carbon is between water and nitrogen, although it flows through the system like water. Its residence time is 20 years.

The differences in the mean residence times between different water pools in the ecosystem are small, only one magnitude between soil and wood. Carbon compartment pools have bigger variation because the gaseous CO_2 pool in the soil is very dynamic; the residence time is only a day. The humus storage in the soil is extremely stable: its residence time is millennia. The variation in residence times is about six magnitudes. Nitrogen has also big variation in the residence times but smaller than carbon. The fastest pool is ammonium ions in the soil: they have residence time about 30 days. The residence time of organic nitrogen in the ecosystem is in the range of millennia. Thus, the variation in residence times is one magnitude smaller than in the case of carbon.

The variations in the fluxes, amounts and residence times of carbon, water and nitrogen are large, often several orders of magnitude. The big variation hampers the analysis of ecosystems since several time levels have to be treated simultaneously. The metabolism and the physical phenomena generating the fluxes take place at the volume and time element level. Proper analysis of the fluxes, amounts and residence times is the key to understand the growth of forest ecosystems and their interaction with their surroundings, especially to the response to climate change.

5.6.6 *Ion Fluxes in Forest Ecosystem*

5.6.6.1 Potassium, Calcium and Magnesium

Seventeen (17) elements are generally considered essential nutrients for a proper metabolism of plants (Mengel and Kirkby 2001; Taiz and Zeiger 2002). Carbon, oxygen and hydrogen are taken up by plants directly from the atmosphere or from water molecules, whereas the other nutrients are primarily taken up as mineral ions from soil solution by roots. The ions enter the soil solution either by deposition from the atmosphere or weathering of soil minerals and leave the forest ecosystem by water runoff. These flux are small compared to the rather large amounts stored in the vegetation and soil organic matter. Along with nitrogen (N), phosphorus (P) and sulphur (S), potassium (K), calcium (Ca) and magnesium (Mg) are important macronutrients taken up by plants in relatively large amounts (about 100 mg m^{-2} $year^{-1}$ or more). They are present as cations (K^{+1}, Ca^{2+}, Mg^{2+}) in soil solution and circulate broadly in a similar way between vegetation and forest soil.

The fluxes and pools of nutrient ions in a boreal forest ecosystem were quantified at SMEAR II station by combining own measured data with data from other studies (Figs. 5.38 and 5.39). The amount and elemental concentrations in bulk precipitation, throughfall, litterfall, soil water and on soil particle surface were measured, except for soil pools of P and S that were estimated using current literature. Biomass and biomass accumulation were based on biomass inventory, and we obtained the elemental composition of biomass from Saarela et al. (2005) and Berg and McClaugherty (2003). We utilised for litter nutrient concentrations measurements by Johansson et al. (1995). Plant uptake, growth, mineralisation, canopy interception, weathering and nutrient balance were calculated using measurements and the mass balance. We assumed that nutrients storages in the foliage, soil organic matter, soil particle surfaces and bark were in the steady state in timescale of 1 year.

The two weirs constructed to the small watersheds (Chap. 9) around SMEAR II station allowed the measurement of total runoff (surface plus subsurface flow) from them. The soil at the hill around the station is very shallow (on average about 0.6 m) because the ice during the last glaciation pushed the soil from Hyytiälä to south. New mineral soil material was released from the bedrock by the action of ice. These deposits were initially under water at the end of glaciation and were redistributed by flowing water with thickest deposits accumulating at lower elevations in the landscape. With time, they became dry land due to the isostatic rebound of Earth crust. The bedrock under the shallow soil is solid granite with such a low water permeability, and it is considered practically waterproof. This allows quantitative measurements of runoff from watersheds by the weirs.

The large fluxes between vegetation and soil dominate the flows of potassium and calcium and to lesser extent those of magnesium, in the forest ecosystem around SMEAR II station (Fig. 5.38). The vegetation takes up 600, 2,000 and 180 mg m^{-2} $year^{-1}$ of K, Ca and Mg, respectively. The litterfall returns the macronutrients largely back into the pool of soil organic matter with fluxes of 250, 1,300 and

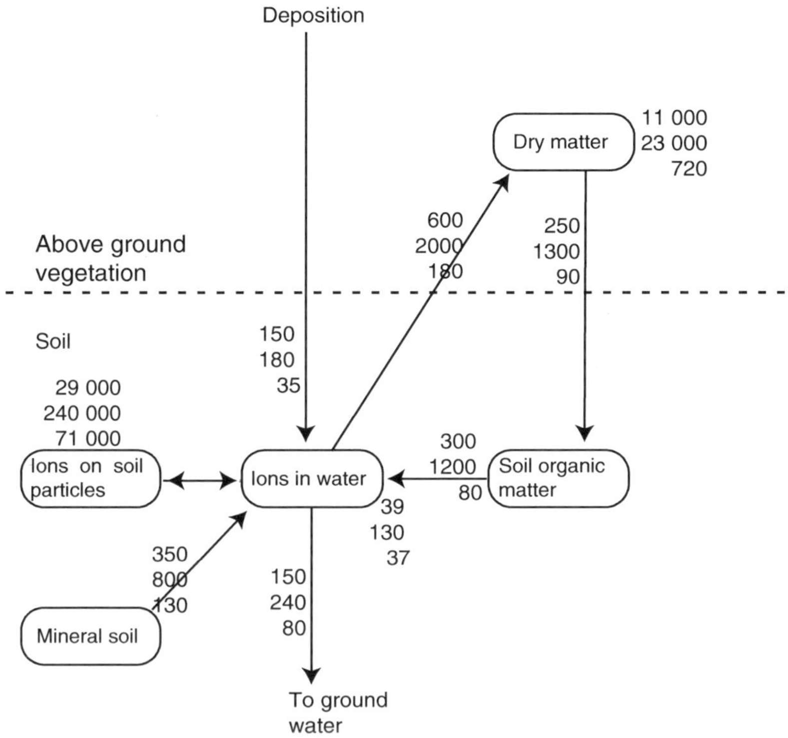

Fig. 5.38 The fluxes (mg m^{-2} year^{-1}) and pools (mg m^{-2}) of potassium, calcium and magnesium in the forest ecosystem around SMEAR II measuring station. The numbers on the *top*, *middle* and *bottom* of lists stand for K, Ca and Mg, respectively

90 mg m^{-2} year^{-1} for K, Ca and Mg, respectively. The decomposition of soil organic matter (mineralisation) releases potassium, calcium and magnesium ions into soil water. If we assume the amounts of nutrients in soil organic matter do not change, the release of these cations by the decomposition of soil organic matter equals the litter input of K, Ca and Mg into soil.

Most cations are bound to soil particle surfaces by a mechanism called cation exchange (Bohn et al. 2001; Blume et al. 2010). This is important as it determines largely the availability and mobility of nutrients in soil. Cation exchange consists of reversible transfer between the ions sorbed on the solid surfaces and those free in soil solution. Exchangeable cations are bound to the surfaces relatively loosely by electrostatic forces and can be replaced by the cations in soil solution. The cation exchange reactions are dependent on such factors as the cation exchange capacity, the accessibility of surfaces, the ion concentrations and the properties of cations. For the macronutrient cations discussed here, the tendency to be bound on the exchange sites decreases in the order Ca, Mg and K. Due to the electrostatic nature of cation exchange, the differences in the ease of exchange are mostly explained by the valence and size of ions: multivalent and larger cations are bound to the surfaces more strongly than monovalent and small cations (Bohn et al. 2001; Blume et al. 2010).

The deposition flux of cations is rather small, only 150, 180 and 36 mg m^{-2} $year^{-1}$ for K, Ca and Mg, respectively. In comparison, the runoff is somewhat larger with fluxes amounting to 150, 240 and 80 mg m^{-2} $year^{-1}$ for K, Ca and Mg, respectively. The data indicate stationary amounts of K and a slow depletion of Ca and Mg in the current ecosystem around SMEAR II station. This is broadly in line with the results by Ukonmaanaho and Starr (2002) who reported negative input–output budgets (depletion) for all elements, except for nitrogen and acidity (H^+) with a positive budget and sulphur with a negative or positive budget depending on the site.

The rate of weathering is the only unknown flux changing the amounts of potassium, calcium and magnesium in soil solution and on the surfaces of soil particles. If we further assume the amounts in soil solution and on the particle surfaces do not change with time, the mass balance gives an estimate for the ion fluxes released by weathering of mineral soil. In this way, we estimate the weathering fluxes of 350, 800 and 130 mg m^{-2} $year^{-1}$ for K, Ca and Mg, respectively.

The assumption of constant ion concentrations in soil is probably approximately valid for the time period of a few decades. However, during longer time periods, the amounts of ions stored in soil changes, as evidenced by the development of current, quite large pools of nutrient ions in soil. As a consequence, the above estimates indicating small weathering of K, Ca and Mg can be considered conservative estimates. These rates are in the range of past and current weathering rates determined in other studies for boreal podzol soils (Land and Öhlander 2000; Olsson and Melkerud 2000; Melkerud et al. 2003).

The annual outflows of potassium, calcium and magnesium are small when compared with the mobile ion pools in the forest ecosystem, especially those adsorbed on soil particles. Thus, their lifetimes in a forest ecosystem are rather large. Potassium circulates about 250 years, and Ca and Mg about 1,000 years, between soil and vegetation before leaching out of the forest ecosystem.

5.6.6.2 Phosphorus and Sulphur

Phosphorus and sulphur are essential nutrients for proper metabolism of vegetation and microbes (Mengel and Kirkby 2001; Taiz and Zeiger 2002). Phosphorus (P) is a vital component of many important phosphate- or pyrophosphate-containing biochemical molecules such as DNA, RNA, ATP, various coenzymes, phytin and the phospholipidic membranes of cells and cellular organelles. Sulphur (S) is part of thiol group (–SH) containing amino acids (cysteine, methionine) that are important building blocks of proteins. Moreover, the reactions of thiol groups in the cysteine–cystine, glutathione and lipoic acid redox systems give rise to disulfide bonds that stabilise the tertiary and quaternary structure (three-dimensional conformation) of enzyme proteins. Sulphur is also part of coenzyme A and the vitamins thiamine and biotin. Both phosphorus and sulphur belong to macronutrients taken up by plants in relatively large amounts (about 100 mg m^{-2} or more).

The chemical behaviours of P and S in soil differ quite considerably from that of K, Ca and Mg, because both P and S exist as phosphate ($H_2PO_4^{-1}$, HPO_4^{-2}) and sulphate (HSO_4^-, SO_4^{2-}) anions, respectively, in the soil solution. In addition, large amounts of P and S are also present in various organic forms.

Even if some variable-charge components (especially organic matter and Al and Fe oxides) become more positive with decreasing pH, the solid phase of boreal soils remains net negatively charged even at low pH. For this reason, phosphate and sulphate cannot be effectively adsorbed on soil particle surfaces merely by electrostatic anion exchange. However, phosphate can be adsorbed strongly, and sulphate less strongly, by ligand exchange (Hingston et al. 1967) on the Al and Fe oxide surfaces. In this mechanism, phosphate and sulphate ions replace hydroxyl (–OH) groups or water molecules from the inner coordination sphere around the metal ion and form a chemical bond with the metal ion. However, organic molecules may also bind to the oxide surfaces and compete with inorganic ligands for sorption sites. The extent to which this occurs in practice is still partly an open question, but current research suggests it may be smaller than previously anticipated (Borggaard et al. 2005; Guppy et al. 2005; Achat et al. 2011).

In podzol soils, the sorption of both phosphate and sulphate is abundant in the B-horizons enriched with Al and Fe oxides (Kaiser and Kaupenjohann 1998; Gustafsson et al. 2000; Väänänen et al. 2008). The mechanisms and extent of phosphorus and sulphur retention by soils have been recently discussed in more detail by McGechan and Lewis (2002a, b), Peltovuori (2006), Sokolova and Alekseeva (2008), Väänänen (2008) and Scherer (2009).

Due to weaker adsorption, dissolved sulphate is more abundant in soil solution and can move more freely in soil compared with phosphate. Although the adsorption of sulphate is generally weak, it is increased by the acidity of soil, as both anion and ligand exchange mechanisms become more efficient with decreasing pH (Sokolova and Alekseeva 2008; Scherer 2009).

Besides mineral P forms, some 20–80% of phosphorus in podzolic forest soils may be in organic forms (Compton and Cole 1998; Cade-Menun et al. 2000). Especially O-horizons are rich in organic P. In practice, the contents of organic P and S forms in soil are only rarely analysed due to measurement problems. However, the elemental ratios of C, N, P and S in soil organic matter are rather constant. The C/P ratio in soil organic matter is usually around 500 but may vary in the range 100–1,000 (Cade-Menun et al. 2000; Blume et al. 2010). The C/S ratio in soil organic matter is about 200, varying in the range 60–400 (Blume et al. 2010). The ratio S/N is about 0.1 (Nilsson et al. 2001). These ratios allow the estimation of organic P and S pools in soil, if the amounts of C or N are known.

The large fluxes between vegetation and soil dominate the flows of phosphorus. The vegetation takes up about 110 mg m^{-2} $year^{-1}$ of P, and the litter fall returns some 65 mg m^{-2} $year^{-1}$ back into the pool of soil organic matter. These P fluxes between plant and soil are an order of magnitude larger than the small fluxes of P deposition and outflow of about 7 mg m^{-2} $year^{-1}$. In contrast, the plant uptake and

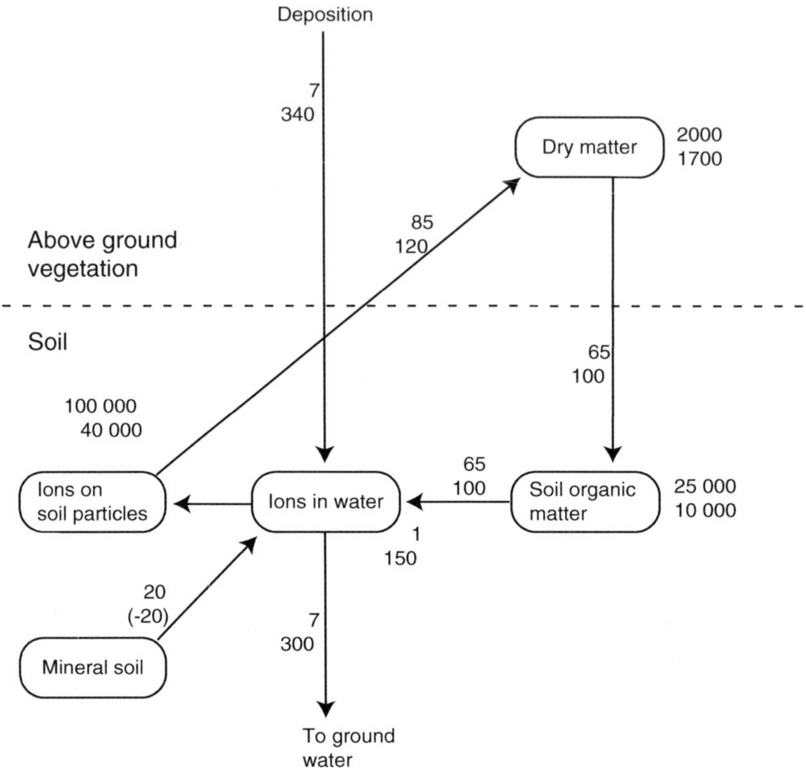

Fig. 5.39 The fluxes (mg m^{-2} $year^{-1}$) and pools (mg m^{-2}) of phosphate-P and sulphate-S in the forest ecosystem around SMEAR II. The numbers on the *top* and *bottom* stand for P and S, respectively

litterfall fluxes for S are larger than those for P. Nevertheless, they are only about half the fluxes of S deposition and outflow (300–340 mg m^{-2} $year^{-1}$).

The data indicate that the phosphorus is not accumulating in the system (deposition equals the runoff). The very small leaching of P (7 mg m^{-2} $year^{-1}$) is of similar magnitude as that determined in another study with another method by Piirainen et al. (2007). In contrast, a small accumulation of sulphur seems currently going on as the deposition exceeds the runoff. This is also in line with the results by Ukonmaanaho and Starr (2002) who reported accumulation of sulphur at some sites.

Based on the same assumptions as used earlier for cations, we estimate a small weathering rate of phosphorus amounting to 20 mg m^{-2} $year^{-1}$ from mineral soil (Fig. 5.39). This suggests that the weathering rate of P is somewhat slower than those of cations (Fig. 5.38). This is in partial agreement with the results by Olsson and Melkerud (2000) showing that the mean long-term weathering rate of P was slowest of all elements studied, even if for Hyytiälä site they reported a small gain of phosphorus (negative weathering of about –7 mg m^{-2} $year^{-1}$). In contrast, our data suggest some precipitation of sulphur into mineral form (negative weathering of –20 mg m^{-2} $year^{-1}$). This is unlikely to be true. Instead, a small change in the amount of sulphur stored on soil oxide surfaces or soil organic matter is more likely.

The annual outflow of phosphorus is very small when compared with the mobile ion pools in the forest ecosystem, especially those in soil. Accordingly, the average life time of phosphate ion in a boreal forest ecosystem is very long, as it will circulate about 10,000–20,000 years between soil and vegetation before leaching out. Even if the outflow of sulphate is comparatively much larger, its circulation time in a boreal forest is still quite long of about 200 years.

References

Achat DL, Augusto L, Morel C, Bakker MR (2011) Predicting available phosphate ions from physical-chemical soil properties in acidic sandy soils under pine forests. J Soil Sediment 11:452–466

Ågren GI (1996) Nitrogen productivity or photosynthesis minus respiration to calculate plant growth. Oikos 76:529–535

Amthor JS, Chen JM, Clein JS, Frolking SE, Goulden ML, Grant RF, Kimball JS, King AW, McGuire AD, Nikolov NT, Potter CS, Wang S, Wofsy SC (2001) Boreal forest CO_2 exchange and evapotranspiration predicted by nine ecosystem process models: intermodel comparisons and relationships to field measurements. J Geophys Res 106:33623–33648

Anfodillo T, Carraro V, Carrer M, Fior C, Rossi S (2006) Convergent tapering of xylem conduits in different woody species. New Phytol 169:279–290

Baldocchi DD, Vogel CA (1996) Energy and CO_2 flux densities above and below a temperate broad-leaved forest and a boreal pine forest. Tree Physiol 16:5–16

Baldocchi DD, Vogel CA, Hall B (1997) Seasonal variation of energy and water vapor exchange rates above and below a boreal Jack pine forest canopy. J Geophys Res 102(D4):28939–28951

Berg B, McClaugherty C (2003) Plant litter: decomposition, humus formation, carbon sequestration. Springer, Heidelberg/Berlin

Bergh J, Ross E, Linder S (1998) Climatic factors controlling the productivity of Norway spruce: a-model-based analysis. For Ecol Manag 110:127–139

Berninger F, Coll L, Vanninen P, Mäkelä A, Palmroth S, Nikinmaa E (2005) Effects of tree size and position on pipe model ratios in Scots pine. Can J For Res 35:1294–1305

Blume H-P, Brümmer GW, Horn R, Kandeler E, Kögel-Knabner I, Kretzshmar R, Stahr K, Wilke B-M (2010) Scheffer/Schachtschabel Lehrbuch der Bodenkunde, 16th edn. Spektrum, Heidelberg

Bohn HL, McNeal BL, Myer RA, O'Connor GA (2001) Soil chemistry, 3rd edn. Wiley, New York

Bonan GB (2008) Forests and climate change: forcings, feedbacks, and the climate benefits of forests. Science 320:1444–1449

Bond B, Kavanagh K (1997) Stomatal behavior of four woody species in relation to leaf-specific hydraulic conductance and threshold water potential. Tree Physiol 19:503–510

Boone RD, Nadelhoffer KJ, Canary JD, Kaye JP (1998) Roots exert a strong influence on the temperature sensitivity of soil respiration. Nature 396:570–572

Borggaard OK, Raben-Lange B, Gimsing AL, Strobel BW (2005) Influence of humic substances on phosphate adsorption by aluminium and iron oxides. Geoderma 127:270–279

Brough DW, Jones HG, Grace J (1986) Diurnal changes in the water-content of stems of apple trees as influenced by irrigation. Plant Cell Environ 9:1–7

Buchmann N (2000) Biotic and abiotic factors controlling soil respiration rates in *Picea abies* stands. Soil Biol Biochem 32:1625–1635

Cade-Menun BJ, Berch SM, Preston CM, Lavkulich LM (2000) Phosphorus forms and related soil chemistry of Podzolic soils on northern Vancouver Island. I. A comparison of two forest types. Can J For Res 30:1714–1725
Caraglio Y, Barthélémy D (1997) Revue critique des termes relatifs à la croissance et à la ramification des tiges des végétaux vasculaires. In: Bouchon J, de Reffye P, Barthélémy D (eds) Modélisation et simulation de l'architecture des végétaux. INRA Éditions, Versailles
Čermák J, Nadezhdina N (1998) Sapwood as the scaling parameter–defining according to xylem water content or radial pattern of sap flow? Ann For Sci 55:509–521
Čermák J, Kucera J, Bauerle WL, Phillips N, Hinckley TM (2007) Tree water storage and its diurnal dynamics related to sap flow and changes in stem volume in old-growth Douglas-fir trees. Tree Physiol 27:181–198
Choat B, Jansen S, Zwieniecki MA, Smets E, Holbrook NM (2004) Changes in pit membrane porosity due to deflection and stretching: the role of vestured pits. J Exp Bot 55:1569–1575
Cochard H, Hölttä T, Herbette S, Delzon S, Mencuccini M (2009) New insights into the mechanisms of water-stress induced cavitation in conifers. Plant Physiol 151:949–954
Compton JE, Cole DW (1998) Phosphorus cycling and soil P fractions in Douglas-fir and red alder stands. For Ecol Manag 110:101–112
Constantin J, Grelle A, Ibrom A, Morgenstern K (1999) Flux partitioning between understorey and overstorey in a boreal spruce/pine forest determined by the Eddy Covariance method. Agr For Meteorol 98(99):629–643
Davidson EA, Belk E, Boone RD (1998) Soil water content and temperature as independent or confounded factors controlling soil respiration in a temperate mixed hardwood forest. Glob Change Biol 4:217–227
Dawson TE (1993) Hydraulic lift and water use by plants: implications for water balance, performance and plant-plant interactions. Oecologia 95:565–574
Debenedetti P (1996) Metastable liquids. Princeton University Press, Princeton
Desutter TM, Sauer TJ, Parkin TB, Heitman, JL (2008) A subsurface, closed-loop system for soil carbon dioxide and its application to the gradient efflux approach. Soil Sci Soc Am J 72:126–134
Dickson RE, Isebrands JG (1991) Leaves as regulators of stress responses. In: Mooney HA, Winner WE, Pell EJ (eds) Response of plants to multiple stresses. Academic, San Diego
Domec J-C, Lachenbruch B, Meinzer FC, Woodruff DR, Warren JM, McCulloh KA (2008) Maximum height in a conifer is associated with conflicting requirements for xylem design. Proc Natl Acad Sci USA 105:12069–12074
Doran JW, Mielke LN, Stamatiadis S (1988) Microbial activity and N cycling as regulated by soil water-filled pore space. In: Proceedings of 11th ISTRO conference, Edinburgh
Drake JE, Gallet-Budynek A, Hofmockel KS et al (2011) Increases in the flux of carbon below ground stimulate nitrogen uptake and sustain the long-term enhancement of forest productivity under elevated CO_2. Ecol Lett 14:349–357
Drebs A, Nordlund A, Karlsson P, Helminen J, Rissanen P (2002) Climatological statistics of Finland 1971–2000, climatological statistics of Finland 2001. Finnish Meteorol Institute, Helsinki
Duursma R, Mäkelä A (2007) Summary models for light interception and light-use efficiency of non-homogeneous canopies. Tree Physiol 27:859–870
Duursma RA, Marshall JD, Robinson AP, Pangle RE (2007) Description and test of a simple process-based model of forest growth for mixed-species stands. Ecol Model 203:297–311
Falge E, Baldocchi D, Olson RJ et al (34 authors) (2001) Gap filling strategies for defensible annual sums of net ecosystem exchange. Agr For Meteorol 107:43–69
Fang C, Moncrieff JP (1999) A model for soil CO_2 production and transport 1: model development. Agr For Meteorol 95:225–236
Fernandez IJ, Kosian PA (1987) Soil air carbon dioxide concentrations in a New England spruce-fir forest. Soil Sci Soc Am J 51:261–263

Flechard CR, Nemitz E, Smith RI, Fowler D, Vermeulen AT, Bleeker A, Erisman JW, Simpson D, Zhang L, Tang YS, Sutton MA (2011) Dry deposition of reactive nitrogen to European ecosystems: a comparison of inferential models across the NitroEurope network. Atmos Chem Phys 11:2703–2728

Fontaine S, Barot S, Barré P et al (2007) Stability of organic carbon in deep soil layers controlled by fresh carbon supply. Nature 450:277–280

Freijer JI, Leffelaar PA (1996) Adapted Fick's law applied to soil respiration. Water Resour Res 32:791–800

Génard M, Fishman S, Vercambre G, Huguet J-G, Bussi C, Besset J, Habib R (2001) A biophysical analysis of stem and root diameter variations in woody plants. Plant Physiol 126:188–202

Glinski JW, Stepniewski W (1985) Soil aeration and its role for plants. CRC Press, Boca Raton

Goulet J, Messier C, Nikinmaa E (2000) Effect of branch position and light availability on shoot growth of understory sugar maple and yellow birch sapling. Can J Bot 78:1077–1085

Granier A (1987) Evaluation of transpiration in a Douglas-fir stand by means of sap flow measurements. Tree Physiol 3:309–320

Greaves JR, Carter EG (1920) Influence of moisture on the bacterial activities of the soil. Soil Sci 10:361–387

Guppy CN, Menzies NW, Moody PW, Blamey FPC (2005) Competitive sorption reactions between phosphorus and organic matter in soil: a review. Aust J Soil Res 43:189–202

Gustafsson JP, van Hees P, Starr M, Karltun E, Lundström U (2000) Partitioning of base cations and sulphate between solid and dissolved phases in three podzolised forest soils. Geoderma 94:311–333

Hacke UG, Sperry JS, Pittermann J (2004) Analysis of circular bordered pit function. II. Gymnosperm tracheids with torus-margo pit membranes. Am J Bot 91:386–400

Hacke UG, Sperry JS, Pockman WP, Davis SD, McCulloh KA (2001) Trends in wood density and structure are linked to prevention of xylem implosion by negative pressure. Oecologia 126:457–461

Hallé F, Oldeman RAA, Tomlinson PB (1978) Tropical trees and forests: an architectural analysis. Springer, Berlin

Hari P, Heikinheimo P, Mäkelä A, Kaipiainen L, Korpilahti E, Salmela J (1986) Trees as a water transport system. Silva Fenn 20:205–210

Hari P, Keronen P, Bäck J, Altimir N, Linkosalo T, Pohja T, Kulmala M, Vesala T (1999) An improvement of the method for calibrating measurements of photosynthetic CO_2 flux. Plant Cell Environ 22:1297–1301

Heinonsalo J, Pumpanen J, Rasilo T, Hurme K-R, Ilvesniemi H (2010) Carbon partitioning in ectomycorrhizal Scots pine seedlings. Soil Biol Biochem 42:1614–1623

Hillel D (1982) Introduction to soil physics. Academic Press, New York

Hingston FJ, Atkinson RJ, Posner AM, Quirk JP (1967) Specific adsorption of anions. Nature 215:1459–1461

Högberg P, Read DJ (2006) Towards a more plant physiological perspective on soil ecology. Trends Ecol Evol 21:548–554

Holbrook NM, Zwieniecki MA (1999) Embolism repair and xylem tension: do we need a miracle? Plant Physiol 120:7–10

Höltta T, Kolari P (2009) Interpretation of stem CO_2 efflux measurements. Tree Physiol 29:1447–1456

Hölttä T, Vesala T, Perämäki MK, Nikinmaa E (2002) Relationships between embolism, stem water tension, and diameter changes. J Theor Biol 215:23–38

Hölttä T, Vesala T, Perämäki M, Nikinmaa E (2006a) Refilling of embolised conduits as a consequence of "Munch water" circulation. Funct Plant Biol 33:949–959

Hölttä T, Vesala T, Sevanto S, Perämäki M, Nikinmaa E (2006b) Modeling xylem and phloem water flows in trees according to cohesion theory and Münch hypothesis. Trees Struct Funct 20:67–78

Hölttä T, Cochard H, Nikinmaa E, Mencuccini M (2009) Capacitive effect of cavitation in xylem conduits. Plant Cell Environ 32:10–21

Horn HS (2000) Twigs, trees, and the dynamics of carbon in the landscape. In: Brown JH, West GB (eds) Scaling in biology. Oxford University Press, New York

Huber B (1935) Die physiologische Bedeutung der Ring- und Zerstreutporigkeit. Ber Deut Bot Ges 53:711–719

Irvine J, Grace J (1997) Continuous measurements of water tensions in the xylem of trees based on the elastic properties of wood. Planta 202:455–461

Jassal R, Black A, Novak M, Morgenstern K, Nesic Z, Gaumont-Guay D (2005) Relationship between soil CO_2 concentrations and forest-floor CO_2 effluxes. Agr For Meteorol 130:176–192

Johansson MB, Berg B, Meentemeyer V (1995) Litter mass loss rates in the late stages of decomposition in a climatic transect of pine forests. Long-term decomposition in a Scots pine forest. Can J Bot 73:1509–1521

Kähkönen MA, Wittmann C, Kurola J, Ilvesniemi H, Salkinoja-Salonen MS (2001) Microbial activity of boreal forest soil in a cold climate. Bor Environ Res 6:19–28

Kaipiainen L, Hari P (1985) Consistencies in the structure of Scots pine. In: Tigerstedt PMA, Puttonen P, Koski V (eds) Crop physiology of forest trees. Helsinki University Press, Helsinki

Kaiser K, Kaupenjohann M (1998) Influence of the soil solution composition on retention and release of sulfate in acid forest soils. Water Air Soil Poll 101:363–376

Kärkkäinen M (2007) Puun rakenne ja ominaisuudet. Metsäkustannus Oy, Helsinki (in Finnish)

Kelliher FM, Lloyd J, Arneth A, Byers JN, McSeveny TM, Milukova I, Grigoriev S, Panfyorov M, Sogatchev A, Varlargin A, Ziegler W, Bauer G, Schulze ED (1998) Evaporation from a central Siberian pine forest. J Hydrol 205:279–296

Kennedy RE, Turner DP, Cohen WB, Guzy M (2006) A method to efficiently apply a biogeochemical model to a landscape. Landsc Ecol 21:213–224

Klepper B, Browning VD, Taylor HM (1971) Stem diameter in relation to plant water status. Plant Physiol 48:683–685

Knoblauch M, van Bel AJE (1998) Sieve tubes in action. Plant Cell 10:35–50

Koch GW, Sillet SC, Jennings GM, Davis SD (2004) The limits to tree height. Nature 428:851–854

Koehler B, Zehe E, Corre MD, Veldkamp E (2010) An inverse analysis reveals limitations of the soil-CO_2 profile method to calculate CO_2 production and efflux for well-structured soils. Biogeosciences 7:2311–2325

Kolari P, Pumpanen J, Kulmala L, Ilvesniemi H, Nikinmaa E, Grönholm T, Hari P (2006) Forest floor vegetation plays an important role in photosynthetic production of boreal forests. For Ecol Manag 221:241–248

Kolari P, Lappalainen HK, Hänninen H, Hari P (2007) Relationship between temperature and the seasonal course of photosynthesis in Scots pine at northern timberline and in southern boreal zone. Tellus 59B:542–552

Kolari P, Kulmala L, Pumpanen J, Launiainen S, Ilvesniemi H, Hari P, Nikinmaa E (2009) CO_2 exchange and component CO_2 fluxes of a boreal Scots pine forest. Bor Environ Res 14:761–778

Kottek M, Grieser J, Beck C, Bruno R, Rubel F (2006) World map of the Köppen-Geiger climate classification updated. Meteorol Z 15:259–263

Kramer K, Leinonen I, Bartelink HH et al (28 authors) (2002) Evaluation of six process-based forest growth models using eddy-covariance measurements of CO_2 and H_2O fluxes at six forest sites in Europe. Glob Change Biol 8:213–230

Kulmala L, Launiainen S, Pumpanen J, Lankreijer H, Lindroth A, Hari P, Vesala T (2008) H_2O and CO_2 fluxes at the floor of a boreal pine forest. Tellus 60B:167–178

Kulmala L, Pumpanen J, Hari P, Vesala T (2011) Photosynthesis of ground vegetation in different aged pine forests: effect of environmental factors predicted with a process-based model. J Veg Sci 22:96–110

Kuzyakov Y (2010) Priming effects: interactions between living and dead organic matter. Soil Biol Biochem 42:1363–1371

Land M, Öhlander B (2000) Chemical weathering rates, erosion rates and mobility of major and trace elements in a boreal granitic till. Aquat Geochem 6:436–460

Landsberg JJ, Waring RH (1997) A generalised model of forest productivity using simplified concepts of radiation-use efficiency, carbon balance and partitioning. For Ecol Manag 95: 209–228
Lang A (1979) A relay mechanism for phloem translocation. Ann Bot 44:141–145
Lang A (1983) Turgor-regulated translocation. Plant Cell Environ 6:683–689
Launiainen S (2010) Seasonal and inter-annual variability of energy exchange above a boreal Scots pine forest. Biogeosciences 7:3915–3940
Launiainen S, Rinne J, Pumpanen J, Kulmala L, Kolari P, Keronen P et al (2005) Eddy covariance measurements of CO_2 and sensible and latent heat fluxes during a full year in a boreal pine forest trunk-space. Bor Environ Res 10:569–588
Launiainen S, Katul GG, Kolari P, Vesala T, Hari P (2011) Empirical and optimal stomatal controls on leaf and ecosystem level CO_2 and H_2O exchange rates. Agr For Meteorol 151:1672–1689
Law BE, Waring RH, Anthoni PM, Aber JD (2000) Measurements of gross and net ecosystem productivity and water vapour exchange of a *Pinus ponderosa* ecosystem, and an evaluation of two generalized models. Glob Change Biol 6:155–168
Liesche J, He H-X, Grimm B, Schulz A, Kühn C (2010) Recycling of *Solanum* sucrose transporters expressed in yeast, tobacco, and in mature phloem sieve elements. Mol Plant 3:1064–1074
Linn DM, Doran JW (1984) Effect of water-filled pore space on carbon dioxide and nitrous oxide production in tilled and nontilled soils. Soil Sci Soc Am J 48:1267–1272
Liski J, Palosuo T, Peltoniemi M, Sievänen R (2005) Carbon and decomposition model Yasso for forest soils. Ecol Model 189:168–182
Long JN, Smith FW (1988) Leaf area – sapwood area relations of lodgepole pine as influenced by stand density and site index. Can For Res 18:247–250
Magnusson T (1995) Relationships between soil properties and the soil atmosphere in Swedish forest soils. Scan J For Res 10:209–217
Mäkelä A (1988) Models of pine stand development: an eco-physiological systems analysis. Univ Helsinki, Dept Silviculture, Res Notes 62:1–267
Mäkelä A, Valentine HT (2006) The quarter-power scaling does not infer size-invariant hydraulic resistance in trees. J Theo Bio 243:283–285
Mäkelä A, Hari P, Berninger F, Hänninen H, Nikinmaa E (2004) Acclimation of photosynthetic capacity in Scots pine to the annual cycle of temperature. Tree Physiol 24:369–376
Mäkelä A, Kolari P, Karimäki J, Nikinmaa E, Perämäki M, Hari P (2006) Modelling five years of weather-driven variation of GPP in a boreal forest. Agr For Meterol 139:382–398
Mäkelä A, Pulkkinen M, Kolari P, Lagergren F, Berbigier P, Lindroth A, Loustau D, Nikinmaa E, Vesala T, Hari P (2008) Developing an empirical model of stand GPP with the LUE approach: analysis of eddy covariance data at five contrasting conifer sites in Europe. Glob Change Biol 14:98–108
Maselli F, Barbati A, Chiesi M, Chirici G, Corona P (2006) Use of remotely sensed and ancillary data for estimating forest gross primary productivity in Italy. Remote Sens Environ 100:563–575
McCulloh KA, Sperry JS, Adler FR (2003) Water transport in plants obeys Murray's law. Nature 421:939–942
McGechan MB, Lewis DR (2002a) Sorption of phosphorus by soil, Part 1: Principles, equations and models. Biosyst Eng 82:1–24
McGechan MB, Lewis DR (2002b) Sorption of phosphorus by soil, Part 2: Measurement methods, results and model parameter values. Biosyst Eng 82:115–130
McMurtrie RE, Gholz HL, Linder S, Gower ST (1994) Climatic factors controlling the productivity of pine stands: a model-based analysis. Ecol Bull 43:173–188
McQueen JC, Minchin PEH, Thorpe MR, Silvester WB (2005) Short-term storage of carbohydrate in stem tissue of apple (*Malus domestica*), a woody perennial: evidence for involvement of the apoplast. Funct Plant Biol 32:1027–1031
Mecke M, Ilvesniemi H (1999) Near-saturated hydraulic conductivity and water retention in coarse podzol profiles. Scand For Res 14:391–401

Medlyn B, Barrett D, Landsberg J, Sands P, Clement R (2003) Conversion of canopy intercepted radiation to photosynthate: review of modelling approaches for regional scales. Funct Plant Biol 30:153–169
Meinzer FC, Clearwater MJ, Goldstein G (2001) Water transport in trees: current perspectives, new insights and some controversies. Env Exp Bot 45:239–262
Melcher PJ, Goldstein G, Meinzer FC, Yount DE, Jones TJ, Holbrook NM, Huang CX (2001) Water relations of coastal and estuarine *Rhizophora mangle*: xylem pressure potential dynamics of embolism formation and repair. Oecologia 126:182–192
Melkerud P-A, Bain DC, Olsson MT (2003) Historical weathering based on chemical analyses of two Spodosols in southern Sweden. Water Air Soil Poll Focus 3:49–61
Mencuccini M, Hölttä T (2010) The significance of phloem transport for the speed of link between canopy photosynthesis and belowground respiration. New Phytol 185:189–203
Mencuccini M, Hölttä T, Petit G, Magnani F (2007) Sanio's laws revisited. Size-dependent changes in xylem architecture of trees. Ecol Lett 10:1084–1093
Mengel K, Kirkby EA (2001) Principles of plant nutrition, 5th edn. Kluwer, Dordrecht
Milburn JA (1996) Sap ascent on vascular plants: challengers to the cohesion theory ignore the significance if immature xylem and recycling of Münch water. Ann Bot 78:399–407
Millington RJ, Quirk JP (1961) Permeability of porous solids. Trans Faraday Soc 57:1–8
Milner EM (2010) Magnificent trees of Britain. National Trust Books, London
Minchin PEH, Thorpe MR (1984) Apoplastic phloem unloading in the stem of bean. J Exp Bot 35:538–550
Minchin PEH, Thorpe MR (1987) Measurement of unloading and reloading of photoassimilate within the stem of bean. J Exp Bot 38:211–220
Moldrup P, Olesen T, Gamst J, Schjønning P, Yamaguchi T, Rolston DE (2000a) Predicting the gas diffusion coefficient in repacked soil: water-induced linear reduction model. Soil Sci Soc Am J 64:1588–1594
Moldrup P, Olesen T, Schjønning P, Yamaguchi T, Rolston DE (2000b) Predicting the gas diffusion coefficient in undisturbed soil from water characteristics. Soil Sci Soc Am J 64:94–100
Moncrieff JB, Fang C (1999) A model for soil CO_2 production and transport 2: application to a Florida Pinus elliottii plantation. Agric For Meteorol 95:237–256
Nikinmaa E (1992) Analyses of the growth of Scots Pine; matching structure with function. Acta For Fenn 235:1–68
Nikinmaa E, Kaipiainen L, Mäkinen M, Ross J, Sasonova T (1996) Geographical variation in the regularities of woody structure and water transport. In: Hari P, Ross J, Mecke M (eds) Production process of Scots Pine: geographical variation and models. Acta For Fenn 254: 49–78
Nikinmaa E, Goulet J, Messier C, Sievänen R, Perttunen J, Lehtonen M (2003) Shoot growth and crown development; the effect of crown position in 3D simulations. Tree Physiol 23:129–136
Nilsson SI, Andersson S, Valeur I, Persson T, Bergholm J, Wirén A (2001) Influence of dolomite lime on leaching and storage of C, N and S in a Spodosol under Norway spruce (*Picea abies* (L.) Karst.). For Ecol Manag 146:55–73
Nobel PS (1991) Physiochemical and environmental plant physiology. Academic, San Diego
Olsson MT, Melkerud P-A (2000) Weathering in three podzolized pedons on glacial deposits in northern Sweden and central Finland. Geoderma 94:149–161
Patrick JW (1997) Phloem unloading: sieve element unloading and post-sieve element transport. Annu Rev Plant Phys 48:191–222
Paul MJ, Foyer CH (2001) Sink regulation of photosynthesis. J Exp Bot 52:1383–1400
Peltovuori T (2006) Phosphorus in agricultural soils of Finland – characterization of reserves and retention in mineral soil profiles. Dissertation, University of Helsinki, Pro Terra 26
Perämäki M, Nikinmaa E, Sevanto S, Ilvesniemi H, Siivola E, Hari P, Vesala T (2001) Tree stem diameter variation and transpiration in Scots pine; an analysis using dynamic sap flow model. Tree Physiol 21:889–897
Perämäki M, Vesala T, Nikinmaa E (2005) Modeling the dynamics of pressure propagation and diameter variation in tree. Tree Physiol 25:1091–1099

Petit G, Anfodillo T, Mencuccini M (2008) Tapering of xylem conduits and hydraulic limitations in sycamore (*Acer pseudoplatanus*) trees. New Phytol 177:653–664

Phillips NG, Ryan MG, Bond BJ, McDovell NG, Hinckley TM, Čermák J (2003) Reliance on stored water increases with tree size in three species in the Pacific Northwest. Tree Physiol 23:237–245

Phillips RP, Finzi AC, Bernhardt ES (2011) Enhanced root exudation induces microbial feedbacks to N cycling in a pine forest under long-term CO_2 fumigation. Ecol Lett 14:187–194

Piirainen S, Finér L, Mannerkoski H, Starr M (2007) Carbon, nitrogen and phosphorus leaching after site preparation at a boreal forest clear-cut area. Forest Ecol Manage 243:10–18

Priestley CBH, Taylor RJ (1972) On the assessment of surface heat flux and evaporation using large scale parameters. Mon Weather Rev 100:81–92

Pumpanen J, Ilvesniemi H, Keronen P, Nissinen A, Pohja T, Vesala T, Hari P (2001) An open chamber system for measuring soil surface CO_2 efflux: analysis of error sources related to the chamber system. J Geophys Res 106(D8):7985–7992

Pumpanen JS, Heinonsalo J, Rasilo T, Hurme K-R, Ilvesniemi H (2009) Carbon balance and allocation of assimilated CO_2 in Scots pine, Norway spruce and Silver birch seedlings determined with gas exchange measurements and ^{14}C pulse labelling in laboratory conditions. Trees Struct Funct 23:611–621

Rannik Ü, Kolari P, Vesala T, Hari P (2006) Uncertainties in measurement and modelling of net ecosystem exchange of a forest ecosystem at different time scales. Agric For Meteorol 138:244–257

Room PM, Maillette L, Hanan JS (1994) Module and metamer dynamics and virtual plants. Adv Ecol Res 25:105–157

Ruehr NK, Offermann CA, Gessler A, Winkler JB, Ferrio JP, Buchmann N, Barnard RL (2009) Drought effects on allocation of recent carbon: from beech leaves to soil CO_2 efflux. New Phytol 184:950–961

Running S, Gower S (1991) FOREST-BGC, a general model of forest ecosystem processes for regional applications. II dynamic carbon allocation and nitrogen budgets. Tree Physiol 9:147–160

Running SW, Hunt ER (1993) Generalization of a forest ecosystem process model for other biomes, BIOME-BGC, and an application for global-scale models. In: Ehleringer JR, Field CB (eds) Scaling physiological processes: leaf to globe. Academic, San Diego

Ryan MJ, Yoder B (1997) Hydraulic limits to tree height and tree growth. Bioscience 47:235–242

Saarela KE, Harju L, Rajander J, Lill JO, Heselius SJ, Lindroos A, Mattsson K (2005) Elemental analyses of pine bark and wood in an environmental study. Sci Total Environ 343:231–241

Salleo S, Lo Gullo M, Trifilo P, Nardini A (2004) New evidence for a role of vessel-associated cells and phloem in the rapid xylem refilling of cavitated stems of *Laurus nobilis*. Plant Cell Environ 27:1065–1076

Saranpää P (1990) Heartwood formation in stems of *Pinus sylvestris* L. Publ Dept Bot, Univ Helsinki 14:1–22

Scherer HW (2009) Sulfur in soils. J Plant Nutr Soil Sci 172:326–335

Schlentner RE, van Cleve K (1985) Relationships between CO_2 evolution from soil, substrate temperature, and substrate moisture in four mature forest types in interior Alaska. Can J For Res 15:97–106

Scholander PF, Hammel HT, Bradstreet ED, Hemmingsen EA (1965) Sap pressure in vascular plants. Science 148:339–346

Schulze ED, Cermak J, Matyssek R, Penka M, Zimmerman R, Vasicek F, Gries W, Kuèera J (1985) Canopy transpiration and water fluxes in the xylem of the trunk of Larix and Picea trees – a comparison of xylem flow, porometer and cuvette measurements. Oecologia 66:475–483

Seneviratne SI, Lüthi D, Litschi M, Schär C (2006) Land–atmosphere coupling and climate change in Europe. Nature 443:205–209

Sevanto S, Vesala T, Perämäki M, Nikinmaa E (2002) Time lags for xylem and stem diameter variations in a Scots pine tree. Plant Cell Environ 25:1071–1077

Sevanto S, Vesala T, Perämäki M, Nikinmaa E (2003) Sugar transport together with environmental conditions controls time lags between xylem and stem diameter changes. Plant Cell Environ 26:1257–1265

Sevanto S, Hölttä T, Hirsikko A, Vesala T, Nikinmaa E (2005) Determination of thermal expansion of green wood and the accuracy of tree stem diameter variation measurements. Bor Environ Res 10:437–445

Shinozaki K, Yoda K, Hozumi K, Kira T (1964) A quantitative analysis of plant form – the pipe model theory. I. Basic analyses. Jpn J Ecol 14:97–105

Simmoneau T, Habib R, Goutouly JP, Huguet JG (1993) Diurnal changes in stem diameter depend upon variations in water content: direct evidence in peach trees. J Exp Bot 44:615–621

Simojoki A, Garcia H, Pihlatie M, Pumpanen J, Kurola J, Salkinoja-Salonen M, Hari P (2008) Environmental factors in soil. In: Hari P, Kulmala L (eds) Boreal forest and climate change, vol 34, Advances in global change research. Springer, Dordrecht

Šimůnek J, Suarez DL (1993) Modeling of carbon dioxide transport and production in soil 1. Model development. Water Resour Res 29:487–497

Skopp J, Jawson MD, Doran JW (1990) Steady-state aerobic microbial activity as a function of soil water content. Soil Sci Soc Am J 54:1619–1625

Smith ML, Ollinger SV, Martin ME, Aber JD, Hallett RA, Goodale CL (2002) Direct estimation of aboveground forest productivity through hyperspectral remote sensing of canopy nitrogen. Ecol Appl 12:1286–1302

Sokolova TA, Alekseeva SA (2008) Adsorption of sulfate ions by soils (a review). Eurasian Soil Sci 41:140–148

Solantie R (1981) Lumipeitteen vesiarvon vuotuinen maksimi ja sen ajankohta. Vesitalous 5, Helsinki (in Finnish)

Stenberg P (1996) Simulations of the effects of shoot structure and orientation on vertical gradients in intercepted light by conifer canopies. Tree Physiol 16:99–108

Steppe K, De Pauw DJW, Lemeur R, Vanrolleghem PA (2006) A mathematical model linking tree sap flow dynamics to daily stem diameter fluctuations and radial stem growth. Tree Physiol 26:257–273

Stevens GC, Perkins AL (1992) The branching habits and life history of woody plants. Am Nat 139:267–275

Suarez DL, Šimůnek J (1993) Modeling of carbon dioxide transport and production in soil 2. Parameter selection, sensitivity analysis and comparison of model predictions to field data. Water Resour Res 29:499–513

Taiz L, Zeiger E (2002) Plant physiology, 3rd edn. Sinauer Associates, Sunderland

Tang J, Baldicchi DD, Qi Y, Xu L (2003) Assessing soil CO_2 efflux using continuous measurements of CO_2 profiles in soils with small solid-state sensors. Agric For Meteorol 118:207–220

Thompson MV (2006) Phloem: the long and the short of it. Trends Plant Sci 11:26–32

Thompson MV, Zwieniecki MA (2005) The role of potassium in long distance transport in plants. In: Holbrook NM, Zwieniecki MA (eds) Vascular transport in plants. Elsevier Academic Press, Boston

Thornton PE, Law BE, Gholz HL, Clark KL, Falge E, Ellsworth DS, Goldstein AH, Monson RK, Hollinger D, Falk M, Chen J, Sparks JP (2002) Modeling and measuring the effects of disturbance history and climate on carbon and water budgets in evergreen needleleaf forests. Agric For Meteorol 113:185–222

Troeh FR, Jabro JD, Kirkham D (1982) Gaseous diffusion equations for porous materials. Geoderma 27:239–253

Turcu VE, Jones SB, Or D (2005) Continuous soil carbon dioxide and oxygen measurements and estimation of gradient-based gaseous flux. Vadose Zone J 4:1161–1169

Turgeon R (2010) The puzzle of phloem pressure. Plant Physiol 154:578–581

Turner DP, Urbanski S, Bremer D, Wofsy SC, Meyers T, Gower ST, Gregory M (2003) A cross-biome comparison of daily light use efficiency for gross primary production. Glob Change Biol 9:383–395

Tyree MT, Sperry JS (1988) Do woody plants operate near the point of catastrophic xylem dysfunction caused by dynamic water stress? Answers from a model. Plant Physiol 88:574–580
Tyree MT, Yang S (1990) Water-storage capacity of *Thuja*, *Tsuga* and *Acer* stems measured by dehydration isotherms. Planta 182:420–426
Tyree MT, Zimmermann MH (2002) Xylem structure and the ascent of sap. Springer, New York
Tyree MT, Salleo S, Nardini A, Lo Gullo MA, Mosca R (1999) Refilling of embolized vessels in young stems of laurel: do we need a new paradigm? Plant Physiol 120:11–21
Ukonmaanaho L, Starr M (2002) Major nutrients and acidity: budgets and trends at four remote boreal stands in Finland during the 1990s. Sci Total Environ 297:21–41
Väänänen R (2008) Phosphorus retention in forest soils and the functioning of buffer zones used in forestry. Dissertation, University of Helsinki. Dissertationes Forestales 60, 42 p
Väänänen R, Hristov J, Tanskanen N, Hartikainen H, Nieminen M, Ilvesniemi H (2008) Phosphorus sorption properties in podzolic forest soils and soil solution phosphorus concentration in undisturbed and disturbed soil profiles. Boreal Environ Res 13:553–567
Vakkilainen P (1986) Haihdunta. In: Mustonen S (ed) Sovellettu hydrologia. Mäntän kirjapaino, Mänttä (in Finnish)
van Dijk AIJM, Dolman AJ, Schulze ED (2005) Radiation, temperature, and leaf area explain ecosystem carbon fluxes in boreal and temperate European forests. Glob Biogeochem Cy 19:GB2029
Vesala T, Hölttä T, Perämäki M, Nikinmaa E (2003) Refilling of hydraulically isolated embolised vessels: model calculations. Ann Bot 91:419–428
Waring RH, Running SW (1978) Sapwood water storage: its contribution to transpiration and effect on water conductance through the stems of old-growth Douglas fir. Plant Cell Environ 1:131–140
Waring RH, Whitehead D, Jarvis PG (1979) The contribution of stored water to transpiration in Scots pine. Plant Cell Environ 2:309–317
Waring RH, Schroeder PE, Oren R (1982) Application of pipe model theory to predict canopy leaf area. Can J For Res 12:556–560
Watson M, Casper BB (1984) Morphogenetic constraints on patterns of carbon distribution in plants. Ann Rev Ecol Syst 15:233–258
Weitz JS, Ogle K, Horn HS (2006) Ontogenetically stable hydraulic design in woody plants. Funct Ecol 20:191–199
West GB, Brown JH, Enquist BJ (1999) A general model for the structure and allometry of plant vascular systems. Nature 400:664–667
Whitehead D, Livingston NJ, Kelliher FM, Hogan KP, Pepin S, McSeveny TM, Byers JN (1996) Response of transpiration and photosynthesis to a transient change in illumination foliage area for a *Pinus radiata* D. Don tree. Plant Cell Environ 19:949–957
Widén B, Majdi H (2001) Soil CO_2 efflux and root respiration at three sites in a mixed pine and spruce forest: seasonal and diurnal variation. Can J For Res 31:786–796
Williams M, Rastetter EB, Fernandes DN et al (1996) Modeling the soil–plant–atmosphere continuum in a *Quercus–Acer* stand at Harvard Forest: the regulation of stomatal conductance by light, nitrogen and soil/plant hydraulic properties. Plant Cell Environ 19:911–927
Wilson KB, Hanson PJ, Mulholland PJ, Baldocchi DD, Wullschleger SD (2001) A comparison of methods for determining forest evapotranspiration and its components: sap-flow, soil water budget, eddy covariance and catchment water balance. Agric For Meteorol 106:153–168
Wood WW, Petraitis MJ (1984) Origin and distribution of carbon dioxide in the unsaturated zone of the southern high plains of Texas. Water Resour Res 20:1193–1208
Woodruff DR, Bond BJ, Meinzer FC (2004) Does turgor limit growth in tall trees? Plant Cell Environ 27:229–236
Yuan W, Liu S, Gua Z, Guo Z, Tieszen LL, Baldocchi D, Bernhofer C, Gholz H, Goldstein AH, Goulden ML, Hollinger DY, Hu Y, Law BE, Stoy PC, Vesala T, Wofsy SC (2007) Deriving a light use efficiency model from eddy covariance flux data for predicting daily gross primary production across biomes. Agric For Meteorol 143:189–207

Zimmermann MH (1983) Xylem structure and the ascent of sap. Springer, Berlin
Zweifel R, Item H, Häsler R (2001) Link between diurnal stem radius changes and tree water relations. Tree Physiol 21:869–877
Zwieniecki MA, Holbrook NM (2009) Confronting Maxwell's demon: biophysics of xylem embolism repair. Trends Plant Sci 14:530–534
Zwieniecki MA, Melcher PJ, Holbrook NM (2001) Hydraulic properties of individual xylem vessels of *Fraxinus americana*. J Exp Bot 52:257–264

Chapter 6
Structural Regularities in Trees

Pertti Hari, Mikko Havimo, Juho Aalto, Pauliina Schiestl-Aalto, Eero Nikinmaa, Anna Lintunen, Tuomo Kalliokoski, Heljä-Sisko Helmisaari, and Inge Stupak

Contents

Abstract Several processes, such as photosynthesis and water and nutrient uptake, are simultaneously running in plants. Fluxes of processed products, such as sugars, water and nutrient ions connect the processes with each other. Highly specialised structures have developed in evolution for each process and transport phenomenon. According to the basic idea 7 of the cover theory (Chap. 2), these structures are effective, and basic idea 9 says that the biochemical regulation systems play an important role in the formation of the structures. The water transport within trees generates regularities in the structure of woody components.

Keywords Regularities in tree structure • Evolution • Biochemical regulation system • Water transport • Nitrogen concentration

P. Hari (✉)
Department of Forest Sciences, University of Helsinki, P.O. Box 27, 00014 University of Helsinki, Helsinki, Finland
e-mail: pertti.hari@helsinki.fi

P. Hari et al. (eds.), *Physical and Physiological Forest Ecology*,
DOI 10.1007/978-94-007-5603-8_6,

6.1 Functional Basis of the Regularities in Tree Structure

Pertti Hari, Mikko Havimo, Juho Aalto, Pauliina Schiestl-Aalto, and Eero Nikinmaa

The lifetime of plant cells is rather short, usually only some years, and the senescent ones must be replaced. The metabolism of living cells consumes energy for growth and maintenance respiration. In addition to energy, plant growth requires material for cell walls, membranes and other structures.

The material and energy fluxes between plants and their surroundings are characteristic features for ecosystems. The materials enter the vegetation as small molecules or ions. Metabolism synthesises numerous new compounds that have very complicated structure and large molecular weights.

The routes of material and energy in the metabolism and in the transport within a tree are numerous and versatile. There are, however, some metabolic and transport phenomena that dominate. We can identify these as metabolic or transport tasks:

1. Photosynthesis begins the journey of carbon in plants.
2. Water uptake from soil makes water available for transpiration.
3. Nutrient uptake, with the products of photosynthesis, offers necessary building materials for functional substances.
4. Respiration provides energy for the metabolism of cells.
5. New cells are formed in growth.
6. Transport of water and sugars within trees connects the roots, the stem, the branches and the needles with each other.

These six main metabolic and transport tasks have their own highly specialised structures. Leaves are specialised for photosynthesis. There are complex structures in the mesophyll cells to capture light energy and convert it into a chemical form as sugars, which are the main raw material for growth and source of energy for metabolism.

Fine roots have special structures for nutrient and water uptake. Osmosis, caused by high sugar concentration in the endodermis cells in the roots, generates water flow into the roots via aquaporins, and membrane pumps move nutrient ions into root cells (Sect. 4.2.5). Nitrogen plays a crucial role in the nutrient and water uptake since aquaporins and membrane pumps are proteins.

Several phenomena, such as membrane penetration or synthesis of substances, require energy that is obtained from ATP. Mitochondria produce ATP from ADP using sugar and other small carbon molecules as source of energy. Simultaneously CO_2 is released. The energy in the form of ATP is of vital importance for the cell; without ATP production and release of CO_2, the cell dies.

P. Hari (✉)
Department of Forest Sciences, University of Helsinki, P.O. Box 27, 00014 University of Helsinki, Helsinki, Finland
e-mail: pertti.hari@helsinki.fi

The growth of new cells is a long chain of synthesis of new substances. In cell growth, cellulose and lignin are needed for cell walls, lipids for membranes. These substances form the passive structure of a cell that is by itself unable to perform metabolic actions. Functional substances, enzymes, membrane pumps and pigments, form the tools in the cells. These substances are needed for biochemical reactions, membrane penetration and capturing light energy. The nitrogen content of functional substances is high and that explains the important role of nitrogen as nutrient.

The long distances between the intake of material and its consumption call the need for transport within plants. Vegetation, especially trees, has specialised structures for the various transport tasks within individuals. According to the basic idea 7 in the cover theory, the transport structures are efficient in performing their task, and the properties of the transport structure can be studied, at least in the case of crucial tasks, as a solution to the general problem of optimisation.

There is a clear water transport system in the trees: chains of interconnected tracheids run in coniferous trees from roots to leaves via coarse roots, stems and branches. Transpiration from leaves generates water tension in leaves that pulls water through the tracheid chains from roots.

Sugars are transported in a system of highly active cells that form the phloem which connects the leaves with all other parts of the plant. The high sugar content in phloem cells generates pressure that pushes water and the dissolved sugars from leaves to everywhere needed, to the last end of the roots. The water flows in the xylem and in phloem are strongly coupled with each other.

The metabolic and transport tasks are strongly interconnected:

1. The water uptake and transport are connected with transpiration and photosynthesis.
2. Photosynthesis and sugar transport provide sugars for respiration and growth.
3. Nitrogen uptake and transport are needed for the synthesis of functional substances into the cells.

The performance of each structure in their metabolic or transport tasks should match with the needs of the next step in the task chain. The amount of the structure and its fine structure determine the performance of the structures in their metabolic or transport tasks. Plants can increase or decrease the performance of metabolic or transport structures by allocating sugars or nitrogen to the growth of cells in the structures. The increase in the amounts of sugars or nitrogen consumed to a structure will increase its performance, and in the case of excessive consumption, the sugars or nitrogen could produce more benefit if they were used in the construction of some more crucial structure. Thus, proper allocation is of vital importance for the functioning of trees (Figs. 6.1 and 6.2).

A powerful biochemical regulation system to allocate sugars and nitrogen to the structures of different metabolic or transport tasks has emerged during evolution, resulting in effective couplings between the tasks. In this way, emergent regularities in tree structure are formed during growth, and detecting these regularities is an important key to understand tree behaviour. The regular behaviour of photosynthesis

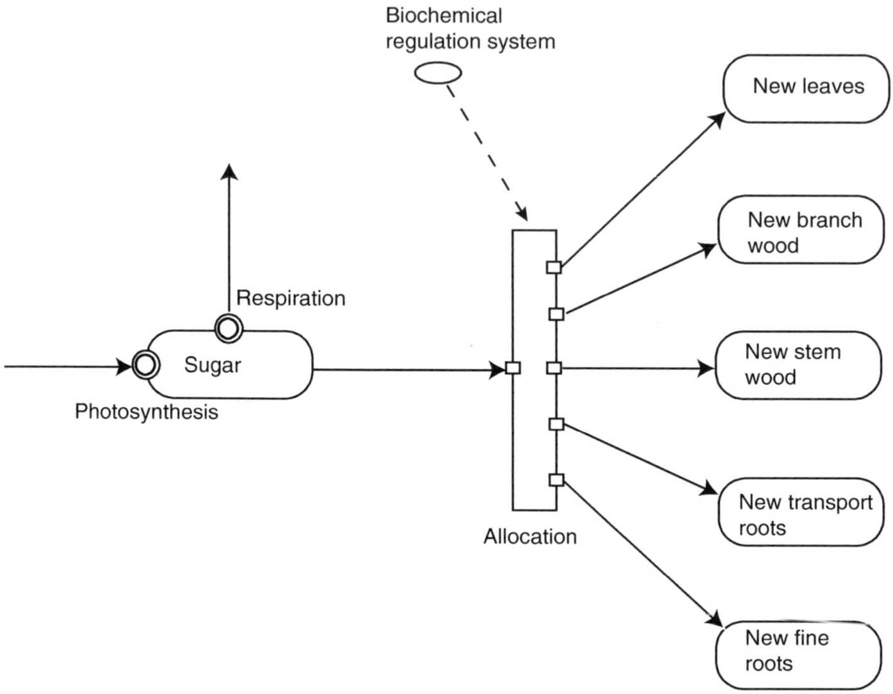

Fig. 6.1 Visualisation of allocation of carbon compounds to new leaves, new water transport system in branches, stem, transport roots and fine roots under the control of the biochemical regulation system. We introduce a new symbol, the rectangular, for allocation. The other symbols are introduced in Fig. 2.3

in midsummer (Sect. 4.2.2.2) indicates balance between the water availability for transpiration and water transport in the stem, and the regular behaviour of growth shows sufficient availability of sugars (Sect. 4.2.7).

The plants meet the allocation problem also within metabolic tasks since in a tree the leaves grow on numerous branches, each of them having meristems for growth. The extinction of light in the canopy and variation in the length of the water transport system are additional complications in the solution of the spatial allocation problem. Especially, the water should flow smoothly without any disturbances throughout the long journey from roots to leaves.

The pine trees lay down buds in late summer and early autumn at the ends of each twig for growth during next summer. There are primordia – essentially embryos of each needle – in the bud, and thus, the number of the new needles is preprogrammed in the buds (Fig. 6.3). The development of a pine bud during a growing season can be seen in the photos taken from a bud and from the shoot developing from the bud during the growing season (Fig. 6.4). The bud first

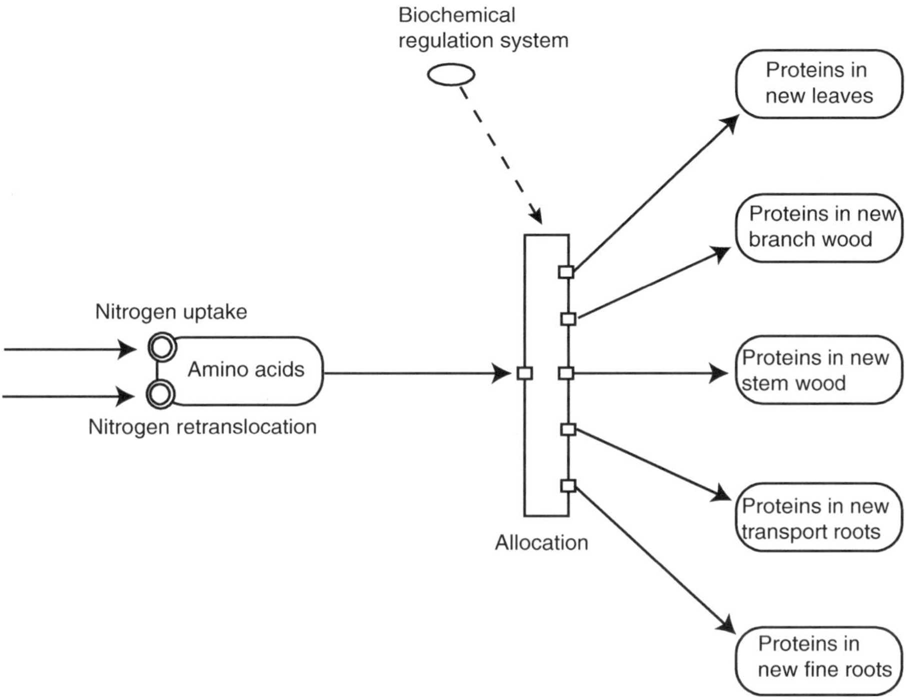

Fig. 6.2 Visualisation of allocation of nitrogen compounds to new leaves, new water transport system in branches, stem, transport roots and fine roots under the control of the biochemical regulation system

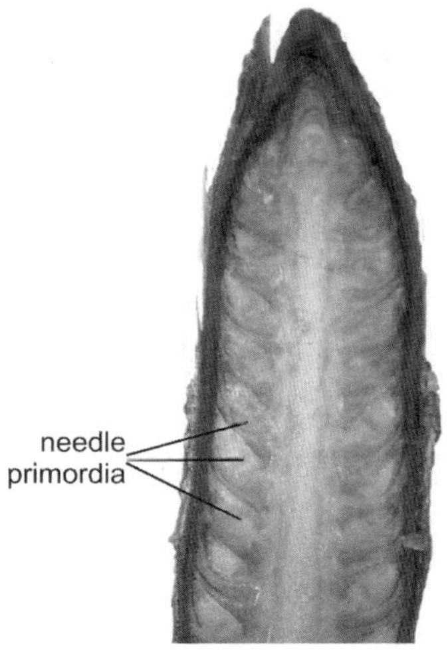

Fig. 6.3 Cross section of a Scots pine bud in autumn. There are primordia of each needle in the bud to grow in the next spring

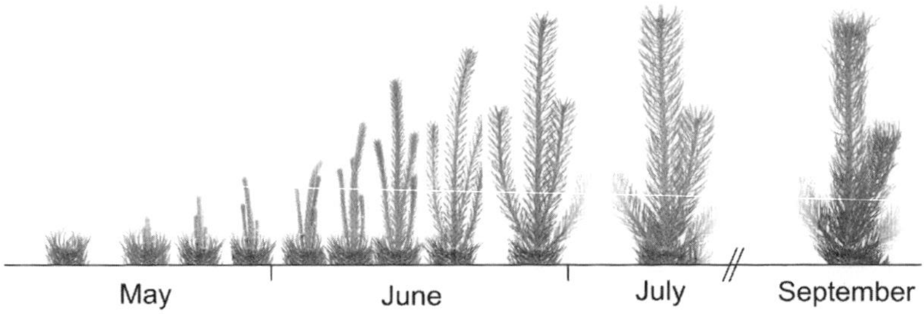

Fig. 6.4 Scots pine shoots developing from the buds during summer

elongates, and thereafter the primordia of the needles grow to the final needle size during midsummer. The growth is preprogrammed in the bud, and the weather conditions during growing summer have only minor modifying effect. Thus, the solution of the allocation problem is to a great extent in the buds formed in late summer, before the actual growth in spring.

6.2 Tree Structure as Balanced Water Transport System

Pertti Hari, Mikko Havimo, Anna Lintunen, Tuomo Kalliokoski, and Eero Nikinmaa

The action of the biochemical regulation system generates emergent regularities in xylem structure from fine roots to the needles, since the xylem tissue in coarse roots, stems, branches and twigs has to transport water needed for transpiration. A mature pine tree consumes a major part of its annual photosynthetic production in the build-up of the xylem structures for water transport. Proper allocation of sugars between the coarse roots, stem, branches and twigs is of vital importance for the success of the tree. In addition, the performance of the water transport systems has to be sufficient to replace water lost by transpiration, whilst the fine roots must be sufficient to access the required amount of water.

The tradition to study regularities in tree structure is old. Already Leonardo da Vinci was interested especially in the structure of branches and stem (MacCurdy 2002). In modern times, Sinozaki et al. (1964) discovered that there is linear relationship between cross-sectional area of a branch and its leaf mass. Hari et al. (1986) studied the tree structure as a balanced water transport system.

P. Hari (✉)
Department of Forest Sciences, University of Helsinki, P.O. Box 27, 00014 University of Helsinki, Helsinki, Finland
e-mail: pertti.hari@helsinki.fi

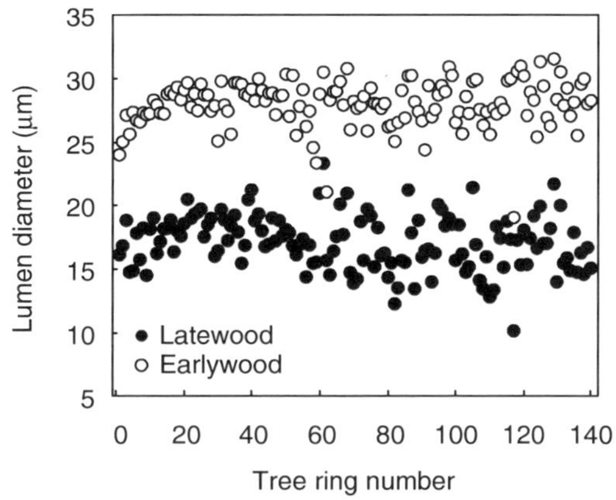

Fig. 6.5 Radial lumen diameter in Scots pine as function of the tree ring number from pith to bark in breast height. The figure is based on data described in Havimo et al. (2009)

6.2.1 Water Transport in Xylem

Fine roots take water from the soil, and needles transpire the water to the atmosphere. The distance between fine roots and needles is often tens of metres (the tallest trees are 100 m). Water is transported in the xylem of coarse roots, stem and branches. Tension pulls water through xylem into the needles, and since water is under tension, in a metastable state, the xylem has to have highly specialised structure. In coniferous trees, xylem consists of tracheids and rays. An annual ring in the xylem is commonly divided into earlywood and latewood. The main function of earlywood tracheids is water transport; they have wide lumens and narrow cell walls. Latewood tracheids have wide cell walls and small lumens, which confer good mechanical properties but only small water-conducting capacity. Main function of latewood tracheids is to give mechanical support to the above growing biomass.

Tracheid dimensions vary somewhat depending on the position in the trunk. Their length in *Pinus sylvestris* varies from 1 to 3 mm (Rautiainen and Alen 2009). The width of tracheids is circa 30 μm, several orders of magnitude smaller than the length (Havimo et al. 2009). Cell wall thickness varies between earlywood and latewood. On average, the wall thickness is 2.0 μm in earlywood and 3.7 μm in latewood.

Tracheids are dead cells, which consist of lignified cell wall and hollow lumen. The tracheid lumen is connected to the lumens of neighbouring tracheids through small pores called pits (see Fig. 5.1). The lumen diameter determines, to a great extent, the water transport capacity of xylem. Figure 6.5 shows the lumen diameter from pith to bark in a Scots pine trunk. In latewood, the lumen diameter is considerably smaller than in earlywood, and it is quite constant from pith to bark. In earlywood, the diameter increases rapidly in juvenile wood, which in Scots pine is approximately 20 innermost annual rings. After the rapid increase, in mature wood, the lumen diameter is relatively constant, although there is some variation in the measured values.

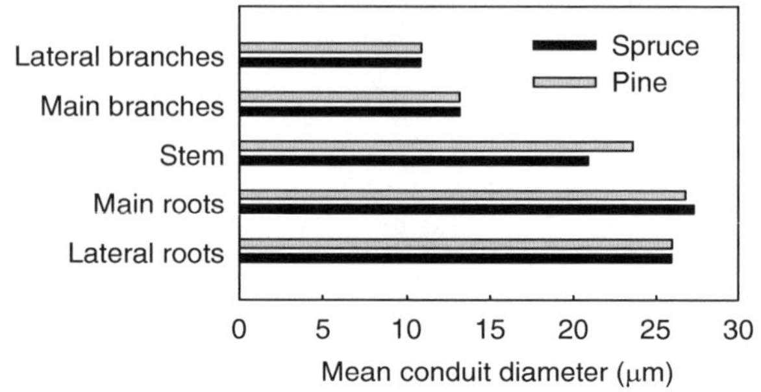

Fig. 6.6 Mean conduit diameter in Norway spruce and Scots pine from lateral roots to lateral branches. The figure is based on measurements described in Lintunen and Kalliokoski (2010)

According to Lintunen and Kalliokoski (2010), the lumen diameter decreases from roots to the stem and from the stem to the branches (Fig. 6.6). In other words, the water transport capacity of an individual tracheid decreases from root to the needle. This is compensated by the increase in the number of tracheids, which increases from the roots to the top (Lintunen and Kalliokoski 2010).

In young trees, all annual rings in xylem conduct water. When the tree grows, the oldest annual rings transform into heartwood. Heartwood does not conduct water, and in some tree species, such as Scots pine, the heartwood tracheids are filled with air.

We pay special attention to the following aspects of the water transport system:

1. Heartwood is a non-water-conducting tissue, and it should be excluded from the water transport capacity.
2. Roots, stem and branches have their own water transport capacities and are treated separately.
3. The amount of xylem tissue varies between roots, stem and branches.
4. The roots, stem and branches form a balanced water transport system.

6.2.1.1 Basic Concept and Ideas

Water transport from soil to needles is essential for photosynthesis. Trees invest a large share of their annual photosynthetic production for growth of new water transport tissue. In order to understand the allocation of sugars produced in photosynthesis and the development of tree growth, we need a theoretical understanding of the regularities in the water transport system.

Water transport capacity of roots, stem and branches is an emergent property that describes the water transport ability of a cross-sectional area of sapwood. It arises from the tracheid dimensions and the amount of latewood and earlywood tracheids in annual rings. There is a considerable variation in the tracheid dimensions. The largest differences are between roots, stem and branches.

The allocation of photosynthetic products to the water transporting tissue plays a key role in the growth of a tree. There are emergent regularities in the water transport system in different parts of a tree, generated by the action of the biochemical regulation system. Here we focus on the differences of xylem properties between roots, stem and branches.

Basic idea W1: The coarse roots, stem, branches and twigs form an efficient and balanced water transport system.

The action of the biochemical regulation system generates emergent regularities in the water transport system.

Basic idea W2: The water transport capacities of coarse roots, stem and branches match with each other.

The action of the biochemical regulation system takes care also of the balance between water uptake, transport and transpiration.

Basic idea W3: The water uptake by fine roots matches the transport capacity of the root, and the transpiration of the leaves matches the water transport capacity of the branch on which the leaves are attached.

6.2.1.2 Theoretical Model

The lumen diameter is rather stable in the stem from the edge of the pith to the bark as can been seen from Fig. 6.5, and consequently, the water transport capacity is horizontally stable in the xylem if we exclude the juvenile wood. Thus, we can use the cross-sectional area of xylem sapwood as a measure of the water transport capacity of xylem tissue. The observation by Lintunen and Kalliokoski (2010, Fig. 5.6) indicates that the xylem per cross-sectional area in roots transports water more effectively than in stem and correspondingly in stem and branches.

The basic ideas can, now, be formulated more precisely utilising sapwood area as measure of water transport capacity. Let A_c denote the sapwood area in coarse roots, A_s in stem and A_b the sum of sapwood areas in branches. The basic assumption W2 can now be expressed utilising the relationship between cross-sectional area and water transport capacity of woody tissue:

$$a_r A_r = a_s A_s = a_b A_b, \tag{6.1}$$

where a_r, a_s and a_b are parameters, which take into account differences in water conductivity in coarse roots, stems and branches.

If we assume that needle mass on a branch determines the rate of transpiration, and the required water transport capacity, we can formulate the basic idea W3 as linear relationship between needle mass and water transport capacity of the branch. Let M_n^i denote the needle mass in the branch i and $A_b{}^i$ its sapwood area. Thus,

$$M_n^i = b_n A_b^i, \tag{6.2}$$

where b_n is a parameter.

If we assume that fine root mass on a root characterises its water uptake, then the basic idea W3 results in a linear relationship between the water transport capacity of a coarse root and its fine root mass. Let M_r^i denote the fine root mass on the root i and A_r^i its sapwood area:

$$M_r^i = b_r A_r^i, \tag{6.3}$$

where b_r is a parameter.

6.2.1.3 Field Measurements

For testing the above developed theory, we measured 15 pine trees near the city of Petrozavodsk in Karelia, in the former Soviet Union, in the summer of 1983. We felled the trees and measured the trunk diameter at the crown base. Also the diameters of all branches in the trees were measured.

Under-bark diameters were measured, but the heartwood area was not explicitly determined. Instead we assumed that there is no heartwood in the crosscutting of stem below lowest living branch. The assumption was based on observation by Nikinmaa et al. (1996). According to Nikinmaa et al. (1996), the number of heartwood tree rings equals with the number of the senescent whorls above the point in consideration. We applied the same argument also on branch level, and we assumed that the branch is sapwood at the point where all above twigs carry living needles (Fig. 6.7).

We picked the needles from 25 branches and dried them in oven for 24 h at 100°C. We also excavated the fine roots attached to 16 coarse roots in the sandy soil (it had a low content of organic matter). Sand grains adhered to the roots, and this is why we measured the fine root mass as loss of ignition.

We paid no attention to the determination of measurement noise in the year 1983 assuming it to be small. However, the heartwood may cause systematic measuring errors in the sapwood areas. The separation of fine roots from the soil organic matter is highly problematic, and it generates systematic errors and noise in the measurements.

6.2.1.4 Tests of the Theory of Tree Structure as Balanced Water Transport System with Field Data

Determination of the parameter values. We determined the values of the parameters with standard linear regression procedures.

Characteristic features in tree structure. In balanced water transport systems, there should be regularities in the water transport capacities of roots, stems and branches. According to Eq. 6.1, we expect linear relationships between the sapwood areas in roots, stems and branches. The relationships between coarse root area and stem sapwood area (Fig. 6.8a), stem sapwood area and the sum of the areas of the branches in a tree (Fig. 6.8b and c) are indeed linear as we expected. In addition, there is also a similar linear relationship within branches (Fig. 6.8d).

The fine roots take up water and nutrients that should be transported to the crown of the tree. From Eq. 6.3 we expect a linear relationship between the fine root mass

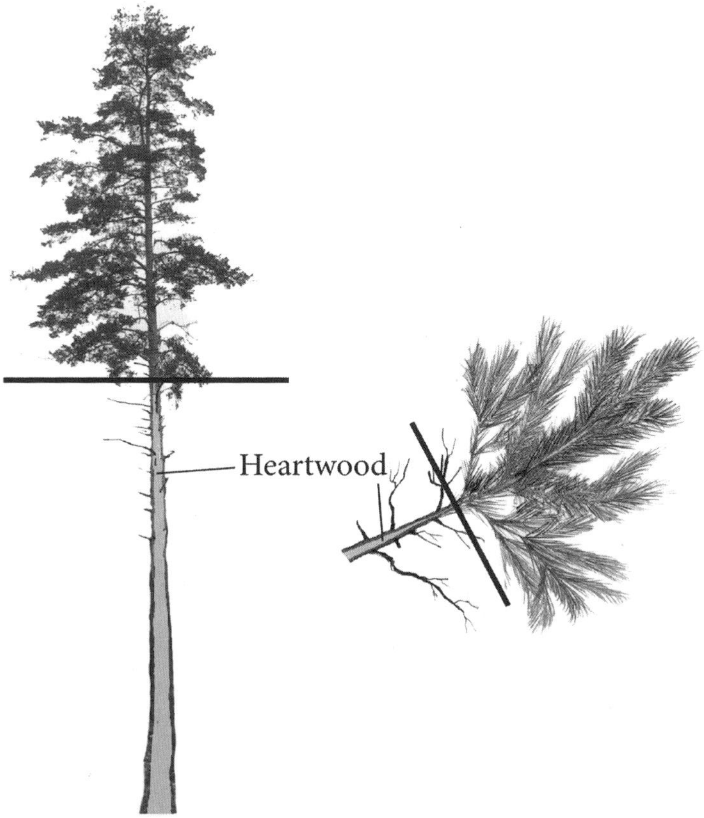

Fig. 6.7 The heartwood reaches up to the height of the lowest living branch, and above it the stem is sapwood (Nikinmaa et al. 1996). The proper measuring point cross-sectional area of stem and branch sapwood is indicated with a *line*

in a root and its sapwood area. Our measurements north of Petrozavodsk indeed show a strong linear relationship (Fig. 6.9a).

The needles transpire large amounts of water. This quantity should be capable of being transported by the sapwood in branches. According to Eq. 6.2, we expect a linear relationship between the sapwood area in a branch and its needle mass. Figure 6.9b shows a clear linear regression between the sapwood area in a branch and its needle mass.

Adequacy of the model structure. There seems to be no systematic deviation from the linear model, and the residuals are within the likely measuring precision (although this was not exactly determined). The scatter in the regression between fine root mass and coarse root area is rather large. This is natural since the determination of fine root mass is problematic.

Explaining power of the theory of tree structure as a balanced water transport system. The dominating task of sapwood is to transport water from fine roots to

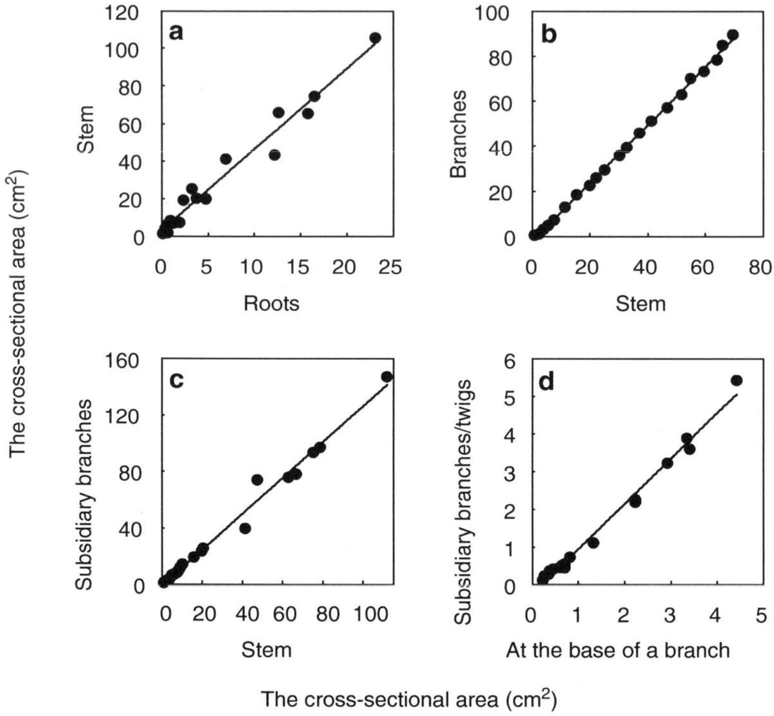

Fig. 6.8 Relationship between the cross-sectional area of a stem at the height of its lowest living branch and the sum of the cross-sectional area of its roots (**a**) and relationship within a tree between the cross-sectional area of a stem at varying heights and the sum of the cross-sectional areas of its branches above the height in consideration (**b**). Relationship in 15 trees between the cross-sectional area of a stem at the height of its lowest living branch and sum of the cross-sectional areas of the branches above the measuring point (**c**) and relationship between the cross-sectional area at the base of a branch and the sum of the cross-sectional areas of its subsidiary branches/twigs (**d**)

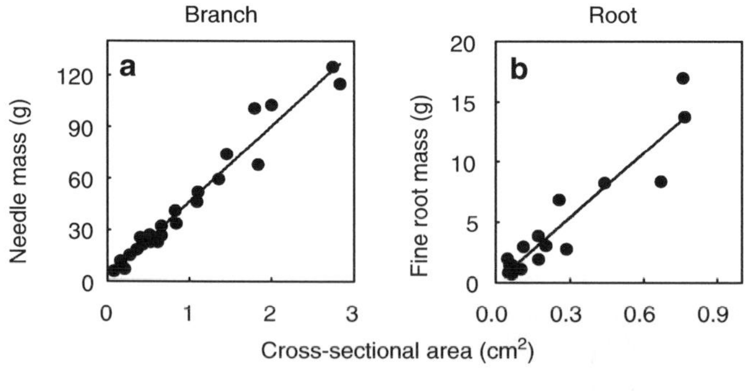

Fig. 6.9 Relationship of needle mass of branch on cross-sectional area of the branch below the oldest living subsidiary branch in a branch (**a**) and relationship between cross-sectional area of a root and its fine root mass (**b**)

needles. Although there are differences in the tracheids, most of the sapwood can be considered as a homogeneous porous medium having the same water transport properties within roots, stems and branches. Thus, the sapwood area is a good measure of the water transport capacity of the water pipe formed by the tracheids in xylem.

The mean conduit diameter decreases when moving from soil to the top of the tree; thus, the water transport capacity per unit area is bigger in the roots than in the branches (Fig. 6.6). This difference in the wood fine structure explains the fact that there is less water-conducting area in roots than in stem.

The linear models between sapwood areas explained 96–99.8% of the variance in the sapwood area, 95% for needle mass and 88% for fine root mass.

Comparison with other theories. The idea, presented by Sinozaki et al. (1964), that there is linear relationship between wood area and needle mass in a tree has met criticism because there is rather large scatter in the relationship between wood area at breast height and the needle mass in the tree. The wood area at breast height consists normally both heartwood and sapwood. In large trees, heartwood can even dominate sapwood area. Thus, the heartwood evidently smears the regression between sapwood area and needle mass.

Conclusion. The theory of tree structure as a balanced water transport system was able to explain the characteristic features in the tree structure. There were very regular relationships between sapwood areas within the water transport system and between fine roots and sapwood area in coarse roots transporting the water from the fine roots and between needle mass and the sapwood area in branches feeding the needles with water. The theory of tree structure as balanced water transport system gained clear corroboration in the test with field data. However, the low number of measurements reduces the corroboration.

6.3 Nitrogen Concentration in Leaves, Wood and Fine Roots in Scots Pine

Heljä-Sisko Helmisaari and Inge Stupak

Functional Role of Nitrogen

Metabolism and the biochemical regulation system have the demanding task in growth of building up living cells mainly from simple carbon compounds, sugars

H.-S. Helmisaari (✉)
Department of Forest Sciences, University of Helsinki, P.O. Box 27, 00014 University of Helsinki, Helsinki, Finland
e-mail: helja-sisko.helmisaari@helsinki.fi

and from nitrogen ions and amino acids. The cells have complicated geometrical and chemical structures. Cellulose and lignin form the cell walls, and lipids the membranes, whilst starch is the main storage compound in most species, and functional substances carry out the metabolism. The elemental composition of cellulose, lignin, starch and lipids is rather simple, only carbon, oxygen and hydrogen atoms. In contrast, nitrogen plays an important role in the composition of the functional substances since they are mainly proteins. These nitrogen-rich macromolecules, proteins, enable metabolism throughout the plant. The nitrogen concentration in proteins is about 15–16% with protein nitrogen representing 70% of the total nitrogen in, for example, needles (Näsholm and Ericsson 1990). Typically, the bulk N-content of pine needles is 1–2%.

The tissues have different tasks: needles synthesise sugars, fine roots with their mycorrhizae take up water and nutrients and woody structures do the transport. Each task requires a specific cell structure and specific functional substances. Needles have a large array of functional substances from pigments to membrane pumps and enzymes enabling photosynthesis, whilst fine roots have specialised structures for membrane penetration, and wood consists of tracheids for water flow with only a small proportion of living cells, the medullary rays. Evidently, each tissue has its specific nitrogen concentration reflecting its cell structure and the tissue-specific functional substances. For example, in one Scots pine stand in Sweden, the nitrogen concentration in stem wood, branches, stem bark, coarse roots and needles was 0.9, 2.1, 3.5, 5.9 and 13.4 mg g^{-1}, respectively (Ladanai et al. 2007).

The proper allocation of nitrogen to different tissue types and also within each tissue type has been crucial for the success of pine trees during evolution, and effective regulation of the allocation pattern has occurred by natural selection during evolution (basic idea 10), leading to regularities in the nitrogen concentrations between tissue types and within each tissue type.

Retranslocation

Retranslocation is a regulatory mechanism, which causes nutrients to redistribute from older needles to current year needles. There is also evidence that retranslocation takes place during formation of heartwood from the senescing wood towards the sapwood (Meerts 2002) and further to the inner bark (Rochon et al. 1998). As a result, nitrogen concentrations increase from heartwood to sapwood and from outer bark to inner bark as well as vertically upwards in the inner bark (Helmisaari and Siltala 1989; Meerts 2002).

Retranslocation of nutrients depends on their mobility within the phloem. N, P, K and Mg are mobile in the phloem, whereas Ca is relatively immobile (Helmisaari 1992a; Schleppi et al. 2000). Retranslocation of mobile nutrients thus helps maintain adequate concentrations in the youngest and the most active tissues, that is, in the photosynthesising current needles or mycorrhizal fine root tips active in nutrient uptake. For nonmobile nutrients, the pattern is usually the opposite. They accumulate with the concentration being highest in the older tissues.

Genetic Adaptation to Nutrient Availability

Variation in nutrient concentrations can also have a genetic origin. In a provenance trial situated in a common garden in Poland, Oleksyn et al. (2002) reported that needle nitrogen concentration was on average 4–8% higher in the northern population group than in the southern group. They suggested that under northern conditions, where nutrient availability is often limited as a result of interactions between temperature, litter quality and its mineralisation, the higher foliage concentrations of macronutrients can be an adaptive feature enhancing plant's metabolic activity in their native habitats. They reason that the observed difference could reflect genetic differences in mechanisms responsible for nutrient uptake, transport and accumulation, or it could reflect genetic differences in nutrient storage capacity.

Spatial Variation in Nitrogen Concentrations

There is variation in the nitrogen concentrations of needles in regional, local, stand, tree, compartment and tissue levels, whilst the temporal variation can be decadal, annual, seasonal, diurnal or even shorter. The factors affecting the spatial and temporal variation are the same: external variation in the environmental factors (e.g. soil moisture and temperature), atmospheric deposition and leaching from the soil and internal variation in the physiological processes within trees.

At stand level, the between-stand spatial variation is related to differences in soil fertility and climate as well as stand characteristics, for example, stand age and density. The variation between trees of the same stand reflects the within-stand variation in soil characteristics, the competitive position of the tree within the stand as well as genetic factors.

The tree compartment is a part of the tree where the tissues have a structure and function that is distinctive and different from the rest of the tissues, usually meaning needles, branches, stem wood, stem bark, coarse root and fine root compartments. The variation in nitrogen concentration between tree compartments reflects the structure and metabolic functions of the compartment.

Nitrogen concentration has a narrower range in foliage than in fine roots and in the ectomycorrhizal fine root tips (Fig. 6.10a, b, Helmisaari et al. 2007, 2009). The high concentrations of nitrogen in the ectomycorrhizal fine root tips (Fig. 6.10a) reflect the metabolic activity of the tissues. Högberg et al. (1998) reported a gradient in Norway spruce fine root nitrogen concentrations from 2.1% in central Europe to 1.1% in northern Sweden, whilst the range of foliage concentrations was 1.5–0.7% on the same sites (Bauer et al. 1997).

We have good biological reasons to expect that the nitrogen content of the needles reflects the nitrogen availability at the site. However, our needle nitrogen concentration measurements in Finland indicate that nitrogen concentration is rather constant, around 1.2% (Helmisaari et al. 2007, 2009). In addition, we could detect

Fig. 6.10 Relationship between stand mean nitrogen (N) concentration of ectomycorrhizal root tips and the C/N ratio of the organic layer ($n = 10$ stands) (**a**) and nitrogen concentration of needles and the C/N ratio of the organic layer ($n = 16$ stands) (**b**) (data from Helmisaari et al. 2009; *asterisks* indicate the significance of the p values: $^*p < 0.05$ and $^{**}p < 0.01$)

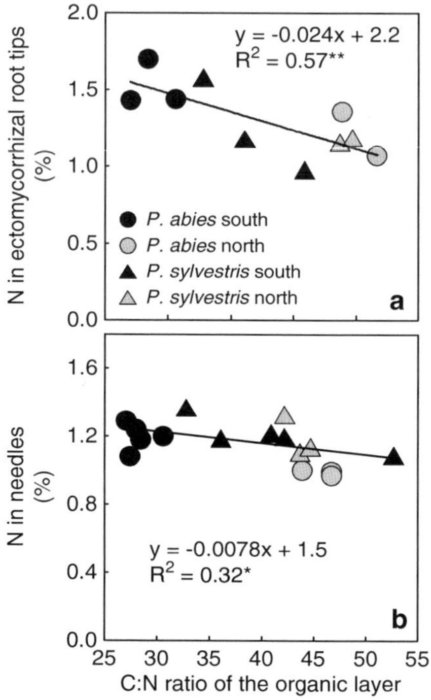

only a weak trend reflecting the nitrogen availability in the soil. This result indicates the power of the biochemical regulation to cope with the environment and adjust the needle nitrogen concentration to genetically determined target value. In fact, tree tissue nitrogen concentrations largely reflect the pattern of carbon allocation within the whole tree. According to Helmisaari et al. (2009), under nitrogen deficiency, trees regulate their needle nitrogen concentrations by reutilising the nitrogen stored in older needles and by allocating less carbon to stem growth but more into fine root and mycorrhiza growth in order to increase uptake. Needle nitrogen concentrations may reveal a change in soil nutrient availability only in conditions with a strong deficiency, whilst a more moderate long-term increase or decrease in the nitrogen availability may be detected by studying carbon allocation to tree compartments such as stem diameter or volume growth (Luiro et al. 2010).

Nitrogen is taken up from the soil as ammonium and nitrate (the latter is rarely significant in less fertile soils on which pines are found). Ammonium and nitrate are reduced and/or assimilated for the biosynthesis of nitrogen-containing metabolites. When returned to the soil in litter, also the organic forms of nitrogen can be taken up. Isotope studies in the field suggest that amino acid uptake is widespread in temperate and boreal forests (Näsholm et al. 1998; Gallet-Budynek et al. 2009; Averill and Finzi 2011). The capacity of ectomycorrhizal fungi to produce the enzymes involved in the degradation of organic matter and to mobilise organic forms of nitrogen is well documented (Näsholm et al. 1998; Read and Perez-Moreno 2003; Lindahl et al.

2005). On a gradient from northern Sweden into central Europe, Taylor et al. (2000) reported that the characteristic fungal species of northern sites which are able to utilise organic nitrogen, decreased in number as the availability of mineral nitrogen increased towards the south. Also, organic forms of nitrogen became the dominant source for hardwood and coniferous tree species with increasing altitude (Averill and Finzi 2011).

Temporal Variation in Nitrogen Concentrations

Year-to-year variation in the environmental conditions and biomass production may influence the variation of needle concentrations more than the supply of nutrients at the site (Merilä and Derome 2008). A large part of the variation can be described by biomass changes as differences in the dry weight of tissues explain the nitrogen concentration variation within the canopy or between years.

Nitrogen concentrations also vary between seasons. This variation is related to the annual metabolic cycle, the internal retranslocation and uptake, controlled by the demand (i.e. the sink strength of growing meristems). The concentrations of N, P and K in current year needles of Scots pine decrease in spring and early summer and increase in autumn; they are at their maximum during winter, as a result of both late-season nutrient uptake and retranslocation from senesced, older needles (Helmisaari 1990, 1992a, b). Retranslocated nutrients are stored within the tree over winter and then transported to growing tissues at the beginning of the growing season. Senescing needles retranslocate 60–90% of their N, P and K content before abscission. Stand age may influence these seasonal changes; Helmisaari (1990) found a tendency towards seasonal changes being less in older stands.

Seasonal variation and retranslocation among needles of different age must be taken into account when choosing a sampling time and age class for detecting nutrient deficiencies. Nitrogen concentrations vary between needle age classes (Helmisaari 1992c), which have a different spatial position on the branch but also temporarily a different history of the environmental conditions prevailing when they were formed and grew. Thus, variation in nitrogen concentrations observed in time interacts with variation in space. Needle age classes are often classified as current (C) and first-year needles (C + 1), second-year needles (C + 2), etc., with current year needles being those emerging at shoot elongation in spring and summer. When new shoots appear, current needles become first-year needles, etc. For analysing the stand nutritional status, current needles are commonly sampled in winter. This means that needle nitrogen concentrations are measured at their temporal maximum and at their spatial maximum within the crown.

Even if sampling would take place in early summer at spatial minimum, needle nutrient concentrations would reveal a change in soil nutrient availability only in conditions with a strong deficiency or nutrient addition. A more moderate long-term increase or decrease in the availability of a limiting nutrient could be detected by studying carbon allocation to tree components.

Also, retranslocation helps to maintain adequate foliage concentrations of mobile nutrients, especially nitrogen, for the synthesis of functional substances to be used in the actively photosynthesising current needles and fine roots active in nutrient uptake.

References

Averill C, Finzi A (2011) Increasing plant use of organic nitrogen with elevation is reflected in nitrogen uptake rates and ecosystem $\delta^{15}N$. Ecology 92:883–891

Bauer G, Schulze E-D, Mund M (1997) Nutrient contents and concentrations in relation to growth of *Picea abies* and *Fagus sylvatica* along a European transect. Tree Physiol 17:777–786

Gallet-Budynek A, Brzostek E, Rodgers VL, Talbot JM, Hyzy S (2009) Intact amino acid uptake by northern hardwood and conifer trees. Oecologia 160:129–138

Hari P, Heikinheimo P, Mäkelä A, Kaipiainen L, Korpilahti E, Salmela J (1986) Trees as water transport system. Silva Fenn 20:205–210

Havimo M, Rikala J, Sirviö J, Sipi M (2009) Tracheid cross-sectional dimensions in Scots Pine (*Pinus sylvestris*) – distributions and comparison with Norway Spruce (*Picea abies*). Silva Fenn 43:681–688

Helmisaari H-S (1990) Temporal variation in nutrient concentrations of *Pinus sylvestris* needles. Scand J For Res 5:177–193

Helmisaari H-S (1992a) Nutrient retranslocation within the foliage of *Pinus sylvestris*. Tree Physiol 10:45–58

Helmisaari H-S (1992b) Nutrient retranslocation in three *Pinus sylvestris* stands. For Ecol Manag 51:347–367

Helmisaari H-S (1992c) Spatial and age-related variation in nutrient concentrations of *Pinus sylvestris* needles. Silva Fenn 26:145–153

Helmisaari H-S, Siltala T (1989) Variation in nutrient concentrations of *Pinus sylvestris* stems. Scand J For Res 4:443–451

Helmisaari H-S, Derome J, Nöjd P, Kukkola M (2007) Fine root biomass in relation to site and stand characteristics in Norway spruce and Scots pine stands. Tree Physiol 27:1493–1504

Helmisaari H-S, Ostonen I, Lõhmus K, Derome J, Lindroos A-J, Merilä P, Nöjd P (2009) Ectomycorrhizal root tips in relation to site and stand characteristics in Norway spruce and Scots pine stands in boreal forests. Tree Physiol 29:445–456

Högberg P, Högbom L, Schinkel H (1998) Nitrogen-related root variables of trees along a N-deposition gradient in Europe. Tree Physiol 18:823–828

Ladanai S, Ågren GI, Hyvönen R, Lundkvist H (2007) Nitrogen budgets for Scots pine and Norway spruce ecosystems 12 and 7 years after ending of long-term fertilization. For Ecol Manag 238:130–140

Lindahl BO, Finlay RD, Cairney JWG (2005) Enzymatic activities of mycelia in mycorrhizal fungal communities. In: Dighton J, White JF, Oudemans P (eds) The fungal community: its organization and role in the ecosystem, 3rd edn. CRC Press, Boca Raton

Lintunen A, Kalliokoski T (2010) The effect of tree architecture on conduit diameter and frequency from small distal roots to branch tips in *Betula pendula*, *Picea abies* and *Pinus sylvestris*. Tree Physiol 30:1433–1447

Luiro J, Kukkola M, Saarsalmi A, Tamminen P, Helmisaari H-S (2010) Logging residue removal after thinning in boreal forests: long- term impact on the nutrient status of Norway spruce and Scots pine needles. Tree Physiol 30:78–88

MacCurdy E (2002) The notebooks of Leonardo da Vinci. Definitive edition in one volume. Konecky & Konecky, Old Saybrook

Meerts P (2002) Mineral nutrient concentrations in sapwood and heartwood: a literature review. Ann For Sci 59:713–722

Merilä P, Derome J (2008) Relationships between needle nutrient composition in Scots pine and Norway spruce stands and the respective concentrations in the organic layer and in percolation water. Boreal Environ Res 13 (Suppl B):35–47

Näsholm T, Ericsson A (1990) Seasonal changes in amino acids, proteins and total nitrogen in needles of fertilised Scots pine trees. Tree Physiol 6:267–281

Näsholm T, Ekblad A, Nordin A, Giesler R, Högberf M, Högberg P (1998) Boreal forest plants take up organic nitrogen. Nature 392:914–916

Nikinmaa E, Kaipiainen L,Mäkinen M, Ross J. SazonovaT (1996) Geographical variation in the regularities in the woody structure and water transport. In: Hari P, Ross J, Mecke M (eds) Production process of Scots pine: Geographical variation and models. Acta Forestalia Fennica 254:49–78

Oleksyn J, Reich PB, Zytkowiak R, Karolewsko P, Tjoelker MG (2002) Needle nutrient in geographically diverse *Pinus sylvestris* L. populations. Ann For Sci 59:1–18

Rautiainen R, Alen R (2009) Variations in fiber length within a first-thinning Scots pine (*Pinus sylvestris*) stem. Cellulose 16:349–355

Read DJ, Perez-Moreno J (2003) Mycorrhizas and nutrient cycling in ecosystems – a journey towards relevance? New Phytol 157:475–492

Rochon P, Paré D, Messier C (1998) Development of an improved model estimating the nutrient content of the bole of four boreal tree species. Can J For Res 28:37–43

Schleppi P, Tobler L, Bucher JB, Wyttenbach A (2000) Multivariate interpretation of the foliar chemical composition of Norway spruce (*Picea abies*). Plant Soil 219:251–262

Sinozaki K, Yoda K, Kira T (1964) A quantitative analysis of plant form – the pipe model theory I. Basic analyses. Jpn J Ecol 14:97–105

Taylor A, Martin F, Read DFJ (2000) Fungal diversity in ectomycorrhizal communities of Norway spruce [*Picea abies* (L.) Karst] and Beech (*Fagus sylvatica* L.) along north–south transects in Europe. Ecol Stud 142:43–365

Chapter 7
Dynamics of Carbon and Nitrogen Fluxes and Pools in Forest Ecosystem

Pertti Hari, Mikko Havimo, Kourosh Kabiri Koupaei, Kalev Jõgiste, Ahto Kangur, Mirja Salkinoja-Salonen, Tuomas Aakala, Juho Aalto, Pauliina Schiestl-Aalto, Jari Liski, and Eero Nikinmaa

Contents

Abstract Metabolism of trees, ground vegetation and microbes generate carbon and nitrogen fluxes in forest ecosystems. Carbon flows through the system, and nitrogen circulates between vegetation and soil. Trees synthesise sugars in photosynthesis and take nitrogen from soil. The biochemical regulation system allocates the annual amounts of synthesised sugars and nitrogen taken up to the growth of needles, wood and fine roots. The regularities in tree structure, generated by the action of the biochemical regulation system, determine the allocation to different tree components.

P. Hari (✉)
Department of Forest Sciences, University of Helsinki, P.O. Box 27, 00014 University of Helsinki, Helsinki, Finland
e-mail: pertti.hari@helsinki.fi

P. Hari et al. (eds.), *Physical and Physiological Forest Ecology*,
DOI 10.1007/978-94-007-5603-8_7,

The ecosystem model MicroForest is based on the theoretical ideas outlined above. It calculates the annual amounts of photosynthesised sugars and the annual nitrogen taken up. The allocation to needles, wood and fine roots is obtained as solution of carbon and nitrogen balance equations. In addition, several rather technical assumptions are needed to describe height growth, stem base enlargement, etc. SMEAR II and I measuring stations provide data for determination of the values of the parameters. MicroForest is able to predict successfully tree growth in six stands near SMEAR II in Finland and in five stands in Estonia.

Keywords Ecosystem • Carbon and nitrogen fluxes • Allocation • Photosynthesis • Decomposition with extracellular enzymes • Field test

7.1 Structural, Metabolic and Evolutionary Basis of Ecosystem Development

We define forest ecosystems as structurally similar functional units consisting of trees, ground vegetation and soil. This definition builds upon the traditional concept of the forest stand, a term used widely in forestry, by including also the soil and ground vegetation, and a strong emphasis on the fluxes of energy and materials. Forest ecosystems are the basic functional units in forests and, given the areas of land they occupy, have an important role in ecology more generally. Trees dominate the material fluxes most of the time during the development of forest ecosystem, as their capacity to photosynthesise is commonly much greater than that of other plants. There are, however, times when ground vegetation has important roles in material and energy fluxes. If the stand development starts from a situation where there are no trees, ground vegetation dominates carbon and nitrogen fluxes. In boreal conditions this situation is common after many kinds of perturbation, for example, fire, insect and fungal damage or violent storm, where almost all trees in the forest die. Before new saplings are large enough, ground vegetation often flourishes and contributes to mass and energy fluxes.

Fire and storms are typical disasters in boreal forests. If they are severe enough, large number of trees dies, but a stand commonly regenerates quite well, and development of the stand starts again. Since in severe disasters almost all trees die, the stand is dominated by a new, even-aged generation of trees. The same occurs after perturbation by humans, especially forestry management practices that harvest all the trees (a practice known as clear cutting or clear fell). After the cutting, the stand is regenerated either naturally or by planting, and trees are also in this case even-aged.

In this chapter, we will focus on the development over time of an even-aged forest ecosystem. Development starts from a situation where there is a young monoculture of Scots pine. At the beginning, trees are 3 years old. As they grow, the canopy starts to close. This changes light conditions for the ground vegetation. After canopy closure, ground vegetation receives relatively small amount of light, and trees dominate photosynthesis and nutrient uptake. Canopy closure also means tightening interactions between individual trees. Smaller trees become shaded by the largest trees, and small trees start to lag behind in growth.

By fixing solar energy into carbon compounds, trees cause a large carbon flux from the atmosphere to the ecosystem. The carbon fixed by trees provides also a major source of energy. When parts of the trees die and fall to ground as litter, microbes start to break down the carbon compounds and utilise them as a raw material for energy and growth.

Understanding the development of a forest ecosystem over the years and decades requires understanding of tree growth, carbon cycling and nitrogen cycling. These all are connected to each other by mass and energy fluxes. Photosynthesis fixes radiation energy of the Sun into chemical energy stored in the chemical bonds of complex carbon compounds. Carbon fluxes in the ecosystem transform therefore both mass and energy and provide an energy source for plants and soil microbes. Nitrogen is an essential element for both plants and microbes, and it cycles between soil and plants. The nitrogen cycle has its own dynamics, which accelerates or reduces growth of trees and ground vegetation.

Growth of trees forms large pools of carbon and nitrogen in the ecosystem, in which these elements are ‘fixed’ in organic compounds. The xylem of the trees is a major pool of carbon but contains only a small amount of nitrogen, perhaps only 1 atom of nitrogen per 300 atoms of carbon. Leaves contain a lot of proteins that are formed from nitrogen, carbon and other elements, and the leaves form an important nitrogen pool in the ecosystem. In this case, the ratio of C to N is around 40. As trees grow, the amount of carbon and nitrogen fixed to tree tissues increases, but since chemical composition differs between tissues, we have to analyse tree structure carefully.

There are regularities in tree structure. For example, amount of sapwood that transports water to needles is connected to the needle mass. These regularities result from evolution, and we will utilise them to analyse tree growth. Trees and other plants are composed of large number of different tissues. We cannot analyse them all, but instead we can make some useful generalisations about their chemical composition. Cell walls are formed from cellulose, and in the case of xylem, they contain also lignin. Cell membranes are mostly lipids, and energy is stored in the cell in the form of starch. All these substances contain carbon, oxygen and hydrogen, which are readily available in water and carbon dioxide.

Nitrogen is in various forms in tree cells. It is found in enzymes, membrane pumps and pigments complexes. The number of protein types in plant cells is vast, but they are all formed from amino acids. In the following analysis, we will lump all nitrogen into one protein type.

The nitrogen concentration in proteins is high, usually 15–17%, but in cellulose, lignin, lipids and starch, it is negligible. Needles and fine roots are metabolically very active: they synthesise new molecules and move molecules and ions through membranes. In contrast, wood is rather a passive structure: it transports water and gives mechanical support to above-growing biomass. This activity difference is reflected in the concentrations of functional substances, being high in needles and fine roots and low in wood.

Nitrogen is not, in contrast to carbon, oxygen and hydrogen, readily available in forest ecosystems. It is often scarce, and its low availability limits tree growth. This is easily seen in fertilisation experiments, where addition of nitrogen as ammonium results in major growth responses.

In the above, we assumed that tree tissue consists of four types of carbon macromolecules: cellulose, lignin, lipids and starch. There is also one nitrogen compound: protein. This simplistic division of chemical compounds is needed in the analysis of tree growth, and it is also used to analyse the breakdown of senescent litter in soil. When needles and branches die, they fall to the ground as a litter, and the chemical composition of litter is similar to the living tissue. The only major exception is the reuse of proteins, which reduces the amount of nitrogen in dead needles and branches.

The senescent leaves, branches, stems and roots feed the pool of carbon compounds in the soil. Litter is the source of raw material and energy for the soil microbes. Microbes emit extracellular enzymes to break the macromolecules, such as cellulose and protein, to smaller compounds. Microbes are not able to take up large macromolecules. Instead, they emit enzymes to their surrounding, which in turn break macromolecules into smaller units. Microbes take up the resulting smaller compounds and utilise the carbon and nitrogen they contain for growth and metabolism. Microbes emit ammonium ions formed in the metabolism of amino acids. These ammonium ions are the dominating source of nitrogen for vegetation.

Growth of trees and microbes requires energy. In forest ecosystems, the Sun is the primary energy source, and the energy of the Sun is carried in the system mainly in the form of carbon compounds. This carbon is taken into the system from the carbon pool of the atmosphere (CO_2), but due to the respiration of vegetation and microbes, carbon is also returned from ecosystem to the atmosphere, mostly as CO_2.

Carbon is returned to the atmosphere in the respiration process of trees and microbes. Respiration can be divided into two classes: maintenance respiration and growth respiration. Even if living cells are not growing, they require some energy for maintenance, and this results in the formation of CO_2. Carbon and nitrogen behave very differently in the material and energy flows in forest ecosystem. Carbon enters vegetation in photosynthesis; thereafter, it is utilised for formation of new cells or for energy needs. Litter formation moves carbon into the pool of organic compounds in the soil, and finally microbes release carbon back into the atmosphere as CO_2 (Fig. 7.1).

The nitrogen cycle in the ecosystem is quite different to the carbon cycle. It is very important since nitrogen is commonly a limiting resource in tree and plant growth. Nitrogen enters vegetation through root uptake as ions. In plants it is converted to amino acids and thereafter utilised for the synthesis of proteins. In dying tissues, part of the nitrogen bound to proteins is converted to amino acids and returned back to the metabolism of the tree, where it is used to synthesise new proteins. Trees cannot circulate all nitrogen back to metabolism. Part of nitrogen in the dying tissues is lost, but finally it enters the soil pool of nitrogen. Soil microbes break proteins in the senescent tissues into ammonium ions. Trees can uptake these

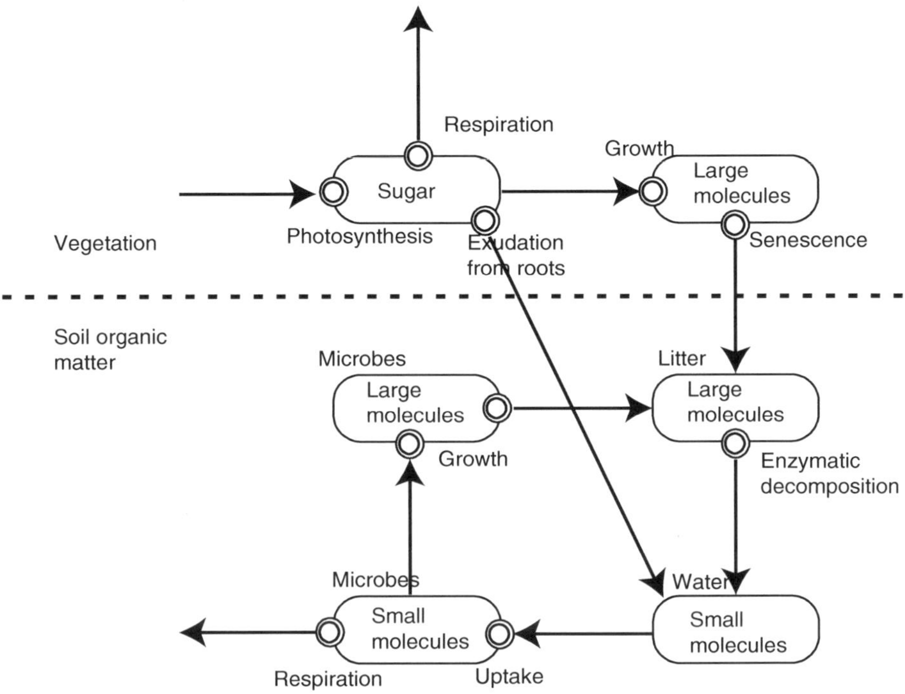

Fig. 7.1 Fluxes of carbon compounds in the forest ecosystem from atmosphere via leaves and from microbes back to the atmosphere. *Boxes* indicate amounts, *arrows* flow and *double rings* conversion of carbon compounds

ions, and the nitrogen cycling in the ecosystems starts again. Some nitrogen enters to the ecosystem from its surroundings, for example, as an anthropogenic deposition of ammonium or nitrate, but this flux is in most cases quite small.

Nitrogen in the proteins of senescent tissues is either returned to the metabolism or added to the soil nitrogen pool. The nitrogen returned to the metabolism is used in the synthesis of new functional substances. The proteins which end to the soil are broken to ammonium ions by microbes. Thus, nitrogen is circulating in forest ecosystems (Fig. 7.2). The nitrogen ion fluxes between a forest ecosystem and its surroundings are small. This makes the ammonium ion flux from microbes important for the stand growth. The results, obtained previously in this book, are very useful in the analysis of carbon and nitrogen fluxes and accumulation into forest ecosystems.

Trees are outcomes of evolution that has resulted in efficient structures and metabolism. The regulation of the formation of the structure and metabolism plays an important role in the growth of trees. There are regularities in tree structures: they contribute to efficient transport within trees, to efficient use of resources available and to effective metabolism. The annual cycle dominates the regulation of metabolism: trees are inactive in winter and active in summer. The structure of trees changes slowly due to formation of new cells in branches, stems and roots.

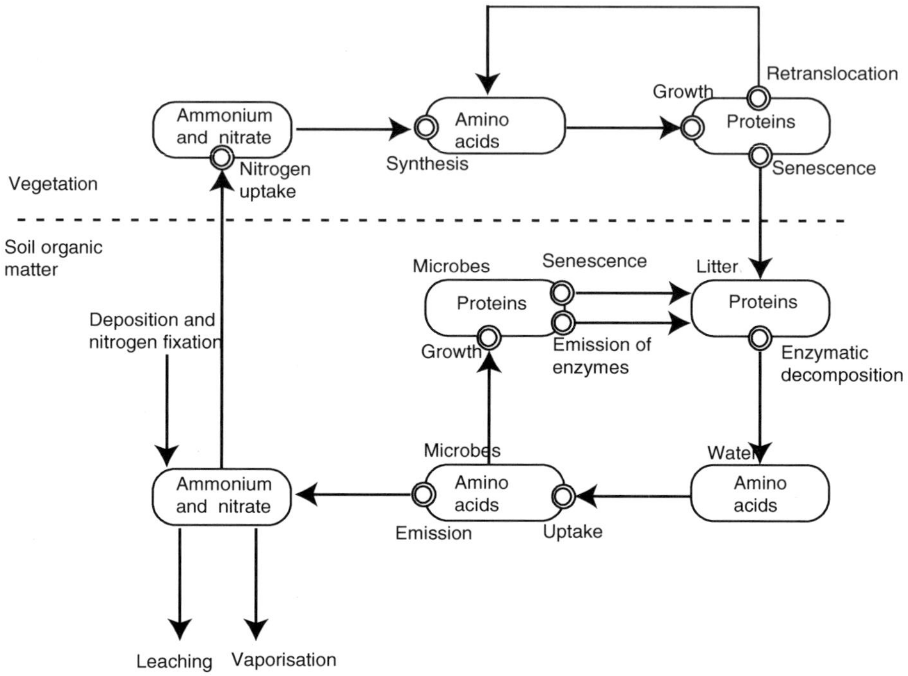

Fig. 7.2 Circulation of nitrogen compounds in forest ecosystem from available nitrogen in the soil into the metabolism of trees and microbes back to available nitrogen. *Boxes* indicate amounts, *arrows* flow and *double rings* conversion of nitrogen compounds

Analysis of forest ecosystems requires a description of plant structure. We need a description that can link the carbon and nitrogen fluxes occurring in the ecosystem to the growth of trees. Plant cells are complex structures, and a large number of different chemicals can be found in living cells. For modelling the stand development and mass fluxes in the stand, a simplified scheme of tree structure has to be developed.

Our simplified tree structure is hierarchical. At the lowest level, there are chemical compounds, which are basic building blocks of different tree tissues. Trees can be divided roughly into three different tissue types: needles, water-transporting tissue and fine roots. The chemical composition differs between tissue types, requiring sophisticated allocation mechanisms in tree growth. Water-transporting structures can be divided into branches, stem wood and transport roots. Later we will introduce a concept of whorl, which is a basic tree building block in our analysis. It is more an artificial functional concept than a real structural part of tree.

We use only five compounds in the analysis: cellulose, lignin, starch, lipids and proteins. They reflect the chemical structure of many plant cell types in a very rough way. Cell wall is formed by cellulose and lignin and cell membrane by lipids. Plant stores energy in the form of starch, and proteins are the primary actors in the cell, for example, as enzymes catalysing chemical reactions.

The chemical compounds used here to analyse stands development are all macromolecules. They are made from smaller building blocks. For example, cellulose is formed from glucose units, and the cellulose molecule can contain tens of thousands glucose units. Another example is proteins, which are formed from amino acids. The important aspect in all these macromolecules is that they can be broken into smaller units by extracellular enzymes emitted by microbes. Microbes utilise the resulting sugar amino acids in their metabolism.

Plants synthesise cellulose, lipids, lignin and starch from carbon, hydrogen and oxygen, which are readily available in CO_2 or water. Proteins contain also these three elements, and in addition nitrogen. Nitrogen is a scarce nutrient in most boreal forests, and the analysis of nitrogen fluxes in the forest ecosystem gives interesting insight to the process accelerating or decelerating stand growth.

The senescent leaves, branches, stems and roots feed the pool of carbon compounds in the soil. Litter is the source of raw material and energy for the microbes in the soil. Microbes emit extracellular enzymes to break the macromolecules in litter, such as cellulose, lignin and protein to smaller compounds. Microbes are not able to take up large macromolecules, but they emit enzymes to their surrounding, which, in turn, break macromolecules. Microbes take in the smaller carbon compounds and utilise the obtained carbon and nitrogen for growth and metabolism. Microbes emit ammonium ions formed in the metabolism of amino acids.

7.1.1 Basic Ideas

We have studied rather different processes in forest ecosystems, ranging from photosynthesis to the microbial decomposition of soil organic matter. We can now condense our understanding into basic ideas describing the most important phenomena in the development of ecosystem, especially in the accumulation of carbon and nitrogen. Our vision dealing with the connections between ecological entities and their environment can now be expressed in more concrete form.

Basic idea ES1: Carbon, nitrogen, water and energy fluxes connect the forest ecosystem with its environment.

The forest ecosystems are extremely versatile, and they include very different types of structures and processes. We should treat simultaneously the whole system: large trees, microbes in the soil and tissue elements. Thus, we need clear structure for the analysis of forest ecosystems.

Basic idea ES2: Forest ecosystems are hierarchical.

We will focus on three levels in the hierarchy, which are combinations of spatial and temporal elements. The levels are the following:

1. Space and time element
2. Individual and 1 year
3. Whole ecosystem over the lifetime of the dominant trees

The fundamental metabolic and physical processes occur on space and time element level, because then the affecting environmental factors are spatially and temporally constant. For example, the area fed by one single stoma with CO_2 is the spatial element in photosynthesis. The single stoma is selected as a spatial element, since the spatial variation in photosynthetic active radiation is strong, and even a neighbouring stomata can obtain different amount of radiation. Transport of CO_2 by diffusion within leaves is unable to smear the spatial variation. Temporal element is less than a second, since diffusion of CO_2 into a stoma takes about 1 s. Thus, photosynthesis takes place in the timescale of a second.

The knowledge about the space and time element level should and can be utilised also at the annual and ecosystem level. Reductionist method, that is, integration over space and time, can be used for climbing to a more aggregated level. The role of emergent properties, when utilising detailed knowledge, should be analysed, and if emergent properties are important, they should be included into the theoretical thinking.

Basic idea ES3: Emergent properties in the metabolism and in the structure are important aspects in the formation of bigger entities from detailed knowledge.

Annual photosynthetic production, allocation of carbon and nitrogen, interactions between trees and nitrogen cycling in the ecosystem should be analysed as emergent properties and properly introduced into the reductionist approach.

Plant cells have strong walls that are mainly formed by cellulose and lignin. Cell membranes, formed by lipids, control the flow of material in the cell and between cells.

Basic idea ES4: The metabolism in living cells is based on functional substances. Thus, nitrogen is a crucial nutrient in boreal forests since the functional substances in the cells, such as enzymes and membrane pumps, are nitrogen derivatives. The metabolism in living cells is based on functional substances.

The nitrogen content of proteins is high, 15–17%. The nitrogen in plant cells is needed for the synthesis of proteins for functional substances. Cellulose, lignin and lipids are carbon, oxygen and hydrogen compounds.

Vegetation synthesises macromolecules from available raw material, that is, simple compounds or ions from the environment.

Basic idea ES5: Sun is the primary energy source for the forest ecosystem. The ecosystem fixes solar radiation energy in photosynthesis, and this energy is carried in the ecosystem in the form of chemical energy bound to various carbon compounds. Senescent plant tissues contain various carbon macromolecules that provide energy for soil microbes.

Nitrogen plays an important role in ecosystems, because it is the vital component in all functional substances. The amount of available nitrogen is always low compared to the nitrogen required in growth. The scarcity of nitrogen makes its cycling in the ecosystem highly important.

Basic idea ES6: Nitrogen cycling in the ecosystem has a great influence on the development of the stand. Vegetation takes nitrogen ions from soil and synthesises amino acids and later proteins to construct enzymes, membrane pumps and pigment complexes. In addition, the reuse of nitrogen from senescent cells is an important source of nitrogen.

The regulation system of growth has the demanding task to allocate carbon and nitrogen for growth of new cells to get efficient use of the available sugars and nitrogen. The observed regularities in the tree structure and in the nitrogen concentrations in different parts of trees indicate successful operation of the regulation system.

Basic idea ES7: The action of the biochemical regulation system of growth generates emergent regularities in tree structure, and these structural regularities determine allocation in trees.

Individuals form the ecosystem, and the interactions connect the individuals with each other. The interactions are an emergent property when moving towards to more aggregated level.

Basic idea ES8: The most important interaction between trees in the ecosystem is shading. Trees shade each other, causing reduction in the photosynthesis of individual trees. Shading is not a constant factor, but evolves as trees grow bigger, and becomes especially important when canopy of the stand closes.

An important aspect in the functioning of the biochemical regulation system is the formation of buds in late summer and early autumn, and, in this way, it controls the growth during the following summer.

Basic idea ES9: The trees operate on an annual timescale.

Coniferous trees form a new whorl every spring; it grows annually and finally dies. The needles in the whorl transpire, requiring water uptake and transport. Growth requires nitrogen for the synthesis of functional substances in the new cells.

Basic idea ES10: The whorls are the functional units of a tree. We include the water-transport structures in branches, stem, transport roots and fine roots into the whorl as suggested by Watson and Casper (1984 and Chapter 6).

Note that 'whorl' in the sense described here differs from the concrete whorl in a real pine tree consisting of needles, twigs and branches. The abstract whorl, used here for our analysis, contains water-transport tissue down to roots and fine roots that provide water and nutrients for the needles.

The reuse of water pipes and nitrogen from senescent needles is an important source of transport capacity and nitrogen for a whorl. There is no senescence in the top whorls of a coniferous tree, and the reuse of water pipes has no role in the top whorls. This is an additional complication for the action of growth regulation system.

Basic idea ES11: The growth regulation system allocates resources from lower parts of the crown to the top of the tree.

Ground vegetation plays a minor role in the material and energy fluxes in later phases of succession of forest ecosystems. However, ground vegetation dominates the fluxes in the early phase of succession. This gives an important role to ground vegetation in the ecosystem dynamics.

Basic idea ES12: There are no fundamental differences in the basic processes of ground vegetation and trees. They photosynthesise and take up nitrogen and use the obtained resources in an efficient way for maintenance and growth.

Dead needles, branches and other components of vegetation enter the pool of soil organic matter, and their chemical character is the same as during their lifetime. The microbes in the soil utilise the litter as raw material for their metabolism.

Basic idea ES13: Soil microbes decompose macromolecules like cellulose or proteins. Microbes emit extracellular enzymes for macromolecule breakdown and take up the released small molecules.

The ammonium ion released in the nitrogen metabolism of microbes is important for nutrient cycling in the ecosystem.

The processes in a forest ecosystem generate concentration, pressure and temperature differences. These cause material and energy flows in the ecosystems. The flow of water is the largest mass flow in the system. This large flow has only a slight connection to plant metabolism, since most of the flow is caused by transpiration. Carbon forms the second biggest mass flow. Carbon plays a key role in the metabolism and is also an important raw material. Nitrogen flows are rather small but important. They provide raw material for synthesis of proteins that are necessary for all metabolic reactions.

The carbon cycle is driven by photosynthesis. During the stand development and tree growth, carbon fixed in photosynthesis is bound to long-lasting tree structures. For example, carbon can be bound in stem and roots for decades, and in some extreme cases even for centuries. The carbon bound to the living trees forms an important pool. This pool is not stable: tree growth increases the pool, and the senescence of tree parts decreases it.

Basic idea ES14: Metabolic processes and senescence generate carbon flow through the forest ecosystems and nitrogen cycling in the ecosystems.

The basic ideas outline the most important phenomena in forest ecosystems, generating accumulation of carbon and nitrogen into the system. However, several aspects have to be specified, and additional assumptions have to be introduced to obtain a quantitative description of the material and energy fluxes in a forest ecosystem.

7.2 Quantification of Carbon and Nitrogen Dynamics: Ecosystem Model MicroForest

The next step in the analysis is to formulate the rather general basic ideas as equations. The equations are combined into a model called MicroForest. Quantitative results can be obtained by simulating the phenomena outlined above. In this way, we can get exact predictions that can be tested against field measurements.

Models can be divided into classes according to how they work. For example, in classical physics differential equation models are common, whereas in forest

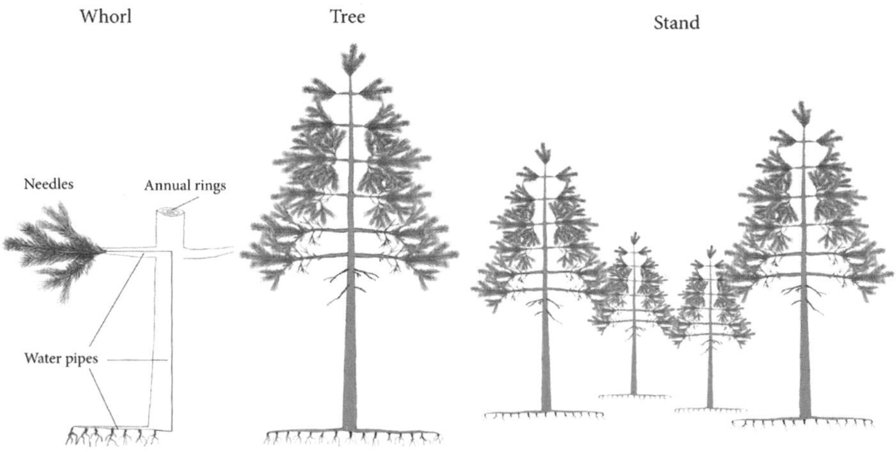

Fig. 7.3 Hierarchical description of stand structure in MicroForest. Most detailed level is a whorl including needles, water pipes and fine roots. Intermediate level is tree obtained as sum of whorls. Most aggregated level is stand formed by size classes

mensuration and in many biological disciplines, statistical models are mainly used. Basic principles of MicroForest are in many aspects similar to the models used in physics. MicroForest contains state variables and initial state as in the differential equation models. However, we cannot solve the difference equations in MicroForest with the methods of differential calculus; instead, we have to use computer simulations.

The simulation with MicroForest starts from an initial state of the forest ecosystem. Thereafter we determine the annual growths and senescence utilising the material and energy fluxes and structural regularities. We obtain the state a year later by adding the growth increments and senescence to the initial state. Thereafter we can again determine the annual growths and senescence and add them to the state, thus obtaining the new state a year later. This procedure continues until the maximum age is reached in the simulation.

7.2.1 Trees

7.2.1.1 Structure

The modelling of tree and stand growth requires a description of tree structure. The description needs to reflect the allocation of resources within a single tree and differentiation of trees within a stand (Hari et al. 1985). We describe the stand structure hierarchically: the most detailed level is the whorl, the tree being the intermediate level and the stand the most aggregated (Fig. 7.3). Growth and senescence change tree structure slowly during the lifetime of the dominant trees, but on the annual level, a clear structure can be found.

Growth of the tree is derived as the sum of growths of individual whorls. We first model the growth of individual whorls. Pine trees form annually a whorl on the top of the tree, where new branches and a leader shoot are formed. Other branches elongate, new needles grow on them and the annual rings are formed in stem and branches. Tree growth dynamics are driven by the growth dynamics of the individual whorls.

A stand is a population of different trees. We divide the trees into size classes according to their diameter. The mean tree in each size class, together with the number of trees in the class, describes the stand.

The needles, woody structures and fine roots can be described with masses of carbon and nitrogen. This is not, however, sufficient since the chemical structure of wood differs rather considerably from that of needles and fine roots.

Plants synthesise macromolecules, such as proteins and cellulose, from sugars and nutrients (basic idea ES5). Proteins are the active components in living cells; they catalyse biochemical reactions and transport substances through membranes (basic idea ES4). Cellulose and lignin form the cell wall, lipids the membranes, starch acts as reserve and proteins are functional substances (Sect. 4.1). In addition, the lifetimes in the decay process of litter are macromolecule-specific. Thus, all structures are treated with the five macromolecules both in trees and soil.

We describe the stand as a discrete system where masses and dimensions can be assigned to whorls. The whorls, as mentioned before, are the fundamental units in our thinking. We form a tree by combining it from individual whorls. We obtain stand by combing trees in the size classes. Dividing the stand in this manner to small units, which are then combined into more aggregated units, is somewhat abstract and rough, but it is still very helpful in the analysis of stand development.

The abstract whorl used in the analysis is an independent and fully functional unit of a tree. Whorl contains all necessary parts for its functioning: needles, water-transporting tissue and fine roots.

For analysis, we have to develop a notation for tracking the masses of different whorls. For example, analysing needle mass M_n requires division into size class, whorl and year. Let $M_n(i, j, k)$ denote the needle mass of size class i, whorl j and year k. The unit of needle mass is gram of dry matter. We follow this notation hereafter. Other whorl components expressed with this notation are mass of branches $M_b(i, j, k)$, mass of stem $M_s(i, j, k)$, mass of transport root $M_c(i, j, k)$ and mass of fine roots $M_r(i, j, k)$. In addition to mass, height and diameter of the trunk and length and diameter of branches are also included into the analysis.

7.2.1.2 Photosynthesis

Photosynthesis is the energy capturing process in the stand, and photosynthesis rate gives the instantaneous rate of fixing of energy. The photosynthesis rate depends on environmental factors, especially light, which can vary greatly even in small timescales. Rate of photosynthesis depends also on the efficiency of the photosynthetic functional substances. The efficiency develops during the growing season: it is low in spring, grows slowly to maximum in middle summer and then starts to decline towards winter.

According to the basic idea ES3, important emergent properties are involved when moving to the more aggregated levels in a forest ecosystem. According to the basic idea ES9, the trees operate at an annual timescale. The great temporal variability in rate of photosynthesis has to be aggregated to obtain the annual photosynthetic production. The annual production can then be divided between growth and respiration.

Annual photosynthesis can be integrated from the specific rate of photosynthesis. Let x be a given point in the stand and $p(x, t)$ be the specific rate of photosynthesis. The annual photosynthetic production $P(x, i)$ during the year i is then integral

$$P(x,\ i) = \int_{t_b^i}^{t_e^i} p(x,t)\ \mathrm{d}t, \tag{7.1}$$

where $t_\mathrm{b}{}^i$ is the beginning instant of photosynthesis in the year k and $t_\mathrm{e}{}^i$ is the end of photosynthetically active season during the year k.

Reduction of photosynthesis caused by shading is the most important interaction in a forest ecosystem when moving from tree to ecosystem level (basic idea ES8). Therefore, we need a description of this interaction. Needles above a given point in the canopy decrease light intensity. The great temporal variation in the degree of interaction is noise caused by the movement of leaves above and the travelling of Sun over the sky. Thus, also, interactions are very variable in time and space.

We define the annual degree of interaction at a given point as the ratio between photosynthetic production at a point and photosynthetic production at the top of the canopy. The former depends on shading, and the latter gives photosynthesis in unshaded conditions. Let $i_P(x, t)$ denote the annual degree of interaction at the point x during the year k and P_o the photosynthetic production at the top of the canopy. The annual degree of integration $i_P(x, t)$ is

$$i_P(\mathrm{x}_{ij}, k) = \frac{P(\mathrm{x}_{ij}, k)}{P_o}. \tag{7.2}$$

The annual photosynthetic production of a whorl is the starting point for analysis of the development of tree structure (basic idea ES5). The photosynthetic rate, that is, the rate of formation of sugars, is determined by environmental factors and the annual cycle of the trees (Sect. 4.2.1, Hari and Mäkelä 2003; Mäkelä et al. 2004). The annual photosynthetic production is obtained by integrating the photosynthetic rate over the photosynthetically active period.

The reduction of photosynthetic production of a whorl by shading is important for understanding of canopy development and differentiation of size classes. A link between the shading needles above the whorl and its photosynthetic production is needed. Let $P_W(i, j, k)$ (g[CO_2]) denote the photosynthetic production of the *j*th whorl in size class *i* during year *k*. We can see with simple calculations that the degree of photosynthetic interaction at whorl level can be utilised to get the production of the *j*th whorl in size class *i* during year *k*

$$P_W(i, j, k) = P_o\, i_P(x_{ij}, k)\, M_n(i, j, k), \tag{7.3}$$

where x_{ij} is the middle point of the whorl.

The shading reduces light inside the canopy, and consequently the degree of interaction decreases towards the lower parts of the canopy. There are two components in the shading: a tree shades other trees, and it also shades needles within the same tree. The two sources of shading have to be combined. Let $M_S(i, j, k)$ denote the shading needle mass of the whorl of the *j*th whorl in size class *i* during year *k*.

The shading determines the degree of interaction, which is approximated as follows:

$$i_P(i, j, k) = f_S(M_S(i, j, k)) = \frac{1}{1 + (a_{S\,2}\, M_S(i, j, k))^{0.8}}. \tag{7.4}$$

When Eqs. 7.3 and 7.4 are combined, we get

$$P_w(i, j, k) = P_o \frac{M_n(i, j, k)}{1 + (a_{S\,2}\, M_S(i, j, k))^{0.8}}. \tag{7.5}$$

This equation combines the structure of the stand with photosynthetic production of a whorl. It is an aggregated formula that utilises the understanding of photosynthetic process and extinction of light in the canopy.

7.2.1.3 Respiration

Most biochemical reactions demand energy. On the molecular level energy in a cell is provided by the conversion of energy-rich ATP to ADP (basic idea ES5). The conversion of ATP to ADP has to be reversed in order to maintain the energy resources in the cell. The reverse reaction consumes sugars and releases CO_2. This process is called maintenance respiration. It can be divided into classes, depending on where in the tree structure it takes place.

On a given year *k*, the maintenance respiration $R(i, j, k)$ of a *j*th whorl in the *i*th size class can be summed from the respiration of needles, branches, stem, transport roots and fine roots belonging to the whorl *j*. We assume that maintenance respiration is proportional to living mass of each tissue. Each tissue type has its own respiration rate, which we present with parameter a_{xr}, where x is the type of the tissue. Let M_n be the mass of needles, M_b mass of branches, M_s mass of stem, M_t mass of transport roots and M_r mass of fine roots. We get then

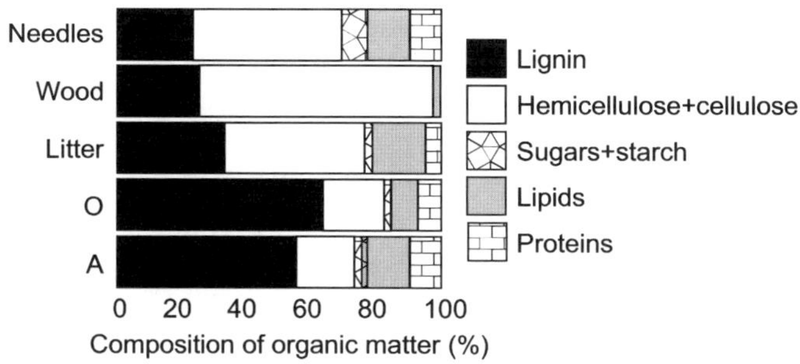

Fig. 7.4 The composition of needles, wood, needle litter, and soil organic matter in the O and A layers in Scots pine trees and soil layers at SMEAR II

$$R(i,j,k) = a_{\mathrm{n\,r}}\,M_{\mathrm{n}}(i,j,k) + a_{\mathrm{b\,r}}\,M_{\mathrm{b}}(i,j,k) + a_{\mathrm{s\,r}}\,M_{\mathrm{s}}(i,j,k) + a_{\mathrm{c\,r}}\,M_{\mathrm{c}}(i,j,k) + a_{\mathrm{r\,r}}\,M_{\mathrm{r}}(i,j,k). \tag{7.6}$$

Apart from maintenance respiration, energy is needed for tissue growth, where new macromolecules are synthesised. This is proportional to the amount of new tissue formed and is called growth respiration. In material transformations, considerable weight loss occurs, part of which is due to growth respiration. We use the following equivalencies: 1 kg of sugars is used in the synthesis of 0.7 kg protein, 0.8 kg cellulose, 0.44 kg lignin and 0.28 kg lipids (Penning de Vries et al. 1974).

The macromolecule concentrations are tissue type specific (Fig. 7.4). Needles and fine roots have the same concentrations. The characteristic feature of wood is low protein and lipid concentrations. The protein concentration of needles and fine roots is 8.9%, starch 7.9%, cellulose 46.6%, lignin 33.1% and lipids 16.9%. In wood the concentrations are: proteins 0.16%, starch 0.34%, cellulose and hemicellulose 72.1%, lignin 25.7% and lipids 01.7%. When the tissue-type-specific concentrations and mass loss are combined with the synthesis of macromolecules, then the conversion coefficients $a_{\mathrm{n\,gr}}$, $a_{\mathrm{w\,gr}}$ and $a_{\mathrm{r\,gr}}$ are obtained (Penning de Vries et al. 1974).

7.2.1.4 Conservation of Carbon

The regulation system of the tree allocates sugars produced in photosynthesis to growth of productive parts and water-transport system. Needles and fine roots are productive parts, and branches, stem and transport roots form the water-transport system. The allocation has to maintain a balance between productive parts and water-transport system. For example, the water-transport capacity of stem has to match the mass of needles supported by the transport system. The regulation system maintains this balance, resulting in regularities in the tree structure (basic idea ES7). These regularities play an important role in the development of tree structure during the stand development.

The whorls are the functional units of a tree, that is, the sugars formed and nitrogen taken up by the whorl are used to the growth and maintenance of the whorl itself (basic idea ES10). However, the regulation system allocates also sugars and nitrogen to the top of the tree (basic idea ES11).

Tree growth takes place at an annual timescale (basic idea ES9). Productive parts and the water-transport system form an apparatus that requires balance between each part. However, tree parts do not grow simultaneously. In Scots pine, shoots elongate from early May to early July, after which needles grow to their full size. Simultaneously trunk and branches grow in diameter, and finally fine roots grow mainly in the early autumn. We cannot treat this stepwise growth in time periods shorter than one growing season. For the carbon and nitrogen balances, we will sum up the annual production of sugars and intake of nitrogen, and then divide it between the growth of productive parts and water-transport system. This procedure is the operationalisation of basic idea 9.

Let $G_n(i, j, k)$, (g(dry matter)) denote the growth of needles in the whorl during the year k in the jth whorl in the ith size class. The lifetime of sapwood is long when compared to that of needles, and this has to be introduced into the analysis. When needles die (on average at the age of 3 years in Scots pine growing in southern Finland), the sapwood used for water transport to the dying needles is released for reuse. Let $G_{Ab}(i, j, k)$, (cm^2) denote the growth of sapwood area in branches, $G_{As}(i, j, k)$, (cm^2) in stem, and $G_{At}(i, j, k)$, (cm^2) in transport roots.

Water transport is the key to understanding the regularities in tree structure (Chap. 6). This fact is introduced into our analysis of tree structure with the requirement that the area of new sapwood fulfils the need of water-transport capacity for the growing needles. Thus,

$$G_{Ab}(i, j, k) = a_b \, (G_n(i, j, k) - G_n(i, j, k - 3)), \tag{7.7}$$

$$G_{As}(i, j, k) = a_s \, (G_n(i, j, k) - G_n(i, j, k - 3)), \tag{7.8}$$

$$G_{At}(i, j, k) = a_t \, (G_n(i, j, k) - G_n(i, j, k - 3)), \tag{7.9}$$

where parameters a_b, a_s and a_r describe the sapwood requirement per unit leaf mass for branches, stem and coarse roots. If needle mass in the whorl is decreasing, then the extra water-transport capacity is lost.

Pine trees have a swelling at the stem base that is not explained by water transport. The basal swelling is mostly needed to give mechanical support against wind and snow loads. Let $\Delta A_E(h, i, j, k)$ (cm^2) denote the additional stem area at height h grown during year k utilising sugars from whorl j in the ith size class. A rough approximation is used to introduce the stem base expansion

$$\Delta A_E(h, i, j, k) = \begin{cases} 0, & \textit{if } h > h(i,\ k)/10 \\ a_{b1} \, (1 - \frac{h/10}{h(i,\ k)}) \, G_{As}(i.j, k), & \textit{if } h < h(i,\ k)/10 \end{cases}, \tag{7.10}$$

where $a_{b\,l}$ is a parameter.

Let $M_B(i, j, k)$, (g(dry weight)) denote the mass of growth of stem base expansion supporting the jth whorl in the ith size class during the kth year. $M_B(i, j, k)$ is

$$M_B(i, j, k) = d_S \int_0^{h(i,k)/10} \Delta A_E(h, i, j, k)\mathrm{d}h, \tag{7.11}$$

where d_S is wood density.

Let $M_T(i, j, k)$ (g(dry weight)) denote the mass of the growth of the water-transport system supporting the jth whorl in the ith size class during the kth year. It is formed by the growths in the branches, stem and transport roots. Let $l_b(i, j, k)$, (cm) denote the mean length of branches in the jth whorl during the kth year, $h(i, j)$, (cm) the height of the jth whorl in the ith size class and $b(i, j, k)$, (cm) the length of transport roots. Then, the mass of the growth of the water-transport system is

$$M_T(i, j, k) = d_b\, l_b(i, j, k)\, G_{Ab}(i, j, k) + d_s\, h(i, j)\, G_{As}(i, j, k) + d_t\, b_t(i, j, k)\, G_{At}(i, j, k). \tag{7.12}$$

The parameters d_b, d_s and d_t describe wood density in branches, stem and transport roots.

The top of a tree is unable to grow with its own photosynthetic production, because the young whorls do not have reuse of the water-transporting tissue. Living needles survive to the age of 3 years in Scots pine in southern Finland. In whorls where there are dying needles, the water-transporting capacity of these needles is released and can be used to transport water to new needles. In the young, uppermost whorls, there are no dying needles, and thus no reuse of water-transport capacity. To compensate for this, the lower whorls allocate sugars to the three uppermost whorls (basic idea ES11).

From all the photosynthetic products, the biochemical regulation system allocates a share that depends on tree size for the development of the top of the tree. Let $T_A(i, j, k)$ (g(CO_2)) denote the amount of carbohydrates allocated for the top of the tree from the jth whorl in the ith size class during the kth year. It is approximated with the following function:

$$T_A(i, j, k) = \frac{a_{t1}\,(P_w(i, j, k) - R(i, j, k))}{1 + h(i, k)/a_{t2}}, \tag{7.13}$$

where $a_{t\,1}$ and $a_{t\,2}$ are parameters.

The preceding equations describe growth relative to other biomass compartments. To obtain the actual growth, additional constrains are needed. The carbon balance equation has been the cornerstone of process-based growth models since the work by De Wit et al. in 1970. It is mass conservation applied at annual and tree level (basic idea ES9). The carbon balance equation states that all sugars formed during a year are used for growth, for production of ATP resulting in maintenance respiration and for the top of the tree. The carbon balance equation is

$$P_w(i,j,k) - R(i,j,k) - T_A(i,j,k) = a_{n\,gr}\, G_n(i,j,k) + a_{w\,gr}\, M_T(i,j,k) + a_{r\,gr}\, M_r(i,j,k). \quad (7.14)$$

The parameters $a_{n\,gr}$, $a_{w\,gr}$ and $a_{r\,gr}$ are chemical conversion coefficients from sugars to needle, wood and fine root tissues. Conservation of mass and energy was frequently used to derive equations in Chap. 4. The carbon balance equation is an application of the same principle at both whorl and annual levels.

7.2.1.5 Conservation of Nitrogen

Forest stands take up nitrogen from soil mainly through the fine roots (basic idea ES6). Spatiotemporal complexity of the soil system, the root growth dynamic with microbial (mycorrhizal) interaction and the short lifetime of ammonium ion in soil (Chap. 5.6) render the quantitative description of the system very complex. However, there are some general phenomena which can be used for modelling the system. Available nitrogen content and amount of fine roots seem to determine the annual uptake. Another useful idea is the retranslocation of nitrogen from senescent tissues (basic idea ES6). It is an important source of nitrogen for the bigger trees. The role of foliar nitrogen exchange with the atmosphere is an open question, but in this study the nitrogen oxide emissions and depositions are considered negligible (Hari et al. 2003).

The lifetime of cells varies from one growing season to decades. Fine roots are active only for about one growing season, and needles photosynthesise for 3 years. Parenchyma cells in the xylem have long lifetime: they can reach the age of decades.

Internal circulation of nitrogen within a tree is an important source of nitrogen (Sect. 4.2.4). When a cell dies, a proportion of proteins is broken down to amino acids. These are transported away and later utilised in the synthesis of new proteins. The internal circulation is reflected in the nitrogen content of needle litter: it is less than half of that in active needles (Helmisaari 1992).

All tissues have their specific protein content, and thus they require specific amounts of nitrogen for growth (e.g. Mohren 1987, Chap. 4). Water-transporting tissue contains only few living cells and has low nitrogen content. Therefore, the nitrogen needed for growth of branches, stem and transporting roots is quite small compared to nitrogen required for needles and fine roots. These two tissue types contain many living cells, and they have high protein content.

All tissues have their specific protein content, and thus they require of specific amount of nitrogen for growth (e.g. Mohren 1987). Water-transporting tissue contains only few living cells and has low nitrogen content. Therefore, the nitrogen needed for growth of branches, stem and transporting roots is quite small compared to nitrogen required for needles and fine roots. These two tissue types contain many living cells, and they have high protein content.

Balance equations form the basis for dynamic process models. In the previous chapter, we presented a balance equation for carbon. We can formulate similar balance equations for nitrogen by using a few basic principles: (1) nitrogen uptake depends on the mass of fine roots and availability of nitrogen for roots, and (2) nitrogen in the needles is reused. We omit the reuse of nitrogen in dying woody tissue and fine roots, partly because these reuse processes are poorly understood. In woody tissue, the concentration of nitrogen is very low, and omitting its reuse causes negligible errors.

The balance equation states that the amount of nitrogen needed for growth equals the sum of uptake and reuse of nitrogen. Water-transporting tissues in branches, stem and roots are xylem cells, and we can lump them into a single term M_T which describes the mass of wood in water-transporting structures:

$$n_n\, G_n(i,j,k) + n_w\, M_T(i,j,k) + n_r\, M_r(i,j,k) = u\, M_r(i,j,k)\, N_a(k) + c\, n_n\, G_n(i,j,k-3), \tag{7.15}$$

where N_a(k), (g(N) m^{-2}) is the amount of nitrogen in the soil that is available to the tree in year k and n_n, n_w and n_r, (g(N) g(dry weight)$^{-1}$) are nitrogen concentrations of needles, wood and fine roots, u is nutrient uptake and c is a reuse parameter. Again, the nitrogen balance equation is an application of the conservation principle at whorl and annual levels.

7.2.1.6 Solution of Allocation Problem

The regularities in tree structure have to be fulfilled to obtain efficient use of sugars and nitrogen for tree growth. This is a very demanding task for the growth regulation system. According to basic idea ES7, the regulation system is able to solve the allocation problem. The action of the regulation system is introduced into MicroForest by the solution of the structural constraints (Eqs. 7.8–7.10), carbon (Eq. 7.15) and nitrogen (Eq. 7.16) balance equations for each whorl in the tree. The five equations contain five unknowns: (1) needle growth, (2) growth of sapwood area in branches, (3) stem, (4) coarse roots and (5) growth of fine roots. These unknown variables can be solved from the equations. Solution of the five equations results in an allocation of carbon and nitrogen within the tree under the control of the regulation system (Figs. 6.1 and 6.2).

There is no senescence in the three uppermost whorls of a pine tree when the maximum needle age is 3 years. This is why the release of water-transport capacity and reuse of nitrogen play no role in the allocation of carbohydrates for the treetops. Instead, sugars from lower parts of the crown have to be used (basic idea ES11). This additional photosynthetic production, collected from the photosynthesis of lower whorls, is divided in equal shares between the top whorls. The growth terms in the top whorls can be solved from structural regularities and balance equations with slight modifications.

The solution of the five equations, introducing structural regularities and carbon and nitrogen balance, results in allocation of sugars and nitrogen to the growth of needles, fine roots and the water-transport system.

7.2.1.7 Annual Needle Masses and Diameters

The needle mass in a whorl in the *i*th size class in the *j*th whorl during year *k* is increased by growth and decreased by senescence of needles born in year *k*–3. When these are combined, we get the needle mass

$$M_{\mathrm{n}}(i, j, k+1) = M_{\mathrm{n}}(i, j, k) + G_{\mathrm{n}}(i, j, k) - G_{\mathrm{n}}(i, j, k-3). \tag{7.16}$$

So far we have been dealing with abstract whorls, which contain the water-transport system from needles to transport roots. These abstract whorls can be summed up to form the complete tree. The area of the annual tree ring in each stem segment is determined by the water transport for the needles in the above whorls. Let $A(i, j, k)$, (cm^2) denote the cross-sectional area of the stem segment formed in the *i*th size class in the *j*th whorl at age *k*. The cross-sectional area of the stem at the top of the tree is zero, thus $A(i, j, k) = 0$. The area, $A(i, j, k)$ is obtained recursively by adding new water pipes for the above whorls on the area during the previous year:

$$A(i, j, k) = A(i, j, k-1) + \sum_{jj=j+1}^{k} G_{A\,s}(i,\ jj,\ k). \tag{7.17}$$

The diameter of each segment of the stem can now be determined from the relationship between the area and diameter of a cylinder. Let $d(i, j, k)$, (cm) denote the diameter of the segment formed during the *k*th year in the *i*th size class at age *k*:

$$d(i, j, k) = 2\sqrt{A(i, j, k) / \pi}. \tag{7.18}$$

7.2.1.8 Height Growth

The biochemical regulation system accelerates height growth in trees growing in the shade cast by bigger trees. This is very easy to understand from the evolutionary point of view, since by accelerating the height growth, trees can, in some cases, escape the shade. Some trees will be suppressed when the stand grows. The biochemical regulation system of a suppressed tree accelerates height growth, and in this way a tree 'tries' to maintain its position in the stand. We need to operationalise the action of the regulation system. For this we define a term annual degree of interaction of a tree. It is the ratio of photosynthetic production of the tree divided by the hypothetical photosynthetic production of the same tree in unshaded conditions. Let $I_{pi}(i, k)$ denote the mean annual degree of interaction of the tree i during the year k. We get

$$I_{\mathrm{pi}}(i,k) = \frac{\sum_{j=i}^{k} i_{\mathrm{p}}(i,j,k)\ M_{\mathrm{n}}(i,j,k)}{\sum_{j=i}^{k} M_{\mathrm{n}}(i,j,k)}, \tag{7.19}$$

where $i_p(i, j, k)$ is the degree of interaction of the jth whorl in the ith size class during the year k.

The height growth of trees is rather problematic to derive from functional knowledge, and more evolutionary arguments have been used (Mäkelä 1985, 1988). The length and diameter are mechanistically related (Ylinen 1952), but also a simple empirical linear relationship for open-grown trees has been found (Ek 1971). Ek (1971), and Mäkelä and Sievänen (1992) have observed that for open-grown pines the ratio between height and diameter growth is constant. The shading accelerates height growth in stands, and this acceleration is relative to diameter growth (Mäkelä and Vanninen 1998; Ilomäki et al. 2003). The height growth of the tree in the ith size class during the year k, $\Delta h(i, k)$, (cm), is assumed to be

$$\Delta h(i,\ k) = a_{\mathrm{h}\,1}\ \Delta r(i,\ k)\ (1 + a_{\mathrm{h}\,2}(1 - I_{\mathrm{Pi}})^4), \tag{7.20}$$

where $a_{h\,1}$ and $a_{h\,2}$ are parameters, Δr (cm) is radial growth at stem base, and IT^i the mean interaction defined in Eq. 7.22.

The length growth of branches is derived from a geometrical principle. We assume that the needles fill space with constant volume density, ρ_n, (g(dry weight) cm^{-3}) at the outer surface of the crown at each whorl. Let $L_{\mathrm{B}}(i, j, k)$ denote the mean length of a branch in the ith size class in the jth whorl during year k. The above assumption determines the length growth of branches, $\Delta L_{\mathrm{B}}(i, j, k)$, (cm), and it is the solution of the following equation:

$$\frac{G_{\mathrm{n}}(i,j,k)}{2\,\pi\ \Delta L_{\mathrm{B}}(i,j,k)\ L_{\mathrm{B}}(i,j,k)\ \Delta h(i,\ k)} = \rho_{\mathrm{n}}. \tag{7.21}$$

7.2.1.9 Senescence of Branches, Needles and Fine Roots

Needles become senescent and die at the age of 3 years. The lifetime of fine roots is still shorter, they are annual. Proteins in dying tissues are decomposed to amino acids and transported away for synthesis of proteins in growing tissues (basic idea ES6). A new whorl is formed annually at the top of the tree. In following years, it grows its needles, branches, water-transport system and fine roots by utilising its own photosynthesis. When the stand grows, the branches become more and more shaded, and the whorl photosynthesis is reduced. This results in a slow decline of the needle mass in the whorl, which in turn causes shorter branch growth. We assume that the whorl finally dies when its extension growth is less than 1 cm.

When a stand grows, the smaller trees become more and more shaded. The shading becomes especially strong after canopy closure. Canopy closure causes a declining trend in photosynthesis per gram of needles in the tree. Thus, the carbohydrates available for growth have also a declining trend, which is reflected in the growth of the tree. We assume that the probability of a tree dying is proportional to the relative decrease in its needle mass.

7.2.2 Growth of Ground Vegetation

The forest floor is covered by ground vegetation, which usually consists of dwarf shrubs, herbs and mosses. The living biomass of the ground vegetation is often rather small, but when the trees are not reducing light too strongly, the annual growth of new leaves per unit area is of the same magnitude as that of trees. As a rough approximation, the chemical composition of ground vegetation can be assumed to be close to that of tree leaves. The ground vegetation acts in similar way to the trees (basic idea ES12). Due to short transport distances, the dwarf shrubs and herbs do not allocate considerable amounts of resources to the water-transport system.

We model ground vegetation in a similar way as trees. The most essential difference is that the water-transport system can be omitted. The effect of shading is introduced in a rough way. We assumed that the annual photosynthetic production of ground vegetation, P_G, g(CO_2) m^{-2}, is

$$P_G\,(M_{n\,G}) = p_g\; f_s\,(M_{sn})\; M_{n\,G}\left(1 - \frac{M_{n\,G}}{M_{n\,G} - g_1}\right), \tag{7.22}$$

where $M_{n\,G}$, (g(dry weight)) is the leaf mass of ground vegetation, $M_{s\,n}$ the needle mass of the tree stand and function f_s is defined in Eq. 7.4 and p_g and g_1 are parameters.

The leaf and fine root growths are solved from carbon and nitrogen balance equations. We assume, however, that the annual leaves require only the amount of nitrogen that is in the leaf litter. The balance equations are

$$(P_G - R_G) = a_{n\,gr}\; M_{n\,G} + a_{r\,gr} M_{r\,G}, \tag{7.23}$$

$$(1 - c)\;\; n_n\; M_{n\,G} + n_r\; M_{r\,G} = u\; M_{r\,G}\; N_a(k), \tag{7.24}$$

where $a_{n\,go}$ and $a_{r\,go}$ are conversion coefficients from sugars to dry matter and n_n and n_r are nitrogen concentrations in leaves and roots, c is the nutrient recycling and u a nutrient uptake parameter.

7.2.3 Carbon and Nitrogen Dynamics in Soil

Litter from trees and ground vegetation enters the pool of organic matter in soil. The large biopolymers in the litter are not available for microbes or trees. Microbes decompose the macromolecules in the organic matter with extracellular enzymes (basic idea ES13) (Linkins et al. 1990; Criquet 2002; Waldrop et al. 2004). Large biopolymers (proteins, lipids, starch, cellulose and lignin) cleave to form small molecules such as sugars, amino acids, fatty acids and various other aliphatic and aromatic compounds (Tomme et al. 1995; Eivazi and Bayan 1996; Saito et al. 2003; Pavel et al. 2004; Wittmann et al. 2004). Microbes take up these small molecules (basic idea ES13). In addition, small amounts of humic substances are formed. The decomposition of organic matter in soil is a slow process, requiring decades (Sect. 5.6). The lifetime of humus is very long and varies according to its chemical properties (Liski et al. 1998). The fluxes connected with humic substances are, however, so small that we neglect them from the carbon dynamics.

Most of the nitrogen available for trees is loosely adsorbed on the surface of soil particles as ammonium ions (Sect. 5.6). When microbes use amino acids to produce ATP, they emit ammonium ion. The lifetime of ammonium in the soil is short, only about 30 days (Sect. 5.6).

Litter fall is the main carbon and nitrogen input into forest soil, while nutrient uptake causes the main nitrogen flow out of the soil. In addition, deposition and leaching of nitrogen to ground water and emission of nitric gases to atmosphere connect the forest soil with its environment. The microbes decompose larger molecules in the litter fall with extracellular enzymes. Figure 7.5 shows carbon compound fluxes between the main pools in the soil. Circulation of nitrogen within soil (Fig. 7.6) is an additional complication in the soil nitrogen fluxes compared to carbon fluxes.

We developed a quantitative description of the carbon and nitrogen fluxes in forest soil (Fig. 7.6). As a first approximation, simple model structures and either linear or multiplicative dependencies are used. In order to capture the qualitative difference between the rather permanent, structural carbohydrates and readily usable smaller carbon-based molecules, we divide the organic matter into categories of large and small molecules.

We analyse five types of biopolymers (proteins, lipids, starch, cellulose and lignin). Each biopolymer has its own specific microbe that emits an enzyme capable of breaking the biopolymer. The following symbols are used for the state variables describing the amounts of carbon-rich substances in soil:

$C_{\mathrm{L\,n}}$ = carbon compounds in soil formed by large molecules of type n, (g m^{-2})
$C_{\mathrm{S\,n}}$ = carbon compounds in the soil formed by small molecules of type n, (g m^{-2})
$M_{\mathrm{M\,n}}$ = mass of microbes emitting enzyme to break molecules of type n, (g m^{-2})
E_{n} = enzymes breaking large carbon molecules of type n, (g m^{-2})
$C_{\mathrm{SM\,n}}$ = mass of free small carbon compounds of type n in microbes, (g m^{-2})

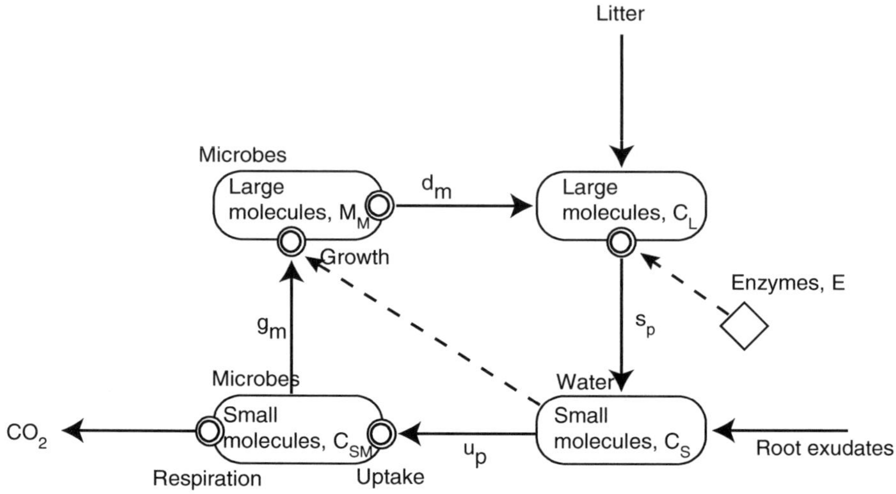

Fig. 7.5 Carbon compound fluxes in forest soil starting with litter fall via enzymatic splitting of large molecules to release in atmosphere in microbial metabolism. *Boxes* indicate amounts, *arrows* (*solid line*) flow, *double circles* process, *arrows* (*dashed line*) effect on process

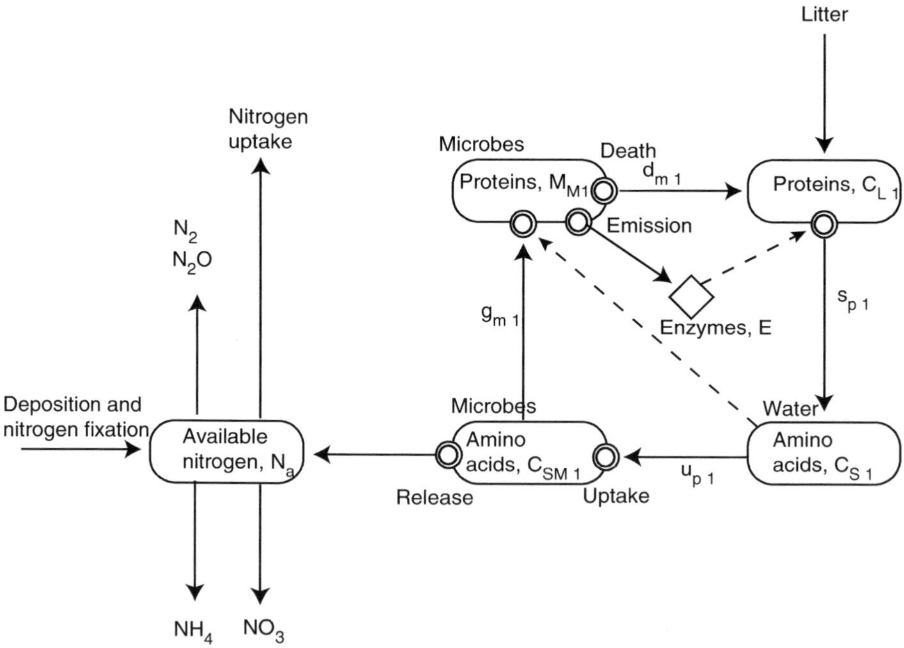

Fig. 7.6 Fluxes of nitrogen compounds loop in forest soil. *Boxes* indicate amounts, *arrows* (*solid line*) flow, *double circles* process, *arrows* (*dashed line*) effect on process

Table 7.1 The connection between the index and molecule types

Index	Small molecules	Large molecules	Enzyme
1	Amino acids	Proteins	Peptidase
2	Sugars	Starch	Amylase
3	Fatty acids	Lipids	Esterase
4	Sugar	Cellulose	Glucanase
5	Aliphatic acids	Lignin	Ligninase

The connection between the index and molecule types is presented in Table 7.1.

The structure of the equations for flows can be seen from Fig. 7.5 where we utilise the conservation of mass as presented in Chap. 2.

The enzymes emitted by microbes split carbon macromolecules. Splitting of large carbon molecules of type n, s_{pn}:

$$s_{\mathrm{pn}} = b_{1\mathrm{n}}\ E_{\mathrm{n}}\ C_{\mathrm{Ln}}. \tag{7.25}$$

Uptake of small molecules of type n by microbes u_{pn}:

$$u_{\mathrm{pn}} = b_{2\,\mathrm{n}}\ C_{\mathrm{Sn}}. \tag{7.26}$$

Growth of microbes emitting enzyme-splitting macromolecules of type n, g_{mn}:

$$g_{\mathrm{mn}} = b_{4\,\mathrm{n}}\ C_{\mathrm{Sn}}\ M_{\mathrm{Mn}}. \tag{7.27}$$

Death of microbes emitting enzyme-splitting macromolecules of type n, d_{mn}:

$$d_{\mathrm{mn}} = b_{5\,\mathrm{n}}\ M_{\mathrm{Mn}}. \tag{7.28}$$

The masses of microbes are treated in terms of the five chemical groups.

The protein and amino acid fluxes in the soil include additional aspects connected with the circulation of nitrogen-rich compounds in the soil. The uptake of amino acids by microbes of type n, u_{pn}:

$$u_{\mathrm{pn}} = c_1\ \frac{M_{\mathrm{Mn}}}{\sum_{j=1}^{5} M_{\mathrm{Mj}}}\ C_{S\ 1}. \tag{7.29}$$

Production of an enzyme that splits macromolecules of type n, p_{en}:

$$p_{\mathrm{en}} - c_2\ C_{\mathrm{SM}\ 1}\ \frac{M_{\mathrm{M\,n}}}{\sum_{j=1}^{5} \mathrm{M}_{\mathrm{M\,j}}}. \tag{7.30}$$

Release of ammonium n_{r}:

$$n_{\mathrm{r}} = c_3\ C_{\mathrm{SM}\ 1}. \tag{7.31}$$

Microbial respiration r_m:

$$r_m = c_4 \sum_{j=1}^{5} e_j \, C_{SM\,j}, \tag{7.32}$$

where chemical constants e_j convert different carbon compounds to CO_2.

Enzymes in the soil that break down macromolecules are vulnerable to the action of the enzyme that decompose proteins. This is the reason for enzymes having limited lifetime in the soil. Destruction of the enzyme-splitting macromolecules of type n, d_{en} is

$$d_{en} = c_5 \, E_1 \, E_n. \tag{7.33}$$

The differential equations for the state variables are based on the in- and outflows as shown in Chap. 2, Eq. 2.3), consider as an example the small carbon compounds of type n:

$$\frac{d \; C_{Sn}}{dt} = s_{pn} - u_{pn}. \tag{7.34}$$

The nitrogen in the soil that is available to plants, N_a, plays a key role in connecting trees and soil. It is obtained from the flows changing the amount of available nitrogen

$$\frac{dN_a}{dt} = -u \sum_{i=1}^{m} \sum_{j=1}^{k} M_r(i,j,k) N_a(k) - u \, M_{rG} \, N_a(k) + n_r + d_{ep} + n_f - l_e - e_v + sp1, \tag{7.35}$$

where d_{ep} is nitrogen deposition, n_f nitrogen fixation, l_e is nitrogen leaching, e_v is nitrogen vaporisation as NO or N_2O and the sum over size classes and whorls is the nutrient uptake term in Eq. 7.16 for trees.

7.2.4 Connections Between Trees, Ground Vegetation and Soil

The ecosystem model MicroForest combines the three functional components of forest ecosystem: trees, ground vegetation and soil. The model describes the carbon and nitrogen fluxes in the forest at an annual timescale and can simulate how the fluxes behave as the stand develops from the sapling stage.

Processes generate material fluxes between the pools and also between the pools and the surroundings of the forest. The knowledge of the processes at volume and time element level can explain the changes in the pools. In this way we can obtain an understanding of the development of forest ecosystems. However, we pay special emphasis on the emergent features in the ecosystem, when moving to a more aggregated level. We consider explicitly annual photosynthetic production, regularities in tree structure, annual interactions between trees and nitrogen cycling in the ecosystem as emergent properties that are important for the development of the ecosystem.

The carbon and nitrogen fluxes connect the main components of the forest ecosystem, that is, trees, ground vegetation and soil. The trees and ground vegetation take carbon from the atmosphere and nitrogen from soil, and in turn trees and vegetation feed the soil with large carbon compounds and with proteins in the form of litter. Respiration of trees and microbes release carbon to the atmosphere in the form of CO_2. Microbes also release nitrogen from proteins by breaking down the large protein molecules in litter and utilising the resulting small amino acid molecules in their metabolism, and they emit ammonium ions into soil solution. Trees take up the ammonium ions, and in this way, the nitrogen in litter becomes available to trees.

Nitrogen cycling is an important emergent feature at ecosystem level. Before current large anthropogenic nitrogen deposition, the nitrogen released in decomposition of proteins was the main source of nitrogen. It determined the fertility of the site. The present large nitrogen deposition has changed, to some extent, the situation in areas close to industrialised centres. Often, deposited anthropogenic nitrogen is an important raw material for protein synthesis.

The large pools in living trees and in the soil dominate the carbon and nitrogen dynamics in forest ecosystems. The ground vegetation plays a minor role in the fluxes when the canopy is closed. The needle mass of trees is low during early stages of stand development, and then the ground vegetation is important for carbon and nitrogen fluxes.

We call the combined model describing the development of carbon and nitrogen fluxes and pools in a forest ecosystem as MicroForest. The first part of the name of the model refers to the important role of microbes in the dynamics of nitrogen and carbon in forest ecosystems. MicroForest calculates all carbon and nitrogen pools in terms of cellulose, lignin, lipids, starch and proteins. In addition, tree characteristics, such as diameter and height, are modelled. MicroForest provides a large number of predictions that can be tested with field measurements.

7.3 Carbon and Nitrogen Dynamics in Forest Ecosystem around SMEAR II

7.3.1 Field Measurements

SMEAR II (Station for Measuring Ecosystem-Atmosphere Relations) was planned and implemented to measure material and energy fluxes and pools and processes generating the fluxes in a forest ecosystem (Chap. 9). Since MicroForest and SMEAR II deal with carbon and nitrogen fluxes and pools, and processes generating these fluxes, the model and the measurement speak the same 'language'. Therefore, the data obtained at SMEAR II are very useful in developing and testing MicroForest.

Development of MicroForest requires stand data, such as annual heights and diameters of trees. These data were obtained from the SMEAR II stand in autumn 2001, when the stand was 40 years old. The stand consists almost purely of Scots pine. Annual heights were determined from whorl heights. These are easily identifiable in Scots pine. Only very old whorls, which have born at age of 6 or before, are somewhat difficult to identify.

Annual diameters were measured from discs cut from stump height. Diameters are also difficult to measure from three oldest annual rings, since these are often narrow and the border between the rings can be unclear. In younger rings the diameter measurement is in most cases very accurate.

The measurements include always systematic errors and noise that should be taken into consideration when using the data. Annual carbon fluxes are used to evaluate the performance of MicroForest. The accuracy and precision of soil CO_2 efflux measurements are rather poor, about 0.5 kg (CO_2) ha^{-1} yr^{-1}, canopy photosynthesis is more accurate and precise about 0.2 kg (CO_2) ha^{-1} yr^{-1} and the precision and accuracy of the CO_2 flux of the ecosystem is close to that of soil efflux about 0.6 kg (CO_2) ha^{-1} yr^{-1}.

As mentioned above, the retrospective measurement of annual diameters and tree heights is rather easy, when height of trees is above 1 m and diameter above 1 cm. However, detection of the whorl and annual tree ring borders becomes difficult when the seedlings are below 0.5-m tall. This fact causes disturbing systematic error and measuring noise in the initial state of the stand, and causes some problems, since the initial state is the only place in the model, where information from stand and tree structure are needed.

7.3.2 Parameter Values

There are 54 parameters to be determined for MicroForest model (Appendix 2). The measurements at SMEAR II are the backbone, when values for the parameters are determined. We applied five approaches to determine the values: (1) chemical information, (2) direct measurements of structure, process rates or chemical composition at SMEAR II, (3) required static behaviour of some state variables, (4) literature values and (5) estimation from the behaviour of the model. We tried to minimise the number of parameters estimated from the behaviour of the model MicroForest.

Photosynthesis is the key process in plant growth and metabolism, since it produces raw material and energy in form of sugars, which are consumed in growth and metabolism. Therefore, the most important parameter influencing stand behaviour is the annual photosynthetic production in unshaded conditions. We utilised the measurements at SMEAR II for determining the photosynthetic production in unshaded conditions.

Another key process in the plant is the conversion of sugars, formed in photosynthesis, into tissue material such as needles, xylem and fine roots. MicroForest includes also these conversion processes. We used the conversion coefficients according to Penning de Vries (1975). For the conversion parameters, we have to know also chemical composition of needles and wood. These were measured from the SMEAR II stand.

The values of the parameters in the chemical composition of needles and wood were based on measurements of samples from the stand. The chemical composition of fine roots was assumed to be the same as that of needles.

Similarly as is the case with trees, we have to determine parameter values for forest soil. However, the useful information in literature, where soil parameters could be determined, is very limited, since the traditional measurements do not differentiate soil organic matter into proteins, cellulose, lignin, starch and lipids. Some useful information can still be found in literature.

The pools of macro carbon compounds are very large, indicating slow turnover rates (Liski et al. 1998), and in contrast, the pools of small carbon compounds are small. The parameter values related to the decomposition process, that is, those parameter values that were connected with microbes and extracellular enzymes, were adjusted so that the pools of organic matter were close to observed values at SMEAR II (Fig. 5.35 and 5.37) and also in such way that the pools in soil remained static. The chemical properties of microbes were based on the literature.

There are a number of parameters, which cannot be measured directly or obtained from the literature. In some cases parameters are difficult to measure: there may not be proper measurement devices, methods or resources for obtaining these values. These parameters may be measured in the future, when devices and methods have developed. However, it is not a good strategy to wait with the model for years for obtaining additional information to determine the values of the parameters.

Another group of parameters, which cannot be derived from measurements or literature, are slow changes occurring in nature. The proper determination of slow changes requires continuous monitoring for years and a long time series. Monitoring and time series construction may take decades or even centuries. Some time series are available from measurement stations such as SMEAR II, but compared to the duration of the slow phenomena, our measuring series are very short. If we want these values from measurements, we should have measured them already 50 or 100 years ago.

Parameters that cannot be found from the literature or measured can be estimated from the model behaviour. The values of seven parameters were fitted to give acceptable model behaviour. The four criteria were (1) the parameters on the effect of shading on canopy photosynthesis and fine root respiration (including root exudates) were fitted to yield annual photosynthetic production and soil CO_2 efflux measured with soil chamber and Eddy Covariance at SMEAR II; (2) the stem base enlargement parameters and allocation to the top of the crown were selected

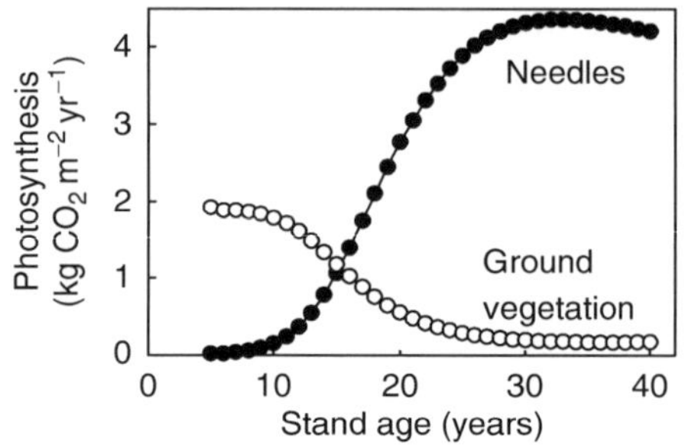

Fig. 7.7 Simulated annual photosynthetic production of tree needles and ground vegetation as function of time

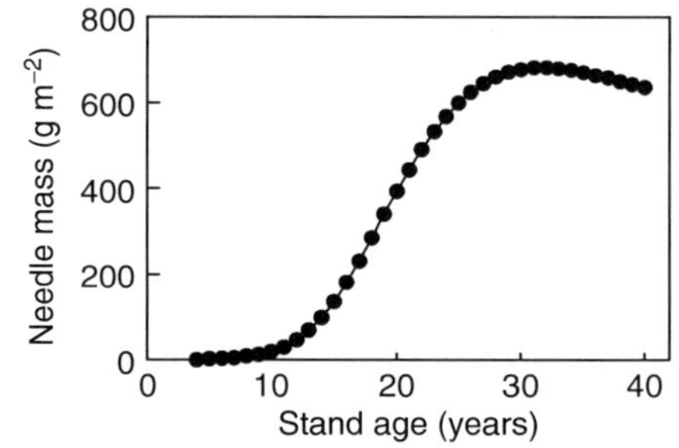

Fig. 7.8 Simulated needle mass in the stand as function of time

so that the model produced the observed form of the stem base and crown; (3) nitrogen uptake efficiency was estimated so that the model produced reasonable fine root mass and (4) the effect of shading on height growth was adjusted to produce measured height growth patterns at SMEAR II.

We introduced the influence of increased nitrogen deposition during the last decades by increasing it from initially 1–5 kg ha^{-1} at the age of 40 years. In addition, we included nitrogen fixation into the approximation of nitrogen deposition. We assumed that the leaching of nitrogen from the stand is negligible, which is in accordance with observations at our field sites.

7.3.3 Simulation of the Forest Ecosystem around SMEAR II

Photosynthesis drives the carbon and nitrogen cycles not only within plants but also in the whole ecosystem. The development of CO_2 flow from atmosphere to trees and ground vegetation and accumulation into the ecosystem (Fig. 7.7) are closely connected with the dynamics of needle mass in the stand (Fig. 7.8). The low needle mass of trees, in the early phase of stand development, provides only weak photosynthetic production, but at the same time, the unshaded ground vegetation fixes considerable amounts of CO_2. Later during the stand development,

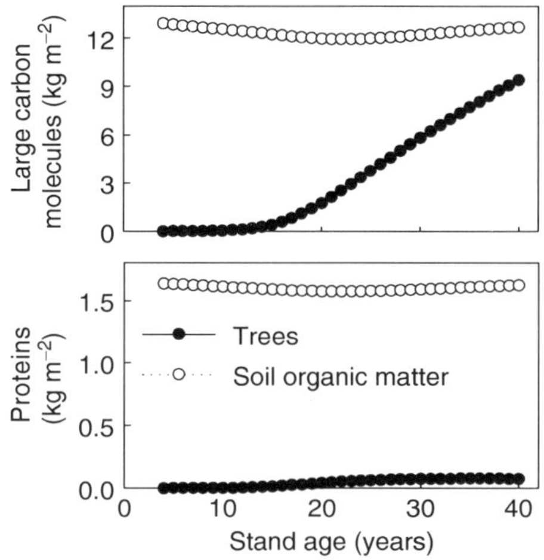

Fig. 7.9 Simulated development of carbon compounds (*upper panel*) and proteins (*lower panel*) in trees and soil

the situation is reversed: tree foliage fixes more carbon but also shades the ground vegetation, which then photosynthesises very little.

The simulated development of the stand around SMEAR II shows that carbon accumulates into the ecosystem throughout the period of stand development, but nitrogen cycles in the system with rather constant rates (Fig. 7.9). A major part of the biomass in the growing stand is accumulated into the xylem of trees. There is also a small increasing trend in the mass of soil organic matter, as the branch litter accumulates into soil (Fig. 7.9).

The input of protein-rich litter does not vary much during the rotation time. After the regeneration of the stand, young trees are unable to shade ground vegetation or take a large proportion of nutrients. Ground vegetation develops rapidly and consumes a large proportion of nitrogen released in decomposition. Later on, the gradual development of trees suppresses ground vegetation and increases the nitrogen uptake of trees. Nitrogen taken up by trees is used to grow more needles, fine roots and other active tissues, where concentration of proteins is high. The life span of these tissues is rather short and nitrogen returns to the soil within a couple of years. Only nitrogen-poor wood accumulates year after year, and finally the growth of the stand results in a large nitrogen pool in living trees. The vegetation is able to take up all the nitrogen that is released in decomposition; thus, the amount of plant-available nitrogen in soil is low (Fig. 5.37).

The released nitrogen from organic matter is sufficient for maintaining the high biomass growth in trees, as the average concentration of nitrogen in tree tissue is lower than in ground vegetation. Over the rotation, a small accumulation of nitrogen in wood can be observed (Fig. 7.9), but it does not affect plant-available nitrogen. This is because in the simulations we assume, following the trend of anthropogenic emissions during the last century, an increasing nitrogen input from atmosphere to soil.

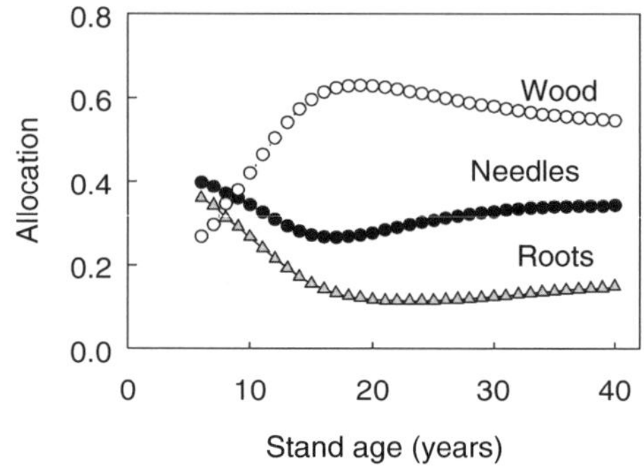

Fig. 7.10 Simulated development of allocation to wood, needles and roots. The allocation is the solution of the five equations describing structural regularities, and carbon and nitrogen balance

MicroForest allocates carbon and nitrogen between different tree parts dynamically (Fig. 7.10). In other words, the amount of nitrogen and carbon used in the growth of needle, fine root and xylem masses is calculated each year and is based on the carbon fixed in photosynthesis and on nitrogen acquired by roots. These available resources are allocated to tree parts by the biochemical regulation system that takes in consideration the regularities in tree structure. The amount of needles has to correspond to the water-transport capacity in woody structures, and the pool of available nitrogen has to be sufficient to synthesise all needed proteins in new tissues. In practice we have mass balance equations for nitrogen and carbon. When we solve these equations, we obtain the allocation of nitrogen and carbon to different parts of the tree, which, in turn, results the tree growth at the given year.

Since the allocation of resources is a dynamic phenomenon, we get some interesting results. In the beginning of stand rotation, when trees are small, they allocate resources mostly to needles and fine roots. As the tree grows in height, the water-transport distance increases, and more resources have to be allocated to the construction of the water-transport system. During the first 20 years, there is a change in the allocation from needles and fine roots to the water-transporting xylem.

There are small differences in initial values of the height, diameter and needle mass between the trees in different size classes. The seedlings grow exponentially during the first 10 years. Thereafter, the shading begins to reduce the photosynthesis, especially in the smaller size classes. This results in reduced growth and leads to differentiation of size classes (Fig. 7.11). The differences in annual photosynthetic production generate the differentiation of the size classes into dominated and suppressed trees in simulations.

The flux of CO_2 from soil to atmosphere (Fig. 7.12) is more static during stand development than the CO_2 fixing by the photosynthesis of trees. Respiration of tree and ground vegetation roots, and also root exudates, dominates CO_2 flux. The microbial respiration depends on litter production and root exudates and remains very stable over the rotation. Together all these tendencies cause forests to be carbon sources when young but to become sinks as they mature.

Nitrogen fluxes from the surrounding environment into forest ecosystems are small, at least when compared to the amount of nitrogen circulating within

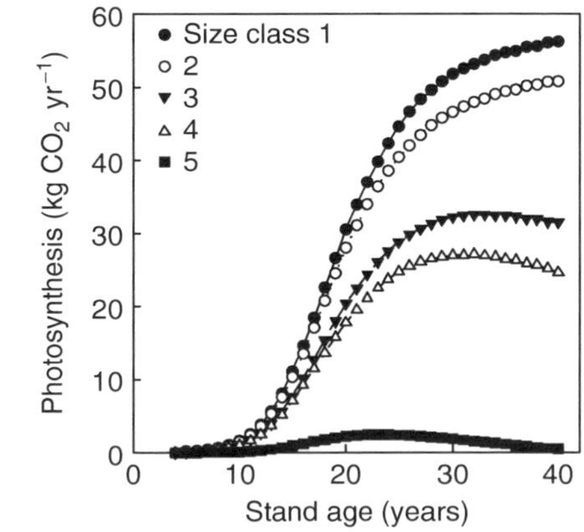

Fig. 7.11 Simulated development of photosynthesis in the five size classes

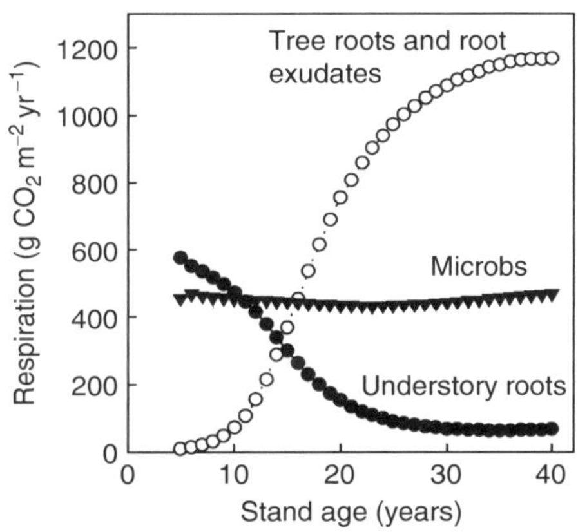

Fig. 7.12 Simulated annual CO_2 fluxes generated by respiration of tree roots and root exudates, microbes and roots of understory vegetation

the system. Leaves of the trees form one of the largest nitrogen pools in the ecosystem. The amount of leaves in the stand is rather stable after canopy closure. Then, when the amount of leaves has stabilised to a level characteristic to the site, the flow of nitrogen into the vegetation, generated by root uptake, and flow out, by senescent leaves and fine roots, are close to each other. This is why the nitrogen release into a plant-available form by microbes from proteins in the soil is very important for tree growth. The large protein pool in the soil filters out the effect of small variations in the litter input into the soil and the release of ammonium ions, and nitrogen uptake is rather stable during stand development. The ground vegetation has an important role in taking up nitrogen released in decomposition at the early phase of stand development.

The performance analysis of the model against measured tree growth at the SMEAR II station shows good agreement with measurements. The simulated height developments are rather close to the measured ones in each size class: the fit is demonstrated in Fig. 7.13. The fits between the observed and modelled height and diameter developments in the four biggest size classes are rather similar to

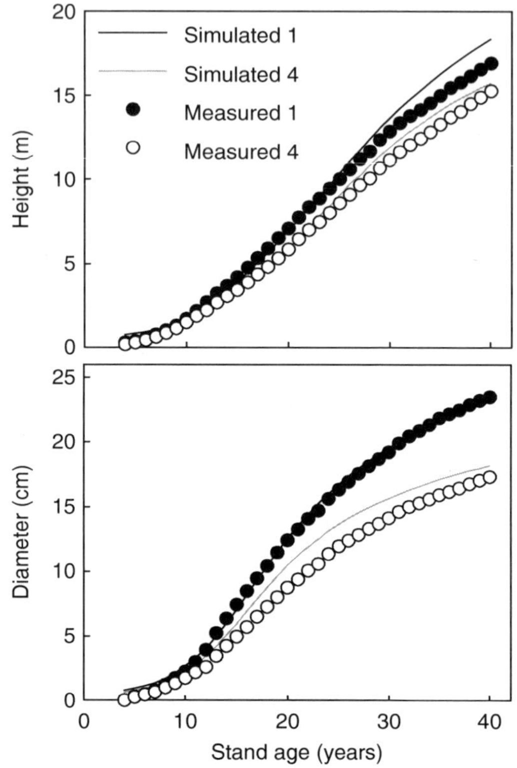

Fig. 7.13 Comparison of measured and simulated height (*upper panel*) and diameter (*lower panel*) developments in the size classes 1 and 4 at SMEAR II

that shown in Fig. 7.13. However, for the smallest size class, MicroForest gives smaller heights and diameters than the observed ones are. On the other hand, the measured and simulated stand volumes are rather close to each other. This is because small trees, in which the discrepancy between model and measurements is largest, contribute only very little to the stand volume.

7.3.4 Evaluation of the Simulations at SMEAR II

Our theory of carbon and nitrogen accumulation in the forest ecosystem is based on the analysis of several processes in the forest. These processes generate carbon and nitrogen fluxes in the ecosystems. Another fundamental idea in our theory is utilisation of the regularities in tree structure. Our theory includes several novel aspects, especially the chemical properties of living cells in vegetation and in soil, nitrogen as a vital component in the functional substances (enzymes, membrane pumps and pigments complexes), regularities in tree structure, ground vegetation and intensive and extensive field measurements.

In reality, living cells include thousands of different chemical compounds. This versatility can be, however, systemised by introducing groups of chemical compounds that have common features. Most of the compounds are macromolecules formed by interconnected basic units of small and simple carbon compounds, like sugars and fatty and amino acids. In the model, we have treated five dominating macro compounds: starch, cellulose, lignin, lipids and proteins. The synthesis of the basic building blocks, and also the construction of the macromolecules, requires very specific enzymes. These enzymes, in turn, require nitrogen that is the basic building block of amino acids.

Cellulose, lignin and lipids form the passive cell walls in a living plant cell, that is, they form the skeleton of the cell. The metabolism is based on the action of functional substances, which synthesise and move molecules and ions through membranes. Functional substances are also needed to capture energy in photosynthesis. The reverse phenomena occur in soil when microbes utilise solar energy, stored in the senescent litter, in their metabolism. Microbes also utilise a functional substance to decompose the macromolecules with extracellular enzymes. The decomposition requires specific enzymes emitted by microbes.

The macromolecule groups are also important for the accumulation of carbon compound in the forest soil. Each macromolecule requires specific decomposing enzyme, and they have different lifetimes in the soil. The decomposition is a very slow process, and the accumulation of carbon macromolecules in forest soil is an important phenomenon. Proteins originating from the functional substances in litter play a very important role in the nitrogen dynamics of the ecosystem.

Description has dominated the research of the structure of trees. For example, we know the fine details of cells, the geometry of leaves and trunks and chemical properties of different tissues. However, although tree structure has been described well, the connection between structure and tree metabolism is missing to a great extent. Another aspect, which is often omitted in previous studies, is the regularities in tree structure. There are some philosophical problems in the analysis of the regularities, since it requires evolutionary argumentation and optimality considerations. However, connections between metabolism and the structure needed in the metabolism have been in the focus of evolution. We may assume that good, even optimal, solutions have developed in evolution during many hundreds of generations. These regularities in tree structure are very fundamental properties of forests.

Trees are facing the challenge of running several interconnected processes simultaneously, such as nutrient uptake and photosynthesis. In addition, trees have to transport chemical substances between processes, for example, when water and nitrogen are transported from roots to needles for photosynthesis. The transport system between the tissues is formed on yearly basis. The raw material, sugars and nitrogen, has to be allocated to different metabolic tasks and to the transport system. The proper allocation of sugars and nitrogen has been the focus of evolution, since proper allocation greatly contributes to the success of individuals.

If we think that the biochemical regulation system of allocation is able simultaneously to take into consideration (1) the structural regularities and (2) the resources available for growth, then the allocation can be found as the solution of mass balance equation. This idea was introduced already in the 1980s (Hari et al. 1985; Valentine 1985; Mäkelä and Hari 1986). The idea has been implemented in carbon balance models of stand growth, but these lack the important nitrogen cycling in the ecosystem. We have expanded the thinking behind carbon balance models to include also nitrogen and implemented nitrogen with the aid of functional substances and the nitrogen balance. In our model, functional substances connect metabolism and nitrogen. This expansion, in turn, requires the treatment of soil proteins but provides interesting insight to the cycling of nitrogen between plants and soil.

Allocation evidently varies during the growing season. For example, in Scots pine, the radial growth begins in the spring, whereas needle growth takes place at midsummer. The biochemical regulation system has a demanding task to build a balanced functional system in this situation, where tree parts grow sequentially. Radial growth produces new water-transporting xylem tissue, and the amount of this tissue has to correspond to the amount of needles. The observed regularities in tree structure indicate that the regulation system is able to solve the allocation problem, and the allocation to needles, water-transport system and fine roots is an emergent property at annual and tree levels. The allocation takes place every year, and this also implies that ecosystem dynamics should be treated on yearly basis.

The differences in annual growth in forest ecosystems in the same region are large. In addition, nitrogen fertilisation increases growth very clearly in most ecosystems, especially in nitrogen-poor boreal forests. The allocation to roots explains these phenomena in our theory. On nutrient-poor sites, trees react to the small availability of nitrogen by allocating more to the root system and to root exudates. Any change in nitrogen availability will change the allocation between roots and trunk, and this results in large growth responses in trunk volume.

Ground vegetation has gained rather little attention in the research of forest ecosystems, evidently because it often plays a minor role in the mature phase of stand development. In an old stand, ground vegetation is scarce or even missing, and trees dominate material and energy fluxes between the ecosystem and its surroundings. The situation is totally different in the early phase of stand development, when trees are young and small. Then trees contribute very little to the material and energy fluxes, and the ground vegetation dominates fluxes. Therefore, if ground vegetation is omitted, an important aspect of the ecosystem dynamics is missing.

We have utilised a rather long tradition of dynamic carbon-based models of Scots pine stands (Hari et al. 1982, 1985, 2008; Mäkelä and Hari 1986; Nikinmaa 1992; Perttunen et al. 1996; Mäkelä 1997, 2002; Nissinen and Hari 1998; Mäkelä and Mäkinen 2003). We introduced explicitly the structural regularities as the basis of carbon allocation. We expand this thinking also to nitrogen in MicroForest. This approach leads to very dynamic allocation to roots that is a major factor in the model behaviour.

Traditionally, decomposition models have used aggregated soil organic matter types, such as rapidly, slowly and extremely slowly decaying organic matter, for which they have produced aggregated decomposition rates. Some models have been only interested in carbon release (e.g. Liski et al. 1998), while in others, also nitrogen release is considered (e.g. Aber et al. 1991; Rastetter et al. 1991; Running and Gover 1991; Johnson 1999). The problem with this approach of rapidly, slowly and extremely slow soil organic matter types is that there is no means to measure the state variables describing the soil properties. Therefore, testing of the model is also extremely problematic.

The crucial role of extracellular enzymes in the decomposing of soil organic matter was realised about 10 years ago. Thereafter, a rather rapid development in soil microbiology has taken place (Linkins et al. 1990; Criquet 2002; Waldrop et al. 2004). MicroForest is the first dynamic model that attempts to base decomposition on the extracellular enzymes. In principle, these developments offer very good possibilities for studying the nitrogen cycle in the natural environment. We have divided soil organic matter into chemical compounds such as cellulose, lignin and proteins. The treatment of soil compounds according to their chemical structure allows efficient use of quantitative methods, since the amounts of all soil chemicals can, at least in principle, be directly measured. This allows also testing of the model in the field. This is in agreement with Hartmut Bossel's idea that models should be built in such a manner that they consist of state variables that have real-world counterparts (Bossel 1991). The same is not true, if soil organic matter is divided hypothetically into rapidly, slowly and extremely slowly decaying components, since one cannot measure these variables directly. This causes methodological and practical problems.

The ecosystem model MicroForest and the measuring system SMEAR II have both been constructed to analyse and measure material and energy fluxes in a forest ecosystem. They are based on the same vision dealing with the most important aspects in forest ecosystems. SMEAR II provides information that can be easily utilised in the determination of the parameter values and in testing of the model. Most of the parameter values in the tree component were determined with direct measurements, with literature and general chemical information; only seven parameter values had to be estimated from the behaviour of the model. As such, this number is still too high. However, most of the values are such that they can, in principle, be measured, at least in the future. As we reach the limits of our understanding, a close collaboration between modelling and design of measurements becomes apparent. Complex systems require modelling tools that help to understand dynamics of these systems. However, modelling that cannot be evaluated against measurements is closer to computer games than science.

The measurements at SMEAR II cover all measurable carbon and nitrogen fluxes between the forest ecosystem and its surroundings. Measurements also cover carbon and nitrogen pools in trees and soil. The simulations with MicroForest are in agreement with these measurements. The only exception is that the simulated soil CO_2 efflux remains somewhat low. Our description of the carbon and nitrogen balances of the ecosystem with its surroundings are within the margins of the present measuring accuracy and precision.

Global climate change, increasing CO_2, increasing temperature and enhanced nitrogen deposition, will all have major effects upon boreal forests. Evidently, photosynthesis is accelerating, the photosynthetic active period is expanding, decomposition of soil organic matter is becoming more rapid and nitrogen pool in soil is increasing. The causal effects of increasing temperature and atmospheric CO_2 on processes in forest ecosystems can easily be introduced into the annual photosynthetic production and on decomposition of soil organic matter. In addition, MicroForest uses nitrogen deposition as input and can therefore simulate the effect of anthropogenic nitrogen emissions. All these features make MicroForest a powerful tool to analyse the effect of climate change on boreal forests as will be shown in Chap. 8.

MicroForest is an operational summary of the regularities in the processes and structure in the forest ecosystem discussed in the previous chapters. The theory of carbon and nitrogen dynamics in forest ecosystems gained corroboration. However, the test was not a severe one since our data deal with one single stand around SMEAR II measuring tower.

7.4 Test of MicroForest Against Data on Tree Growth

7.4.1 First Test of MicroForest

In the work outlined above, we utilised the versatile measurements at SMEAR II as estimation data, and consequently, we had no real test data. This is an evident shortcoming. However, SMEAR II measurements are unique, covering 15 years. The versatile measurements require rather a complicated system of instrumentation and recording, and thus, we cannot obtain all the additional test data that we might wish for. However, to test a complicated ecosystem model, we can focus on a specific component of the ecosystem. We can measure easily and retrospectively the time series of diameters and heights in coniferous stands. This enables creation of new additional test data sets.

MicroForest should be able to predict the stand development from its initial state in all even-aged pine stands as it did in the stand around SMEAR II measuring tower. To avoid the bias in the test caused by estimation of the parameter values, the parameters in MicroForest should have the values as presented in Appendix 3. The model MicroForest was constructed utilising measurements at SMEAR II. This raises the question: can MicroForest be applied to Scots pine stands growing in other locations than near SMEAR II?

We selected six even-aged Scots pine stands near SMEAR II measuring station to test the performance of MicroForest in the same climatic conditions as the estimation stand. In addition, we selected five stands in Estonia at the southern border of boreal forests about 300–400 km south from Hyytiälä to test the MicroForest. The altitude in the selected stands varies between 32 and 56 m above the sea level in Estonia, and near SMEAR II, the altitude is about 170 m asl. Selected stands represent Scots pine-dominated hemiboreal coniferous forests on sandy soil with different site fertilities. The climatic conditions are rather similar in all locations. Mean annual temperature is 4–6°C. Annual precipitation is 500–750 mm, about 40–80 mm of which falls as snow. The length of the growing season (daily air temperature above 5°C) is between 170 and 180 days. The three stands in Vihterpalu in the north-western part of Estonia grow on poor sandy soil. The stands were regenerated by planting after forest fire. The stand in Lahemaa in northern Estonia is also planted on poor sandy soil. The stand in Järvselja in south-eastern Estonia grows also on sandy soil, but the site is rather fertile. The ages of the stands varied from 32 to 61. For detailed description of the Estonian stands, see Köster et al. (2008).

We established a sample plot consisting about 200 Scots pines in each stand. Trees damaged by moose (*Alces alces*) or snow were excluded. Measured trees were divided into five size classes by breast-height diameter in such a way that size classes from biggest to smallest consisted of 5, 15, 30, 30 and 20% of the trees. We sampled six trees randomly in each size class, and a sample disc was cut on stump height. Discs were photographed, and widths of annual rings were measured with an image analysis programme. Havimo et al. (2008) described the test stands in more detail.

Initial state. MicroForest simulates stand development starting from an initial state. Numerous components of the initial state have only minor effect on stand dynamics. For example, the amount of cellulose in the soil in the beginning of the simulation period has only a minor effect on the stand dynamics. For those initial state components, which have only a small effect on the stand dynamics, we can apply values determined for the stand around SMEAR II. The site-specific components of the initial state, which has to be determined separately for each stand, consists of (1) needle masses, (2) tree heights and diameters at stump height, (3) number of trees in each size class and (4) amount of proteins in the soil.

We obtained the initial values for heights and diameters at stump height in each size class from the measured of heights and diameters, when the trees were 3–6 years old. The mortality is very low in young Scots pine stands, and we assumed that the number of trees in each size class did not change during the simulations. We determined the initial needle masses with empirical regression between sapwood area and needle mass.

The initial value for the protein pool in the soil is problematic since we have no measurements available. The only option is to estimate the amount of proteins in the soil at the initial state from the model behaviour, that is, estimate it from the simulations. We used the measured diameter in the second biggest size class at the end of the measurements as criterion to determine the initial state of the proteins in soil. We varied the initial protein pool until the simulated and observed diameters in the second size class at the end of the measuring period were equal.

Estimation of the parameter values. To obtain severe test of MicroForest, we estimated no parameter value whatsoever from the measurements in the test stands.

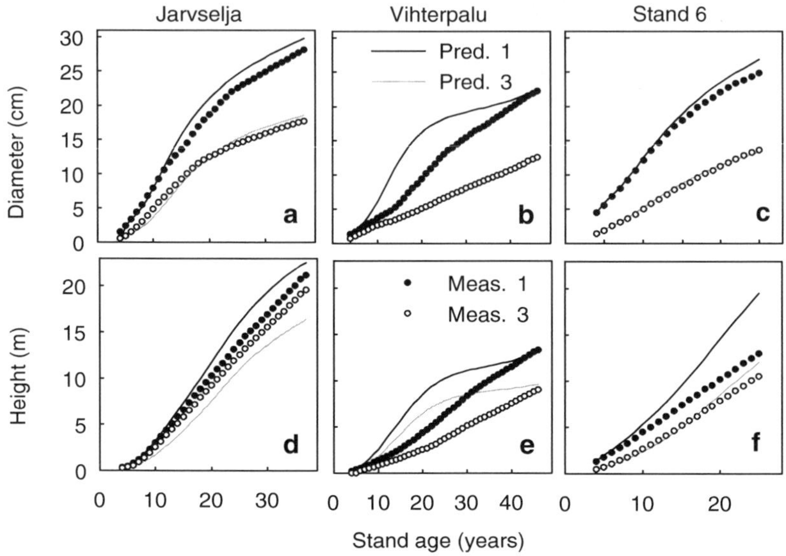

Fig. 7.14 Predicted (Pred.) and measured (Meas.) diameter and height in Järvselja (**a**, **d**) and Vihterpalu (**b**, **e**), Estonia and Stand 6 (**c**, **f**) near SMEAR II in southern Finland

Performance of MicroForest in the first test. The predicted diameter growth patterns were rather close to the measured ones in the test stands near SMEAR II (Fig. 7.14c), although there is variation in the fit between stands. The predictions of height developments were less successful, and the discrepancies between measured and predicted heights are in some cases rather large (Fig. 7.14f).

The predictions by MicroForest in Estonia were rather successful on fertile site in Järvselja (Fig. 7.14a and d). However, on poor sites, the measured and predicted diameter and height patterns were rather different (Fig. 7.14b and e). Also the mean patterns differed from each other; thus, there is a systematic discrepancy between the predictions and measurements in Estonia.

When studying the model behaviour, we found that the reason for the overestimation of growth on poor sites in the early phase of stand development was in the soil nitrogen dynamics. The nitrogen availability was too large at the age of 10–20 years, and thereafter too small. The model was recovering from the nitrogen shortage at the end of the study period. The nitrogen dynamics in the soil explains the failure in the predictions of growth.

The outcome of the first test was rather encouraging: with parameter values from SMEAR II, MicroForest predicted stand development in Hyytiälä and in Estonia if the site was rather fertile. In contrast, MicroForest was unable to predict stand growth properly on poor sites in Estonia before canopy closure.

7.4.2 Second Test of MicroForest

We found in the first test of MicroForest that the model is able to predict the growths on fertile sites rather well, but on poor sites, especially in Estonia, the measured and predicted growth patterns were rather different. We analysed carefully the simulation runs, and found that the ground vegetation was growing very little on poor sites. Also the nitrogen uptake by ground vegetation was low, which resulted in high-nitrogen availability and excessive growth before 20 years age. Evidently, the parameter values describing the behaviour of ground vegetation contribute to this systematic error in the model behaviour.

We used static behaviour of the soil macromolecule pools for estimating parameter values of organic matter decomposition. We expanded this criterion also to the behaviour of ammonium in the soil. We found that the parameter values for ground vegetation were such that on poor sites ammonium ions accumulated in the soil and the ammonium behaviour became rather dynamic. This resulted in excessive growth before canopy closure. We reestimated the parameter values describing the behaviour of ground vegetation in such a way that the ammonium concentration remained stable also on poor sites. We did not utilise tree growths in the estimation. In other words, in the second version of the model we require that ammonium concentration in the soil is rather constant over the rotation time, and if trees cannot consume all available ammonium, ground vegetation may use this extra ammonium. We obtained in reestimation $p_g = 22$ g(sugar) g(dry weight)$^{-1}$ and $g_l = 190$ g(dry weight) m^{-2}, Eq. 7.25.

Estimation of the parameter values. Similarly to the first test run, we estimated no parameter value whatsoever in the model MicroForest from the test stands.

Determination of the initial state. We utilised the measured diameters and heights at 4 years to determine the initial diameters and heights. We determined the initial value of the protein pool in the soil (the fertility of the site) utilising the last measured diameter in the second biggest size class, that is, we varied the initial protein pool in soil in such a way that the measured and modelled diameters were equal.

Prediction of characteristic features in diameter and height growths. The tree diameter has a clear pattern as a function of age. First, diameter increases rapidly, but later, there is a slow declining trend in growth. MicroForest was able to predict this characteristic feature. However, there is a rather weak tendency that the predicted decline begins earlier and is stronger in the prediction than the observed one in the smaller size classes (Figs. 7.15 and 7.16). The height growth pattern is more linear, but slowing down can be seen also in height development. Again, slowing down is clearer in the prediction than in the measurements in the smallest size classes (Figs. 7.15 and 7.16). MicroForest was able to predict the observed diameter and height patterns in all measured stands near SMEAR II and in Estonia rather well

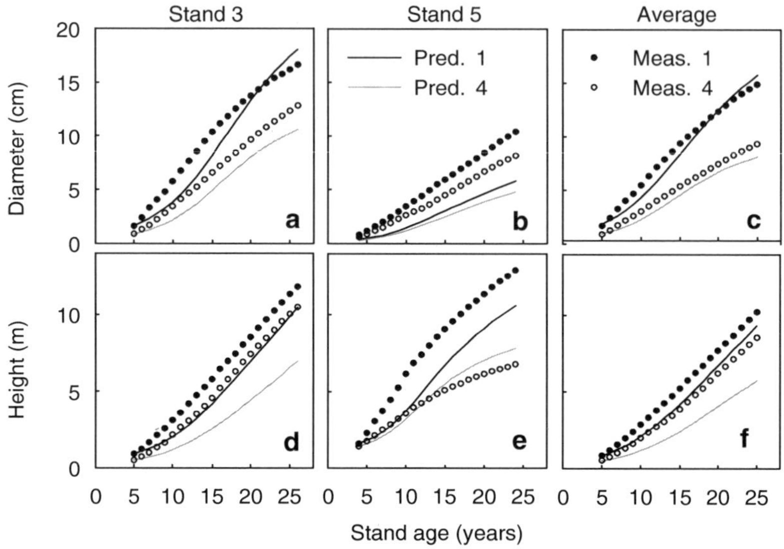

Fig. 7.15 Predicted (Pred.) and measured (Meas.) height and diameter for first and fourth size classes in stand number 3 (*Myrtillus* type site, 2,000 stems ha^{-1}, best fit), in stand number 5, (*Vaccinium* type site, 4,400 stems ha^{-1}, poorest fit) and in average in the test performed in Hyytiälä

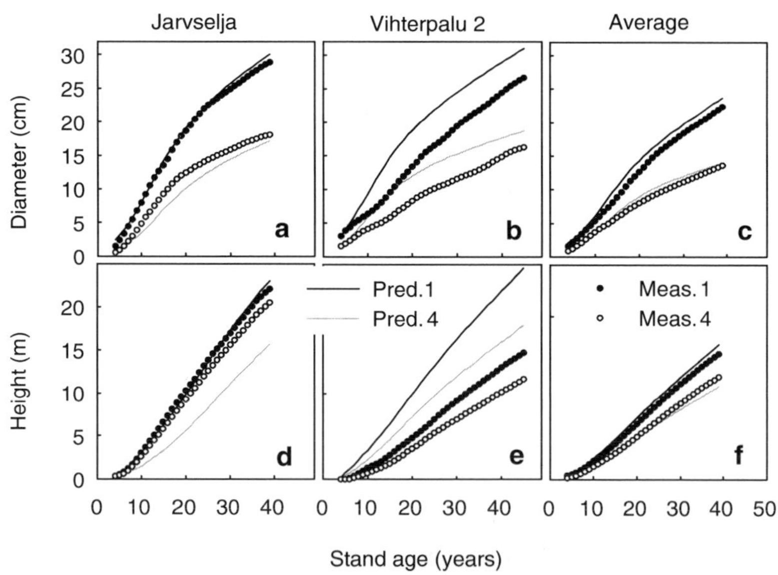

Fig. 7.16 Predicted (Pred.) and measured (Meas.) diameters and heights at two sites in Estonia (Jävselja, the best fit, and Vihterpalu 2, the poorest fit) and the average of all the sites in Estonia

(Figs. 7.15 and 7.16). However, the timing of this slowing down of growth takes place in some simulations too early in the smallest size class, and the prediction of diameters was more successful than for heights.

There is stand-specific variation in the fits between the measured and predicted diameters and heights. When we take the means of size classes over the stands, we can see more clearly the systematic failures in predictions by MicroForest. Figures 7.15 and 7.16 indicate that MicroForest predicts successfully the mean diameters in the stands near SMEAR II and in Estonia. The height predictions show discrepancy near SMEAR II, but in Estonia, also the height predictions are successful.

Rather, considerable size differences are characteristic for Scots pine stands. Our measurements indicate that the size classes observed at the age 25–50 years are formed already when the diameters of the seedlings are about 1 cm. MicroForest was able to predict the diameters and heights in each size class and the differentiation into the size classes from the small differences in the initial state.

Adequacy of the model structure. The strong role of theory is the characteristic feature of the model MicroForest. We base the equations on theoretical ideas. The carbon and nitrogen fluxes in forest ecosystems response to the prevailing environmental factors and great temporal variations are characteristic for the fluxes. There are, however, emergent properties at annual level such as annual photosynthetic production, annual nitrogen uptake and regularities in tree structure. MicroForest is constructed for working at the annual level, and the emergent annual properties have an important role in it. Carbon and nitrogen balance equations play the key role in the allocation of carbon and nitrogen. In this way, the understanding of the fluxes gained in Chaps. 4 and 5 is utilised at the annual level. The growth of Scots pine trees during next summer is programmed into the buds in August, and the growth is a realisation of the programme. The structure of MicroForest is in general agreement with this behaviour of Scots pine trees.

The diameter predictions are based on the core of the model MicroForest, that is, we obtain them in the simulations from the solution of the carbon and nitrogen balance equations. On the contrary, the height growths are based on the product of a varying empirical coefficient and the simulated diameter increments. Thus, the theoretical basis for height growth is rather poor. The rather large discrepancies between predicted and observed heights reflect the weak theoretical basis.

There are only weak systematic features in the residuals when the fertility of the site and stand density differ considerably from those at SMEAR II stand (Figs. 7.17 and 7.18). The residuals, as function of the initial state, reveal clear systematic behaviour when the initial diameter is less than 0.5 cm (Figs. 7.17 and 7.18). This is related to the determination of the tree sizes for the initial state. Annual ring width measurements are not reliable for four or five innermost annual rings, where the detection of ring boundary is often ambiguous.

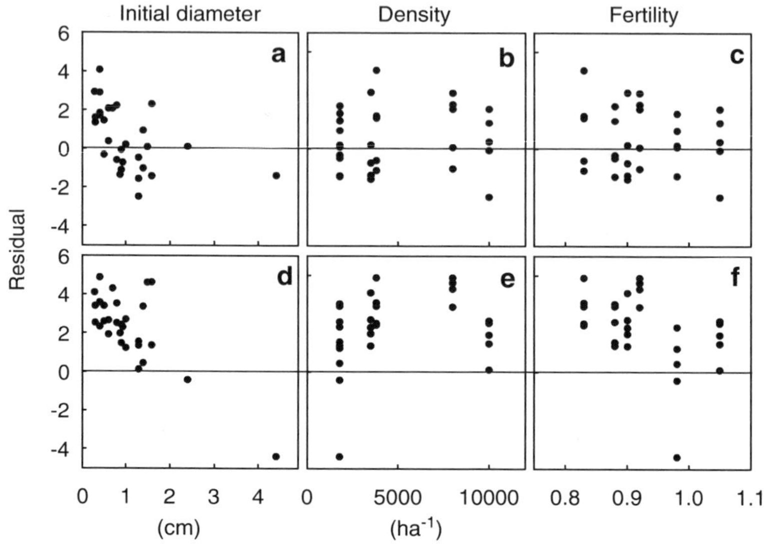

Fig. 7.17 Residuals of final diameter (*upper panels*) and final height (*lower panels*) against initial diameter, density and fertility in Hyytiälä

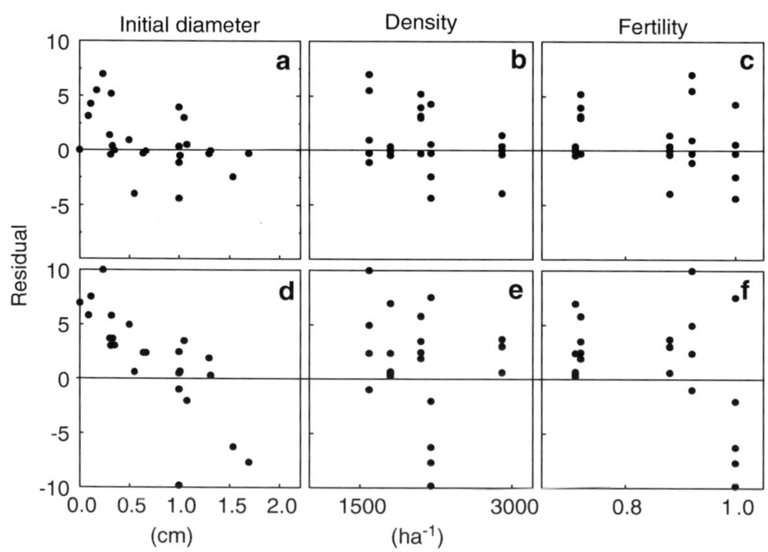

Fig. 7.18 Residuals of final diameter (*upper panels*) and final height (*lower panels*) against initial diameter, density and fertility in Estonia

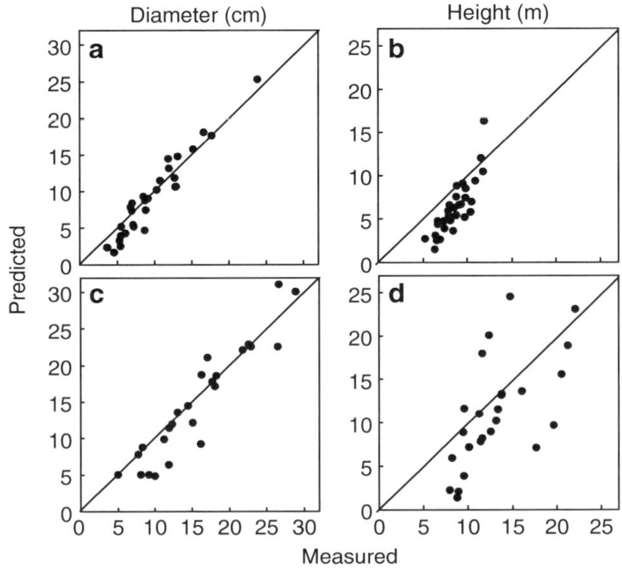

Fig. 7.19 Measured and modelled final diameter and height in Hyytiälä (*upper panels*) and in Estonia (*lower panels*)

Explaining power. The model MicroForest explains the development of carbon and nitrogen pools in the ecosystem with the carbon and nitrogen fluxes and with the structural regularities in the trees. These general principles together with more technical assumptions were able to predict the relevant features of the ecosystem dynamics, especially the growths of trees in size classes starting from small seedlings, indicating very high explaining and prediction power of the theory.

Although there are some systematic features in the residuals, MicroForest was able to explain the effect of site fertility and stand density in totally different conditions as the model was constructed and tuned.

The relationships between observed and predicted diameters at the end of the simulation period in trees from SMEAR II and from Estonia (Fig. 7.19) are close to each other. MicroForest can explain 90% of the variance in diameters in the test stand near Smear II station; in Estonia, the proportions are little lower 84% in diameter. The prediction of final heights (Fig. 7.19) is less successful but still rather good. The proportion of explained variance is near SMEAR II 53% and in Estonia 47%.

Comparison with other theories. MicroForest has several unique features: (1) it treats the main chemical properties of the trees, (2) the soil organic matter is treated utilising macromolecule classes, (3) it utilises emergent regularities in tree structure, (4) it includes ground vegetation and (5) it gives simultaneous solution of carbon and nitrogen balance equations. It represents a new development of the old carbon balance models of stand growth (de Wit et al. 1970).

We do not know any comparable tests of ecosystem theories starting from small seedlings and using data that is not utilised in the determination of the values of the parameters in the model.

Conclusion: MicroForest passed successfully all four aspects of a severe test and gained clear corroboration.

References

Aber J, Mellilo J, Nadelhofer J, Pastor J, Boone D (1991) Factors controlling nitrogen cycling and nitrogen saturation in northern temperate forest ecosystem. Ecol Appl 1:303–315

Bossel H (1991) Modelling forest dynamics: moving from description to explanation. For Ecol Manag 42:129–142

Criquet S (2002) Measurement and characterization of cellulase activity in sclerophyllous forest litter. J Microbiol Method 50:165–173

De Wit CT, Brouwer R, Penning de Vries FWT (1970) The simulation of photosynthetic systems. In: Setlik I (ed) Prediction and measurement of photosynthetic production. Proceedings of the IBP/PP Technical Meeting Trebon 1969, Pudoc, Wageningen. Centre for Agricultural Publishing and Documentation, Wageningen

Eivazi F, Bayan MR (1996) Effects of long-term prescribed burning on the activity of select soil enzymes in an oak-hickory forest. Can J For Res 2:1799–1804

Ek AR (1971) Size-age relationships for open grown red pine, vol 156, Forest research notes. University of Wisconsin, Madison

Hari P, Mäkelä A (2003) Annual pattern of photosynthesis of Scots pine in the boreal zone. Tree Physiol 23:145–155

Hari P, Kellomäki S, Mäkelä A, Ilonen P, Kanninen M, Korpilaahti E, Nygren M (1982) Dynamics of early development of tree stand. Acta For Fenn 177:1–39

Hari P, Kaipiainen L, Korpilahti E, Mäkelä A, Nilsson T, Oker-Blom P, Ross J, Salminen R (1985) Structure, radiation and photosynthetic production in coniferous stands, vol 54, Department of silvicultural research notes. University of Helsinki, Helsinki, pp 1–233

Hari P, Raivonen M, Vesala T, Munger JW, Pilegaard K, Kulmala M (2003) Ultraviolet light and leaf emission of NOx. Nature 422:134

Hari P, Salkinoja-Salonen M, Liski J, Simojoki A, Kolari P, Pumpanen E, Kähkönen M, Aakala T, Havimo M, Kivekäs R, Nikinmaa E (2008) Growth and development of forest ecosystems; the MicroForest model. In: Hari P, Kulmala L (eds) Boreal forest and climate change, vol 34, Advances in global change research. Springer, Dordrecht

Havimo M, Kivekäs R, Aalto J, Schiestl P, Hari P (2008) Test with six stands near SMEAR II. In: Hari P, Kulmala L (eds) Boreal forest and climate change, vol 34, Advances in global change research. Springer, Dordrecht

Helmisaari H-S (1992) Nutrient retranslocation within the foliage of *Pinus sylvestris*. Tree Physiol 10:45–58

Ilomäki S, Mäkelä A, Nikinmaa E (2003) Crown rise due to competition drives biomass allocation in silver birch (*Betula pendula* L.). Can J For Res 33:2395–2404

Johnson D (1999) Simulated nitrogen cycling response to elevated CO_2 in *Pinus taeda* and mixed deciduous forests. Tree Physiol 19:321–327

Köster K, Kangur A, Hari P, Jõgiste K (2008) Test in Estonia at the southern border of the boreal zone. In: Hari P, Kulmala L (eds) Boreal forest and climate change, vol 34, Advances in global change research. Springer, Dordrecht

Linkins AE, Sinsabaugh RL, NcClugherty CA, Melillo JM (1990) Comparison of cellulase activity in decomposing leaves in a hardwood forest and woodland stream. Soil Biol Biochem 22: 423–425
Liski J, Ilvesniemi H, Mäkelä A, Starr M (1998) Model analysis of the effects of soil age, fires and harvesting on the carbon storage of boreal forest soils. Eur J For Sci 49:407–416
Mäkelä A (1985) Differential games in evolutionary theory: height growth strategies of trees. Theor Popul Biol 27:239–267
Mäkelä A (1988) Models of pine stand development: an eco-physiological systems analysis, vol 62, Department of silvicultural research notes. University of Helsinki, Helsinki, pp 1–267
Mäkelä A (1997) A carbon balance model of growth and self-pruning in trees based on structural relationships. For Sci 43:7–24
Mäkelä A (2002) Derivation of stem taper from the pipe theory in a carbon balance framework. Tree Physiol 22:891–905
Mäkelä A, Hari P (1986) Stand growth model based on carbon uptake and allocation in individual trees. Ecol Model 33:205–229
Mäkelä A, Mäkinen H (2003) Generating 3D sawlogs with a process-based growth model. For Ecol Manag 184:337–354
Mäkelä A, Sievänen R (1992) Height growth strategies in open-grown trees. J Theor Biol 159: 443–467
Mäkelä A, Vanninen P (1998) Impacts of size and competition on tree form and distribution of above ground biomass in Scots pine. Can J For Res 28:216–227
Mäkelä A, Hari P, Berninger F, Hänninen H, Nikinmaa E (2004) Acclimation of photosynthetic capacity in Scots pine to the annual cycle of temperature. Tree Physiol 24:369–376
Mohren GMJ (1987) Simulation of forest growth, applied to Douglas fir stand in the Netherlands. Dissertation, Agriculture University of Wageningen
Nikinmaa E (1992) Analyses of the growth of Scots pine; matching structure with function. Acta For Fenn 235:1–68
Nissinen A, Hari P (1998) Effects of nitrogen deposition on tree growth and soil nutrients in boreal Scots pine stands. Environ Pollut 102:61–68
Pavel R, Doyle J, Steinberger Y (2004) Seasonal patterns of cellulase concentration in desert soil. Soil Biol Biochem 36:549–554
Penning de Vries FWT (1975) The cost of maintenance processes in plant cells. Ann Bot 39:77–92
Penning de Vries FWT, Brunsting AHM, van Laar HH (1974) Products, requirements and efficiency of biosynthesis: a quantitative approach. J Theor Biol 45:339–377
Perttunen J, Sievänen R, Nikinmaa E, Salminen H, Saarenmaa H, Väkevä J (1996) LIGNUM: a tree model based on simple structural units. Ann Bot 77:87–98
Rastetter E, Ryan M, Shaver G, Mellilo J, Nadelhoffer K, Hobbie J, Aber J (1991) A general biochemical model describing the responses of the C and N cycles in terrestrial ecosystems to changes in CO_2, climate and N deposition. Tree Physiol 9:101–126
Running S, Gover S (1991) FOREST-BGC, A general model of forest ecosystem processes for regional applications. II Dynamic carbon allocation and nitrogen budgets. Tree Physiol 9: 147–160
Saito T, Hong P, Kato K, Okazaki M, Inagaki H, Maeda S, Yokogava Y (2003) Purification and characterization of an extracellular laccase of fungus (family Chaetomiaceae) isolated from soil. Enzym Microb Technol 33:520–526
Tomme P, Warren AJ, Gilkes NR (1995) Cellulose hydrolysis by bacteria and fungi. Adv Microb Physiol 37:2–81
Valentine HT (1985) Tree-growth models: derivations employing the pipe model theory. J Theor Biol 117:579–584
Waldrop MP, Zak DR, Sinsabaugh RL (2004) Microbial community response to nitrogen deposition in northern forest ecosystems. Soil Biol Biochem 36:1443–1451

Watson M, Casper BB (1984) Morphogenetic constraints on patterns of carbon distribution in plants. Annu Rev Ecol Syst 15:233–258
Wittmann C, Kähkönen MA, Ilvesniemi H, Kurola J, Salkinoja-Salonen MS (2004) Areal activities and stratification of hydrolytic enzymes involved in the biochemical cycles of carbon, nitrogen, sulphur and phosphorus in podsolized boreal forest soils. Soil Biol Biochem 36:425–433
Ylinen A (1952) Über die mechanische Schaftformtheorie der Bäume. Silv Fenn 76:1–52

Chapter 8
How to Utilise the Knowledge of Causal Responses?

Pertti Hari, Mikko Havimo, Heljä-Sisko Helmisaari, Liisa Kulmala, Eero Nikinmaa, Timo Vesala, Jouni Räisänen, Tuukka Petäjä, Erkki Siivola, Heikki Tuomenvirta, Jaana Bäck, John Grace, Federico Magnani, Twan van Noije, Jukka Pumpanen, David Stevenson, Markku Kulmala, Sampo Smolander, Ilona Riipinen, and Miikka dal Maso

Contents

Abstract Our physical and physiological theory provides causal explanations of various phenomena in forests. This causal nature of the theory enables versatile applications in forestry and in the research of the interactions between climate change and forests. We treat the effects of thinnings and whole-tree harvesting on wood production and the responses of forest ecosystem to nitrogen deposition in more detail. The forests react to the increasing CO_2 concentration and also to temperature increase generating feedbacks from forests to climate change. The changes in the carbon storages in forest ecosystems and in the emission of volatile organic compounds are evidently the most important feedbacks from forest ecosystems to the climate change.

P. Hari (✉)
Department of Forest Sciences, University of Helsinki, P.O. Box 27, 00014 University of Helsinki, Helsinki, Finland
e-mail: pertti.hari@helsinki.fi

P. Hari et al. (eds.), *Physical and Physiological Forest Ecology*,
DOI 10.1007/978-94-007-5603-8_8,

Keywords Forestry • Thinning • Whole-tree harvesting • Climate change • Response of forest ecosystem • Feedback to climate change

8.1 Introduction

Our theories provide causal explanations of many phenomena in forest ecosystems based on metabolism of living organisms, on transport of material and energy and on structural regularities developed during evolution. These causal explanations enable exploration of a wide range of applications in forestry and facilitate research on the effects of climate change on forests.

The extinction of light in the canopy conveys the principal interaction between trees. This opens a fruitful way to explain the growth of dominating and suppressed trees in a stand and also the effect of stand density on growth of individual trees. Nitrogen in many chemical forms within trees is removed from the stand by the practice of whole-tree harvesting, thus reducing the amount of reactive nitrogen in the remaining stand and reducing the tree growth in the stand. Anthropogenic nitrogen deposition increases the amount of reactive nitrogen in forest ecosystems resulting in a slow increase in the fertility of the sites.

The present climate change effects on processes in forest ecosystems, that is, increasing CO_2, enhance photosynthesis while increasing temperature accelerates decomposition of proteins in the soil, and at the same time, nitrogen deposition increases reactive nitrogen in forests. All these changes increase forest growth. The changes in the processes are also reflected into the feedbacks from forests to climate change since they affect the atmospheric CO_2 and aerosol concentrations.

8.2 Applications to Forestry

Pertti Hari, Mikko Havimo, Heljä-Sisko Helmisaari, Liisa Kulmala, and Eero Nikinmaa

8.2.1 *The Effect of Thinnings on Wood Production*

The influence of stand density on timber dimensions and production is one of the main questions to optimise in forest management. The large but fluctuating price difference between that of sawn wood and pulpwood guarantees that the question

P. Hari (✉)
Department of Forest Sciences, University of Helsinki, P.O. Box 27, 00014 University of Helsinki, Helsinki, Finland
e-mail: pertti.hari@helsinki.fi

remains a valid driver for management decisions. The increasing use of wood as a renewable source of energy is expanding again the usage and dimensions of forest products, and it is expected that market prices will change. These considerations will also influence optimal management regimes, not to mention the impact that the changing climate will have.

The main approach to study the effects of thinnings on forest growth has been observational, relying on permanent and semipermanent sample plots. Permanent sample plots have been established at varying site fertilities, tree species, species mixtures, etc., and valuable results have been obtained. However, the interpretation of the results is problematic since by necessity the experiments are long-lasting, and during the passage of time, the relevant research questions may well have changed, hence making the results somewhat redundant.

Statistical methods have been the main tool in growth studies, and regression models have been developed to link directly the tree dimensions and stand level variables with growth. These models give rather good predictions of growth in the time span of a couple of decades. The statistical approach has also been used to study the effect of stand density on growth. The idea there is that the stand growth conditions and competition between trees can be described with stand level variables such as dominant height and stand density, and the response of any individual tree would be reflected in its own dimensions. However, there are both theoretical and methodological problems involved, particularly when simulating the influence of actions that drastically alter the stand conditions such as thinning.

Theoretically, it is problematic to explain how stand basal area or any other tree characteristic of tree stems has a causal effect on growth; the observed correlations are only indicators of the connections between the stand dimensions and growth under the specified conditions. The explaining factors in a causal regression model should be independent of each other when the model is used as causal explanation. However, the stand characteristics are strongly correlated. This correlation of the variables can be utilised in the growth analysis under regular conditions, but during thinning, the nature of the correlation changes, and this makes the use of regression models in the analysis of the effects of thinning and initial density on tree growth very problematic. To be able to account for the thinning impacts with regression type analysis, we would need to have a very large set of treatments that would cover the varying initial densities in different ages, treatment strengths and growth responses, which is experimentally costly and requires observations that are very difficult to carry out. Even in this case, we are facing the problem of connections between explaining variables that are problematic for the causal use of the regression model.

In the true causal approach, the growth results from the underlying combination of environmental conditions, the metabolic and structural state of the trees and the different processes leading to tree growth. The interactions between trees reduce the growth of individual trees from what it could be if those trees were grown without competition. The main mechanisms to decrease growth of individual trees are the extinction of light in the canopy and the consequent reduced photosynthesis, especially in the lower parts of the crown, and also the low availability of nitrogen in the soil. When the stand is thinned, then the needles of the remaining trees receive more light than previously resulting in increased photosynthesis and growth in the remaining trees. In addition, the thinning increases nitrogen availability that contributes to the enhanced growth. These changes also influence the allocation of growth within trees as the time constants for additional light and nitrogen release are different, thus influencing growth allocation between root and shoot growth. Also the improved light climate of the crown influences the allocation of growth between stem and leaves. These causal mechanisms of the effect of thinning on tree growth are built into MicroForest which deals with the carbon and nitrogen fluxes and with the allocation of the obtained resources within the tree. Thus, we can analyse the responses of tree growth to thinning using MicroForest.

8.2.1.1 Simulations

The stand around SMEAR II measuring station (Section 9) is an even-aged and artificially regenerated Scots pine forest. We utilise MicroForest, and we start from the initial state when the seedlings are small, less than 1 m tall, and simulate the stand development for 80 years at varying thinnings. In order to show the potential for management purposes, we treat separately sawn wood and pulpwood.

We used the measured diameters and heights in each size class at the age of 5 years as the initial state of the stand in the simulations. We divided the stand into five size classes, and the shares of the number of stems in the five size classes were 5, 15, 30, 30 and 20% starting from the biggest size class. We used the measured values at SMEAR II as the initial state of the soil organic matter. The number of seedlings in the initial state was 2,000 per ha.

We applied two thinnings from below in the simulations. The thinnings occurred at the ages of 35 and 55 years, a common thinning schedule for Scots pine forest in south Finland. We varied the intensity in both thinnings in such way that the number of remaining trees increased at 5% interval starting from 15 to 100%.

We classified the wood obtained in the thinnings into pulpwood and sawlogs using the commercial minimum dimensions for pulpwood and sawlogs.

8.2.1.2 Results

The thinnings reduce the stand needle mass, especially in the heavy treatments (Fig. 8.1a). The recovery of the needle mass depends both on the thinning intensity and the stand age. If about 50% of the trees in the stand are thinned, the foliage recovers in about 20 years to the value it had before thinning (Fig. 8.1a). In more moderate thinnings, the recovery takes place in about 10 years in the simulations. The annual flux of ammonium ions from soil proteins caused by microbial decomposition, nitrogen deposition and fixation determines the annual amount of proteins synthesised for the growth of new tissues and in this way the growth of needle mass.

The simulated response in photosynthesis is clearly smaller than the reduction in the needle mass. The shading of the remaining trees by neighbouring trees is less

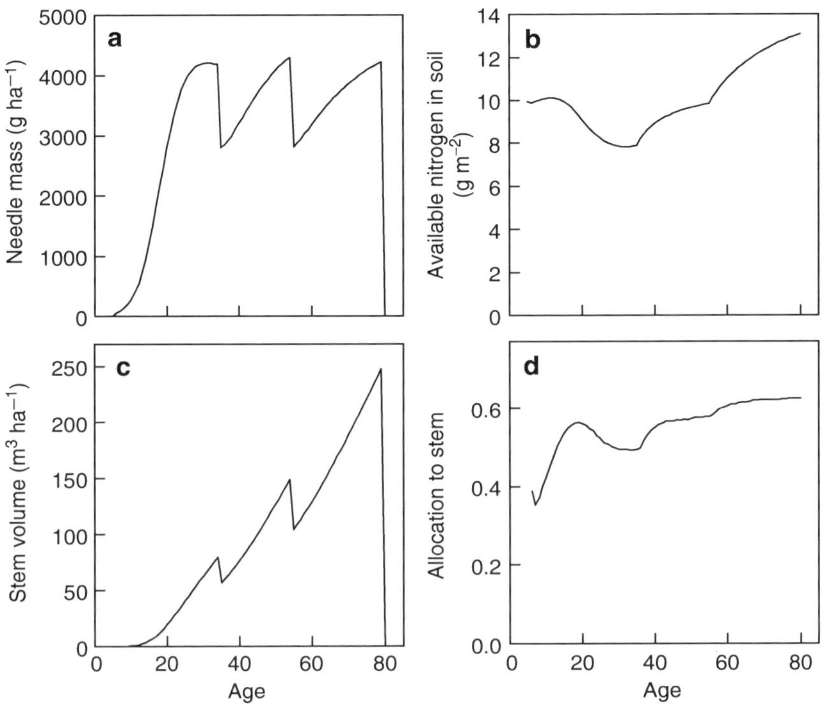

Fig. 8.1 The simulated needle mass (**a**), available nitrogen in the soil (**b**), stand volume (**c**) and allocation to stem (**d**) as function of stand age. Each thinning from below at the age of 35 and 55 years removes 50% of the trees

severe, and so the needles lower in the canopy can photosynthesise more after the thinning than before it. No reduction in the annual amount of photosynthesis was observed immediately after experimental thinning at SMEAR II site in which about 50% of the stem number and about 25% of the stem volume was removed (Vesala et al. 2005). These authors argued that the result was both due to remaining trees and ground vegetation compensating for the decreased foliage and reduction in the root respiration rate associated with lowered biomass.

The thinnings also affected the nitrogen availability in the soil (Fig. 8.1b). When thinnings reduce the simulated needle mass, then the nitrogen uptake of trees is also reduced. The increased growth of ground vegetation is not able to compensate for the reduction in the nitrogen uptake by trees, and the pool of available nitrogen increases slightly. Thus, the thinnings have two effects: increased photosynthetic production and increased nitrogen availability.

The simulated stand volume responded to the thinnings; however, the reduction in volume was clearly smaller than in the number of trees (Fig. 8.1c). This is as expected since the thinnings removed the smaller trees. The volume growth decreased slightly during the first 10 years after thinning, the more as more trees

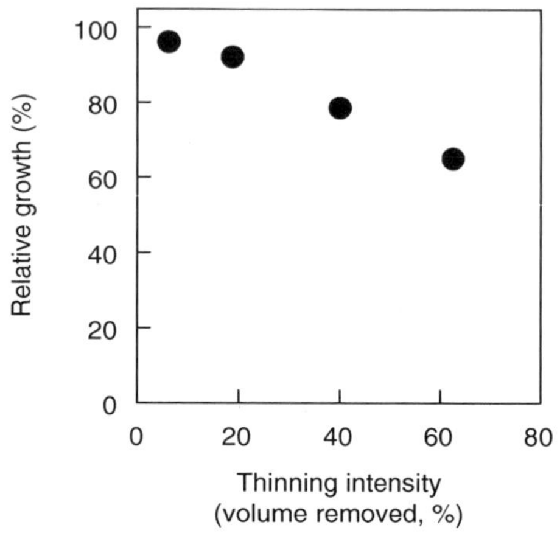

Fig. 8.2 Relative volume growth as function of thinning intensity according to Eriksson and Karlsson (1997)

were removed. After 10–20 years from thinning, the growth recovered completely to the pre-thinning growth rate with exception of the most extreme case of 60% tree removal, in which case the growth was still only about 90% of the pre-thinning value. Thus, although the thinnings reduced the needle mass and the annual photosynthetic production, the trees were able to produce nearly the same amount of wood. The growth immediately after thinning corresponded very closely to observations made in similar age pine stands (Eriksson and Karlsson 1997) (Fig. 8.2). However, the model predicted somewhat better recovery of growth on the long run than was observed.

The allocation of sugars and nitrogen is very demanding task for the biochemical regulation system. The observed regularities in tree structure indicate that trees are able to solve the allocation problem. The thinnings affect allocation in two different ways, the increased nitrogen availability reduces the share of roots and the increasing needle mass per tree in the remaining stand requires more water transport capacity in the stem. The increase in nitrogen availability in soil takes years, and the response of root allocation is also slow. The simulated allocation to stem (Fig. 8.1d) indicates rapid and clear response to the thinnings as we expected from the theoretical argumentation. The increased allocation to stem growth explains the strange result that thinnings reduce annual photosynthetic production, but the volume growth of the stand remains unaltered after recovering from the initial thinning shock.

The simulated response of pulpwood production to thinnings during the rotation period was rather small. Heavy thinnings increased pulpwood production, mainly obtained in first thinning (Fig. 8.3a). The response was, however, rather small, only about 20 m^3.

The thinnings increased slightly the simulated log production up to a value of about 300 m^3 (Fig. 8.3b) if the remaining number of trees was more than 40% of the situation before thinning or the removed volume was more than 30% of the stand volume. The 60% reduction in trees by thinning resulted in only a 30% decrease

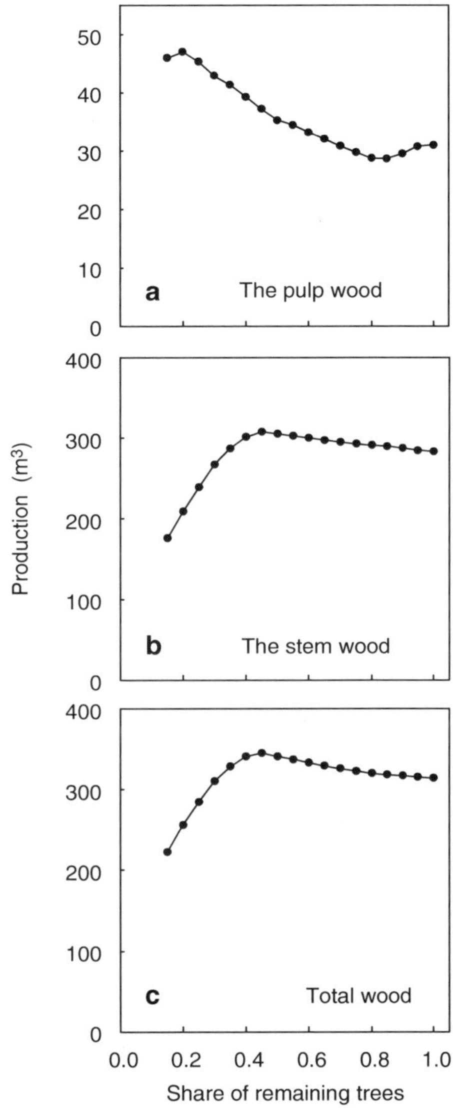

Fig. 8.3 The pulpwood production (**a**), stem wood production (**b**) and total wood production (**c**) during the rotation period as function of the share of remaining trees in the thinnings

in tree volume. Heavier thinnings started to reduce the simulated log production considerably. There is maximum in the total wood production when 40% of the trees remain after thinning (Fig. 8.3c). We have omitted for simplicity the natural mortality in the simulations.

Simulating the stand level responses to management operations has been one of the cornerstones of forest sciences. Determination of optimal rotation periods and timing of intermediate cuttings, not to mention the selection of most appropriate silvicultural methods, require that we are able to estimate properly the forest growth responses. Various different growth simulators have been developed, and overall, they have served well these purposes (Hyytiäinen et al. 2005). The problem arises in using models that are based on empirical growth function when one needs to extrapolate the growth for conditions that have not been covered in the combinations of conditions used to make these models. There is a possibility for a very unrealistic behaviour, and thus quite often artificial limitations are needed to force the model outcome, and the resulting model behaviour does not necessarily correspond to natural behaviour. This poses problems, particularly if one would want to use mathematical tools to analyse new type of optimal solutions for forest management (Niinimäki et al. 2012). Due to the many degrees of freedom that there are when planning and executing intermediate cuttings over the stand rotations, the problem of extrapolation is easily encountered in connection to stand responses to thinning.

The process-based approach that is based on causal relationships, and that we have used here, has the benefit that the growth of trees is always constrained by the material balances, and grossly unreasonable behaviour is not possible once the specific process rates and stand level interactions are satisfactorily described. The good comparability between simulated and observed stand growth and reasonable behaviour in these initial tests to simulate thinning responses indicate that we are quite close to reaching this situation. The growth behaviour immediately after thinning corresponded to observations of long-term thinning experiments (Eriksson and Karlsson 1997). However, our model predicted faster growth recovery on the long run than was observed. One of the reasons may be that in current simulations, we assumed that nitrogen leaching from the site is negligible. Experiments have shown that removal of logging residue from forests may cause considerable long-term growth reduction (Helmisaari et al. 2011). Similarly, nitrogen leaching from abundant logging residue would produce similar outcome when compared to no leaching assumption.

The many degrees of freedom available when performing thinning make it difficult to describe the responses with traditional methods of forest growth and yield. However, it is also a very challenging task for process-based modelling. As previous chapters have indicated, metabolism, transport and growth are highly integrated phenomena and often produce very tightly confined relationships between process rates and structural quantities, properties and dimensions. These features are widely exploited in the model MicroForest. However, at thinning, growth conditions are abruptly changed; they necessarily trigger a chain of acclimations at different temporal and spatial domains in the remaining trees.

Current results suggest that process-based models, such as MicroForest, are very valuable tools among others for planning optimal forest management. They are capable of simulating that type of information that is required for comparison of different management regimes. The very valuable feature of such models is that the material balances of the forests restrict and constrain the simulations, and therefore completely unrealistic behaviour is impossible in new conditions. This makes dynamic models of forest growth based on carbon and nitrogen balance particularly interesting tools for climate change studies.

8.2.2 The Effect of Whole-Tree Harvesting on the Site Fertility

8.2.2.1 Whole-Tree Harvesting in Forestry

The problem of global warming caused by increasing atmospheric CO_2 concentration stresses the importance finding other energy sources than fossil fuels. Bioenergy, especially the burning of wood and other carbon material obtained from forests has been a leading alternative to fossil fuels. The wood in the stems of trees dominates the biomass in forests, but also the masses of branches and leaves are considerable, and the energy yield can be maximised by whole-tree harvesting. This involves removal of all the above-ground biomass for use in power plants.

Wood has been harvested for centuries from forests, and no evident problems have emerged with the nutrients being removed in the logs. On the other hand, the litter collection from forests to agricultural land has had a negative effect on tree growth, documented centuries ago (see Ebermayer 1876). These facts lead us to the question, does whole-tree harvesting reduce the fertility on the site due to export of nutrients in the leaves and branches?

The model MicroForest is based on the carbon and nitrogen fluxes and pools in a forest ecosystem. Thus, it can be used to analyse the changes, generated by whole-tree harvesting, in the various nitrogen pools and also in stand growth. MicroForest also explains the development of the pools and growth enabling understanding of the changes caused by whole-tree harvesting in a forest ecosystem.

The Finnish Forest Research Institute established experiments 20–30 years ago to study the effects of the whole-tree harvesting on stand growth. The 14 experiments represented a range of site types. The experimental design comprised thinning treatments with or without logging residue removal at establishment. The thinning grade was 16–39% of the standing tree volume. Conventional harvesting of commercially exploitable stem wood with bark served as a control, with logging residues left in the stand and evenly distributed on the plots. In the whole-tree harvesting treatment, the residues were manually removed. At thinning, between 21 and 73 kg N ha^{-1} was removed in logging residues in pine stands and between 58 and 129 kg N ha^{-1} in spruce stands (Helmisaari et al. 2011).

Statistical analysis of the whole-tree harvesting experiments indicated that during the first 10-year period after thinning, the average volume increment was significantly lower on the whole-tree harvesting plots in the spruce stands and tended to be lower ($p = 0.06$) in the pine stands, while in the second 10-year period, the growth decrease was significant also in the pine stands. Volume increment of pine stands was 4 and 8% and that of spruce stands 5 and 13% lower on the whole-tree harvesting plots than on the conventional harvesting plots during the first and the second 10-year periods, respectively. The results confirmed the main hypothesis that whole-tree harvesting decreases tree growth in relation to conventional harvesting, and this change especially depends on the amount of nitrogen removed in logging residues (Helmisaari et al. 2011).

Simulations with MicroForest enables deeper insight into the changes in a forest ecosystem, and this is why we made a complementary analysis of the nitrogen fluxes and pools in stands that are treated with whole-tree harvesting. We think that the nitrogen in the harvested needles and small branches may reduce the pool of reactive nitrogen circulating in the ecosystem, and in this way, whole-tree harvesting may generate growth reduction.

8.2.2.2 Analysis of the Whole-Tree Harvesting with MicroForest

We selected the experimental plot 726 of the Finnish Forest Research Institute for closer study with MicroForest. We used the model developed to describe the stand around SMEAR II measuring station and tested and tuned it using data from additional stands nearby SMEAR II and in Estonia. We utilised the values of the parameters obtained from the measurements at SMEAR II, and we applied the model at the experimental plot number 726 without fitting.

The experiment 726 was established when the stand was 24 years old; before that age, no exact information is available dealing with the stand development. We have the diameters of the trees at the age of 24 years. We ordered the trees according the diameters in the first measurement, and we formed five size classes as in the tests nearby SMEAR II and Estonia, 5% of the biggest trees formed the size class 1, next 15% the class 2, next 30% the class 3, next 30% the class 4 and the remaining 20% the smallest size class 5.

The nitrogen deposition has increased during the later part of the twentieth century, and we assumed that the nitrogen deposition was 0.1 g m^{-2} $year^{-1}$ in the beginning of the simulation and grew linearly to the value 0.4 g m^{-2} $year^{-1}$.

The initial state of the trees in each size class and the protein pool in the soil is needed for the simulations, and this is a bit problematic since about 600 numbers are needed to describe the needle masses, stems and branches and proteins in the soil in MicroForest. We have only the measurements of the stem diameters to determine the initial state. The most practical way to obtain the initial values is to generate them with MicroForest. We varied the initial stem diameters at the age of 4 years and the protein pool in the soil in such a way that simulations with MicroForest resulted into the measured diameters in each size class at the age of 24 years. Then the state of the simulation at the age of 24 years formed all the needed initial values as a coherent set of numbers.

We simulated the stand development during the years 1979–2009 and compared the simulations with the measured stem diameters (Fig. 8.4). The model MicroForest predicted the diameters in each size class rather successfully, and further analysis of the effects of whole-tree harvesting on tree growth with MicroForest seems to be justified.

MicroForest calculates all pools of reactive nitrogen in a forest ecosystem. The amount of reactive nitrogen in the soil determines the fertility of the site since the flux of nitrogen generated by the decomposition of proteins determines the amount of synthesised proteins for the new tissues, especially in the leaves and fine roots.

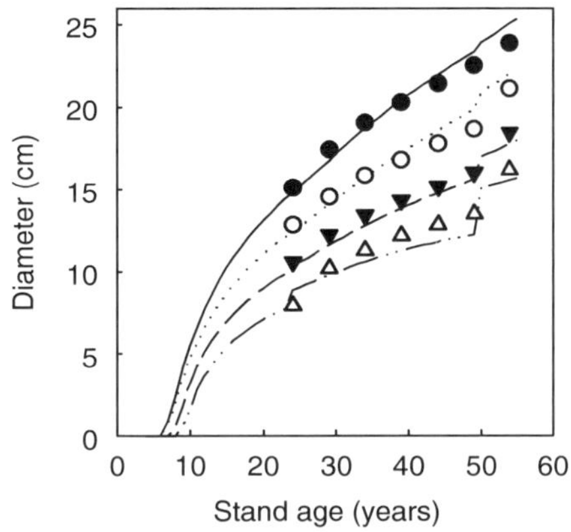

Fig. 8.4 Measured (lines) and predicted (symbols) diameters at the experimental stand 726

The deposition is an additional new source of reactive nitrogen in forest ecosystems. The simulations with MicroForest enable quantification of the reduction in the nitrogen flux from proteins in the soil caused by whole-tree harvesting, and in this way, we obtain also the changes in the growth of the stand.

The proteins in the soil dominate the nitrogen pool in a forest ecosystem; its amount in the simulated initial state at the age of 24 years was 1,300 g m^{-2}. The needle mass changes strongly during the early phase of stand development. When the experiment was established, the simulated needle mass in the stand was stabilising to 400 g m^{-2} containing proteins at about 35 g m^{-2}. The large stem mass in the late phase of stand development makes the protein pool in wood considerable, although the protein concentration is low; this nitrogen is permanently bound into the wood and not circulating in the system. The wood pool was 10 kg m^{-2}, and the protein pool in the wood was 16 g m^{-2} in the beginning of the experiment. The proteins in the bark are missing in MicroForest. The simulated nitrogen flux caused by the decomposition of proteins is 12 g m^{-2} $year^{-1}$ enabling synthesis of 80 g m^{-2} $year^{-1}$ of proteins.

The thinnings at the age of 24 and 49 years of the experiment were rather slight, resulting in about a 35% decrease in the first and 25% in the second thinning in the simulated stand needle mass (Fig. 8.5).

The simulated loss of reactive nitrogen in whole-tree harvesting in the age of 24 years at the experimental plot is 12 g m^{-2}. The observed value of total nitrogen loss was 4.6 g m^{-2} in logging residues and 1.4 g m^{-2} in stem wood. The simulated reduction of the reactive nitrogen in trees and in soil is about 1%.

The simulated effect of whole-tree harvesting on the volume growth was rather small, only about 3% after the first thinning and about 4% after the second one. The comparison with the experimental plots showed a somewhat larger total effect than the empirically measured one. The effect of whole-tree harvesting in this experiment was a 3% reduction in the volume increment during 10–20 years after thinning. During the first 10 years after thinning, there was not yet any effect of whole-tree harvesting.

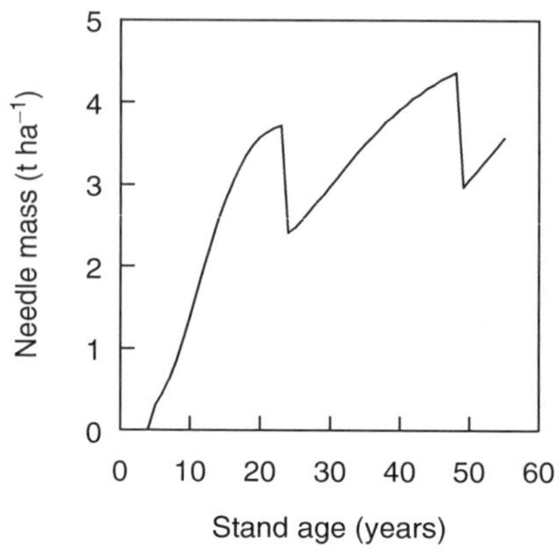

Fig. 8.5 Simulated needle mass in the experimental plot 726

8.2.3 *The Effect of Nitrogen Deposition on Forest Production*

8.2.3.1 Nitrogen Deposition

The nitrogen circulates between vegetation and soil in a forest ecosystem. The nitrogen deposition, fixation, leaching and evaporation generate small nitrogen fluxes between the ecosystem and its surroundings. The natural flux from the atmosphere to forests has evidently been rather small and stable for millions of years, below 0.1 g m^{-2} year^{-1} (1 kg ha^{-1} year^{-1}). The nitrogen fixation is rather weakly known in quantitative terms; however, particularly in northern latitudes, it is also small and of the same magnitude as natural deposition.

The usage of fossil fuels and chemical fertilisers introduced an additional source of reactive nitrogen into the atmosphere. The lifetimes of nitrogen oxides (NO and NO_2) and ammonium in the atmosphere are rather short, only days, and they deposit within some thousands of kilometres from their emission location. Most of the anthropogenic deposition is chemically as ions and readily available for vegetation as a nutrient.

The usage of fossil fuels and chemical fertilisers were rather small before 1950; thereafter, a rapid increase took place, and stabilisation occurred around 1990 in industrialised countries. China and India have dominated the increase in the utilisation of fossil fuels during the last decades. However, increasing traffic continues to be an important and growing nitrogen source also in Europe.

The flux of reactive nitrogen from the atmosphere into forests is rather small when compared with those of water and carbon. There are severe measuring problems involved in the determination of the nitrogen flux from the atmosphere to forests. The wet deposition can be determined from rainwater, but dry deposition is still rather problematic. Small aerosol particles are rich in nitrogen, and they deposit on needles and leaves in forests (Sect. 8.3.6.2) resulting in dry deposition. But that is very problematic to measure. In addition, gaseous nitrogen oxides can deposit directly into leaves (Sect. 4.2.8).

Despite the evident measuring problems, we have rough global picture of the development of nitrogen deposition since 1960 (van Aardenne et al. 2001). The patterns in the development of nitrogen deposition are quite similar in different areas on the globe, a rapid increase, and thereafter, stabilisation reflects the usage of fossil fuels in the industrialised countries. The level of stabile deposition is much higher in the industrialised areas when compared with remote areas far in the north Siberia or Canada. This historical pattern of nitrogen deposition is different in China and India.

The rapid increase in the nitrogen deposition raises the question of the importance of the additional reactive nitrogen in forest ecosystems. The fluxes, and especially nitrogen flux, play an important role in our theory and also in our model MicroForest. Consequently, we use it as a tool to evaluate the role of nitrogen deposition in forests.

8.2.3.2 Simulations of the Effects of Nitrogen Deposition on Forest

We consider in more detail two alternatives: the responses to different increases in nitrogen deposition during the last part of the twentieth century and the response to a step increase in nitrogen deposition.

The patterns of the increase in the nitrogen deposition were similar around the industrialised centres in Europe and America: increase starting around 1950 and stabilisation about 1980 or 1990. We simulate three different deposition patterns:

1. Stable deposition of 0.1 g m^{-2} $year^{-1}$
2. Linear increase from 0.1 to 0.5 g m^{-2} $year^{-1}$ during four decades after 1950 and thereafter stable deposition level
3. Linear increase from 0.1 to 2.0 g m^{-2} $year^{-1}$ during four decades after 1950 and thereafter stable deposition level

The first case refers to the situation before anthropogenic nitrogen deposition, the second one is close to the deposition pattern at SMEAR II station and the third one is possible in Central Europe.

We assumed that the stand was born as even aged in the year 1950 and that the initial state at the age of 4 years was as we measured at Järvselja (Sect. 7.4). We omitted thinnings in our simulations.

The outcome of the simulations is clear. The deposition provides an additional source of nitrogen for the synthesis of proteins, especially in needles and fine roots, and the response is rather large and versatile; the needle mass and stem volume increase (Fig. 8.6). The response saturates at the high deposition and in contrast, the needle mass increases up to the age 80 years in the moderate deposition and the needle mass is rather stable in the case of no anthropogenic deposition. The responses in the stand volume are still larger than in the needle mass.

The responses in needle mass and in photosynthesis are fairly small when compared with the stem volume. The nitrogen deposition generates large changes in allocation, and this change is the key to understand the obtained simulation. The fine

Fig. 8.6 The simulated needle mass (**a**) and stand volume (**b**) as function of stand age at varying nitrogen deposition. We simulated three deposition patterns: (i) stable deposition 0.1 g m^{-2} $year^{-1}$, (ii) linear increase from 0.1 to 0.5 g m^{-2} $year^{-1}$ during four decades and there after stable 0.5 g m^{-2} $year^{-1}$ and (iii) linear increase from 0.1 to 2.0 g m^{-2} $year^{-1}$ during four decades and there after stable 2.0 g m^{-2} $year^{-1}$

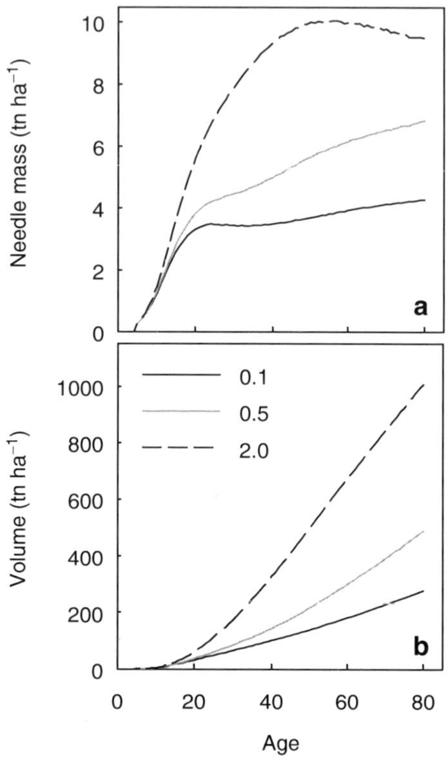

roots and mycorrhiza are the biggest single sink of sugars formed in photosynthesis if the nitrogen deposition is 0.1 g m^{-2} $year^{-1}$, and if the nitrogen deposition is as high as 2 g m^{-2} $year^{-1}$, then the share of roots and mycorrhiza is negligible. This simulation result comes from the fundamental logic of MicroForest. When the deposition increases the availability of nitrogen, the solution of the nitrogen and carbon balance equations results in smaller allocation to roots and mycorrhiza, and the remaining sugars can be used for stem and needle growth.

Our simulations omit the increase in leaching due to the nitrification and gaseous nitrogen losses due to denitrification. Thus, our simulations may result too high response at high deposition.

The simulations that mimic the observed and expected nitrogen deposition indicated that the forests can response with increasing growth to stable deposition for decades. This raises the question, how long are the forests able to accelerate their growth under stable anthropogenic nitrogen deposition?

We simulated the response of forests' ecosystem to step increase in deposition to find the time constant of the response. We assumed stable low nitrogen deposition, that is, 0.1 g m^{-2} $year^{-1}$ for the first 30 years of the ecosystem development, and after the year 30, the deposition was constant either 0.5 or 2 g m^{-2} $year^{-1}$. For comparison, we simulated also the case of stable nitrogen deposition (Fig. 8.7).

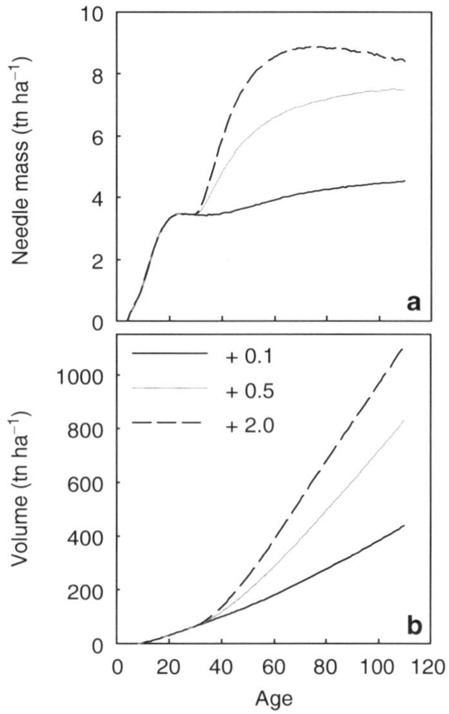

Fig. 8.7 The simulated needle mass (**a**) and stand volume (**b**) as function of stand age at different levels of step increase in nitrogen deposition at the age of 30 years. We applied three increases: (i) from 0.1 to 2.0 g m^{-2}, (ii) from 0.1 to 0.5 g m^{-2} and (iii) stable 0.1 g m^{-2} for comparison

The reactive nitrogen accumulated nearly 50 years into the ecosystem at low anthropogenic deposition, and the release of nitrogen from proteins in soil indicates stabilisation 50 years after the beginning of the additional nitrogen deposition. The stabilisation of the proteins in the soil is reflected also to other components in the system.

At high deposition, the response saturated clearly faster than at low level. On average, the time constant of the response to sudden increase in the nitrogen deposition is roughly about 50 years, at low deposition longer and at high shorter than that. The increased anthropogenic nitrogen deposition stabilised to the present high level some decades ago. Thus, forest ecosystems are still accumulating reactive nitrogen, and the responses are increasing. Deposition provides additional reactive nitrogen that accumulates into forest ecosystems in needles and especially into proteins in the soil. The extra nitrogen enables construction of additional needles, and in this way, the photosynthesis of the ecosystem increases providing additional sugars for growth of wood in stems and branches. Although the nitrogen deposition is small, usually less than 2 g m^{-2} year^{-1}, its accumulation into forest ecosystems generates large effects.

We omit in our simulations the changes in leaching and gaseous release from the forest soil. The concentrations of ammonium and nitrate ions in soil solution will increase due to the nitrogen deposition, and the seepage water will carry nitrogen into ground water. In addition, the nitrification (Sect. 4.4) and denitrification will generate outflow of N_2 and N_2O. The omitting of leaching and gaseous outflow of nitrogen generates bias in our simulations, and the obtained values are evidently too large. However, the deposition levels as observed at SMEAR II have not yet caused any measurable loss of accumulated nitrogen, and in this respect, the result can be considered more realistic. Indeed, the growth of Finnish forest has doubled from 1950s. Much of this change can be contributed to silviculture and changing age structure, but certainly, there has also been fertilisation effect from nitrogen deposition as well.

8.3 Climate Change and Forests

8.3.1 Physical Background of Climate Change

Timo Vesala, Jouni Räisänen, Tuukka Petäjä, Erkki Siivola, and Pertti Hari

The solar radiation energy (1.37 kW m^{-2} through a tangential plane above the atmosphere, Sect. 3.2) generates processes and transport of entities on the planetary surface and in the atmosphere. The processes, principally the absorption and emission of radiation, evaporation, condensation, freezing and melting of water, convert the energy into different forms giving rise to concentration, temperature and density differences that generate material and energy flows within the atmosphere and between the atmosphere and the surface of the globe. These processes and material and energy flows determine the climate on the globe. The conservation of mass, energy and momentum plays a key role in the combining of different phenomena. The behaviour of the atmosphere can be concisely expressed in seven differential equations. The increasing concentrations of greenhouse gases reduce the thermal radiation of the globe into the space generating climate change. The global climate models, based on the seven atmospheric equations, are able to produce the present climate and also the increasing trend in temperature.

Our approach, when we analysed the behaviour of trees, forest soil and forest ecosystem in Chap. 5, was based on the processes generating concentration, temperature and pressure differences that give rise to material and energy flows within trees and ecosystem and between ecosystem and its environment. The conservation of mass and energy plays an important role in the combination of different phenomena. Evidently, our approach in Chap. 5 is very similar to that applied in the meteorology when the behaviour of the atmosphere is analysed.

T. Vesala (✉)
Department of Physics, University of Helsinki, P.O. Box 48,
00014 University of Helsinki, Helsinki, Finland
e-mail: timo.vesala@helsinki.fi

Several processes convert solar energy in the atmosphere into other forms. Absorption in the atmosphere and on the Earth's surface transforms this radiation energy into heat warming the atmosphere. Emission of thermal radiation and transpiration and evaporation processes cool and condensation of water vapour releases heat. Heat exchange tends to level temperature differences. Pressure and temperature differences generate vertical and horizontal flows in the atmosphere. Conservation of mass, energy and momentum forms the basis for the quantitative analysis of the behaviour of the atmosphere (see Basic idea 9, Chap. 2).

8.3.1.1 Phase Transitions

First-order phase transitions, like freezing-melting and condensation-evaporation of water, are associated with heat of transition. The phase transitions are important for the behaviour of the atmosphere since they release and absorb energy affecting distribution and transport of it. *Evaporation* needs external energy which is transported from surroundings, and the surface temperature with evaporating water tends to be lower than the ambient one. Correspondingly, in *condensation*, heat is released and transported to surroundings, and the surface temperature is elevated. When phase transitions occur, they must be taken into account in any considerations of energy balances and exchange. Transpiration is physical evaporation from wetted interfaces along the pores in the cell walls of mesophyll, epidermal and guard cells, via stomatal openings to the atmosphere, lowering the leaf temperature. Conversely, a leaf can gain latent heat if dew or frost condenses onto it (see Nobel 2005). Note that evaporation and condensation are reverse processes, and the amount of heat required to vaporise liquid water is the same as the released heat if the same amount of water vapour is condensed to liquid. The latent heat of vaporisation of water is 2,260 kJ kg^{-1} and the same amount is released in condensation.

Freezing and melting are other relevant examples of phase transitions, and, similarly to evaporation/condensation, external energy is needed for melting and the same amount is released in freezing. The latent heat of freezing of water is much smaller than the heat of vaporisation, only 333 kJ kg^{-1}.

8.3.1.2 Electromagnetic Radiation

The atmosphere and surfaces of solid objects absorb and emit electromagnetic radiation. Conservation of energy means that absorption and emission are always connected with temperature changes, which has strong consequences to the behaviour of the atmosphere. All materials emit radiation since there is a tendency for the atoms and molecules to return spontaneously to lower energy states (see Bird et al. 2002), and in this process energy is emitted. Because the emitted radiation results from changes in the electronic, vibrational and rotational states of the atoms and molecules, the radiation is distributed over a range of wavelengths.

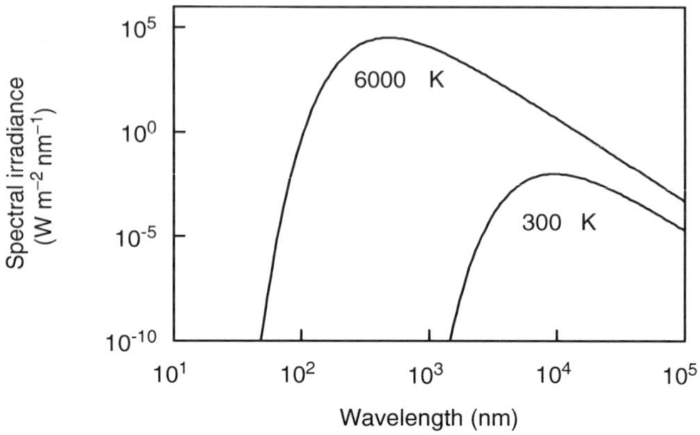

Fig. 8.8 Black-body spectra for 6,000 and 300 K. The temperatures correspond to surface temperature of the Sun and the Earth, respectively. The spectrum of black body at 300 K is invisible in the linear scale

Let us consider first a so-called *black body*, a perfectly efficient emitter, which has a surface emitting hemispherically isotropic radiation at all frequencies. Many natural objects are an approximation to this idealised body. Irradiance $I_{E,\lambda}$ as a function of wavelength (λ) and surface temperature (T) can be approximated with Planck's law of black-body radiation (e.g. Bird et al. 2002):

$$I_{E,\lambda}(\lambda,\ T) = \frac{2\,\pi\,h\,c^2}{\lambda^5} \cdot \frac{1}{e^{c\,h/k\,\lambda\,T} - 1}, \tag{8.1}$$

where c, h and k are speed of light in vacuum (2.9979×10^8 m s^{-1}), Planck's constant (6.626×10^{-34} J s) and Boltzmann's constant (1.381×10^{-23} J mol^{-1} K^{-1}), respectively. Planck's law is one of the triumphs of explaining observed phenomena by using quantum mechanics. It can be derived by applying quantum statistics to a photon gas (or generally to a system of elementary particles called bosons) in a cavity, the photons obeying Bose-Einstein statistics (Bird et al. 2002).

Theoretical spectral irradiance of two black bodies of 6,000 and 300 K are depicted in Fig. 8.8. The selected temperatures are close to the surface temperatures of the Sun and the Earth or any bodies on Earth. The Planck distribution predicts the entire energy versus wavelength curve and the shift of the maximum towards shorter wavelengths at higher temperatures. In fact, the emission spectrum of many bodies at room temperature is very close to 300-K black-body spectrum in Fig. 8.8, and the spectrum of Sun radiation (detected outside the atmosphere) is very close to 6,000-K black-body spectrum (Figs. 8.8 and 8.9) The solar spectrum has a maximum in visible light wavelengths, whereas terrestrial radiation emitted into space has a maximum approximately at 10 μm (infrared).

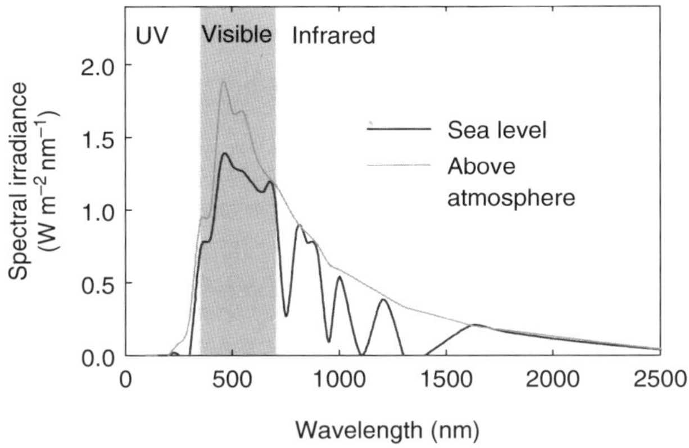

Fig. 8.9 A simplified wavelength distribution of radiative energy from the Sun outside the Earth's atmosphere and at sea level

For non-black surfaces, the emitted energy flux is

$$I_E = e\,\sigma\,T^4, \tag{8.2}$$

where e is the *emissivity* and must be evaluated at temperature T. Unoxidised, clean, metallic surfaces have very low emissivities (like 0.02 for polished copper), whereas most nonmetals and metallic oxides have emissivities above 0.8 at room temperature, and natural/biological surfaces have e of nearly 1.0. The emissivity increases with increasing temperature for nearly all materials. Note that the emissivity may depend on the frequency of radiation. If it is <1 but is the same over all frequencies, the body is called grey.

Formula 8.2 has been effectively used in measurements of surface temperatures of bodies, by infrared sensors and cameras. If the surface emissivity is known, the measured energy flux can be readily inverted to the temperature by the formula.

8.3.1.3 Interaction of Radiation and Matter

The functioning of all land ecosystems is driven by radiative energy from the Sun. Light energy directly drives photosynthesis and transpiration in plants and also affects stomatal action and the rates of biological processes, such as respiration, through temperature by warming up the plants. Infrared energy, which is about half of all the incident energy from the Sun, has a significant warming effect and thus increases the tendency of water to be evaporated from leaves in transpiration.

Photons interacting with material are either absorbed, reflected or transmitted through the material. The fraction of incident radiation flux density that is absorbed

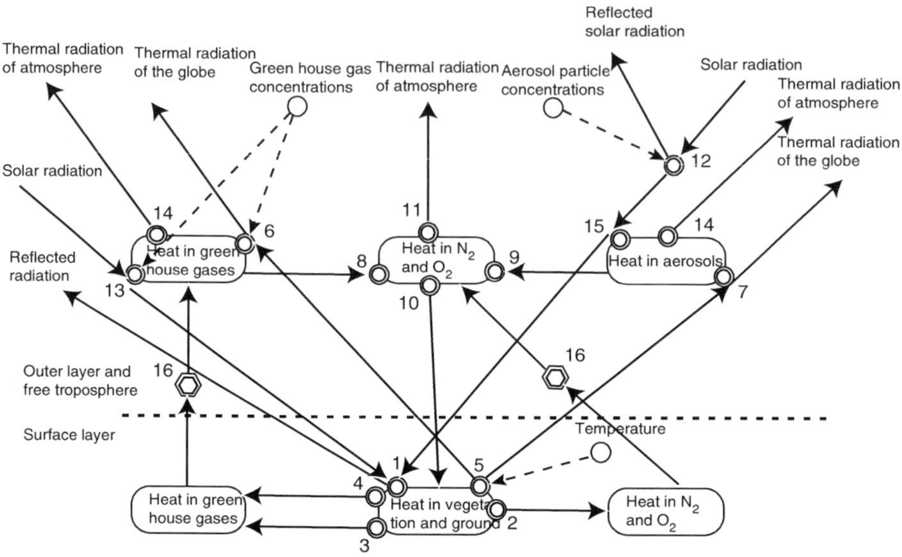

Fig. 8.10 Energy exchange, transport and storages in the atmosphere-biosphere continuum daytime. The atmosphere can be divided into several layers, but here only the surface layer is indicated since ecosystems are located there. The surface layer is located within the atmospheric boundary layer and the boundary layer above the surface layer is called the outer layer (see, e.g. Arya 2001). The free troposphere extends above the boundary layer up to the tropopause. Heat in N_2 and O_2 includes also a contribution from non-greenhouse gases. Note that aerosols include also clouds since physically cloud is aerosol. The *circles* represent environmental factors, *hexagons* physical properties, *double hexagons* transport, and *double circles* processes, *arrows* material/energy flow and *dashed arrows* influence (Fig. 2.3)

by a surface is called absorptivity. Reflectivity is the fraction that is reflected by a surface, and transmissivity is the fraction that is transmitted through a surface. The sum of absorptivity, reflectivity and transmissivity equals 1.

Scattering is a physical process where radiation is forced to deviate from a straight trajectory by the particles in the medium through which the radiation passes. In solid or liquid medium, scattering results in part of the incident radiation being reflected from the surface. Also photons absorbed by the medium or transmitted through it may have undergone scattering in the medium. Because scattering depends on wavelength, the spectral composition of scattered (diffuse) light is altered from that in the Sun's direct beam. The spectrum of diffuse light in a clear sky is shifted towards shorter wavelengths and thus appears blue to the human eye. Under clouds, scattering causes the sky to be greyish. Correspondingly, the spectrum of light inside a tree canopy is different from the spectrum of incoming solar radiation: light transmitted through or reflected from foliage is enriched in green because blue and red wavelengths are absorbed by photosynthetic pigments in the leaves.

8.3.1.4 Heat Transport and Exchange in the Atmosphere and Between Atmosphere and Surface

Here, we discuss on a general level the conservation and transport of energy in the atmosphere. Energy input by solar radiation is the 'engine' driving the processes and the associated transport phenomena in the atmosphere and biosphere. The numbers in the following discussion refer to Fig. 8.10. A fraction of the energy carried by solar radiation penetrates the atmosphere and is absorbed at the surface (1). The heated surface warms the adjacent air layers composed of nitrogen, oxygen and, to a lesser extent, greenhouse gases (GHG) including water vapour (2 and 3). Conduction and convective transport are coupled with evapotranspiration (4), which transforms a fraction of heat from the absorbed radiation to the latent heat associated with vaporised water vapour. The surface emits thermal (long wave) radiation (5). A part of the incident solar radiation is reflected, this fraction being termed the surface albedo. A large fraction of the thermal radiation is absorbed by greenhouse gases (6), clouds (7) (in Fig. 8.10 aerosols include also clouds). Absorbed energy is distributed to the main materials of the air (8 and 9), nitrogen and oxygen and further transported by thermal radiation (10 and 11) and convection. Not all solar radiation penetrates to the surface; some is reflected (12) mainly by clouds and absorbed (13) mainly by gases such as water vapour and ozone and by aerosols (15). The absorbed energy is redistributed among gases and radiated away (14). Convection, that is, bulk flow, transports heat and latent energy vertically and horizontally in the atmosphere. The overall energy balance is maintained by the outward thermal radiation of the atmosphere (11 and 14). This is the key issue in understanding the greenhouse effect. The atmosphere containing the increased concentration of GHGs is more opaque for thermal radiation, and thus the loss of heat by outward thermal radiation occurs effectively from upper atmospheric layers. Upper layers are however colder, and thus the intensity of radiation tends to be reduced. For maintaining the (radiative) energy balance, the temperature is consequently increased at the lower atmosphere.

Figure 8.10 represents the daytime situation when insolation is at high levels. At night and occasionally also on winter days at high latitudes, the solar radiation component is naturally missing or is very low, and thermal radiation (4) may effectively cool the ground surface and surface layer.

8.3.1.5 Conservation Principles

The conservation principles of mass, energy and momentum are fundamental theoretical principles in classical physics as well as in quantum mechanics and the theory of relativity. For phenomena considered in this book, the theory of relativity and quantum mechanics *per se* can be ignored, although they are needed, for example, in fundamental understanding of electromagnetic radiation. The conservation principles in classical physics have been tested since the Newtonian era, and they have passed the tests very successfully. In fact, modern technology can be seen in general as rigorous tests of conservation principles.

The conservation of basic entities sets a strict framework for analysis, which any solution of a problem and a model must follow. An excellent and thorough discussion on conservation principles in the context of transport phenomena is given by Bird et al. (2002). In derivations of conservation equations, one considers the volume element where the amounts of momentum, heat and mass may change, and those changes are balanced by transport of the quantities through the volume boundaries and the rate of production within the volume. The equations are thus derived from a fundamental concept of conservation where the phenomenological transport mechanisms are embedded, and the fluxes are replaced by expressions involving transport properties and gradients of concentration, velocity and temperature. This enables the derivation of formulas that form a closed set of equations for velocity components, temperature, concentration, pressure and density (see Chap. 2, Eq. 2.3).

8.3.1.6 Basic Equations of Atmospheric Models

General circulation models (GCM) are invaluable tools in analysis of past, present and future climate and are thus linked with many themes presented in this book. There is a debate on how much we can trust complicated numerical models. However, it is demonstrated below that in fact the basic formulas are very simple, and they are rigorously based only on conservation principles and transport equations and do not contain any esoteric components. The analysis of atmospheric flows normally utilises some coordinate transforms, but they are just straightforward mathematical treatments.

Present GCMs are based on so-called *hydrostatic primitive equations*, which are normally formulated in Cartesian *pressure coordinates*. At synoptic scale (~1,000 km), the atmospheric flows are very close to incompressible flows, and the condition of hydrostatic balance provides an excellent approximation for the vertical dependence of the pressure field in the real atmosphere (Holton 2004).

The seven differential equations in the Box 8.1 form a full set with as many equations as there are unknowns and in principle they describe the behaviour of the atmosphere very accurately. The atmosphere is divided into grid boxes, and conservation laws are applied for each box. However, as computing resources make it necessary to use a rather coarse grid, small-scale processes, like turbulence and convective clouds, cannot be solved directly. Even their numerical treatment is difficult. The challenge is to calculate 'the source terms' of turbulent friction, diabatic heating and phase transitions accurately enough since they describe the processes occurring in spatial scales smaller than the typical grid size of the numerical model. However, the source terms are very important in simulations of climate and its changes.

Box 8.1

The atmospheric equations in the rotating pressure coordinates are as follows:

Equation for west–east wind u:

$$\frac{\partial u}{\partial t} = -u\frac{\partial u}{\partial x} - v\frac{\partial u}{\partial y} - \omega\frac{\partial u}{\partial p} - \frac{\partial \Phi}{\partial x} + fv + F_x \tag{8.3}$$

Equation for south–north wind v:

$$\frac{\partial v}{\partial t} = -u\frac{\partial v}{\partial x} - v\frac{\partial v}{\partial y} - \omega\frac{\partial v}{\partial p} - \frac{\partial \Phi}{\partial y} - fu + F_y \tag{8.4}$$

Thermodynamic equation (energy conservation):

$$\frac{\partial T}{\partial t} = -u\frac{\partial T}{\partial x} - v\frac{\partial T}{\partial y} - \omega\frac{\partial T}{\partial p} + \omega\frac{RT}{C_p p} + \frac{Q}{C_p} \tag{8.5}$$

Hydrostatic balance equation:

$$\frac{\partial \Phi}{\partial \mathrm{p}} = -\frac{RT}{p} \tag{8.6}$$

Equation of continuity (mass conservation):

$$\frac{\partial u}{\partial x} + \frac{\partial v}{\partial y} + \frac{\partial \omega}{\partial p} = 0 \tag{8.7}$$

Continuity equation (water vapour concentration q):

$$\frac{\partial C_{\mathrm{H_2O}}}{\partial t} = -u\frac{\partial C_{\mathrm{H_2O}}}{\partial x} - v\frac{\partial C_{\mathrm{H_2O}}}{\partial y} - \omega\frac{\partial C_{\mathrm{H_2O}}}{\partial p} + S_{\mathrm{H_2O}} \tag{8.8}$$

Continuity equations for other species A:

$$\frac{\partial C_{\mathrm{A}}}{\partial t} = -u\frac{\partial C_{\mathrm{A}}}{\partial x} - v\frac{\partial C_{\mathrm{A}}}{\partial y} - \omega\frac{\partial C_{\mathrm{A}}}{\partial p} + S_{\mathrm{A}} \tag{8.9}$$

where $\omega = (\mathrm{d}p/\mathrm{d}t)$ is usually called the 'omega' vertical motion, the pressure change following the motion. It plays the role of the vertical wind in the pressure coordinate system. Φ is the geopotential equal to gz where g is the gravity including also the apparent centrifugal acceleration due to rotation of the Earth. The geopotential at the height of z is the work required to raise a unit mass to height z from mean sea level that is the potential energy per unit mass. f is the Coriolis parameter equal to $2\Omega \sin \phi$ where Ω is Earth's angular speed of rotation and ϕ is latitude. F_x and F_y describe the turbulent friction in the atmospheric boundary layer (ABL). R is the gas constant of air. Q involves the diabatic heating, including most importantly radiation and phase transitions of water. $S_{\mathrm{H_2O}}$ describes effects of evaporation and condensation to water vapour concentration.

8.3.2 Mechanisms of Climate Change

Jouni Räisänen and Heikki Tuomenvirta

The increasing concentrations of CO_2 and other greenhouse gases change the behaviour of radiation energy in the atmosphere. Several processes (Sect. 8.3.1) respond to this redistribution of radiation. The above conservation equations of mass, energy and momentum can be solved numerically to study the resulting effects on climate.

Changes in climate occur both as a result of natural variability and as a response to anthropogenic forcing. Some part of the natural variability is forced, that is, caused by external factors such as solar variability and volcanic eruptions. Another part is unforced, associated with the nonlinear internal dynamics of the climate system such as interactions between the atmosphere and the oceans. Because of these natural mechanisms, climate has always varied, on time scales ranging from years to millions of years (Jansen et al. 2007), and it would continue to vary in the future regardless of what mankind is doing. Nevertheless, when we focus on this and the following centuries, the effects of natural variability will likely be secondary when compared with anthropogenic changes in the global climate (Meehl et al. 2007).

Anthropogenic climate change is often referred to simply as *global warming* (e.g. Harvey 2000; Houghton 2004). This refers to the expected tendency of increased greenhouse gas concentrations to warm up the surface and the lower atmosphere in all or nearly all parts of the world. Nevertheless, other anthropogenic forcing mechanisms such as changes in land use and, in particular, increased aerosol concentrations complicate this picture (Forster et al. 2007). Furthermore, although temperature is the climate variable in which the anthropogenic changes are expected to be strongest when compared with natural variability, many other aspects of climate will also change.

The Earth has a *natural greenhouse effect* which helps to keep the surface of our planet much warmer than it would otherwise be. The surface of the Earth emits thermal radiation almost like a black body, with the intensity of radiation increasing with the fourth power of the surface temperature (Jin and Liang 2006; Eq. 8.2). On the other hand, considering the energy budget of the Earth and its atmosphere as a whole, there must be a close balance between the solar radiation absorbed by our planet and the thermal radiation escaping to space. From this condition, it follows that if the atmosphere were transparent to thermal radiation, the average surface temperature of the Earth would be only about −18°C (255 K). The observed mean surface temperature is about 32°C higher. This difference is due

---

J. Räisänen (✉)
Department of Physics, University of Helsinki, P.O. Box 48,
00014 University of Helsinki, Helsinki, Finland
e-mail: jouni.raisanen@helsinki.fi

to the fact that most of the thermal radiation emitted by the surface is absorbed by the atmosphere, and the radiation that actually escapes to space therefore has its origin in atmospheric layers much cooler than the surface. Of the so-called *greenhouse gases* that are responsible for this effect, the most important is water vapour (with a contribution of about 60% in clear-sky conditions), followed by carbon dioxide CO_2 (26%) and ozone O_3 (8%). Methane (CH_4) and nitrous oxide (N_2O) together account for the remaining 6% (Kiehl and Trenberth 1997). Clouds also amplify the greenhouse effect, with a contribution comparable to that of CO_2, but this is more than compensated by their ability to reflect solar radiation to space.

We have very little information about conditions on the globe in the distant geological past. However, some materials can be found which can be dated and which reflect air temperature or gas concentrations characteristic of their formation period. The most valuable sources of information for the most recent 800,000 years are ice cores taken from the Antarctica or Greenland. Small air bubbles in the ice enable the determination of atmospheric CO_2 concentrations when the ice was formed from accumulating snow. In addition, the ratio of stable oxygen isotopes reflects the temperature conditions of the ocean around the Arctic and Antarctic regions.

Ice cores that have been extracted from Antarctica and Greenland indicate that the atmospheric CO_2 concentration has varied in a range from 180 to 300 ppm during the last 800,000 years (Fig. 8.11a). The stable isotopes of oxygen indicate that temperature has varied simultaneously with the atmospheric CO_2 concentration. Periods of low atmospheric carbon dioxide concentration (glacials) have been cooler than those of high concentration (interglacials, including the last 10,000 years) (Fig. 8.11b). This is thought to reflect a positive feedback between carbon cycle and climate. Variations in carbon dioxide concentration have been ultimately driven by changes in climate, but they have also acted to amplify the glacial-interglacial temperature variability.

Changes in atmospheric carbon dioxide concentration during the glacial-interglacial cycles have been slow; they have happened on a time scale of 10,000 years. This tells us that the natural carbon cycle has been closely balanced. Each year, vegetation assimilates about 15% of all the carbon dioxide in the atmosphere by photosynthesis, but nearly the same amount of carbon dioxide is returned to the atmosphere by respiration and decomposition. Similarly, although large amounts of carbon dioxide are dissolved in oceans, especially the northern oceans, this is compensated by outgassing of carbon dioxide back to the atmosphere in other ocean areas. During the past 10,000 years preceding the industrial era, this balance was even more perfect, and the atmospheric CO_2 concentration was stable at about 260–280 ppm. During the same period, temperatures were also relatively stable and warmer than average according to the ice core data.

The slow natural changes in carbon dioxide concentration as seen in the ice core record contrast strongly with the rapid increase during the industrial era and particularly the last few decades (Fig. 8.11a). The concentrations of methane CH_4 and N_2O in the atmosphere have also had a strong increasing trend during the same period.

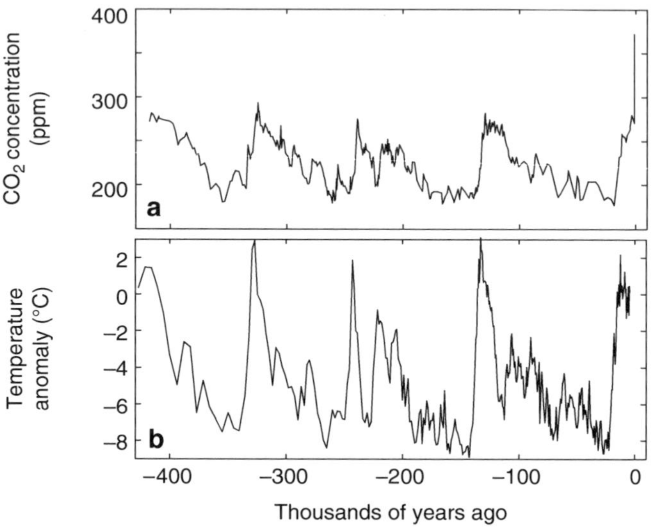

Fig. 8.11 Concentration of CO_2 (**a**) and development of temperature (**b**) during the last 420,000 years. The concentration data are compiled from three sources, (i) Vostock ice cores (Barnola et al. 2003), (ii) Siple station ice core data (Friedli et al. 1986) and (iii) atmospheric measurements (Keeling and Worf 2005). The temperatures are determined from Vostok ice core isotopes as difference from present values (Jouzel et al. 1987, 1993, 1996; Petit 1999)

The absorption of radiation energy by gases depends strongly on the wavelength, the radiation can penetrate a gas without absorption at some and it can be totally absorbed at some other wavelength. The absorption by the atmosphere depends strongly on the wavelength; at visible light (380–740 nm), the absorption is very small, and at some wavelengths in the infrared (740 nm to 300 μm) area, the atmosphere is opaque (Fig. 8.9). The increasing concentrations of CO_2, CH_4 and N_2O enhance the absorption at wavelengths at which the atmosphere is at least partially transparent, and the greenhouse gases can absorb radiation.

Because of human activities, the concentrations of many greenhouse gases have increased and are projected to increase further in the future. The effect that a given change in the atmospheric composition has on the planetary energy budget is commonly quantified by *radiative forcing*. This measures, in approximate terms, the change in the energy balance of the Earth-atmosphere system that a given change in external conditions would induce with no compensating changes in climate (for the exact definition, see Houghton et al. 2001, p. 795). For the increase in CO_2 concentration from pre-industrial time to the year 2005 (280–379 ppm), the radiative forcing was approximately 1.66 W m^{-2} (Forster et al. 2007), and it is currently increasing by about 0.03 W m^{-2} $year^{-1}$. Smaller positive forcing contributions have come from increases in CH_4 (0.48 W m^{-2}), N_2O (0.16 W m^{-2}), various halocarbons (0.34 W m^{-2}) and tropospheric O_3 (0.35 W m^{-2}). With the exception of O_3, all these forcings can be estimated with a good accuracy (±10%) because the radiative properties of gas molecules are well known, and the lifetime of these greenhouse gases (excluding O_3) is long enough to make them almost evenly distributed in the lower atmosphere. All in all, the present-day radiative forcing from anthropogenic increases in greenhouse gas concentrations approaches 3 W m^{-2}. This number does not include changes in water vapour because direct anthropogenic sources of water vapour are negligible when compared with the natural hydrological cycle. However, observations suggest that water vapour is increasing in response to observed atmospheric warming and is thereby providing a positive feedback to ongoing climate change (Trenberth et al. 2007).

Greenhouse gas forcing has, however, been partly compensated by a net negative forcing from increased aerosol loading. The direct effect of anthropogenic aerosols, associated with the scattering and absorption of solar radiation by aerosol particles, varies in sign between different aerosol types, but the net forcing is very likely negative. In its 4th Assessment Report (Forster et al. 2007), the Intergovernmental Panel on Climate Change gives it a best estimate of −0.5 W m^{-2} and a wide estimated uncertainty range of −0.1 to −0.9 W m^{-2}. The anthropogenic increase in aerosols also results in an increase in available cloud condensation nuclei, which has the potential to increase cloud albedo. The negative forcing due to this indirect aerosol effect is estimated as −0.7 W m^{-2}, but with a very wide uncertainty range (−0.3 to −1.8 W m^{-2}). Furthermore, aerosols may also affect the lifetime of clouds and hence the total cloud cover, but this effect is even more poorly quantified.

As the anthropogenic increase in aerosol loading has resulted in a negative radiative forcing, it has most likely slowed down the warming that would have resulted from increases in greenhouse gas concentrations alone. However, the magnitude of the aerosol forcing and hence the degree of this compensation is poorly known (e.g. Andreae et al. 2005). Also in contrast to the more or less evenly distributed greenhouse gas forcing, the aerosol forcing is concentrated in the vicinity of areas with the largest aerosol-generating emissions (e.g. Forster et al. 2007). Thus, greenhouse gas-induced and aerosol-induced climate changes are expected to have partly different geographical distributions. However, these differences are moderated by atmospheric circulation, which tends to spread the effects of even localised forcings to the global scale (e.g. Boer and Yu 2003).

Positive radiative forcing tends to increase, and negative forcing decrease the global mean temperature, but the magnitude of the forcing does not directly tell the magnitude of the temperature response. The latter depends on several feedbacks acting in the climate system, and it therefore needs to be estimated with climate models. However, model simulations suggest that the ratio of the response to the magnitude of forcing is approximately the same for different forcing agents (e.g. Joshi et al. 2003; Hansen et al. 2005).

In the absence of amplifying feedbacks, the global climate would be quite resistant to radiative forcing. For example, a hypothetical doubling of the atmospheric CO_2 concentration, which gives a radiative forcing of about 3.7 W m^{-2}, would result in a global mean warming of only about 1.2°C, once the climate has had sufficient time[1] to reach a new equilibrium (e.g. Hansen et al. 1984). In fact, this number, known as *climate sensitivity,* is most likely higher, but its exact value is very hard to determine. According to the latest estimates, which combine evidence from model simulations and observations (Meehl et al. 2007), it is probably within the range 2–4.5°C.

The water vapour holding capacity of air increases quasi-exponentially with increasing temperature by approximately 7% for each 1°C of warming. Assuming that there will be no large changes in relative humidity, which is supported by both observations and model simulations (Held and Soden 2006), the actual concentration of water vapour will also increase. Because water vapour is the most important greenhouse gas in the atmosphere, the increase in its concentration will result in a strong positive feedback. Another most likely positive feedback is associated with changes in snow and ice cover. With a warming of climate, ice and snow cover will be most likely reduced, which increases the solar radiation absorbed at the surface and therefore acts to amplify the warming. This feedback is particularly important in high latitudes, but its role in amplifying the global mean warming is expected to be smaller than the role of increasing water vapour (Webb et al. 2006).

Clouds are important for the global climate, as they both reflect solar radiation to space (a cooling effect) and absorb thermal radiation emitted by the surface (a warming effect). Changes in cloudiness may thus also result in an important feedback effect. However, the complexity of cloud dynamics and microphysical processes makes this feedback very difficult to determine, although the most recent model estimates suggest that it will also probably be positive (Colman 2003; Webb et al. 2006). As a result, changes in cloudiness appear to be the most important uncertainty in the response of climate to increasing greenhouse gas concentrations (e.g. Webb et al. 2006).

The interaction between biosphere and climate may also result in feedbacks. Focusing, in particular, on the role of boreal forests, there are several possible feedbacks:

1. The model simulations discussed in Sect. 8.3.5 suggest that the growth of boreal forests will increase with increasing temperature. This will cause the trees to sequester more CO_2 from the atmosphere, which represents a negative feedback to the climate system. On the other hand, decomposition of organic matter in the soil is also expected to accelerate with warming. The net effect of boreal forests on the CO_2 concentration will depend on which of these competing feedbacks dominates (Friedlingstein et al. 2006).
2. Model simulations also point to an increase in the leaf area index of the trees in a warmer climate, which has the potential to increase transpiration. This may have multiple effects on climate. Increased transpiration is expected to increase water vapour concentration in the atmosphere, thus amplifying the greenhouse

[1]Due to the large heat capacity of the oceans, this equilibration time is of the order of centuries.

effect. On the other hand, the resulting increase in humidity is also expected to promote formation of low clouds, which will reduce the solar radiation reaching the surface. Finally, an increase in transpiration would leave a smaller fraction of the available radiation energy for heating the surface that, like increases in cloudiness, would tend to reduce the surface temperature.

3. Increased activity of the forest canopy can also lead to an increase in the formation of biogenic aerosols, with implications on the scattering of solar radiation and cloud formation.
4. Increases in the leaf area index and the density of the forest canopy may reduce the surface albedo, particularly during the snow season when the albedo contrast between mostly snow-free tree crowns and open snow-covered land is large. This positive feedback is potentially important, particularly during late winter and early spring when there is still snow on the ground, but a lot of solar radiation is already available. Simulations with coupled climate-vegetation models suggest that this feedback may have played an important role in explaining the relative warmth of the northern hemisphere high-latitude continents during the mid-Holocene 6,000 years ago (e.g. Wohlfahrt et al. 2004).

8.3.3 Observed and Projected Changes in Climate

Jouni Räisänen and Heikki Tuomenvirta

The longest series of direct meteorological observations in the boreal region cover over 200 years. For example, in Stockholm (Moberg et al. 2002) and St. Petersburg (Jones and Lister 2002), air temperature and pressure have been measured since the mid-eighteenth century. However, only from the latter part of the nineteenth century have there been enough stations to calculate the global and boreal zone area-averaged temperatures. For the other parameters describing climatic conditions, such as precipitation amount, data coverage and its accuracy are limited until the twentieth century. Increasingly comprehensive observations are only available, for example, for snow cover and water vapour, during the recent decades. In this section, we will describe succinctly what kind of changes have been observed in the boreal climate.

The global mean temperature has increased 0.8 ± 0.2°C since the late nineteenth century (Trenberth et al. 2007). The warming has been larger over land areas than over the oceans. In the northern hemisphere, the warming has been largest during winter and spring. In the boreal zone, the annual mean temperatures have warmed

J. Räisänen (✉)
Department of Physics, University of Helsinki, P.O. Box 48,
00014 University of Helsinki, Helsinki, Finland
e-mail: jouni.raisanen@helsinki.fi

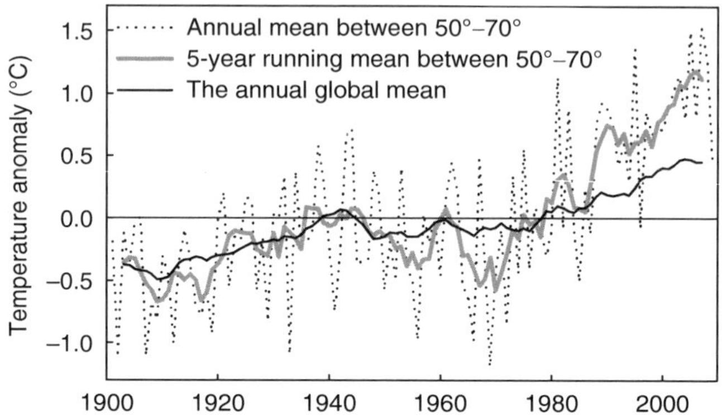

Fig. 8.12 Observed variations in temperature between the years 1901 and 2009. The *dotted line* shows annual temperature anomalies (i.e. differences from the mean value of 1961–1990) averaged over the land areas between 50 and 70°N and the dashed grey line their 5-year running mean. The 5-year running global mean temperature anomalies are shown by a *solid black line*. The data are from NCDC (Smith and Reynolds 2005)

more than in the global mean (Fig. 8.12). The twentieth-century warming has occurred in three phases: warming until the 1940s, followed by a slight cooling and a rapid warming from about the 1970s.

In the boreal zone (land areas between the latitudes 50 and 70°N), mean temperature during the period from May to August (hereafter MJJA) has increased in a similar three-phase behaviour as the annual mean temperature. When expressed as a linear trend, the warming since the beginning of the twentieth century has been about the same as the global annual mean warming (0.07°C/decade or 0.8°C in total). However, the warming since the mid-1970s has been much faster, nearly 0.3°C/decade for the period 1976–2009. Although linear trends depend on the period used in calculation, they allow an easy way to characterise spatial differences. From Fig. 8.13a, b, it can be seen that the warming from the beginning of the twentieth century has not been spatially uniform. There are some limited regions in the boreal zone that show no warming or may even have cooled, but the vast majority of the boreal area shows an increasing temperature trend over the period 1901–2009. The largest positive trends are about +1.5°C per century.

Since the 1970s, MJJA temperature trends are characterised by widespread warming (Fig. 8.13c, d). In parts of Europe and Siberia, the warming has been 0.5–0.7°C per decade, whereas some regions in North America have cooled slightly. Changes in atmospheric circulation modify the regional temperature trends, and natural climatic variability affects the statistical significance of the trends. A majority of the positive trends in Fig. 8.13a are statistically significant at 5% level, but due to the large natural variability and the short time period, just the largest trends in Fig. 8.13c are statistically significant. One clear implication of temperature increase is the lengthening of the growing season which can be seen as advanced bud burst of European rowan (*Sorbus aucuparia*) (Sect. 4.2.6).

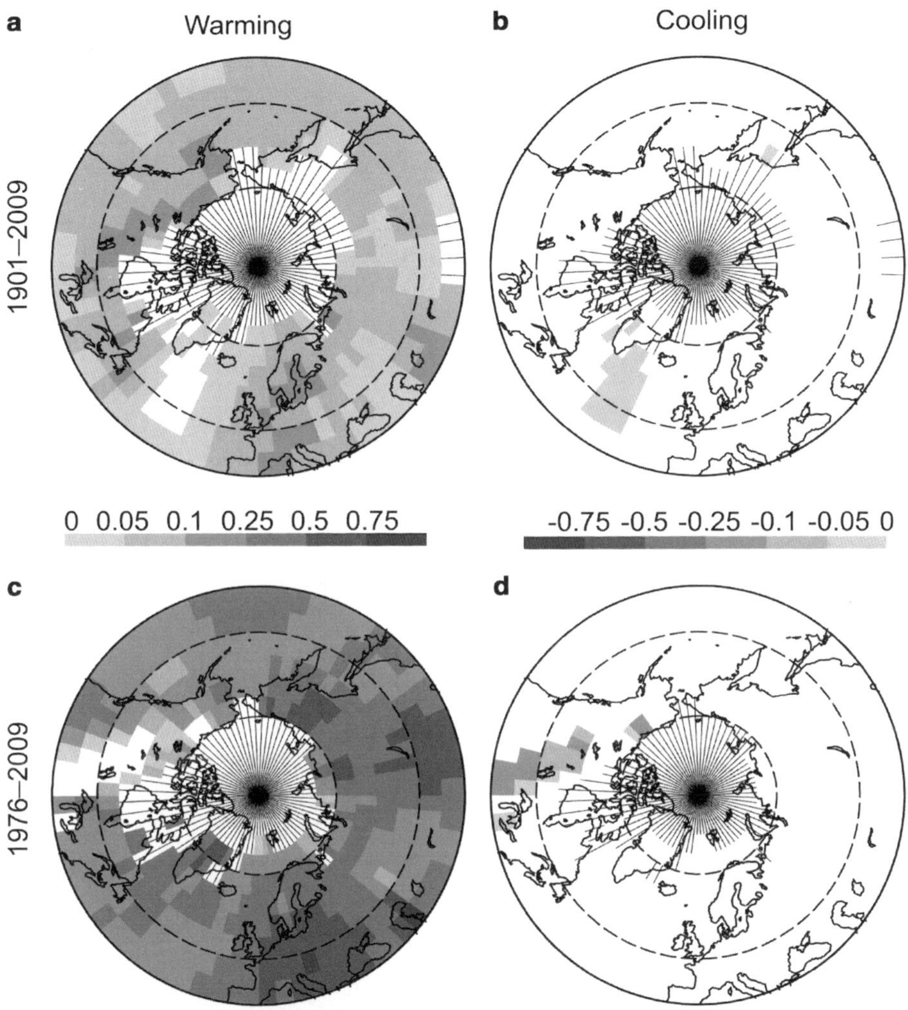

Fig. 8.13 Linear trends of MJJA temperatures (°C/decade) for (**a**)–(**b**) 1901–2009 and (**c**)–(**d**) 1976–2009. Areas with warming are shown in (**a**) and (**c**), areas with cooling in (**b**) and (**d**). Areas with missing data are indicated with north–south-oriented lines. The data are from NCDC (Smith and Reynolds 2005)

Effects of urbanisation on the land-based temperature record are negligible as far as hemispheric- and continental-scale averages are concerned because the very real but local effects are avoided or accounted for in the data sets. Trenberth et al. (2007) estimate the error due to urbanisation to be less than 0.01°C/decade (best estimate 0.006°C/decade) over land areas. Similarly, there may exist somc random like error sources, like changes in observation sites, that tend to cancel out when averaging is done.

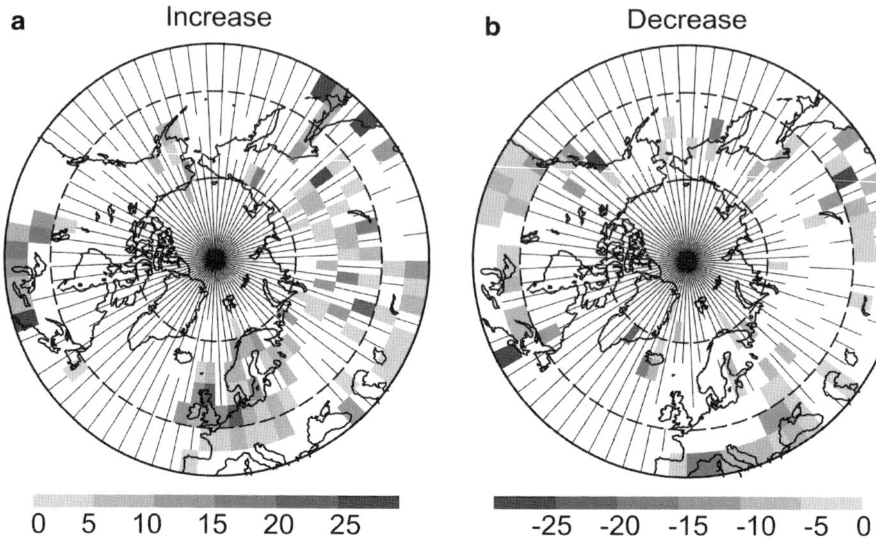

Fig. 8.14 Linear trends of MJJA precipitation sum (mm/decade) for the years 1976–2009. Areas with increasing precipitation are shown in (**a**) and areas with decreasing precipitation in (**b**). Areas with missing data are indicated with north–south-oriented lines The data are from NCDC (Smith and Reynolds 2005)

Area-averaged temperature variations and changes are known with good accuracy during the twentieth century. However, precipitation variations are known with spatially varying accuracy. There are two basic difficulties that cause inaccuracies to precipitation estimates. Firstly, measurements of precipitation suffer from undercatch due to wind and other error sources. Local conditions have a large effect on the measured amount, so any changes, for example, in wind exposure or improvements in the instrument, may break the homogeneity of measurement series. Secondly, global data sets have too few stations to catch the large spatial variability of precipitation; thus, in some regions of the world, the sample is too small to cover the spatially complex phenomena. For these reasons, precipitation trends are known less accurately than temperature trends.

There are several global land precipitation data sets, but they do not show consistent long-term trends over global land areas (Trenberth et al. 2007). In the NCDC data set used in this book, boreal zone annual, as well as MJJA, precipitation shows some increase during the twentieth century (Peterson and Vose 1997). However, the changes have varied in sign within the boreal zone (Fig. 8.14). In most parts of Eurasia, including northern Europe, MJJA precipitation has increased since the 1970s, but there have also been some areas with decrease. In North America, increasing and decreasing trends have been more in balance. From spatially detailed analysis it is known that the pattern of changes can be more complex than shown in Fig. 8.14. For example, the increase in northern Europe breaks into regions of increase and decrease in a more detailed analysis (BACC Author Team 2008). Due to large natural variability, regional trends may lack coherence, and trend analysis is sensitive to selection of the time period.

The moisture holding capacity of the atmosphere increases at a rate of about 7% per °C. Temperature increase alone may therefore alter the hydrological cycle, especially the characteristics of precipitation (amount, frequency, intensity, duration, type). A warmer climate is expected to increase the risk for both drought and floods. Furthermore, changes in precipitation characteristics are complicated by aerosols that affect cloud formation processes. The changes in water that is available for the trees, that is, soil moisture, are determined by changes in precipitation, evapotranspiration and run-off. There are studies utilising observations and model calculations (e.g. Dai et al. 2004; Huntington 2006; Groisman et al. 2007), but a comprehensive view of changes in soil moisture conditions is lacking in the boreal zone.

Intensification of westerlies since the 1960s has increased air flow from oceans to continents, and its effect on precipitation and temperature anomalies can be seen during the winter half-year (Trenberth et al. 2007). Although the most notable climatic changes in the boreal zone have occurred during the winter and spring seasons, some of these changes also affect conditions at the beginning of the growing season that has shifted earlier (Sect. 4.2.6). Snow-covered area has declined in the northern hemisphere, and the reduction has been largest during the months of March and April (Lemke et al. 2007a, b). Earlier snowmelt occurs near the edges of the snow-covered area, thus lengthening the period of evaporation from soil and increasing the risk for drought. In April, the northern hemisphere snow cover extent is strongly correlated with the 40–60°N April temperature reflecting the feedback between snow and temperature.

As noted in the previous subsection, the concentration of greenhouse gases and aerosols has increased markedly during the industrial period. The globally averaged net radiative forcing since 1750, due to both human-induced warming and cooling effects, is estimated to be $+1.6$ W m^{-2}, with uncertainty range from $+0.6$ to $+2.4$ W m^{-2} (Forster et al. 2007). Hegerl et al. (2007) conclude that most of the observed increase in global mean temperature during the last 50 years is very likely due to the increase of anthropogenic greenhouse gas concentrations. However, in the boreal zone, natural climatic variability is large, and modes of variability such as the Arctic Oscillation (or its 'European' manifestation, the North Atlantic Oscillation) provide large regional differences in the observed climatic changes.

8.3.3.1 Global Climate Models

The primary tool used for understanding past climate changes and making projections of future climate change are three-dimensional global climate models (GCMs), also known as general circulation models. These models attempt to explicitly simulate the main physical phenomena that affect atmospheric weather and ocean circulation, land surface conditions and the state of sea ice and snow cover. The models are built on well-known hydrodynamic and thermodynamic principles that describe the conservation of mass, water, energy and momentum within the climate system, rather than, for example, empirical correlations between temperature and greenhouse gas concentrations. The basic equations used in atmospheric models, that is, two equations for the horizontal wind components, the thermodynamic equation, the hydrostatic balance equation, the continuity equation, and conservation equations for water vapour and other trace substances, were given in the Box 8.1. A similar set of equations is also used in ocean models, with the main difference that the equation for water vapour is replaced by an equation for salinity. This solid physical basis gives a strong reason to think that these models are useful tools for exploring the behaviour of the climate system and its response to changes in external forcing such as increases in greenhouse gas concentrations.

On the other hand, limitations in computing power make it impossible to represent the full complexity of the real world in any model. This, together with the uncertainty associated with future greenhouse gas and aerosol emissions and natural climate variability, makes it impossible to give exact predictions of future climate.

The atmospheric components in current GCMs typically have a grid spacing of 200–300 km in the horizontal direction. In the vertical direction, there are typically about 30 levels between the surface and the model top at 30–50-km height, the spacing of levels increasing from a few hundred metres in the boundary layer to several kilometres in the stratosphere. Resolution of the oceanic model components is similar or slightly better. Phenomena acting on scales smaller than the grid spacing cannot be resolved explicitly. The impact of these phenomena needs to be parameterised, that is, estimated indirectly from the grid-scale conditions simulated by the model.

The most important parameterised phenomena in the atmospheric components of climate models include the transfer of solar and thermal radiation, small-scale mixing in the atmospheric boundary layer, vertical convection and, in particular, the formation of clouds and precipitation. The other model components, representing the ocean, sea ice and land surface, have their own parameterised phenomena, and parameterisations are also needed when describing exchange of heat, water and momentum between the atmosphere and other components. Some of the feedbacks that affect the magnitude and in some cases direction of climate changes are strongly sensitive to the parameterisation of sub-grid-scale phenomena, changes in cloudiness being the most notorious example.

In addition to the parameterisation problem, uncertainty in the results of climate models also arises from the neglect of some potentially important phenomena. For example, climate-induced changes in vegetation have this far been excluded in most model simulations, including those used in Sect. 8.3.3.2 below.

The atmospheric components of GCMs are, except for a coarser horizontal and vertical resolution, very similar to the models used in operational weather forecasting. The raw model output thus consists of daily or sub-daily time series of weather, for each grid point at the surface and higher in the atmosphere. Obviously, however, the daily evolution of weather is not predictable for several decades in the future. Useful information in climate model output therefore resides in the statistical properties (long-term means, measures of variability, etc.) rather than in the daily details of simulation. For a sufficiently strong external forcing such as a large increase in greenhouse gas concentrations, the changes in these statistical properties become large enough to be discernible from the internally generated chaotic variability in the simulations.

GCMs have been developed and are used at a large number of research institutions. An intercomparison between the climate changes simulated by different models provides one way of estimating the uncertainty in future climate changes in the real world. However, this measure should be used with some care. While a disagreement between model simulations demonstrates that at least some of the models give unrealistic projections, an agreement between them does not necessarily prove that the models are right.

8.3.3.2 Projected Climate Changes for the Rest of the Twenty-First Century

Climate modellers often prefer to talk about projections rather than predictions or forecasts of future climate change. This is because the actual evolution of climate will be partly dependent on future human activities, in particular the magnitude of greenhouse gas and aerosol emissions. To characterise the uncertainty associated with emissions, a number of different emission scenarios have been developed. The most widely used set of these have been the IPCC (Intergovernmental Panel on Climate Change) SRES (Special Report on Emissions Scenarios) scenarios (Nakićenović and Swart 2000).

The SRES scenarios are based on alternative but plausible and internally consistent sets of assumptions about the demographic, socioeconomic and technological changes that together determine the evolution of emissions in the future. For a more detailed discussion of these underlying assumptions, the reader is referred to Nakićenović and Swart (2000). The scenarios remain reasonably similar in the early twenty-first century, for example, all of them include an initial increase in CO_2 emissions in the next few decades (Fig. 8.15). Towards the end of the century, the scenarios diverge, with some of them indicating a continuing rapid increase of greenhouse gas emissions driven by the increasing energy demand of the world, and others pointing towards a gradual decline in emissions allowed by reduced population growth, alternative energy sources and a shift towards a less energy-intensive economy. Nevertheless, the concentrations of the most long-lived greenhouse gases, particularly CO_2 and N_2O, keep rising even in those scenarios in which emissions are reduced. The best-estimate CO_2 concentration in the year 2100, as derived by an off-line carbon cycle model, varies from ca. 540–960 ppm between the lowest (B1) and highest (A1FI) SRES scenarios (Houghton et al. 2001), to be compared with the present-day concentration of 390 ppm.

Anthropogenic increases in atmospheric aerosol loading have suppressed greenhouse-induced warming during the twentieth century but will not necessarily

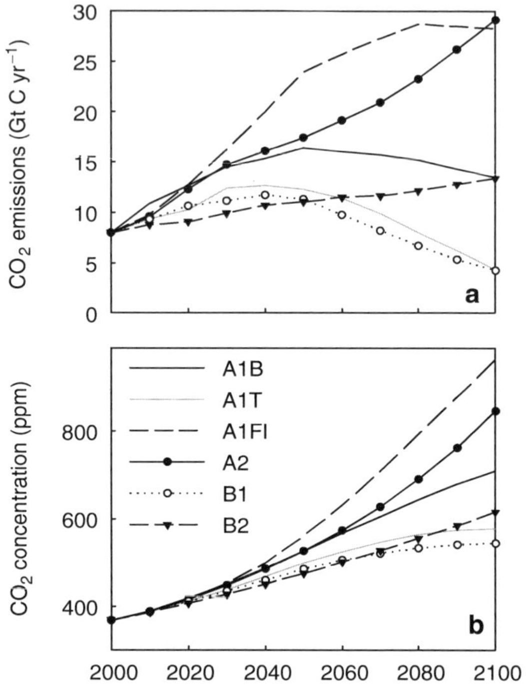

Fig. 8.15 CO_2 emissions (**a**) and the resulting best-estimate CO_2 concentrations (**b**) for the six SRES scenarios

continue to do so in the future. The main contributor to anthropogenic aerosol-induced cooling, SO_2 emissions, is still projected to increase in a global mean sense during the first decades of the twenty-first century in many SRES scenarios, which would suppress warming during this period. By the end of the century, however, the worldwide introduction of cleaner technologies is projected to reduce global SO_2 emissions well below present-day levels. Due to the very short lifetime of tropospheric aerosols, this would result in an immediate decrease in sulphate aerosol concentrations. Thus, greenhouse gas-induced warming is projected to become increasingly dominant over aerosol-induced cooling.

In its Fourth Assessment Report (Meehl et al. 2007), the IPCC estimates that the global mean warming that would occur during this century under the lowest SRES scenario (B1) would be within the range 1.1–2.9°C, with a best estimate of 1.8°C. The corresponding range for the highest scenario (A1FI) is 2.4–6.4°C, with a best estimate of 4.0°C. In both cases, the quoted uncertainty ranges take into account both the variation in warming under the same scenario of greenhouse gas concentrations between different GCMs and the uncertainty in modelling the feedback from climate change to the atmospheric CO_2 concentration. The within-scenario uncertainty associated with the modelling of feedbacks in the climate system is thus comparable with the differences between the lowest and highest emission scenarios. However, this should not obscure the fact that the magnitude of the warming will depend strongly on the magnitude of greenhouse gas emissions.

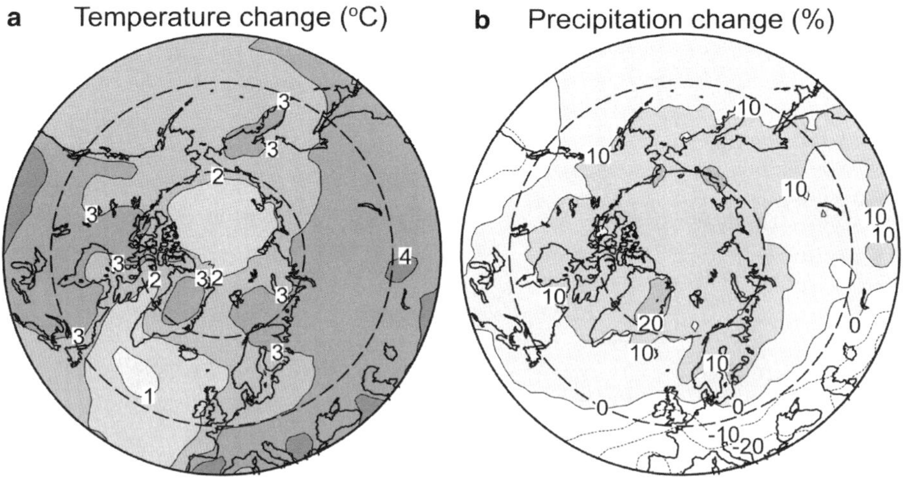

Fig. 8.16 Simulated changes in MJJA mean temperature (°C) (**a**) and precipitation (%) (**b**) from the recent past (1971–2000) to the late twenty-first century (2070–2099), as averaged over 21 climate models. All simulations are based on the SRES A1B emissions scenario. The two *dashed lines* show the latitudes 50 and 70°N

Thus, as mankind has little control of feedbacks in the natural climate system, our primary possibility to alleviate future climate changes is to reduce the emissions of CO_2 and other greenhouse gases.

Below, the typical features and uncertainties of GCM-simulated climate change are illustrated for the SRES A1B scenario, which is, in terms of greenhouse gas emissions, in the midrange of the SRES scenarios (Fig. 8.16). For example, CO_2 emissions are projected to approximately double by 2050, declining slightly thereafter, and CO_2 concentration is projected to rise approximately to 710 ppm by the year 2100. The analysis is based on the results of 21 recent GCMs contributing to the World Climate Research Programme 3rd Coupled Model Intercomparison Project (WCRP CMIP3, Meehl et al. 2007), and climate changes are calculated as differences between the mean values for the years 2070–2099 and 1971–2000. The focus is on the MJJA season that is most important for growth of boreal forests.

Maps of temperature and precipitation change, as averaged over the 21 models, are shown in Fig. 8.16. The average simulated warming in the boreal forest zone is slightly over 3°C, being comparable to but slightly higher than the 21-model annual mean global warming (2.6°C) in the same simulations. The warming is geographically relatively uniform but has, in this season, a tendency to increase from north to south. Precipitation is simulated to increase slightly in almost all of the boreal forest zone but with larger increases in the north than in the south. The opposite patterns of temperature and precipitation change may be physically related via the surface energy balance. The risk that warming would lead to a drying of the soil is larger in the southern areas where the increase in precipitation is smaller and, if the soil becomes sufficiently dry, then the ability of evapotranspiration to

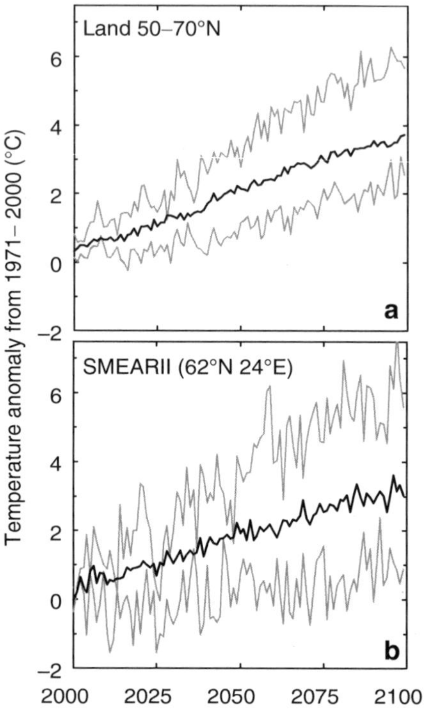

Fig. 8.17 Simulated time series of MJJA mean temperature, as averaged over all land areas at 50–70° N (**a**) and in an individual location (**b**). Temperatures are given as differences from the mean value in the years 1971–2000. In each case, the *central line* represents the mean of 21 climate models and the other *two lines* the individual models with the smallest and the largest warming

cool the surface will be reduced. Nevertheless, as illustrated by Figs. 8.17 and 8.18 below, the geographical variation of these simulated climate changes is relatively small when compared with the variation between individual models.

Climate is expected to warm gradually throughout the twenty-first century, although the warming is punctuated by substantial interannual variability (Fig. 8.17). The details of this variability are not predictable, and they are therefore largely smoothed out when averaging the temperatures over a large number of individual models. However, the differences in the rate of warming between individual models are also substantial and more so for individual grid boxes (Fig. 8.17b) than for averages taken over the whole boreal zone (Fig. 8.17a). Still, all models simulate some warming in practically the whole boreal forest zone.

For precipitation, the signal-to-noise ratio between the simulated anthropogenic climate change and natural variability is much lower than for temperature (Fig. 8.18), particularly when we consider the evolution of climate on the scale of individual grid boxes. Anthropogenic changes in precipitation are thus likely to become more slowly discernible from natural variability than changes in temperature. The variation between the individual models is also larger for precipitation than for temperature changes. Nevertheless, in most parts of the boreal forest zone, a large majority of the models agree on a long-term increase in precipitation.

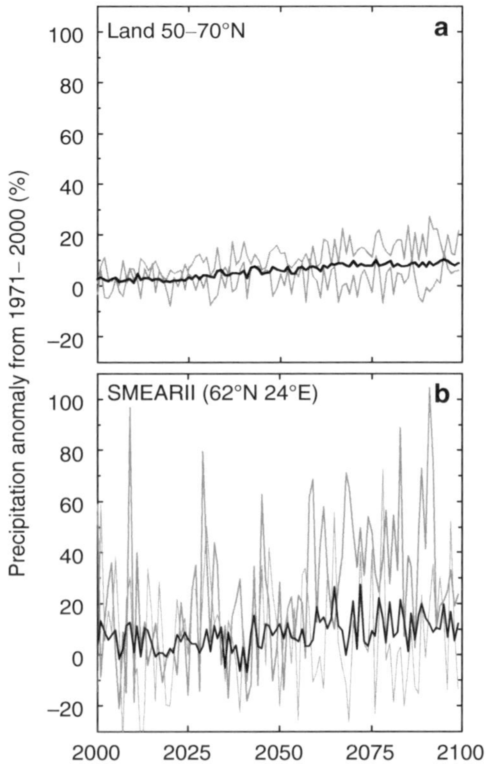

Fig. 8.18 As Fig. 8.17, but for changes in precipitation

8.3.3.3 Acknowledgments

We acknowledge the modelling groups for making their model output available for analysis, the Program for Climate Model Diagnosis and Intercomparison (PCMDI) for collecting and archiving these data and the WCRP's Working Group on Coupled Modelling (WGCM) for organising the model data analysis activity. The WCRP CMIP3 multi-model data set is supported by the Office of Science, US Department of Energy.

8.3.4 Responses of Forest Ecosystems to Climate Change

Pertti Hari, Jaana Bäck, and Eero Nikinmaa

There are several changes in the metabolic processes of trees caused by the climate change. These responses have to be converted to stand level to achieve

P. Hari (✉)
Department of Forest Sciences, University of Helsinki, P.O. Box 27,
00014 University of Helsinki, Helsinki, Finland
e-mail: pertti.hari@helsinki.fi

an understanding of the expected changes in boreal forests. For example, the photosynthetic rate can be assumed to accelerate due to the increased availability of CO_2. Is the growth increased in the same way as photosynthesis or are there internal feedbacks in a tree or in forest ecosystems that modify the response? The ecosystem model MicroForest (Chap. 7) combines several processes and regularities in tree structure to produce a prediction of ecosystem development, with important insights into the responses of boreal forest ecosystems.

The increasing atmospheric CO_2 concentration increases maximum photosynthetic rate and increases the annual amount of sugars available for growth and metabolism. The annual amount of photosynthesis in unshaded conditions is a key parameter in the model MicroForest. This fact opens a possibility to introduce the effect of increasing CO_2 concentration on photosynthesis into model simulations.

We analysed the effect of increasing CO_2 concentration on stand development with SMEAR II stand. The basic run in the in Chap. 7 assumes implicitly stable CO_2 concentration. For comparison, we introduced the effect of increasing CO_2 by assuming that the parameter values are estimated at 380 ppm, that is, this concentration is the reference CO_2, and we assumed a linear relationship between maximum photosynthesis and the atmospheric CO_2 concentration. The simulated stand photosynthesis had, as expected, a clear increasing trend (Fig. 8.19a). In contrast, the simulated stand volume reacted very weakly to the accelerated photosynthesis (Fig. 8.19c): the stand volume was even smaller at the end of the simulation than without increasing CO_2 concentration. This result is caused by the change in allocation to favour roots. As only CO_2 concentration was modified, the release of nitrogen from proteins in soil by microbes did not change and annual nitrogen uptake remained stable. The trees were unable to produce additional functional substances, and this is why the response of the stand to elevated CO_2 is very small. The model assumes no acclimation in foliar nitrogen content. If trees were to acclimate their leaf nitrogen content to the availability of nitrogen, the growth of trees would also have reacted along increasing atmospheric CO_2 concentration.

The expected temperature increase has two effects: it prolongs the metabolically active season and accelerates decomposition of soil organic matter. Extension of the active season increases annual photosynthetic production, and its effects are rather similar to those obtained for increasing CO_2 concentration, that is, the effect is quite small if it is not associated with increased nitrogen uptake from soil.

The decomposition of soil organic matter is slow in boreal forests since the pools of proteins—lignin, cellulose and lipids—are over tenfold when compared with the annual litter input. The model MicroForest has a special sub-model for decomposition of soil organic matter. The value of the parameter b_{1n} in Eq. 7.25 determines the decomposition rate of proteins in the simulations. The effect of increasing temperatures on stand growth was introduced by assuming that the dependence of the value of the parameter b_{1n} for temperature increase ΔT is

$$b_{1\,n}(\Delta T) = {q_{10}}^{\Delta T}\; b^0_{1\,n}, \tag{8.10}$$

where q_{10} is parameter and b^0_{1n} is the parameter value used in the Chap. 7.

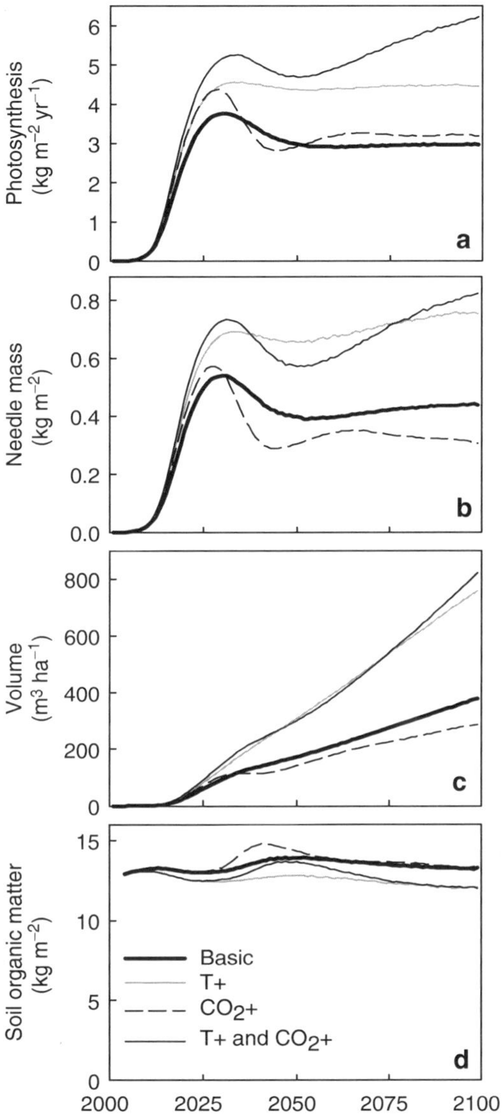

Fig. 8.19 Simulated photosynthesis (expressed in units of sugars) (**a**), needle mass (**b**), volume (**c**) and mass of soil organic matter (dry weight) (**d**) by MicroForest for the years 2000–2100 with different scenarios: no change in temperature or atmospheric CO_2 concentration, increase in temperature (0.4°C/decade), increase in CO_2 concentration (1.5 ppm year^{-1}) and increase both in temperature and CO_2 concentration

Temperature records show warming in southern Finland, although the interannual variation nearly hides the signal. We assumed that temperature will increase 0.4°C/decade which is a higher than the IPCC estimate of warming in the boreal zone. When this temperature increase is combined with the dependence of the value of the parameter on temperature, we get the result that decomposition of soil organic matter will be 20% faster in the year 2020 than in the year 1980 and 70% faster in the year 2100.

The simulations assuming accelerated decomposition of soil organic matter due to temperature increase indicated a strong response. The needle mass increased 70% at the end of the simulation in the year 2100 (Fig. 8.19b) and the annual photosynthetic production of the stand reflected the increase in needle mass (Fig. 8.19a). Annual wood increment responded also strongly (Fig. 8.19c). The increased availability of nitrogen in the soil due to accelerated decomposition of proteins in soil generates these large responses in simulations.

The increasing decomposition of soil organic matter generated large responses in the simulations. It is surprising that changes in the pools of soil organic matter are small, having only a very small declining trend (Fig. 8.19d), quite differently from the prediction of earlier models (e.g. Cox et al. 2000). The accelerated decomposition of proteins in soil enables a clear increase in nitrogen uptake and synthesis of functional substances, especially for leaves and fine roots. The accelerated decomposition is compensated to a great extent by the increased litter input on the soil. Thus, the weak response of soil organic matter can be explained by the compensating effect of increased litter fall.

When the accelerated decomposition due to the temperature increase and increasing CO_2 concentration were combined in the simulation, still larger responses were obtained. The needle mass increased 90%, photosynthesis 110% and stand volume 120%. The large organic molecules in the soil decreased by 10%. This small decrease can be explained with increased litter input that compensates for the faster decomposition. The strong response can be explained with high availability of nitrogen for the synthesis of functional substances and with accelerated photosynthesis due to increased availability of CO_2.

The annual increase in the release of ammonium from proteins in the soil is used to synthesise proteins in leaves, water transport system and fine roots. The lifetime of needles and fine roots is short, and thus nitrogen returns quickly into soil as proteins. In contrast, most of the woody structures endure for the whole lifetime of the trees, and proteins accumulate into woody structures. Thus, effectively, proteins are moved from soil into the pool of woody structures in trees. Since the annual release of nitrogen determines the fertility of the site (Sect. 7.3.3) and the protein concentration in wood is over one magnitude smaller than in soil, the increase in wood mass is about one magnitude larger than the loss in soil organic matter.

Anthropogenic nitrogen deposition is large in the south-western corner of European boreal forests; it can be over tenfold when compared with natural deposition. The leaching and evaporation of nitrogen are very small at SMEAR II (Sect. 5.6), and thus the nitrogen circulates efficiently within the ecosystem. The accumulation is, however, very slow; it can last for centuries. Magnani et al. (2007) reported clear relationship between nitrogen deposition and carbon fixation by ecosystems. Eddy Covariance measurements in stands forming chronosequences were used to obtain the net primary production. The material consists of 20 stands growing in Europe, Siberia and North America. Most of the stands are coniferous but four of them are deciduous. Scots pine is the dominating tree species on six sites (including SMEAR II stand). The analysis by Magnani et al. (2007) resulted in a clear relationship between nitrogen deposition and net primary production as shown in Fig. 8.20a.

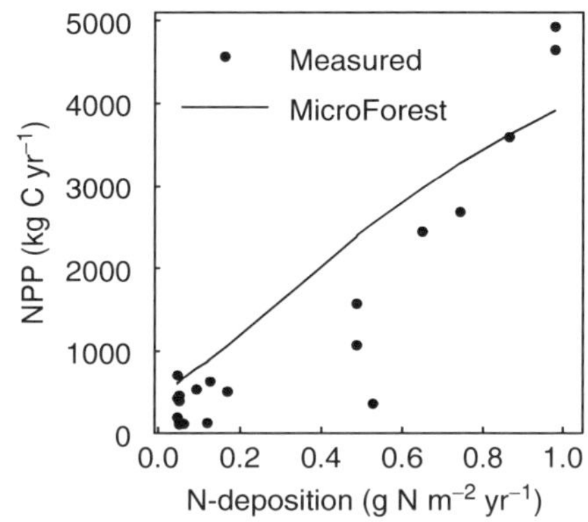

Fig. 8.20 Relationship between nitrogen deposition and net primary production. *Dots* indicate measured values reported by Magnani et al. (2007), and *line* indicates the response simulated with MicroForest

MicroForest is based on carbon and nitrogen fluxes in forest ecosystem, and nitrogen deposition is included as input flux in the ecosystem (Fig. 7.2). We simulate the effect of nitrogen deposition on net primary production with MicroForest using the site-specific nitrogen depositions as in the study by Magnani et al. (2007). We applied linearly increasing nitrogen depositions in such a way that at the end of the simulations at the age of 40 years, the deposition was as used in the study by Magnani et al. (2007). We also determined the net primary production of the stand as the annual increase in wood and soil organic material mass at the end of the simulations. The simulated relationship between nitrogen deposition and net primary production is rather close to that reported by Magnani et al. (2007). However, the shape of the relationship is different, since in the original study the regression is nonlinear and simulations result in linear relationship. The nitrogen impact reported by Magnani et al. (2007) has been argued to be too large. The simulations would, however, suggest that slowly increasing deposition and accumulation into the ecosystem might bring about the observed pattern.

8.3.5 Response of Boreal Forests to Climate Change

Pertti Hari, John Grace, Liisa Kulmala, Federico Magnani, Eero Nikinmaa, Twan van Noije, Jukka Pumpanen, Jouni Räisänen, David Stevenson, Timo Vesala, and Markku Kulmala

Boreal forests are versatile ecosystems, but there are, however, similarities among them. The most important similarity is that the processes are the same. All forests photosynthesise, take up nutrients, grow and finally die. Evidently, also the

P. Hari (✉)
Department of Forest Sciences, University of Helsinki, P.O. Box 27, 00014 University of Helsinki, Helsinki, Finland
e-mail: pertti.hari@helsinki.fi

regularities in the structure are based on water transport and nitrogen allocation. Thus, behind the versatile species, the same processes and structural regularities together with their dependencies on environmental factors govern the flows of material and energy within forests and between forest and the atmosphere. We urgently need understanding of the changes in the material and energy fluxes in forests because the forest soils contain large pools of carbon that may be released into atmosphere with climatic warming.

The boreal forests cover a huge area between 50 and 70°N: their total area is about 1.66×10^9 ha. There are several species forming stands, and the variation in fertility and the length of growing season is large. On the other hand, we have available very detailed information from one stand (SMEAR II) and dynamic model (MicroForest), based on carbon and nitrogen flows and structural regularities, tuned to the SMEAR II stand and tested with stands at the southern border of boreal forests. We need urgently to understand the effects of climate change on boreal forests. Thus, the question arises, can we apply the model MicroForest to all boreal forests?

We are facing two unpalatable alternatives, either we introduce strong assumptions and apply MicroForest or we state that it is premature to say anything about the behaviour of boreal forests under climate change. We think that valuable insights can be gained when applying MicroForest on all boreal forests even though the analysis is rough. The analysis should be tested with new material, and the detected shortcomings should be improved. In this way, we can improve our understanding in the long run. The other alternative is a blind alley and will not result in progress in the field.

Boreal forest has most tendency of all forests to grow as even-aged stands. They are 'born' after some catastrophe, such as clear cut, fire or storm, and they develop until a new catastrophe occurs, often at a rather high age of the stand. The natural fire cycle for the boreal biome is estimated to be approximately 100–150 years (Gauthier et al. 1996). The rotation period in forestry is about 100 years. Also more complicated stand structures exist, but we do not consider them in this first approximation of the changes in the carbon pools in boreal forests.

Boreal forests include several carbon storages that are relevant in the connection of climate change. Let $M_B(t)$ denote the amount of a carbon component in boreal forests as a function of time t and $F_B(t)$ the flow of a compound between the boreal-forests and the atmosphere. The component can be needle mass, amount of wood, amount of proteins, lignin, cellulose, lipids and starch in soil and the flow of CO_2, VOCs or N_2O. Changes in these masses or flows reflect the response of boreal forest to climate change. Let $M_S(t,\ \tau_S)$ denote the masses in a stand at stand age τ as a function of time and $F_S(t,\ \tau_S)$ the fluxes correspondingly. If the boreal forests are formed by even-aged stands, then we can approximate the masses and fluxes by summing the stand-specific values over the whole area. Thus,

$$M_B(t) \approx \sum_{\tau_S=1}^{\infty} A(t, \tau_S)\ M_S(t, \tau_S) \tag{8.11}$$

$$F_B(t) \approx \sum_{\tau_S=1}^{\infty} A(t, \tau_S)\ F_S(t, \tau_S), \tag{8.12}$$

where $A(t, \tau_s)$ is the area of stands having age τ_s in the boreal forests as a function of time t.

We can obtain approximation for any of the masses $M(t, \tau_s)$, at moment t and age τ_s from simulations with MicroForest. The development of the areas of stands at a given age through the centuries is difficult to obtain. This problem can be avoided by assuming that all stands have the same maximum age, often called as rotation time, and that the areas of each stand age classes are the same. Let τ_R denote the rotation time. Then we get

$$M_B(t) \approx \sum_{\tau_S=1}^{\tau_R} \frac{A_B}{\tau_R} M_S(t, \tau_S) \tag{8.13}$$

$$F_B(t) \approx \sum_{\tau_S=1}^{\tau_R} \frac{A_B}{\tau_R} F_S(t, \tau_S), \tag{8.14}$$

where A_B is the total area of boreal forests.

The boreal forests cover a huge area on the northern hemisphere, and the climate change varies within it. The spatial variation in the atmospheric CO_2 concentrations is really small, even the variation within years is larger. The temperature increase has some geographical features that we should introduce into our analysis. The nitrogen oxides have a rather short lifetime in the atmosphere, and they deposit within thousands kilometres from the emission source. This gives a clear geographical deposition pattern since the emissions are strongly concentrated into the industrialised centres in Europe, North America and eastern Asia. We consider that the boreal forests cover the land area between 50 and 70°N, and we divide this area in east–west direction into six zones each of them covering 60°. In this way, we obtain the following areas: Europe, Central Asia, eastern Asia and western, central and eastern North America.

We used the CO_2 concentrations from the SRES A1B emissions scenario, as tabulated in Appendix II.2 of Houghton et al. (2001) and averaged over the ISAM and BERN-CC reference simulations.

We derived the estimates for the twenty-first-century temperature changes in the May–June–July–August (MJJA) season from the SRES A1B climate change simulations included in the Third Coupled Model Intercomparison Project (CMIP3) data set (Meehl et al. 2007). We averaged geographically these changes over land grid boxes in the zone 50–70°N, dividing both North America and Eurasia to three longitude bands. The resulting temperature changes in the twenty-first century are 2.8°C in western, 3.2°C in central, and 3.0°C in eastern North America and 2.9°C in western, 3.5°C in central and 3.0°C in eastern Eurasia. We assumed that the temperature increase started in the year 1965, which is in agreement with measurements (Smith and Reynolds 2005), and it will continue during the twenty-first century according to the scenarios.

We estimated nitrogen deposition to boreal forests in each of the above regions for the time period 1960–2000 using results from the TM4 chemistry-transport model, driven by meteorological data from ERA-40 (European Centre for Medium-Range Weather Forecasts Re-Analysis), as part of the RETRO (REanalysis of the TROpospheric chemical composition over the past 40 years) project (Schultz et al. 2007). Changes in the magnitude and spatial distribution of all the major anthropogenic and natural emissions of oxidised and reduced nitrogen compounds are the main drivers of trends in nitrogen deposition. Prior to 1960, we extrapolate backwards in time using decadal emission estimates from the EDGAR-HYDE data set (van Aardenne et al. 2001), which spans 1890–1990. We calculated for each region, the ratio of the 1960s mean nitrogen deposition from TM4 to the mean 1960s EDGAR-HYDE emissions; we used obtained value to scale the emissions for earlier decades. Prior to 1890, deposition rates were assumed to remain at their 1890s values. Future levels of nitrogen deposition strongly depend upon future emissions and in particular the strength and level of implementation of air quality legislation (Dentener 2006). Here, we assume deposition remains constant beyond 2000.

We assume that the material and energy fluxes in boreal forests will respond to climate change as described in MicroForest, and we apply MicroForest to all boreal forests. In addition, we normalise the results in such a way that the biomass is 4.2×10^7 g ha^{-1} in a year in which the value has been measured for boreal forests in North America (Botkin and Simpson 1990). We analyse the responses of boreal forests to increasing CO_2 to nitrogen deposition and to accelerated decomposition due to temperature increase.

There are rather large differences in the nitrogen deposition between the six areas in boreal forests. Nitrogen deposition is largest near the big industrialised centres and moderate in the central North America and Siberia. The expected temperature increase has more even spatial distribution; the expected increase at 2100 varies from 2.8 to 3.5°C. The winds in the atmosphere mix the emitted CO_2 so effectively that the spatial concentration differences are minimal, and they do not generate spatial features in the responses of boreal forests.

We extended the simulations with MicroForest to cover boreal forests over the period 1750–2100. The nitrogen deposition increases carbon sequestration per hectare over two fold in western Eurasia when compared with east North America. The area of high nitrogen deposition in boreal forests is, however, so small that it reduces the importance of nitrogen deposition in the boreal scale.

When only CO_2 concentration changes in the simulations, then the response begins in the late nineteenth century, accelerates slowly tree growth during the twentieth century and rapidly declines after the year 2000 (Fig. 8.21a). The growth decrease in simulations is caused by accumulation of nitrogen in the soil and the shift in allocation from leaves to roots.

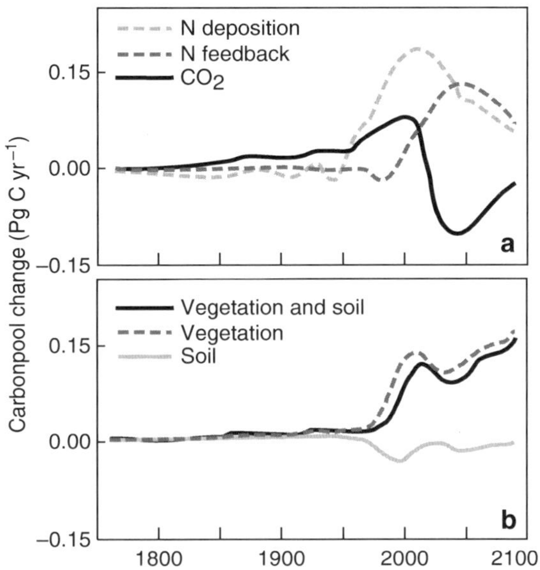

Fig. 8.21 Simulated changes in carbon pool of boreal forests as response to single changing factor (**a**) and as a response to temperature and CO_2 increase (**b**)

Changes in nitrogen availability and the CO_2 increase have strong interaction in carbon sequestration and the response cannot be derived from the single factor simulations. The trees can utilise effectively the additional sugars caused by enhanced photosynthesis associated with increased atmospheric CO_2 concentration to the synthesis of cellulose, lining and lipids for cell walls and membranes when simultaneously enhanced nitrogen availability, either due to accelerated decomposition of proteins or to nitrogen deposition, provides material for protein synthesis to construct the necessary enzymes and membrane pumps in the additional cells (Fig. 8.21b).

The large differences in nitrogen deposition generate geographical features in the responses of boreal forests to climate change. The sequestration increase is in western Europe having high nitrogen deposition, about twofold when compared with eastern North America (Fig. 8.22).

When we add together the responses of the six geographical areas we obtain the carbon sequestration in the boreal forests. According to our simulations, the carbon sequestration in boreal forests started at the beginning of the twentieth century, grew slowly, peaked in the beginning of the twenty-first century to 0.5 Pg (C) year^{-1} and thereafter slowly declined (Fig. 8.23). Enhanced tree growth resulted in the accumulation of wood and other organic matter in boreal forests.

No data of the past carbon stocks in boreal forests are available, and we can only compare our simulations with other results dealing with northern forests around the year 2000. National forest inventories are based on large systematic samples of forest stands. The results provide reliable estimates of time series of several quantities describing forests in Finland. The Finnish forest inventories

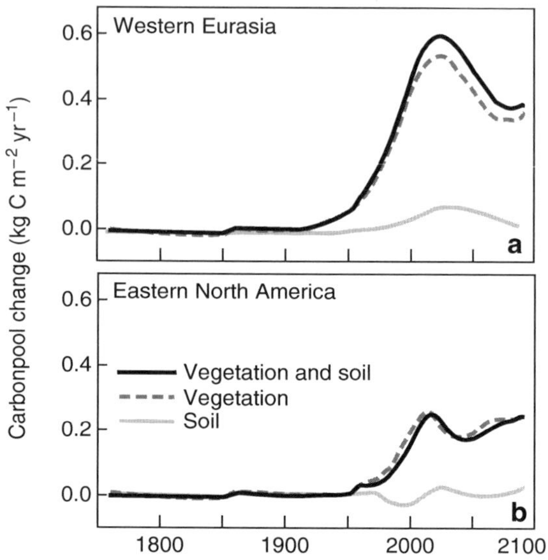

Fig. 8.22 Simulated changes in carbon pool in western Eurasia (**a**), a region of high anthropogenic nitrogen deposition and in eastern North America (**b**), a region of low anthropogenic nitrogen deposition

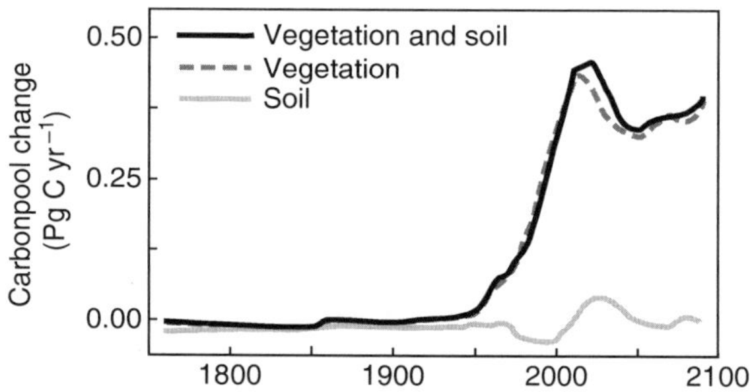

Fig. 8.23 Simulated changes in carbon pools of vegetation, soils and ecosystems in boreal forests during 1750–2100

indicate that annual growth increased from 50 to 100×10^6 m^3 $year^{-1}$ during the period 1950–2000. When the growth increase is converted from volume units to mass of carbon per unit area, we get 0.5 kg (C) m^{-2} $year^{-1}$. The simulations of carbon sequestration by vegetation for western Eurasia in the year 2000 result in 0.52 kg (C) m^{-2} $year^{-1}$, which is close to what is obtained in the Finnish national forest inventories. Drainage of peatlands, changes in forest structure and nitrogen fertilisation have all contributed to the forest growth increase in Finland. Thus, the simulated effect of climate change in carbon sequestration is evidently too large.

The analysis by Myneni et al. (2001) of temperate and boreal forests based on forest inventories and satellite images indicates carbon sequestration of 0.68 Pg (C) year^{-1} by temperate and boreal forests during the period 1981–1999. The result obtained with MicroForest simulations for carbon sequestration by vegetation in boreal forests during the year 2000 is 0.5 Pg (C) year^{-1}. The analysed satellite images cover large areas south of boreal forests, and some northern boreal forests are evidently missing in the material. This discrepancy may explain the difference in the result obtained by Myneni et al. (2001) and simulations with MicroForest.

Stephens et al. (2007) estimated that the carbon sink of northern terrestrial areas in the year 2000 is 1.5 Pg (C) year^{-1} using measured atmospheric CO_2 concentrations and inverse modelling. But there is considerable uncertainty attached to this figure and the northern terrestrial area is rather poorly specified going much further south than in our simulation. Moreover, to obtain comparable results, the amount of wood cut and used in forestry has to be added to the 0.5 Pg (C) year^{-1} found in the MicroForest simulations. According to the official forest statistics by FAO (UN-ECE/FAO), the annual fellings of coniferous trees in the boreal region are 979 million cubic metres, equivalent to about 0.2 Pg (C) year^{-1}; thus, the simulated carbon sink in boreal forests is 0.7 Pg (C) year^{-1}, well within the uncertainty of the result of the inverse modelling.

Four different methods, national forest inventories, remote sensing, atmospheric inversion estimates and MicroForest model, have indicated strong carbon sink and sequestration into boreal forests. The magnitudes of the sink estimates are rather close to each other, although the areas are not exactly comparable with each other. The results based on direct measurements indicate the present situation, but the simulations with MicroForest have the additional advantage that they are able to predict the future carbon sequestration. Moreover, the simulations provide insights into the underlying biological causes. Processes in the boreal forests respond to the changes in the atmospheric properties generating changes in the flows in forest ecosystems. The accelerated decomposition of proteins in the soil, the increased nitrogen deposition from the atmosphere and enhanced photosynthesis due to elevated CO_2 concentration are able to amplify considerably the carbon sequestration in the boreal forests. The development of a simplified version of MicroForest and the introduction of it into GCMs (global climate models) will evidently improve our analysis.

Our results indicate that boreal forests responded strongly to climate change starting from the middle of the twentieth century. They will continue carbon sequestration in the coming decades if there are no increases in disturbance, for example, extensive outbreaks of insects or pathogens. The major influence seems to be through increased rate of nitrogen release from the soil organic matter due to temperature increase. The associated growth enhancement of vegetation seems to more than compensate for the net carbon release from the soil.

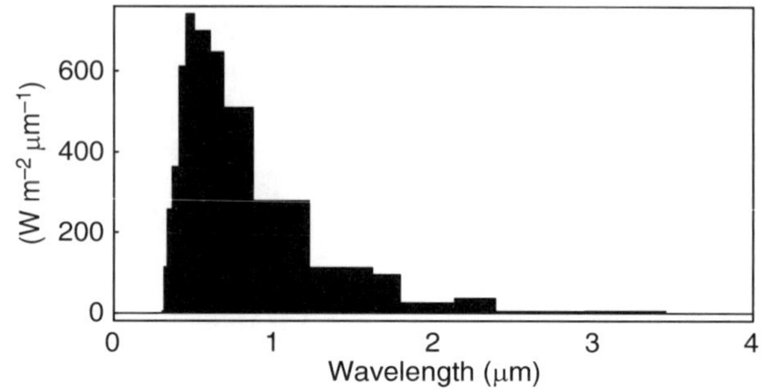

Fig. 8.24 Modelled 25-band spectrum of the radiation penetrating atmosphere, as used in calculating the surface albedo

8.3.6 Feedback from Forests to Climate Change

8.3.6.1 Climatic Effects of Increased Leaf Area: Reduced Surface Albedo and Increased Transpiration

Jouni Räisänen and Sampo Smolander

Green needles absorb strongly solar radiation and convert the radiation energy to heat. In contrast, snow reflects most of the radiation. The temperature increase in a forest stand will accelerate protein decomposition in the soil (Sect. 4.4) which can lead to better forest growth and increasing needle mass in the stand (Sect. 8.3.4). Thus, the increasing needle mass may generate positive feedback to climate change.

The increasing needle mass could result in several climatic feedbacks (Sect. 8.3.2). In this subsection, we focus on two of them: the impact of the leaf area change on surface albedo and its impact on transpiration. After a brief physical discussion of these feedbacks, their potential climatic importance is evaluated in simulations conducted with a state-of-the-art atmospheric general circulation model.

The spectral composition of the incoming below-atmosphere radiation and the spectral albedo of the surface determine the total albedo of a land surface. There are computer codes to determine the spectrum of incoming radiation (Freidenreich and Ramaswamy 1999; Stamnes et al. 1988, 2000). We assumed cloud-free subarctic winter atmospheric conditions (McClatchey et al. 1971) with a visible optical depth of 0.2 for continental aerosols and a solar zenith angle of 60°. The resulting spectra are shown in Fig. 8.24.

Computer models for (1) optical properties of single ice crystals and (2) for the scattering and reflection of light in a layer of snow[2] were used to determine

[2]Available from NASA GSFC FTP ftp://climate1.gsfc.nasa.gov/wiscombe/

J. Räisänen (✉)
Department of Physics, University of Helsinki, P.O. Box 48, 00014 University of Helsinki, Helsinki, Finland
e-mail: jouni.raisanen@helsinki.fi

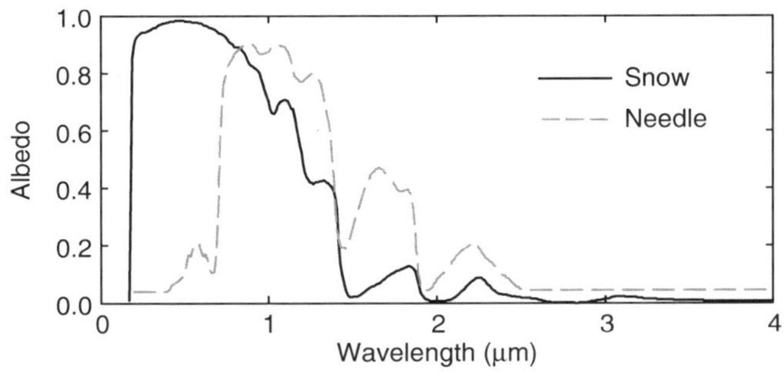

Fig. 8.25 Spectral albedo for snow and conifer needles

the spectral albedo of snow. The snowbed was modelled as an optically thick layer of spherical ice crystals (150 μm diameter). The resulting spectra are shown in Fig. 8.25. When calculating the snow-covered land surface albedo with these spectra, the result is 0.78, close to the value 0.8 used in the ECHAM5 model (Roeckner et al. 2003) as the maximum albedo of unvegetated snow-covered areas.

Coniferous needles absorb, reflect or transmit solar radiation. The energy of the absorbed radiation heats the needle. The LIBERTY model by Dawson et al. (1998) was used to model the fraction of incoming radiation that a needle absorbs, or reflects and transmits, at different wavelengths. The obtained needle spectral albedo is shown in Fig. 8.25. The needle has different albedo at different wavelengths, but if illuminated by the previously modelled skylight, its average albedo is 0.42 (calculated as a weighted average of the needle spectral albedo, with weights from Fig. 8.24), thus rather different from snow.

The spectrum of incoming skylight and the albedo spectrums for snow and conifer needles provide the necessary input to model the albedo of coniferous forests on a snow-covered ground. We used a two-stream canopy radiative transfer model by Ross (1981), modified for coniferous canopies as proposed by Smolander and Stenberg (2005). The model allows for the multiple scattering of radiation inside the forest canopy and between the canopy and the snow-covered ground. The canopy models account for the clumping of needles into coniferous shoots and the resulting effects for small-scale multiple scattering but exclude larger-scale structures (branches and tree crowns) typical of coniferous forest canopies. The lack of tree crowns is a somewhat unrealistic assumption, but in this case, we opted for simplicity. After several model runs for canopies with different leaf area index, the resulting canopy albedo as a function of canopy leaf area is shown in Fig. 8.26. As needles are non-flat objects, the coniferous canopy leaf (needle) area index was defined on a half of the total surface area basis (Chen and Black 1992).

All other factors being the same, transpiration from vegetation should be directly proportional to the transpiring surface area. Thus, for example, a doubling of leaf area should lead to a doubling of transpiration. In reality, this is a gross overestimate. This is because transpiration is limited by the availability of water and, in particular,

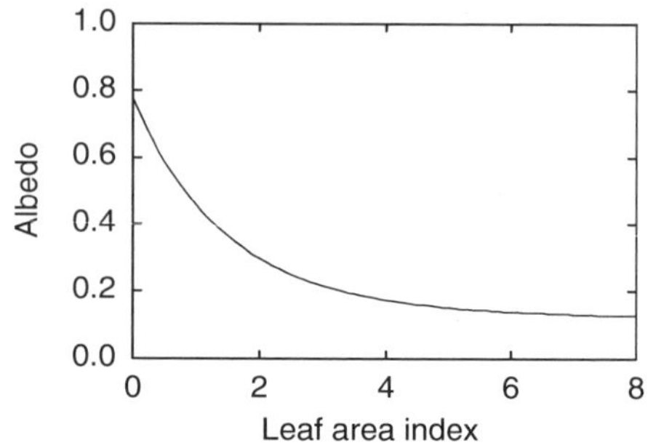

Fig. 8.26 Albedo of coniferous canopy on a snow-covered ground, as a function of canopy leaf area index

by the availability of energy. Furthermore, an initial increase in transpiration tends to increase boundary-layer relative humidity. This reduces the vapour pressure difference between plant stomata and the near-surface air, thus acting as a negative feedback to transpiration. On the other hand, any changes in transpiration may affect other aspects of the hydrological cycle, such as cloudiness and precipitation.

We conducted sensitivity experiments to estimate the impact of the leaf area albedo and transpiration feedbacks in more quantitative terms with the ECHAM5 atmospheric general circulation model (GCM) (Roeckner et al. 2003, 2006) coupled to a simple model of the upper ocean (Box 8.2).

Box 8.2

ECHAM5 is a state-of-the-art atmospheric GCM developed by the Max Planck Institute for Meteorology in Hamburg. For the simulations described here, ECHAM5 was run with a relatively low horizontal resolution (spectral truncation to total wave number 31, corresponding to a grid spacing of approximately 3.75° in latitude and longitude) and with 19 levels in the vertical. The atmospheric model was connected to a simple model of the upper ocean, consisting of a heat balance equation for a 50-m-thick water layer and a prognostic equation for sea ice thickness. To compensate for the lack of ocean currents in the thermodynamic ocean model, flux adjustments (Sausen et al. 1988) were used to keep sea surface temperatures near observed present-day values in the control simulation. An evaluation of the simulated present-day climate with observational data revealed a level of skill well comparable with other global climate models (Räisänen 2007). In a standard experiment with a doubling of the atmospheric CO_2 concentration, the model simulated a global mean warming of about 3°C, in good agreement with other models and the Intergovernmental Panel on Climate Change best estimate of climate sensitivity (Meehl et al. 2007).

In the ECHAM5 standard set-up, the evolution of leaf area follows a prescribed seasonal cycle based on the data set compiled by Hagemann (2002). The growing and dormant season values of leaf area are derived from ecosystem type classifications from the US Geological Survey (2001), and the seasonal cycle between these

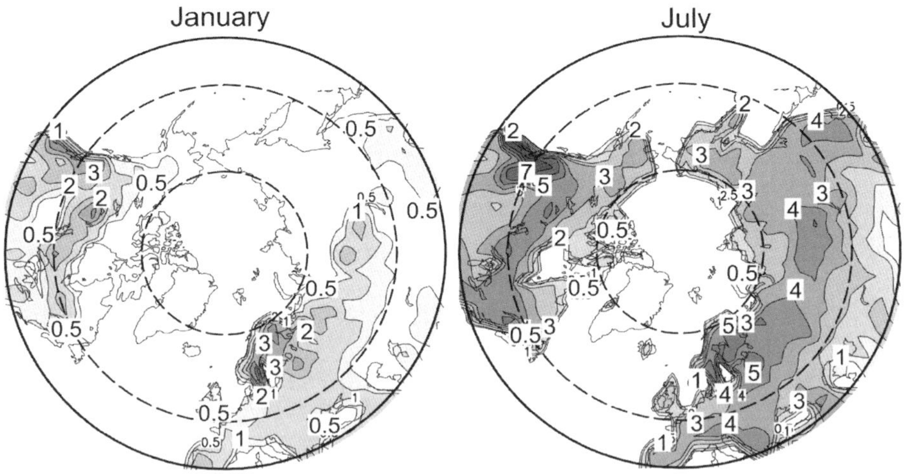

Fig. 8.27 Leaf area index (LAI) in the ECHAM5 control run in January and July

Fig. 8.28 Seasonal cycles of leaf area index (LAI) as averaged over the Eurasian and North American continents at latitudes 50–70° N, in the control simulation (CTRL) and the two sensitivity experiments LAI_PLUS and LAI_DOUBLE

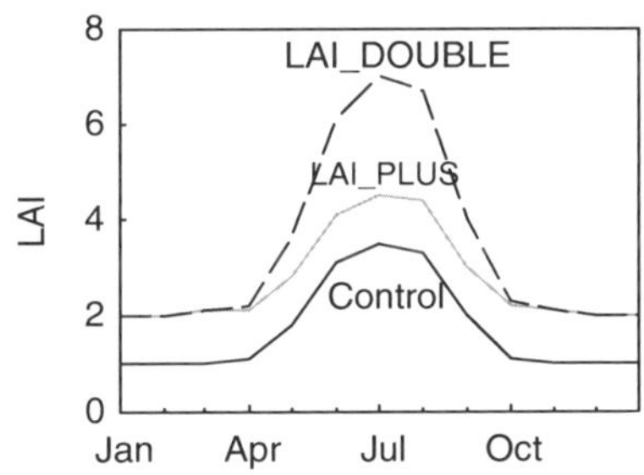

extremes is based at high latitudes on observed monthly mean temperatures. In all months with an observed mean temperature of less than 5°C, leaf area retains its dormant season (winter) value (Fig. 8.27).

To study the sensitivity of the simulated climate to changes in leaf area, two 30-year model simulations with increased leaf area were made and compared with a control simulation (CTRL) with standard leaf area values. In one (LAI_PLUS), a seasonally invariant increase was added to the standard leaf area values at latitudes 50–70°N. The magnitude of this increase was such that the dormant season leaf area was doubled. In the growing season, the absolute increase was equally large, but the relative increase was smaller. In the other experiment (LAI_DOUBLE), leaf area was doubled in each month. All three simulations assumed the same near-present concentrations of greenhouse gases. The seasonal cycles of leaf area in the three simulations, averaged over the Eurasian and North American continents at latitudes 50–70°N, are shown in Fig. 8.28. Note that LAI_PLUS and LAI_DOUBLE are essentially identical from October to April so that they only differ from late spring to early autumn.

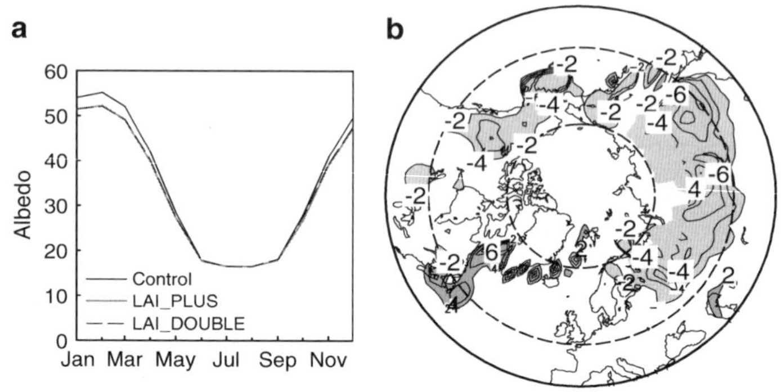

Fig. 8.29 Seasonal cycles of surface albedo, as averaged over the Eurasian and North American continents at latitudes 50–70°N, in the three simulations (**a**). The difference of surface albedo between the simulations LAI_PLUS and CTRL in February–March–April mean surface albedo (**b**). Contour interval 2%, zero contour omitted. *Light* (*dark*) shading indicates areas where the albedo has decreased (increased) by at least 2%

All three simulations were run for 30 years. To avoid eventual spin-up problems in the beginning of the simulations, the mean values shown here were calculated over the last 25 years.

The increase in leaf area affects the simulated climate in ECHAM5 by two primary mechanisms. First, it acts to reduce the surface albedo when there is snow on the ground. Second, it affects the calculation of surface fluxes, including the partitioning of available energy between sensible and latent heat fluxes. Other potential consequences of increased vegetation activity, such as changes in carbon cycle and in the production of biogenic aerosols, are not presented in these simulations.

The average seasonal evolution of the simulated surface albedo in the boreal forest zone is shown in Fig. 8.29a. Because leaf area only affects the albedo in ECHAM5 when there is snow on the ground, there is no difference between the three simulations in summer. In winter and early spring, the average surface albedo is about 3% lower in the simulations LAI_PLUS and LAI_DOUBLE than in CTRL. The geographical distribution of the difference of surface albedo between the simulations LAI_PLUS and CTRL during late winter and early spring (February–March–April) is shown in Fig. 8.29b. Because surface albedo is also affected by changes in snow and ice conditions, the changes are not exclusively limited to those areas where leaf area was increased.

Figure 8.30 shows the relative humidity, total cloudiness, precipitation and temperature differences in the simulations LAI_PLUS and CTRL and LAI_DOUBLE and CTRL, as averaged over the whole boreal forest zone at 50–70°N. The decrease in surface albedo shown in Fig. 8.29 acts to increase the amount of solar radiation absorbed at the surface in winter and spring (Fig. 8.31a). The increase is largest (on the average about 2 W m^{-2}) in March and April when there is much more solar radiation available than earlier in winter.

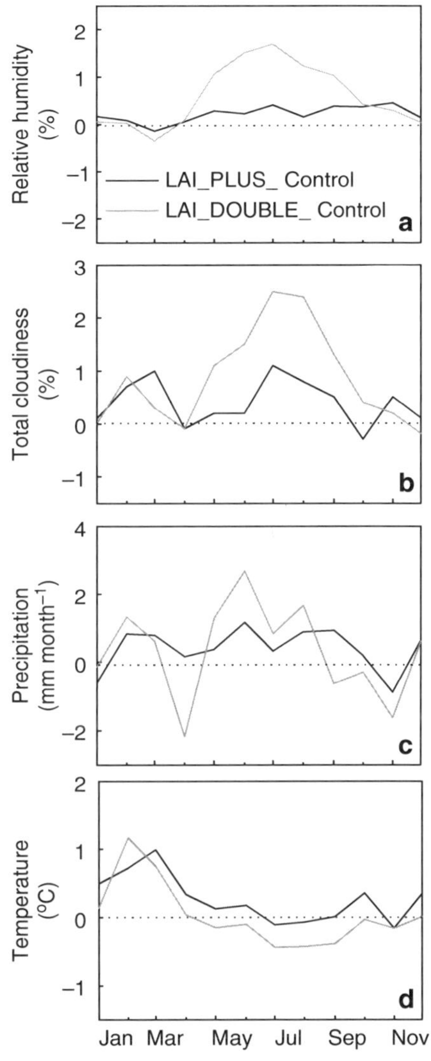

Fig. 8.30 Differences between the simulations LAI_PLUS and CTRL, and LAI_DOUBLE and CTRL in near-surface relative humidity (**a**), total cloudiness (**b**), precipitation (**c**) and surface air temperature (**d**), as averaged over the Eurasian and North American continents at latitudes 50–70°N

In summer, absorbed solar radiation decreases, particularly in the simulation LAI_DOUBLE, in which the increase in leaf area is largest. At the same time, there is an increase in thermal radiation (Fig. 8.31b), although this does not fully compensate the decrease in absorbed solar radiation. Both of these changes appear to be caused by increased cloudiness (Fig. 8.30b) that is most likely a consequence of the leaf area transpiration feedback.

Mean changes in the sensible and latent heat fluxes of the boreal forest zone are shown in Fig. 8.31c, d. For most of the year, the latent heat flux is larger in the simulations LAI_PLUS and LAI_DOUBLE than in CTRL. The sensible heat flux increases relative to CTRL in March and April but decreases in the summer,

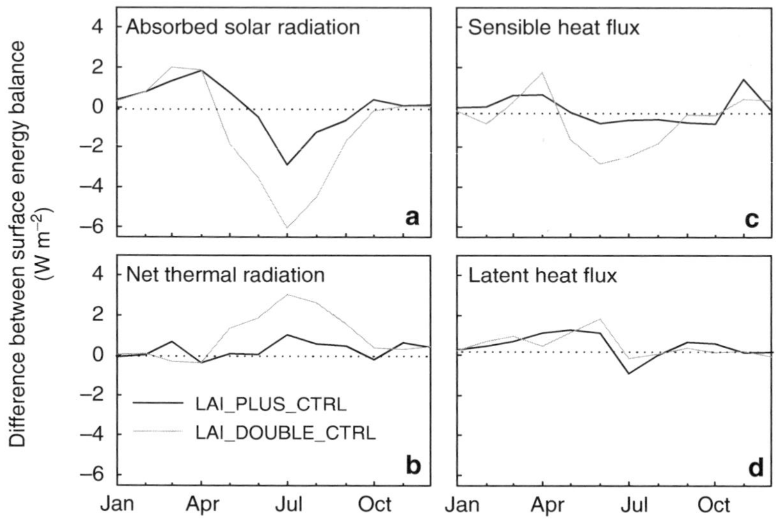

Fig. 8.31 Differences in surface energy balance (W m^{-2}) between the simulations LAI_PLUS and LAI_DOUBLE and the control simulation, as averaged over the Eurasian and North American continents at latitudes 50–70°N. Absorbed solar radiation (**a**), net (downwards minus upwards) thermal radiation (**b**), sensible heat flux (**c**), and latent heat flux (**d**)

particularly in the simulation LAI_DOUBLE. Thus, as expected, the increase in leaf area requires the model to consume a larger fraction of the available radiation energy to the latent heat flux at the expense of the sensible heat flux.

The increase in latent heat flux means an increase in evapotranspiration. However, the increase is relatively modest: in both simulations, LAI_PLUS and LAI_DOUBLE, the annual mean evapotranspiration is only 2% (6 mm $year^{-1}$) higher than in CTRL. The relative change is thus over an order of magnitude smaller than the change of leaf area applied in the experiments. This might partly result from compensation between increasing transpiration and decreasing surface evaporation; unfortunately, this cannot be verified because these two components are not separated in the ECHAM5 output. More importantly, evapotranspiration is limited by the availability of energy.

Another constraint to evapotranspiration is the availability of water. This factor likely limits the simulated increase in evapotranspiration in summer, particularly in the southern parts of the boreal forest zone.

The increase in evapotranspiration and the decrease in sensible heat flux (that reduces the heating of the lower atmosphere) together act to increase the near-surface relative humidity in summer, particularly in the simulation LAI_DOUBLE (Fig. 8.30a). On the other hand, moistening of the surface layer is one of the factors that limit the increase in evapotranspiration. The increase in relative humidity also leads to an increase in cloudiness in summer (Fig. 8.30b). Therefore less solar radiation reaches the surface (Fig. 8.31a), although this is partly compensated by an increase in the atmospheric counterradiation in the thermal part of the

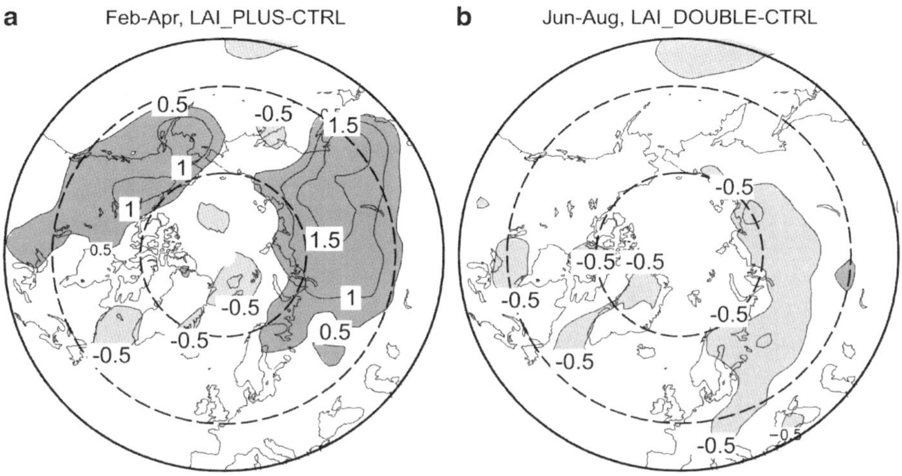

Fig. 8.32 Differences in surface air temperature between the simulations LAI_PLUS and CTRL in February–March–April (**a**) and between the simulations LAI_DOUBLE and CTRL in June–July–August (**b**). Contour interval is 0.5°C; zero contour is omitted. *Dark* (*light*) shading indicates areas where temperature has decreased (increased) by at least 0.5°C

electromagnetic spectrum (Fig. 8.31b). In summary, the changes in the hydrological cycle and in the surface energy balance are strongly interconnected.

The increase in LAI leads to a small overall increase (about 5 mm year^{-1}) of precipitation in the boreal forest zone (Fig. 8.30c). In the simulation LAI_DOUBLE, in particular, the largest increase occurs in summer.

The impacts of increased LAI on simulated surface air temperature are twofold (Figs. 8.30d and 8.32). From January to April, the temperature response is dominated by the decrease in surface albedo. Thus, temperatures are slightly higher in the simulations LAI_PLUS and LAI_DOUBLE than in the control simulation. The largest area mean warming of about 1°C occurs in March in the simulation LAI_DOUBLE and in February in the simulation LAI_PLUS. The warming is greatest in Siberia (Fig. 8.32a). By contrast, the summer mean temperature decreases slightly in most of the boreal forest zone, particularly in the simulation LAI_DOUBLE, at least in part due to the increase in cloudiness (Fig. 8.30b). However, the sign of the summertime temperature changes might be sensitive to the GCM-specific parameterisations of albedo and other surface properties. In an early model study by Bonan et al. (1992), a total removal of boreal forests also led to a cooling in summer, resulting from a much later snowmelt in spring. More recently, Betts et al. (2007) simulated a pronounced cooling in spring but mostly small temperature changes in summer due to historical conversion of North American and Eurasian forests to agricultural land.

Acknowledgments

We thank Dr. Petri Räisänen for the incoming solar radiation computations and Dr. Tristan Quaife for the snow albedo computations.

8.3.6.2 Forests, Aerosols and Climate Change

Ilona Riipinen, Tuukka Petäjä, Pertti Hari, Miikka dal Maso, Jaana Bäck, and Markku Kulmala

The Role of Aerosols in the Atmosphere

Generally, an aerosol is defined as a mixture of a carrier gas and solid or liquid particles suspended in it. In the case of atmospheric aerosols, the carrier gas is air, and the term 'atmospheric aerosol' refers to the mixture of air and *aerosol particles* suspended in the air. When we speak of aerosol concentrations, we refer to the amount (number, surface area or volume/mass) of suspended particles per some unit of carrier gas, usually a volume (like 1 cm^3) of air at atmospheric pressure.

For a particle to be considered an aerosol particle, it needs to remain suspended in air for a significant time. This basically defines a limit for the size of aerosol particles, as larger particles tend to fall to the ground by the pull of gravity. An upper limit for the size can be considered to be a few millimetres, the size of raindrops. Usually, the discussion of aerosol particles is limited to particle sizes smaller than about 100 μm (μm, one millionth of a metre). The lower boundary for the size of aerosol particles is in practise defined by the requirement of a liquid or solid phase (see, e.g. Vehkamaki 2006) and is typically around 1 nm in atmospheric conditions.

Atmospheric aerosols affect the climate in two distinct ways. First, they affect the Earth's radiation budget directly by scattering the solar radiation (direct effect). Second, they act as condensation nuclei for cloud droplets (CCN) having therefore a significant impact on the radiative properties and lifetimes of clouds (indirect aerosol effect). According to the latest report by the Intergovernmental Panel on Climate Change (IPCC 2007), the total climatic effect of atmospheric aerosols is estimated to be cooling: the radiative forcing resulting from the direct effect is estimated to be between -0.1 and -0.9 W m^{-2}, whereas the corresponding estimate for the indirect effect is $-1.8 \ldots -0.3$ W m^{-2}. However, the IPCC also reports that the radiative forcing of the aerosols is currently subject to the largest uncertainties of all individual components in their radiative forcing calculations.

I. Riipinen (✉)
Department of Applied Environmental Science & Bert Bolin Center for Climate Research, Stockholm University, Stockholm, Sweden
e-mail: ilona.riipinen@itm.su.se

In this work, we provide rough estimates on how an increased aerosol number concentration resulting from increased biogenic activity would affect the aerosol radiative forcing.

According to recent global model calculations by Spracklen et al. (2006), approximately 30% of the global particle concentration originates from biogenic secondary aerosol formation events. However, the newly formed particles need to grow fast enough to survive to sizes (ca. 50–100 nm in diameter) at which they can act as cloud condensation nuclei and have a significant climatic effect—otherwise these small particles will be lost by coagulation and deposition processes (see, e.g. Kerminen and Kulmala 2002). Competition between the growth and loss processes is therefore among the main factors determining the magnitude of the particle source provided by the boreal forests.

Aerosol Formation and Growth

Based on their source, atmospheric aerosol particles can be roughly divided into primary and secondary particles. Primary particles, such as dust, sea salt or pollen, enter the atmosphere in the condensed (liquid or solid) phase, whereas secondary particles have been formed from condensable atmospheric vapours. This secondary formation of stable nanosised particles by condensation of gaseous matter is an important source of atmospheric aerosol particles. Such formation often occurs in bursts, called nucleation events, in which new small (a few nanometres in diameter) particles appear and grow to sizes large enough to have climatic importance. Nucleation events have been observed in several different environments around the world, and boreal forests have proven to be an important source of such secondary aerosol particles (see, e.g. Mäkelä et al. 1997; Kulmala et al. 2001, 2004a; Tunved et al. 2006).

An example of a particle formation event taking place at the SMEAR II station in Hyytiälä is shown in Fig. 8.33 (Mäkelä et al. 2000). A basic feature of a particle formation event is an increase in particle number, especially in the smallest observable sizes (3–25 nm). The formed particles then grow for several hours until they have reached sizes of 50–100 nm, where they can act as condensation nuclei for cloud droplets and therefore have a climatic effect. During growth, a part of the freshly formed particles coagulate (i.e. collide and stick) with pre-existing particles, and their number is therefore reduced. The time evolution of the particle size distribution can be obtained by considering the sources and sinks that different dynamical processes provide for the distribution (see, e.g. Seinfeld and Pandis 1998, and Chap. 2).

Atmospheric Aerosol Concentrations

Aerosols absorb, emit and reflect solar radiation as well as thermal radiation of the globe (Sect. 8.3.1). In this way, they contribute to the behaviour of the atmosphere and to the radiation available for photosynthesis. The biogenic volatile organic

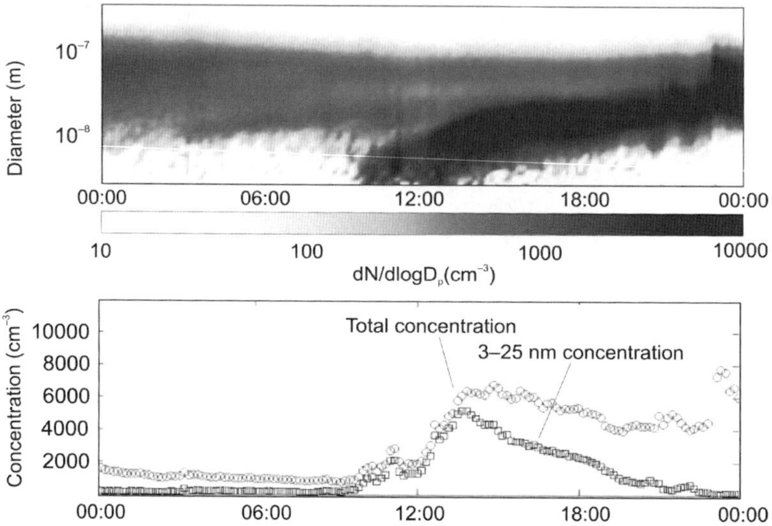

Fig. 8.33 Measured aerosol size distribution data during a particle formation event day. The upper panel shows the time evolution of the particle size distribution, with time on the horizontal axis and particle size on the vertical axis. The colour depicts particle concentrations; the darker the colour, the more particles of that size were observed. The lower panel shows the evolution of the total number concentration (*circles*) and the concentration of the smallest particles (*squares*). One can see that the increase of the particle number is almost exclusively due to formation of the very smallest particles; later, when the particles grow, the small particle number falls below the total number

compounds, BVOCs, emitted by vegetation, effect on the aerosol particle formation and growth. The BVOC emissions react to the climate change, and their contribution to aerosol dynamics will change. Thus, understanding of the formation and growth of aerosol particles is important for the analysis of feedbacks from boreal forests to climate change.

Taking a closer look at the times of particle formation in boreal forest, some clear patterns can be found. The bursts are almost exclusively observed during daytime, at least 2 h after sunrise and on average 3–4 h after sunrise but mostly before noontime. The start time of the bursts typically follows the variation of the time of sunrise (Mäkelä et al. 2000). Incoming solar radiation is found to correlate with the growth rate of the new particles, which leads to speculation that increasing amounts of radiation increase the photochemical production rate of condensing vapour, which increases the formation and growth rates of the particles. These effects naturally increase the probability of particles surviving the early stages of growth until they are detectable.

Probable compounds participating in the atmospheric nucleation are, for example, sulphuric acid, water, ammonia, amines and a variety of organic compounds. In boreal forest, the effect of the organics is clearly seen and a significant part of the condensing vapours is thought to be a mixture of oxidation products of volatile organic compounds (BVOC), which are emitted by trees (e.g. Hakola et al. 2003).

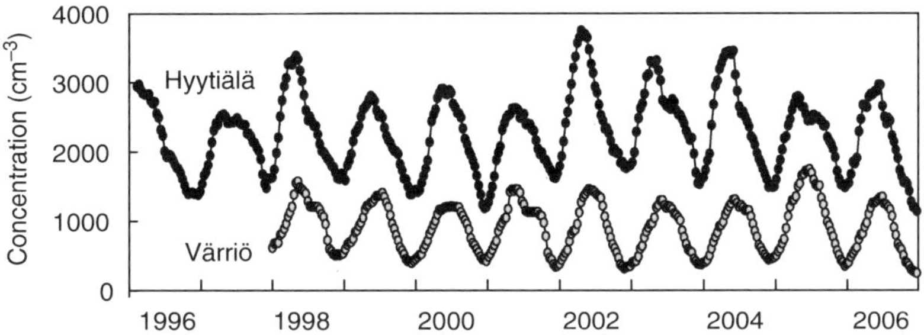

Fig. 8.34 The 4-month running average of the total particle concentration at SMEAR II, southern Finland (Hyytiälä) and at SMEAR I, northern Finland (Värriö). One can easily see the annual variation, with maxima in the summer and minima in the winter. The concentrations are clearly lower at SMEAR I than at SMEAR II

The exact role of condensable organics is not completely understood, but there are clear indications that organics are needed to explain the particle growth rates observed in the boreal forest (Kulmala et al. 2004b). Organics may also affect the very first steps of aerosol formation. However, exact information on, for example, the composition of the smallest atmospheric particles (1–10 nm in diameter) is very difficult to obtain with current commercial instruments.

The condensing vapours affect the aerosol climatic properties in several ways. By making the small particles grow, they increase the atmospheric lifetime of those particles and cause them to reach sizes at which they can act as cloud condensation nuclei. The vapours also naturally affect the composition and therefore the chemical and physical properties of the particles. As the particles deposit on the surfaces, they carry their constituents along with them. Therefore, depositing aerosol particles are also likely to contribute significantly to, for example, the nitrogen deposition.

Total fine particle number concentrations observed in the boreal forests typically range from 10 to 100,000 cm^{-3}, depending on the site, for instance, its distance from roads, cities and other sources of anthropogenic pollution. Figure 8.34 shows an exemplary time series of running 4-month averages of aerosol particle concentrations in the size range of 3–800 nm measured in 1996–2007 at the SMEAR II. The same figure presents also a similar plot for the years 1998–2007 in the cleaner environment of SMEAR I in Värriö. At remote continental sites like Hyytiälä and Värriö most of the particles are typically concentrated in the size range of ca. 50–150 nm. On average, approximately 30% of these, potentially climatically active, particles were originally formed in a nucleation event.

An exemplary plot on the seasonal variation of particle number concentrations in boreal forest is shown in Fig. 8.35, where the 2-week average particle concentration observed at SMEAR II during the year 2000 is plotted. In springtime, a clear maximum in the particle number concentration is observed, and a second, somewhat smaller, maximum is seen in the autumn. The daily variations of particle numbers on a day with clear nucleation and on the other hand on a non-event day are presented in Fig. 8.36. A clear increase in particle concentrations can be observed during new particle formation, compared to a day with no nucleation and growth.

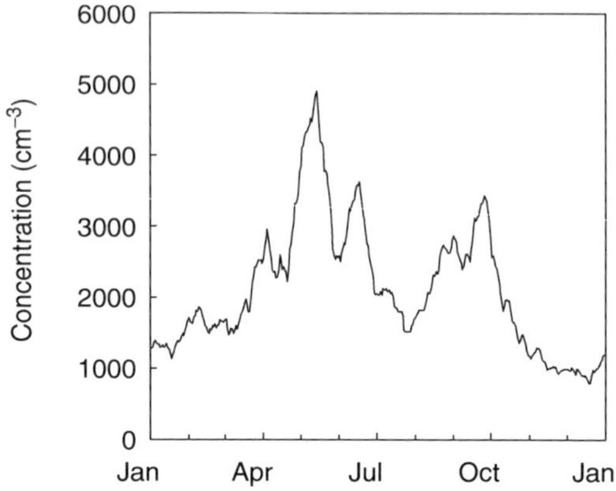

Fig. 8.35 The 2-week running average of the total particle number concentration at SMEAR II, southern Finland, in the year 2000. There is a seasonal variation with maxima in spring and autumn and minima in summer and winter

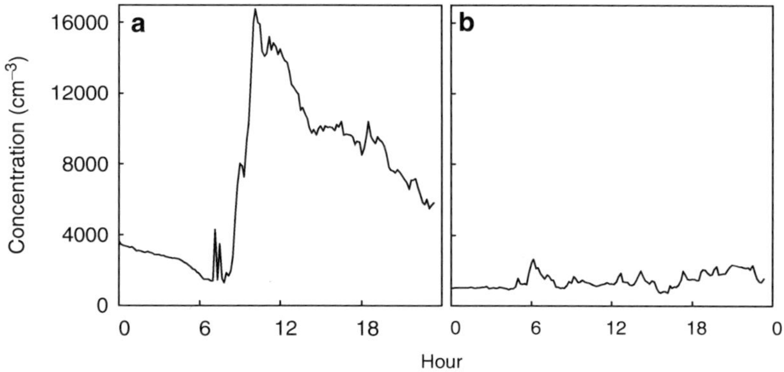

Fig. 8.36 The total particle number concentration during an event day (**a**) a non-event day (**b**) at SMEAR II, southern Finland. During the particle formation period (at ca. 10–11 o'clock) on an event day, the total particle number increases almost 10-fold. The particle number concentration fluctuates around a few thousand particles per cubic centimetre

The vertical profiles of the particle number concentrations vary as a function of, for example, the height of the mixed boundary layer, and the intensity of the possible nucleation in different parts of the troposphere. Figure 8.37 present a vertical concentration profile measured near the SMEAR II station on 3 March 2006 and 13 March 2006 (Laakso et al. 2007). The measurements were taken from a hot air balloon (Fig. 8.38) due to its high lifting capacity and capability to follow air mass. The first of the days shown here corresponds to a day without new particle formation, whereas the latter is a day with a clear nucleation event. The profiles reveal that during new particle formation the particles are clearly concentrated near the surface, whereas on the non-event day no such clear dependence on the measurement height is observed. This indicates that most observed nucleation takes place in the atmospheric boundary layer near the forest canopy, rather than at the

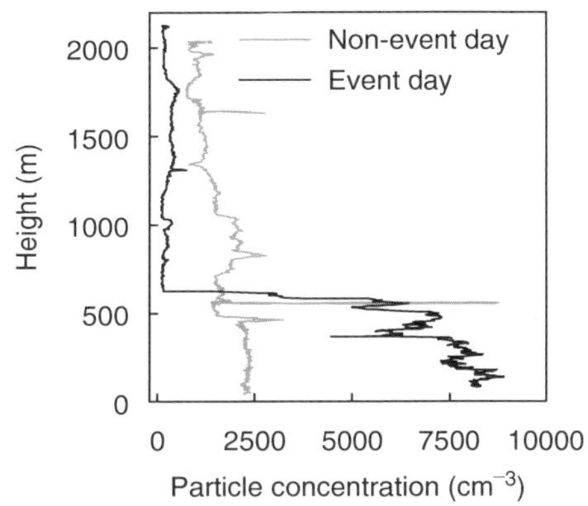

Fig. 8.37 The number concentration of over 10-nm particles as a function of altitude during a non-event day on 3 March 2006 and on an event day on 13 March 2006 at SMEAR II, southern Finland. No significant altitude dependence is observed on the non-event day. The particle number concentration has a clear maximum close to the Earth's surface on the event day

Fig. 8.38 Hot air balloon taking off with eager researchers for boundary-layer profile measurements. The wooden buildings are the dormitory for students and old dining room at Hyytiälä

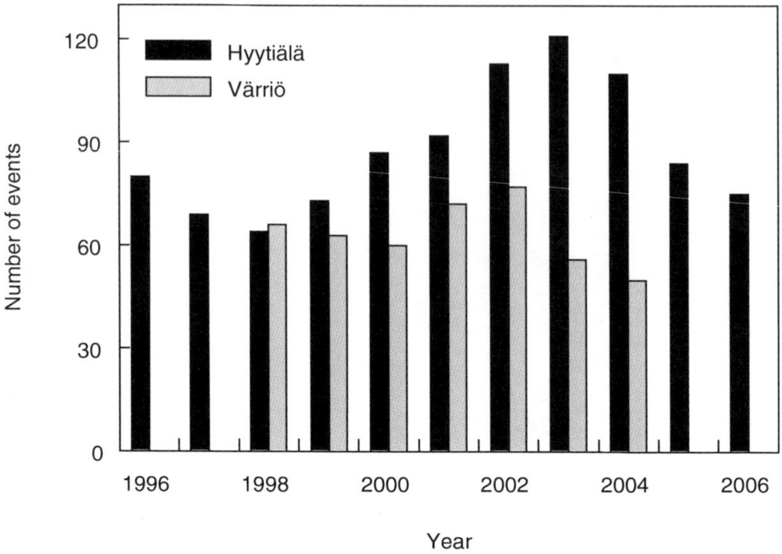

Fig. 8.39 The number of particle formation events at SMEAR I, northern Finland (Värriö) in 1998–2004 and SMEAR II, southern Finland (Hyytiälä) in 1996–2006

higher altitudes. This observation supports the idea that compounds emitted by forest vegetation have an important role in new particle formation, and particularly in particle growth

Annual Cycle of Nucleation Events

Classification and selection of event and non-event days reveals that tropospheric particle formation in the boreal forest is quite common, at least more common than previously assumed. In Hyytiälä, roughly every fourth day is found to contain a particle formation burst, based on the analysis of the particle size distribution data from years 1996 to 2006 (Dal Maso et al. 2005, 2007). For SMEAR I in Värriö, the corresponding frequency is one event every 5 days (Dal Maso et al. 2007). The annual statistics of nucleation event days for Hyytiälä in 1996–2006 and Värriö in 1996–2004 are presented in Fig. 8.39.

On the seasonal scale, two clear peaks in formation frequency are found: one in springtime (March–May) and another centred around September. During these periods, the event fraction in Hyytiälä is typically over 50% of all days and even higher than that if undefined days are removed from the statistics. In contrast, wintertime (November to mid-February) is a very inactive time for particle formation. In summer, a dip in the frequency is observed. The annual distribution of nucleation event days in the more northern Värriö station is similar, with a somewhat less clear minimum in the summer. Summaries of the annual variation of the occurrence of

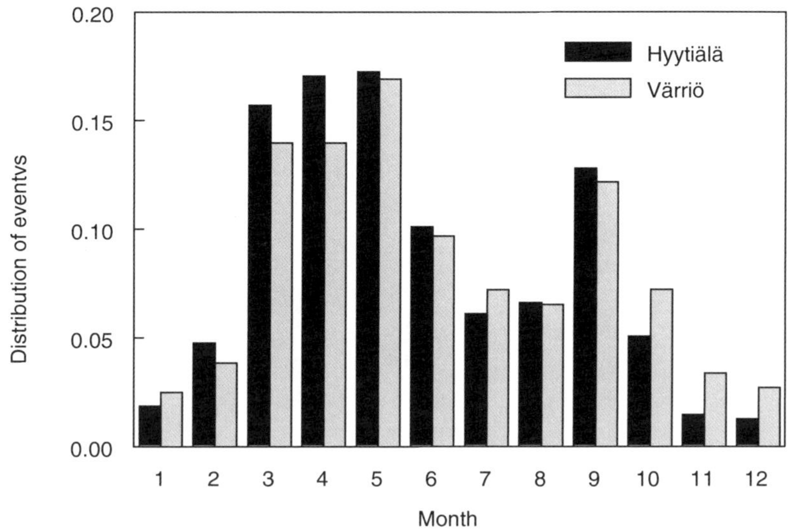

Fig. 8.40 The monthly distribution of particle formation events at SMEAR I and SMEAR II, northern and southern Finland

nucleation events for Hyytiälä and Värriö are presented in Fig. 8.40. The springtime peaks in particle formation are related to the spring recovery of photosynthesis (Sect. 4.2.2), the forest vegetation and related emissions of biogenic organics.

BVOC Emissions Under Changing Climate

The group of the biogenic volatile organic compounds (BVOCs) in the atmosphere-plant interface is very heterogeneous, both in respect of their chemical properties such as volatility and hydrophobicity and in respect of their biogenic origin. The BVOCs are emitted from vegetation as mixtures of compounds, and these mixtures change in time both seasonally and diurnally. BVOCs can be synthesised in both aerial and below-ground plant parts, in floral, vegetative and structural tissues, and they are stored in significant quantities in many plant genera. Their emissions show a conspicuous and compound-specific seasonal pattern. For most compounds, the emission rates are highest in summer and lower, although still significant, in spring and autumn. Even during wintertime, emissions of some BVOCs can be measured (Tarvainen et al. 2005; Hakola et al. 2006).

BVOCs are not found uniformly within the plant. The constituent, defence-related compounds are often located in conifers in resin ducts of needles and resin canals in their trunks. From there, they can be liberated if necessary for defence purposes, for example, after a pathogen or herbivore attack. Another, temporally more fluctuating group of BVOCs is located in, for example, air spaces of photosynthesising tissues and is readily diffused to the boundary layer and further to the atmosphere.

The synthesis of monoterpenes occurs in mesophyll chloroplasts, and the recently synthesised monoterpenes are temporarily located in liquid phase in the apoplast. From the liquid phase, they diffuse to the leaf surface and are emitted into the atmosphere. It is assumed that two independent processes influence emissions from needles to the atmosphere: one is biosynthesis, related to the instantaneous irradiation and mesophyll CO_2 concentration, and the other is temperature-dependent emission from pools. The monoterpene emission processes from Scots pine needles include substrate production, monoterpene biosynthesis, storage, transport within the leaf, and finally emission.

Effects of climate change on ecosystem-level BVOC emissions are produced by many, partially opposite, responses to changing factors, which makes evaluating the net effects at the ecosystem level very difficult (see, e.g. Penuelas and Staudt 2010).

Rising temperatures enhance enzyme activity and accelerate diffusion of BVOCs and thereby increase their volatilisation from tissues. Warming also leads to prolonged growing season and thus longer duration of high emission rates from leafy tissues especially from deciduous trees which normally are considered to be high emitters. Mild drought is reported to increase BVOC emissions, whereas severe drought might result in opposite effects due to substrate or enzyme limitations. Since BVOC biosynthesis is strongly limited by carbon availability, good correlations with photosynthetic rates have been found in many species. The majority of studies report increases in BVOC emissions after N fertilisation or a positive correlation with foliar N levels.

However, short-term exposures to elevated CO_2 often reduce leaf-level isoprenoid emissions, whereas growth in subambient CO_2 tends to increase emissions. Under elevated CO_2, the correlations between the measured emissions and both leaf temperatures and photosynthetic rates become more vague. Emission reductions due to elevated CO_2 have been connected with metabolic competition for intermediates, or downregulation due to lack of energy supply (e.g. Rosenstiel et al. 2003). The instantaneous emission responses may differ from long-term responses at elevated CO_2, and therefore, predictions on emission rates for a time period of years to decades, which are based on short-term experiments, are very problematic (e.g. Niinemets et al. 2010).

We assume that increasing temperature and atmospheric CO_2 concentration accelerates photosynthesis, thus producing more substrates for isoprenoid biosynthesis, and therefore the BVOC emissions increase proportionally to the increase in photosynthesis. Also direct effects of temperature rise to the volatilisation of BVOCs increase their emissions. This assumed behaviour is not in conflict, although not strongly supported by the present experimental results either since the mechanisms of CO_2 influence on BVOC biosynthesis are still ambiguous.

Increasing atmospheric CO_2 concentration accelerates photosynthesis, and the relationship is nearly linear at present atmospheric concentration (Sect. 4.2.2). The acclimation of photosynthesis to changing CO_2 concentration cannot be predicted since it has not been tested in evolution, at least during the last million years. In addition, the increase in needle mass, caused by accelerated decomposition of proteins due to temperature increase, will still enhance the photosynthesis of boreal forests.

Despite of insufficient understanding of the BVOC emission responses to climate change, we know that the synthesis of many compounds is closely linked with photosynthesis at present concentration ranges. Enhanced photosynthetic production, in connection with increasing leaf biomass will strongly influence the BVOC emissions from boreal forest stands in the future climate. Positive effects on emissions are also expected to follow from changes in water and nutrient availability, and increased temperature, in particular for those compounds that are having permanent storage pools in the foliage and trunks, such as monoterpenes. Taking these factors into account, we can roughly assume that the response of BVOC emissions to climatic change is resembling that of photosynthesis.

The photosynthetic production was doubled in simulations with MicroForest at the year 2100. This is why we assume that the BVOC emissions will also double in the year 2100 when compared with present emissions. However, we realise that this is an assumption where potential significant interactions, for example, increased respiratory demands for substrates involved in BVOC biosynthesis (Rosenstiel et al. 2003), may partially cancel the effects of increased photosynthetic production on emissions. In the future, when more field measurements will be available for testing and developing of knowledge, the present emission approximation should be improved.

Climatic Effects of Increasing Aerosols Due to BVOC Emissions from Forests

Condensable vapour concentration affects aerosol particle formation and growth. Anthropogenic emissions of CO_2 change the atmospheric CO_2 concentration, accelerate photosynthesis and consequently potentially also BVOC emissions from boreal forests. The emitted volatile carbon compounds react in the air and lose their volatility and condense to form and grow aerosol particles. Thus, changes in boreal forests influence the aerosols in the air.

Condensation of the oxidation products of volatile organic compounds (BVOCs) emitted by boreal forests seems to explain a significant fraction of the observed 3–25-nm particle growth rates at the SMEAR I and II sites (Kulmala et al. 1998, 2004a; Hirsikko et al. 2005). Besides their evident role in particle growth above 3 nm, it is possible that the organics emitted by the forests affect also the very first steps of the particle formation and growth processes. The atmospheric particle formation takes place close to 2 nm in diameter and the growth rates observed for the 2–3 nm particles (ca. $1.5 \mathrm{nm\,h^{-1}}$, see, e.g. Hirsikko et al. 2005) are typically approximately half of the values reported for particles larger than 3 nm (ca. $3 \mathrm{\,nm\,h^{-1}}$, e.g. Dal Maso et al. 2005).

Because of the links between the BVOCs and the particle formation, an increase in BVOC emissions is likely to result in an increase of the particle source provided by the boreal forests (see also Kulmala et al. 2004c). The change in the produced climatically active aerosol particle number will affect the climatic role of boreal forests: an increase in the cloud condensation nuclei number would lead to more long-lived clouds and therefore add to a negative feedback between climate warming and particle production (IPCC 2007).

To assess this feedback between climate change and boreal forests, we have made rough estimates on the increase of the production of 150-nm particles resulting from doubling the emissions of BVOCs. We assumed that the particle formation rate at 2 nm (J_2) stays constant over time, being approximately 1 cm^{-3} s^{-1}. We then calculated the corresponding formation rates at 3 and 150 nm (J_3 and J_{150}) now and in the future when BVOC emissions are doubled compared to the current situation. The calculations were made taking into account an increase of growth rates caused by an increasing amount of BVOCs, and their effect on the competition of growth and coagulation losses (see Kerminen and Kulmala 2002 for details).

In the estimations, the growth rate of 2–3-nm particles was assumed to be 1.5 nm^{-1}, and the growth rate of particles larger than 3 nm was set to 3 nm^{-1}. We first assumed that the organics start to contribute to growth from the very beginning, that is, from 2 nm, and second that they start to condense on the particles only when they have reached sizes above 3 nm, the latter being the more likely case. The ratios of the 150-nm particle production rates in the future (with doubled BVOC emissions) and at the present time are 130–300% in the case when organics contribute in the growth from 2 nm and 103–110% in the case when the organics contribute to the growth from 3 nm. The obtained ratio in the 150-nm particle production rates ranges from 103 to 300%, which would enhance the climate cooling effect of the forest-produced aerosol at least with the same factor compared to the current radiative forcing (estimated to be between −0.03 and −1.1 W m^{-2}; see Kurtén et al. 2003). However, the response between the aerosol source strength and the radiative transfer is highly nonlinear, so these estimations are subject to significant uncertainties and should be treated as order of magnitude estimations only.

The predicted changes in emissions of volatile organic compounds from forests in response to climate change will increase aerosol formation in these areas. This phenomenon is rather weakly known, and thus, the aerosols increase uncertainty in the expected climate change.

References

Andreae MO, Jones CD, Cox PM (2005) Strong present-day aerosol cooling implies a hot future. Nature 435:1187–1190

Arya SP (2001) Introduction to micrometeorology, 2nd edn. Academic, London

BACC Author Team (2008) Assessment of climate change for the Baltic Sea basin, regional climate studies. Springer, Heidelberg

Barnola J-M, Raynaud D, Lorius C, Barkov NI (2003) Historical CO_2 record from the Vostok ice core. In: Boden TA et al (eds) Trends: a compendium of data on global change. Carbon Dioxide Information Analysis Center, Oak Ridge National Laboratory, U.S. Department of Energy, Oak Ridge

Betts RA, Falloon PD, Goldewijk KK, Ramankutty N (2007) Biogeophysical effects of land use on climate: model simulations of radiative forcing and large-scale temperature change. Agric For Meteorol 142:216–233

Bird RB, Stewart WE, Lightfoot EN (2002) Transport phenomena, 2nd edn. Wiley, New York
Boer GJ, Yu B (2003) Climate sensitivity and response. Clim Dyn 20:415–429
Bonan GB, Pollard D, Thompson SL (1992) Effects of boreal forest vegetation on global climate. Nature 359:716–718
Botkin DB, Simpson L (1990) The first statistically valid estimate of biomass for a large region. Biogeochemistry 9:161–174
Chen JM, Black TA (1992) Defining leaf area index for non-flat leaves. Plant Cell Environ 15:421–429
Colman R (2003) A comparison of climate feedbacks in general circulation models. Clim Dyn 20:865–873
Cox PM, Betts RA, Jones CD, Spall SA, Totterdell IJ (2000) Acceleration of global warming due to carbon-cycle feedbacks in a coupled climate model. Nature 408:184–187
Dai A, Trenberth KE, Qian T (2004) A global data set of Palmer Drought Severity Index for 1870–2002: relationship with soil moisture and effects of surface warming. J Hydrometeorol 5:1117–1130
Dal Maso M, Kulmala M, Riipinen I, Wagner R, Hussein T, Aalto PP, Lehtinen KEJ (2005) Formation and growth of fresh atmospheric aerosols: eight years of size distribution data from SMEAR II, Hyytiälä, Finland. Boreal Environ Res 10:323–336
Dal Maso M, Sogacheva L, Aalto PP, Riipinen I, Komppula M, Tunved P, Korhonen L, Suur-Uski V, Hirsikko A, Kurtén T, Kerminen V-M, Lihavainen H, Viisanen Y, Hansson H-C, Kulmala M (2007) Aerosol size distribution measurements at four Nordic field stations: identification, analysis and trajectory analysis of new particle formation bursts. Tellus B59:350–361
Dawson TP, Curran PJ, Plummer SE (1998) LIBERTY—modelling the effects of leaf biochemical concentration on reflectance spectra. Remote Sens Environ 65:50–60
Dentener F (2006) Global maps of atmospheric nitrogen deposition, 1860, 1993, and 2050. Data set from Oak Ridge National Laboratory, Oak Ridge, Tennessee. http://www.daac.ornl.gov/
Ebermayer E (1876) Die gesamte Lehre von der Waldstreu. Springer, Berlin
Eriksson H, Karlsson K (1997) Olika gallrings och gödslingsregimens effekter på beståndsut vecklingen baserat på långliggande experiment i tall-och granbestånd i Sverige. SLU, Inst Skogsprodukt Rep 42:1–135
Forster P, Ramaswamy V, Artaxo P, Berntsen T, Betts RA, Fahey DW, Haywood J, Lean J, Lowe DC, Myhre G, Nganga J, Prinn R, Raga G, Schulz M, van Dorland R (2007) Changes in atmospheric constituents and in radiative forcing. In: Solomon S, Qin D, Manning M, Chen Z, Marquis M, Averyt KB, Tignor M, Miller HL (eds) Climate change 2007: the physical science basis. Cambridge University Press, Cambridge
Friedli H, Lötscher H, Oeschger H, Siegenthaler U, Stauffer B (1986) Ice core record of $^{13}C/^{12}C$ ratio of atmospheric CO_2 in the past two centuries. Nature 324:237–238
Freidenreich SM, Ramaswamy V (1999) A new multiple-band solar radiative parameterization for general circulation models. J Geophys Res 104:31389–31409
Friedlingstein P, Cox P, Betts R et al (2006) Climate-carbon cycle feedback analysis: results from the C4MIP model intercomparison. J Clim 19:3337–3353
Gauthier S, Bergman A, Bergeron Y (1996) Forest dynamics modelling under natural fire cycles: a tool to define natural mosaic diversity for forest management. Environ Monit Assess 39:417–434
Groisman PY, Sherstyukov BG, Razuvaev VN, Knight RW, Enloe JG, Stroumentova NS, Whitfield PH, Førland E, Hannsen-Bauer I, Tuomenvirta H, Alexandersson H, Mescherskaya AV, Karl TR (2007) Potential forest fire danger over Northern Eurasia: changes during the 20th century. Glob Planet Change 56:371–386
Hagemann S (2002) An improved land surface parameter data set for global and regional climate models. Max-Planck-Institute for Meteorology, Report 336. Hamburg, Germany
Hakola H, Laurila T, Hiltunen V, Hellen H, Keronen P (2003) Seasonal variation of VOC concentrations above a boreal coniferous forest. Atmos Environ 37:1623–1634
Hakola H, Tarvainen V, Bäck J, Ranta H, Bonn B, Rinne J, Kulmala M (2006) Seasonal variation of mono- and sesquiterpene emission rates of Scots pine. Biogeosciences 3:93–101

Hansen J, Lacis A, Rind D, Russell G, Stone P, Fung I, Ruedy R, Lerner J (1984) Climate sensitivity: analysis of feedback mechanisms. Meteorol Monogr 29:130–163
Hansen J, Sato M, Ruedy R et al (2005) Efficacy of climate forcings. J Geophys Res 110:D18104
Harvey LDD (2000) Global warming: the hard science. Pearson Education, Harlow
Hegerl GC, Zwiers FW, Braconnot P, Gillett NP, Luo Y, Marengo J, Nicholls N, Penner JE, Stott PA (2007) Understanding and attributing climate change. In: Solomon S, Qin D, Manning M, Chen Z, Marquis M, Averyt KB, Tignor M, Miller HL (eds) Climate change 2007: the physical science basis. Cambridge University Press, Cambridge
Held IM, Soden BJ (2006) Robust responses of the hydrological cycle to global warming. J Clim 19:5686–5699
Helmisaari H-S, Holt Hanssen K, Jacobson S, Kukkola M, Luiro J, Saarsalmi A, Tamminen P, Tveite B (2011) Logging residue removal after thinning in Nordic boreal forests: long-term impact on tree growth. For Ecol Manag 261:1919–1927
Hirsikko A, Laakso L, Hõrrak U, Aalto PP, Kerminen V-M, Kulmala M (2005) Annual and size dependent variation of growth rates and ion concentrations in boreal forest. Boreal Environ Res 10:357–370
Holton JR (2004) An introduction to dynamic meteorology. Elsevier Academic Press, New York
Houghton J (2004) Global warming: the complete briefing. Cambridge University Press, Cambridge, UK
Houghton JT, Ding Y, Griggs DJ, Noguer M, van der Linden PJ, Xiaosu D (eds) (2001) Climate Change 2001. The scientific basis. Contribution of working group I to the third assessment report of the Intergovernmental Panel on Climate Change (IPCC). Cambridge University Press, Cambridge
Huntington TG (2006) Evidence for intensification of the global water cycle: review and synthesis. J Hydrol 319:83–95
Hyytiäinen K, Tahvonen O, Valsta L (2005) Optimum juvenile density, harvesting, and stand structure in even-aged Scots pine stands. For Sci 51:120–133
IPCC (2007) Contribution of working group I to the fourth assessment report of the Intergovernmental Panel on Climate Change. In: Solomon S, Qin D, Manning M, Chen Z, Marquis M, Averyt K, Tignor MMB, Miller HL (eds) Climate change 2007: the physical science basis. Cambridge University Press, Cambridge
Jansen E, Overpeck J, Briffa KR et al (2007) Paleoclimate. In: Solomon S, Qin D, Manning M, Chen Z, Marquis M, Averyt KB, Tignor M, Miller HL (eds) Climate Change 2007: the physical science basis. Contribution of working group I to the fourth assessment report of the Intergovernmental Panel on Climate Change (IPCC). Cambridge University Press, Cambridge
Jin M, Liang S (2006) An improved land surface emissivity parameter for land surface models using global remote sensing observations. J Clim 19:2867–2881
Jones PD, Lister DH (2002) The daily temperature record for St. Petersburg (1743–1996). Clim Chang 53:253–267
Joshi M, Shine K, Ponater M, Stuber N, Sausen R, Li L (2003) A comparison of climate response to different radiative forcings in three general circulation models: towards an improved metric of climate change. Clim Dyn 20:843–854
Jouzel J, Lorius C, Petit JR, Genthon C, Barkov NI, Kotlyakov VM, Petrov VM (1987) Vostok ice core: a continuous isotope temperature record over the last climatic cycle (160,000 years). Nature 329:403–408
Jouzel J, Barkov NI, Barnola JM, Bender M, Chappellaz J, Genthon C, Kotlyakov VM, Lipenkov V, Lorius C, Petit JR, Raynaud D, Raisbeck G, Ritz C, Sowers T, Stievenard M, Yiou F, Yiou P (1993) Extending the Vostok ice-core record of palaeoclimate to the penultimate glacial period. Nature 364:407–412
Jouzel J, Waelbroeck C, Malaize B, Bender M, Petit JR, Stievenard M, Barkov NI, Barnola JM, King T, Kotlyakov VM, Lipenkov V, Lorius C, Raynaud D, Ritz C, Sowers T (1996) Climatic interpretation of the recently extended Vostok ice records. Clim Dyn 12:513–521
Keeling CD, Whorf TP (2005) Atmospheric CO_2 records from sites in the SIO air sampling network. In: Boden TA et al (eds) Trends: a compendium of data on global change. Carbon

Dioxide Information Analysis Center, Oak Ridge National Laboratory, U.S. Department of Energy, Oak Ridge

Kerminen V-M, Kulmala M (2002) Analytical formulae connecting the "real" and the "apparent" nucleation rate and the nuclei number concentration for atmospheric nucleation events. J Aerosol Sci 33:609–622

Kiehl JT, Trenberth KE (1997) Earth's annual global mean energy budget. Bull Amer Meteor Soc 78:197–208

Kulmala M, Toivonen A, Mäkelä JM, Laaksonen A (1998) Analysis of the growth of nucleation mode particles observed in Boreal forest. Tellus 50B:449–463

Kulmala M, Dal Maso M, Mäkelä JM, Pirjola L, Väkevä M, Aalto P, Miikkulainen P, Hämeri K, O'Dowd CD (2001) On the formation, growth and composition of nucleation mode particles. Tellus 53B:479–490

Kulmala M, Vehkamäki H, Petäjä T, Dal Maso M, Lauri A, Kerminen V-M, Birmili W, McMurry PH (2004a) Formation and growth rates of ultrafine atmospheric particles: a review of observations. J Aerosol Sci 35:143–176

Kulmala M, Laakso L, Lehtinen KEJ, Riipinen I, Dal Maso M, Anttila T, Kerminen V-M, Hõrrak U, Vana M, Tammet H (2004b) Initial steps of aerosol growth. Atmos Chem Phys 4:2553–2560

Kulmala M, Suni T, Lehtinen KEJ, Dal Maso M, Boy M, Reissell A, Rannik Ü, Aalto PP, Keronen P, Hakola H, Bäck J, Hoffmann T, Vesala T, Hari P (2004c) A new feedback mechanism linking forests, aerosols, and climate. Atmos Chem Phys 4:557–562

Kurtén T, Kulmala M, Dal Maso M, Suni T, Reissell A, Vehkamäki H, Hari P, Laaksonen A, Viisanen Y, Vesala T (2003) Estimation of different forest-related contributions to the radiative balance using observations in southern Finland. Boreal Environ Res 8:275–285

Laakso L, Grönholm T, Kulmala L, Haapanala S, Hirsikko A, Lovejoy ER, Kazil J, Kurtén T, Boy M, Nilsson ED, Sogachev A, Riipinen I, Stratmann F, Kulmala M (2007) Hot-air balloon as a platform for boundary layer profile measurements during particle formation. Boreal Env Res 12:279–294

Lemke P, Ren J, Alley R, Allison I, Carrasco J, Flato G, Fuiji Y, Kaser G, Mote P, Thomas R, Zhang T (2007a) Observations: changes in snow, ice and frozen ground. In: Solomon S, Qin D, Manning M, Chen Z, Marquis M, Averyt KB, Tignor M, Miller HL (eds) Climate Change 2007: the physical science basis. Contribution of Working Group I to the Fourth Assessment Report of the Intergovernmental Panel on Climate Change. Cambridge University Press, Cambridge

Lemke P, Ren J, Alley R, Allison I, Carrasco J, Flato G, Fuiji Y, Kaser G, Mote P, Thomas R, Zhang T (2007b) Observations: changes in snow, ice and frozen ground. In: Solomon S, Qin D, Manning M, Chen Z, Marquis M, Averyt KB, Tignor M, Miller HL (eds) Climate Change 2007: the physical science basis. Contribution of working group I to the fourth assessment report of the Intergovernmental Panel on Climate Change (IPCC). Cambridge University Press, Cambridge

Magnani F, Mencuccini M, Borghetti M, Berbigier P, Berninger F, Delzon S, Grelle A, Hari P, Jarvis PG, Kolari P, Kowalski AS, Lankreijer H, Law BE, Lindroth A, Loustau D, Manca G, Moncrieff JB, Rayment M, Tedeschi V, Valentini R, Grace J (2007) The human footprint in the carbon cycle of temperate and boreal forests. Nature 447:848–850

Mäkelä JM, Aalto P, Jokinen V, Pohja T, Nissinen A, Palmroth S, Markkanen T, Seitsonen K, Lihavainen H, Kulmala M (1997) Observations of ultrafine aerosol formation and growth in boreal forest. Geophys Res Lett 24:1219–1222

Mäkelä JM, Dal Maso M, Pirjola L, Keronen P, Laakso L, Kulmala M, Laaksonen A (2000) Characteristics of the atmospheric particle formation events observed at a boreal forest site in southern Finland. Boreal Environ Res 5:299–313

McClatchey RA, Fenn RW, Selby JEA, Volz FE, Garing JS (1971) Optical properties of the atmosphere. Report AFCRL-71-0279. Air Force Geophys Lab, Hanscom Air Force Base, Bedford

Meehl GA, Stocker TF, Collins W, Friedlingstein P, Gaye A, Gregory J, Kitoh A, Knutti R, Murphy J, Noda A, Raper S, Watterson I, Weaver A, Zhao Z-C (2007) Global climate projections. In: Solomon S, Qin D, Manning M, Chen Z, Marquis M, Averyt KB, Tignor M,

Miller HL (eds) Climate Change 2007: the physical science basis. Contribution of working group I to the fourth assessment report of the Intergovernmental Panel on Climate Change (IPCC). Cambridge University Press, Cambridge

Moberg A, Bergström H, Krigsman JR, Svanered O (2002) Daily air temperature and pressure series for Stockholm (1756–1998). Clim Chang 53:171–212

Myneni RB, Dong J, Tucker CJ, Kaufmann RK, Kauppi PE, Liski J, Zhou L, Alexeyev V, Hughes MK (2001) A large carbon sink in the woody biomass of Northern forests. Proc Natl Acad Sci USA 98:14784–14789

Nakićenović N, Swart R (eds) (2000) Emissions scenarios. A special report of working group III of the Intergovernmental Panel on Climate Change. Cambridge University Press, Cambridge

Niinemets U, Arneth A, Kuhn U, Monson RK, Penuelas J, Staudt M (2010) The emission factor of volatile isoprenoids: stress, acclimation, and developmental responses. Biogeosciences 7:2203–2223

Niinimäki S, Tahvonen O, Mäkelä A (2012) Applying a process-based model in Norway spruce management. For Ecol Manag 265:102–115

Nobel PS (2005) Physicochemical & environmental plant physiology, 3rd edn. Academic Press/Elsevier, San Diego

Penuelas J, Staudt M (2010) BVOCs and global change. Trends Plant Sci 15:133–144

Peterson TC, Vose RS (1997) An overview of the Global Historical Climatology Network temperature database. Bull Am Meteorol Soc 78:2837–2848

Petit JR, Jouzel J, Raynaud D, Barkov NI, Barnola J-M, Basile I, Bender M, Chappellaz J, Davis M, Delayque G, Delmotte M, Kotlyakov VM, Legrand M, Lipenkov VY, Lorius C, Pepin L, Ritz C, Saltzman E, Stievenard M (1999) Climate and atmospheric history of the past 420,000 years from the Vostok ice core, Antarctica. Nature 399:429–436

Räisänen J (2007) How reliable are climate models? Tellus 59A:2–29

Roeckner E, Bäuml G, Bonaventura L, Brokopf R, Esch M, Giorgetta M, Hagemann S, Kirchner I, Kornbleuh L, Manzini E, Rhodin A, Schlese U, Schulzweida U, Tomkins A (2003) The atmospheric general circulation model ECHAM5, Part I: Model description. Max-Planck-Inst Meteorol Rep 349, Hamburg

Roeckner E, Brokopf R, Esch M, Giorgetta M, Hagemann S, Kornblueh L, Manzini E, Schlese U, Schulzweida U (2006) Sensitivity of simulated climate to horizontal and vertical resolution in the ECHAM5 atmosphere model. J Clim 19:3771–3791

Rosenstiel TN, Potosnak MJ, Griffin KL, Fall R, Monson R (2003) Increased CO2 uncouples growth from isoprene emission in an agriforest ecosystem. Nature 421:256–259

Ross J (1981) The radiation regime and architecture of plant stands. Kluwer Academic, The Hague

Sausen R, Barthel K, Hasselman K (1988) Coupled ocean–atmosphere models with flux-correction. Clim Dyn 2:145–163

Schultz MG et al (2007) Reanalysis of the tropospheric chemical composition of the past 40 years (RETRO)—a long-term global modelling study of tropospheric chemistry. Final Rep 48/2007, Max Planck Inst Meteorol, Hamburg

Seinfeld JH, Pandis SN (1998) Atmospheric chemistry and physics. Wiley, New York

Smith TM, Reynolds RW (2005) A global merged land and sea surface temperature reconstruction based on historical observations (1880–1997). J Clim 18:2021–2036

Smolander S, Stenberg P (2005) Simple parameterizations of the radiation budget of uniform broadleaved and coniferous canopies. Remote Sens Environ 94:355–363

Spracklen DV, Carslaw KS, Kulmala M, Kerminen V-M, Mann GW, Sihto S-L (2006) The contribution of boundary layer nucleation events to total particle concentrations on regional and global scales. Atmos Chem Phys 6:5631–5648

Stamnes K, Tsay SC, Wiscombe W, Jayaweera K (1988) A numerically stable algorithm for discrete-ordinate-method radiative transfer in multiple scattering and emitting layered media. Appl Optics 27:2502–2509

Stamnes K, Tsay SC, Wiscombe W, Laszlo I (2000) A general-purpose numerically stable computer code for discrete-ordinate-method radiative transfer in scattering and emitting layered media, DISORT Rep v1.1. Stevens Inst Tech, Hoboken

Stephens BB, Gurney K, Tans P, Sweeney C, Peters W, Bruhwiler L, Ciais P, Ramonet M, Bousquet P, Nakazawa T, Aoki S, Machida T, Inoue G, Vinnichenko N, Lloyd J, Jordan A, Heimann M, Shibistova O, Langenfelds RL, Steele LP, Francey RJ, Denning AS (2007) Weak northern and strong tropical land carbon uptake from vertical profiles of atmospheric CO_2. Science 316:1732–1735

Tarvainen V, Hakola H, Hellen H, Bäck J, Hari P, Kulmala M (2005) Temperature and light dependence of the VOC emissions of Scots pine. Atmos Chem Phys 5:6691–6718

Trenberth KE, Jones PD, Ambenje P, Bojariu R, Easterling D, Klein Tank A, Parker D, Rahimzadeh F, Renwick JA, Rusticucci M, Soden B, Zhai P (2007) Observations: surface and atmospheric climate change. In: Solomon S, Qin D, Manning M, Chen Z, Marquis M, Averyt KB, Tignor M, Miller HL (eds) Climate Change 2007: The physical science basis. Contribution of working group I to the fourth assessment report of the Intergovernmental Panel on Climate Change (IPCC). Cambridge University Press, Cambridge

Tunved P, Hansson H-C, Kerminen V-M, Ström J, Dal Maso M, Lihavainen H, Viisanen Y, Aalto PP, Komppula M, Kulmala M (2006) High natural aerosol loading over boreal forests. Science 312:261–263

van Aardenne JA, Dentener FJ, Olivier JGJ, Goldewijk K, Lelieveld J (2001) A 1 deg × 1 deg resolution dataset of historical anthropogenic trace gas emissions for the period 1890–1990. Glob Biogeochem Cycles 15:909–928

Vehkamaki H (2006) Classical nucleation theory in multicomponent systems. Springer, Berlin

Vesala T, Suni T, Rannik U, Keronen P, Markkanen T, Sevanto S, Grönholm T, Smolander S, Kulmala M, Ilvesniemi H, Ojansuu R, Uotila A, Levula J, Mäkelä A, Pumpanen J, Kolari P, Kulmala L, Altimir N, Berninger F, Nikinmaa E, Hari P (2005) Effect of thinning on surface fluxes in a boreal forest. Glob Biogeochem Cycles 19:GB2001

Webb MJ, Senior CA, Sexton DMH, Ingram WJ, Williams KD, Ringer MA, McAvaney BJ, Colman R, Soden BJ, Gudgel R, Knutson T, Emori S, Ogura T, Tsushima Y, Andronova N, Li B, Musat I, Bony S, Taylor KE (2006) On the contribution of local feedback mechanisms to the range of climate sensitivity in two GCM ensembles. Clim Dyn 27:17–38

Wohlfahrt J, Harrison SP, Braconnot P (2004) Synergistic feedbacks between ocean and vegetation on mid- and high-latitude climates during the mid-Holocene. Clim Dyn 22:223–238

Chapter 9
Station for Measuring Ecosystem-Atmosphere Relations: SMEAR

Pertti Hari, Eero Nikinmaa, Toivo Pohja, Erkki Siivola, Jaana Bäck, Timo Vesala, and Markku Kulmala

Contents

Abstract We planned and implemented the SMEAR II and I measuring systems to measure material and energy fluxes between forest ecosystem and its surroundings and within the ecosystem. In addition, we measured the processes generating the fluxes and environmental factors affecting the processes. We constructed SMEAR I in 1991 and SMEAR II in 1994–1996. The mass and energy fluxes play an important role in our physical and physiological theory of forest ecology. The construction principle of SMEAR measuring stations is coherent with our theory, and this is why the data obtained at SMEAR stations have been useful in the development and testing of our theory.

Keywords Field measurements • Digital • Automatic • Material and energy fluxes • Processes • Environmental factors

---

P. Hari (✉)
Department of Forest Sciences, University of Helsinki, P.O. Box 27,
00014 University of Helsinki, Helsinki, Finland
e-mail: pertti.hari@helsinki.fi

P. Hari et al. (eds.), *Physical and Physiological Forest Ecology*,
DOI 10.1007/978-94-007-5603-8_9,

9.1 Short History of Forest Ecological Measurements at the University of Helsinki

As previous chapters have illustrated, advancing ecological understanding of forest productivity and its connections to material and energy fluxes requires intimate connection between theoretical thinking and field measurements. Theories without testing remain speculative arguments while experiments alone do not allow development of logical framework that can explain the observed features in the nature. In the following we give a brief outline on the historical development of the ecosystem measurements that illustrate how the advancement of theory is also linked to the development of instrumentation.

The forest ecological experiments and measurements at the University of Helsinki have long tradition going back to the 1910s. The first field measuring system of photosynthesis and transpiration was purchased in the middle of the 1960s. We first started the measurements with the system in a greenhouse on the roof of the so-called Forest Building in the downtown of Helsinki in the year 1970. After this testing period we moved the system to Hyytiälä, the Forestry Field Station of the University of Helsinki, about 200 km to north from Helsinki in the spring of 1972.

This first measuring system consisted of chambers, a mechanical control unit for opening and closing the chambers, magnetic valves, URAS™ I infrared CO_2 analyser, chart recorders for recording of the readings from gas analyser and environmental sensors on paper. The system was based on analogue techniques since the digital instrumentation was not available at that time. The processing of the results recorded on charts was a laborious and boring task for young researches.

The digitalisation of the system started in 1974 when we introduced Nokia data logger to punch the results on paper tape and to control the magnetic valves. We planned and implemented seven generations of the measuring system during the 1970s and 1980s. Increasing digitalisation and technical development, especially of measuring chambers, characterised the period. We paid special attention to the connection between measurements and the analysis of the obtained results with dynamic models.

Close cooperation developed between physicists and forest ecologists during the 1980s. The nuclear fallout from the nuclear power station accident in Chernobyl in 1986 was abundant in Hyytiälä region, and we wanted to use the radioactive elements from Chernobyl as tracers of material flows in forest ecosystems. This introduced an important novel aspect of material fluxes into our theoretical thinking and dynamic modelling, and physics obtained more weight in our research.

These field measurements since the 1970s and the developed theoretical and technical skills of measuring in the field had an important role when we started to plan advanced forest ecological field measuring station. The plans were implemented in the construction of SMEAR I in Värriö, northern Finland, and especially of SMEAR II in Hyytiälä, southern Finland. However, it really was the widening of the scope of research that made these stations unique. These stations were constructed to add physical understanding into toolbox to solve ecological phenomena and ecological understanding into toolbox to solve questions of atmospheric physics.

9.2 SMEAR I

We constructed SMEAR I (station for measuring ecosystem-atmosphere relations) in the year 1991 about 1 km north from Värriö Research Station in eastern Lapland on the top of Kotovaara hill (67°46′ N, 29°35′ E, 400 m ASL, Fig. 9.1), 200 km north of the Arctic Circle. The station is located in a very remote area, the nearest houses locate at the distance of 10 km and nearest road comes to 8 km from the station.

Extensive forest dieback in Kola Peninsula in the Soviet Union due to air pollution from nickel smelters caused concern of its spreading to the Finnish Lapland in the late 1980s. We constructed the SMEAR I measuring station to study the effects of emissions in Kola on forests and on air quality. Pertti Hari and Markku Kulmala were scientific planners and Toivo Pohja (hardware), Erkki Siivola (electricity) and Tapani Lahti (programming) technical planners and implementers. The experiences gained in the field measurements in Hyytiälä earlier on played an important role in the construction of the SMEAR I station.

SMEAR I includes five components:

1. Instrumented 15-m-tall scaffold including Eddy Covariance measuring system since April 2012
2. Chambers to measure photosynthesis and transpiration
3. Instrumentation for measuring aerosols
4. Gas analysers to monitor SO_2 and O_3
5. Minicomputer and data loggers and serial transmitters to control the system and store the measurements

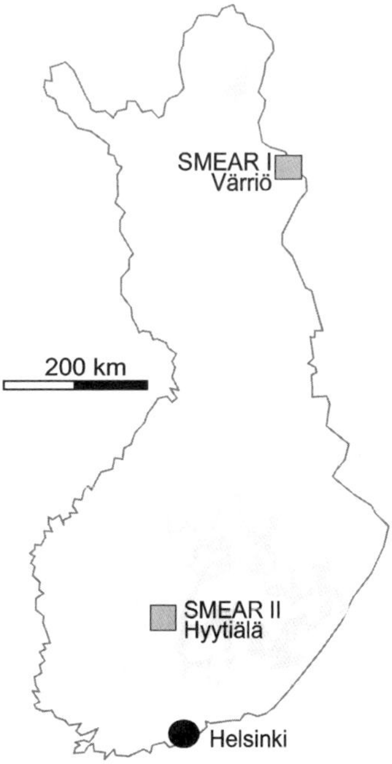

Fig. 9.1 Locations of SMEAR I and SMEAR II stations in Finland

The system is automatic and it is running also in winter even during the polar night. We couple the measuring system off during thunderstorms. In addition, the breaks in electricity cause gaps in the data. The system has worked well and produced good time series of photosynthesis, aerosols, air pollution and environmental factors. Later on the system has been complemented with measurements of tree diameter variations and soil CO_2 concentration.

9.3 SMEAR II

We constructed SMEAR II measuring station about 400 m to East from the Hyytiälä Forestry Field Station (61°46′ N, 24°17′ E, 170 m ASL, Fig. 9.1) during the years 1994–1996. The ecosystem to be measured locates on the top of a small hill in a rather homogeneous Scots pine (*Pinus sylvestris*) stand. The soil is very shallow due to the ice age, only 5–150 cm, mostly about 50 cm. The pine stand was sawn in the year 1962 after prescribed burning. Before the year 1994, the stand was treated as a normal commercial forest including tending of hard woods at the sapling stand stage.

The goal of the novel measuring system was to provide information for studies of forest ecosystem development and for analysis of the role of forests in the atmospheric properties, particularly formation and growth of aerosols and greenhouse gas concentrations. We decided that the material and energy fluxes and the metabolic and physical processes generating the fluxes are in the focus of the measurements.

In forests, the trees and ground vegetation are the engines of the ecosystem: they convert the solar radiation energy to sugars that provide energy and raw material for the metabolism of trees, ground vegetation and microbes in the ecosystem and eventually of animals. In addition, photosynthesis provides material for biogenic volatile organic compounds that are important for aerosol growth in the atmosphere (Sect. 8.3.6.2). We divided the gases into inert and reactive ones. The inert gases, CO_2 and water vapour, are important raw materials for photosynthesis, and they are also important greenhouse gases. The reactive gases such as ozone and NO_x play an important role in air chemistry and also in the growth of aerosols. We visualise our analysis of the important fluxes and processes in the soil, the canopy and outer layer in the atmosphere in Fig. 9.2.

The planning and construction of the SMEAR II station was teamwork. Pertti Hari and Markku Kulmala did general scientific planning and they lead the project. Eero Nikinmaa took care of tree physiology and growth, Hannu Ilvesniemi of soil and Timo Vesala of micro meteorological measurements. Toivo Pohja (hardware), Erkki Siivola (electricity) and Tapani Lahti (programming) were technical planners and implementers.

SMEAR II includes five main components:

1. An instrumented 73-m-high tower (127 m after autumn 2010, Fig. 9.3)
2. Systems to monitor aerosols
3. Instrumentation to monitor tree and soil processes

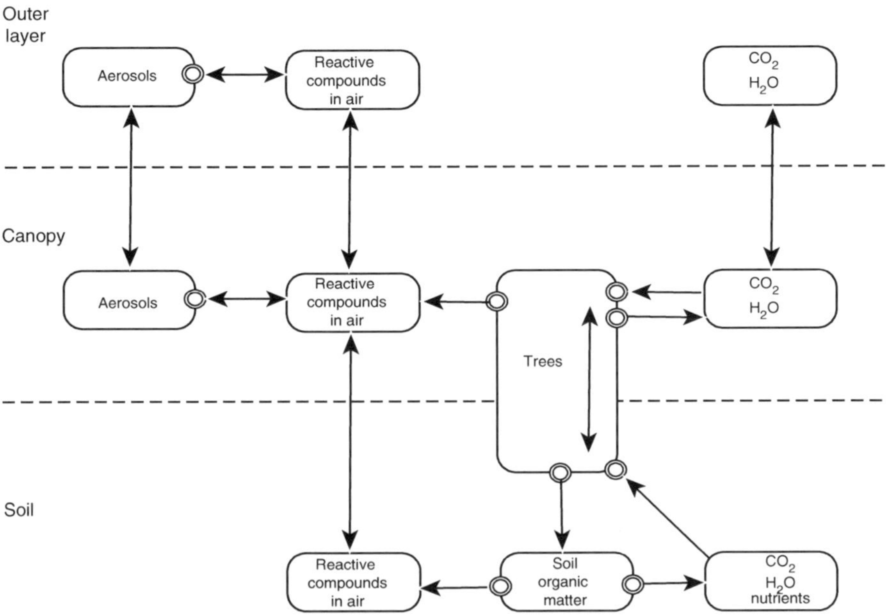

Fig. 9.2 Schematic presentation of the connection between measured fluxes at SMEAR II. We measure all relevant pools and fluxes and processes generating the fluxes. We divide the space into soil, canopy and mixed boundary layer (outer layer) and the measured objects to vegetation (trees), inert and reactive gases and aerosols (Symbols: *Boxes* are amounts, *arrows* fluxes and *double rings* processes)

4. Two instrumented mini catchments
5. Computers, data loggers, serial transmitters, AD converters, magnetic valves, etc. to control the system and to make and store the digital measurements

We monitor profiles of several gases (e.g. CO_2, H_2O, CO, O_3, SO_2, NO, NO_2), profiles of temperature and wind speed, properties of solar and thermal radiation of the stand and fluxes of CO_2, H_2O and O_3 aerosols and several volatile organic compounds between the canopy and the atmosphere. Most of the instrumentation is installed into a wooden cottage (Fig. 9.4) to shelter against moist and low temperatures.

The test measurements started at SMEAR II in August 1995 and the measurements at the station in the winter of 1996. However, as instrumentation is rapidly developing, we have expanded the measuring system continuously, especially aerosol measurements. Electric failures after storm events cause breaks in the measurements; otherwise, the system is always running and measuring also in winter when the temperature may be low, even –30°C. Snow is an additional complication for measurements (Fig. 9.5).

The research of forest ecosystems requires simultaneous measurements of trees and ground vegetation, soil and the atmosphere. It is strange that the novel possibilities opened by the development of instrumentation have been so poorly

Fig. 9.3 The 127-m-high measuring tower at SMEAR II. Photo: J. Aalto

utilised in forest ecological research. SMEAR II is probably the only measuring station that is planned and constructed to measure all relevant aspects of a forests ecosystem that cover both the interest of ecological phenomena and atmospheric physics. The measured time series have been very useful in the testing of our theories (see Chaps. 4, 5 and 7).

At SMEAR II, there are three important measuring methods: chamber measurements, mini watersheds and micrometeorological Eddy Covariance.

9.4 Chamber Measurements

We started chamber measurements of CO_2 exchange in 1972 in Hyytiälä. The chambers were self-made primitive acryl boxes or tubes. The development of the chambers has been characteristic of our fieldwork. Chamber properties have no doubt important influences on the results, particularly when a number of different gases are studied, and we have developed and tested several generations of technical

Fig. 9.4 Measured values are stored and monitored in a cottage at SMEAR II station

Fig. 9.5 Wintery aspect at SMEAR II

solutions. The present tube (Fig. 9.6) and slide (Fig. 9.7) versions are closed during the measurement and they are open the remaining time. When the chamber is closed then the fluxes between leaves and the air in the chamber change the concentrations inside.

Fig. 9.6 The tube version of the measuring chamber when it is closed for measurements. Photo: J. Aalto

The interpretation of the measurements is based on the mass balance for the studied gas in the chamber. Let q denote the flux to be measured, C_c the gas concentration inside the chamber, C_r the gas concentration in the air replacing air drawn into the analyser (often ambient air), q the flow rate of air into the analyser and V_c the volume of the chamber. The mass balance equation (c.f. Eq. 2.3) for the gas in the chamber is

$$\frac{\mathrm{d}\,(V_c C_c)}{\mathrm{d}t} = g - qC_c + qC_r. \tag{9.1}$$

The above mass balance equation can be used in two ways, either assuming steady state conditions and omitting the derivative term or studying concentration changes after closure of the chamber.

Our measured flux is based on the concentration changes after closure of the chamber. Then we obtain the flux g from concentration changes in the cuvette and from the flux through the chamber. We get from the above mass balance equation

Fig. 9.7 An uncovered shoot experiences rain showers and full sunlight in a sliding measuring chamber. The chamber is open in the *upper* photo and closed in the *lower* photo. Photos: J. Aalto

$$g = \frac{V \Delta C_c}{\Delta t} + q\,(C_r - C_c)\,, \tag{9.2}$$

where ΔC_c is the concentration change during the measurement and Δt is the time used for the measurement. The CO_2 concentration decreases clearly during the closure of the chamber 112 at SMEAR II at 12 o'clock, April 2002 (Fig. 9.8). We use for the measurement of CO_2 uptake the period indicated in Fig. 9.8.

Portable gas exchange measuring systems utilise often the steady state approach. Thus, the derivative in Eq. 9.2 is assumed to be zero, and the flux is solved from the simplified equation:

$$g = q\,(C_r - C_c)\,. \tag{9.3}$$

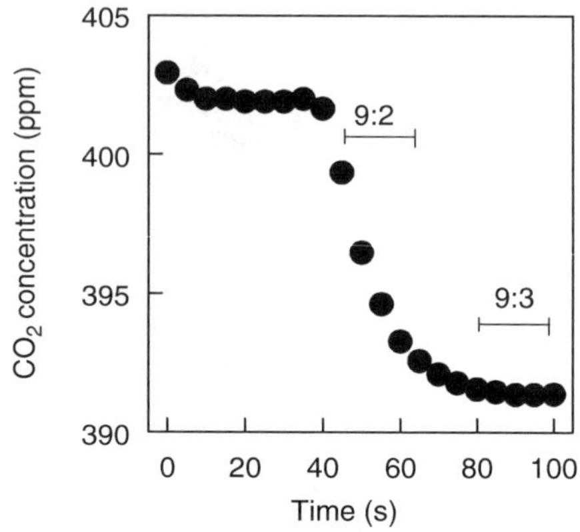

Fig. 9.8 The CO_2 concentration in the measuring chamber as function of time at SMEAR II. The chamber was closed at the time 40 s. The period applicable for calculating the CO_2 exchange with Eqs. 9.2 and 9.3 are indicated with line segments

We have to take several aspects in consideration when selecting technical details of our gas exchange measuring system in the field. Each solution has its own benefits and drawbacks. Satisfactory precision is rather easy to obtain, but accuracy is more problematic since the chambers generate systematic measuring errors due to retention of the gas to be measured on the chamber walls. In addition, the calibration of the gas analysers is rather inaccurate.

The annual photosynthetic production, annual cycle of gas exchange and the effect of drought on photosynthesis and transpiration are in the focus of forest ecological research. Long time series of measured gas exchange are needed to study these vital aspects of photosynthesis. There is variation between leaves and twigs caused by differences in the geometrical structure, in concentrations of the functional substances and in the damage history. The photosynthesis of the leaf or twig reflects the above differences and it varies between objects to be measured. If we measure the same object for longer periods, then the disturbance caused by variation between objects to be measured is missing.

We can arrange the monitoring of gas exchange of a leaf or twig with portable measuring systems. If we use portable measuring system, then the leaves or the twig is permanently closed into the chamber. The leaves react to the chamber climate and acclimate to the unnatural environment, say low CO_2 concentration and strange light climate. This is why we have constructed chambers that are closed only for a short period, and most of the time the chamber is open and the conditions in the chamber are rather close to the ambient ones.

The measuring precision is often rather good in the chamber measurements. Those systems that are closed for measurements and that utilise Eq. 9.2 to determine the gas exchange have the advantage that we can determine the measuring precision. We take seven measurements of CO_2 concentration at 10 s interval at SMEAR I and we can determine CO_2 exchange for six 10 s intervals during each closure. Evidently 10 s is not enough for the determination, but 20 s seems to be rather sufficient for the determination. The regression between the CO_2 exchanges calculated from different periods after chamber closure is rather tight (Fig. 9.9). The measuring noise is about 10% from the CO_2 exchange signal. The noise level is lower if longer measurement

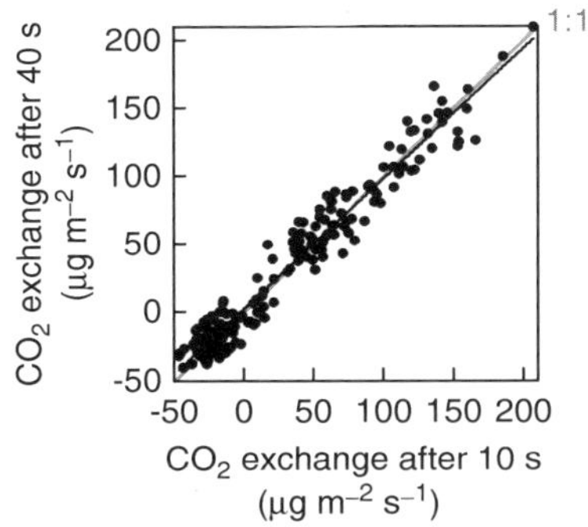

Fig. 9.9 The regression between two measurements of CO_2 exchange during chamber closures measured on 9 August 2011. The *x*-axis is measured CO_2 exchange during the period starting 10 s after closure and the period ends 30 s after closure. The *y*-axis is also measured CO_2 exchange during the period starting 40 s and ending 60 s after the cuvette closure. The *grey line* is 1:1 line whereas the linear regression is marked with a *black line*

periods are used. This fact contributes to the small residuals between predicted and observed CO_2 change observed in Sect. 9.4.

Accuracy is problematic in chamber measurements, because several factors may generate systematic measuring errors. The carbon dioxide concentration in the chamber is in most cases lower than ambient one and consequently the measured CO_2 exchange is too low. This phenomenon is also clear in Fig. 9.9 where the regression line differs from the one-to-one line. Thermal expansion of air may easily cause systematic errors in the flow. The chamber reduces the light falling on the leaves and the humidity in the chamber is also problematic. In this specific case in Fig. 9.9, the systematic error due to the degrease of CO_2 in the chamber is 2%.

9.5 Mini Watersheds

The ice pushed mineral soils from Finland towards Central Europe during the Ice Age, and Finland obtained new layer of mineral soil from the ice when the ice melted about10,000 years ago. Organic matter has accumulated on the topsoil after ice age. Reflecting this pattern, the soil around SMEAR II station is very shallow, only from 5 to 150 cm. The shape of the bedrock underneath is clearly visible to naked eye, there are evident small water sheds bordered by higher elevations.

The bedrock is very solid, and due to short period of subaquatic conditions, there is a layer of silt with very high clay content in it in the bottommost soil layer above the rock surface; thus, we can say that the clay seals the bedrock waterproof. There are two mini catchments (900 and 300 m^2) nearby our measuring cottage. We closed these catchments with a dam, and we constructed measurements of seepage flow and the runoff from the area. The seepage flow and also the elements, ions and compounds in the flow from normal forest soils are impossible to measure. We take regular samples of the outflow water for chemical analysis of ions, nitrogen and carbon compounds. In this way we can determine the fluxes of nutrients, especially nitrogen, and carbon compounds out of the ecosystem in the seepage water flow in the soil. This is a unique possibility in our northern location. We can say that the watersheds are natural lysimeters.

9.6 Eddy Covariance

Consider a hypothetical cubic volume with a fixed size in a homogeneous forest growing on flat land. Horizontal wind carries CO_2 in and out the cubic volume (Fig. 9.10). If the concentrations and temperature are spatially constant, then the wind carries the same amounts of CO_2 in and out through the vertical boundaries of the cube, and these fluxes cancel each other. Rather clear CO_2 concentration differences are characteristic for the vertical direction (Fig. 9.11), and the vertical flow transports considerable amounts of CO_2.

The mass and energy balance (conservation of mass or energy, Chap. 2) states that the change in the amount of CO_2 is caused by the flows in and out together with sources/sinks. Processes consume and produce CO_2 into the hypothetical volume, and horizontal and vertical wind carries in and out CO_2. If the CO_2 concentration is spatially constant, then the horizontal in and out flows of CO_2 compensate each other. If, in addition, the concentrations of CO_2 are stable, then the change term in the mass balance equation of carbon dioxide in the volume can be omitted. Thus, if CO_2 concentration is temporally and horizontally constant, then the vertical flux of CO_2 through the top of the cube equals with their production in the processes in the hypothetical cubic volume (Finnigan et al. 2003).

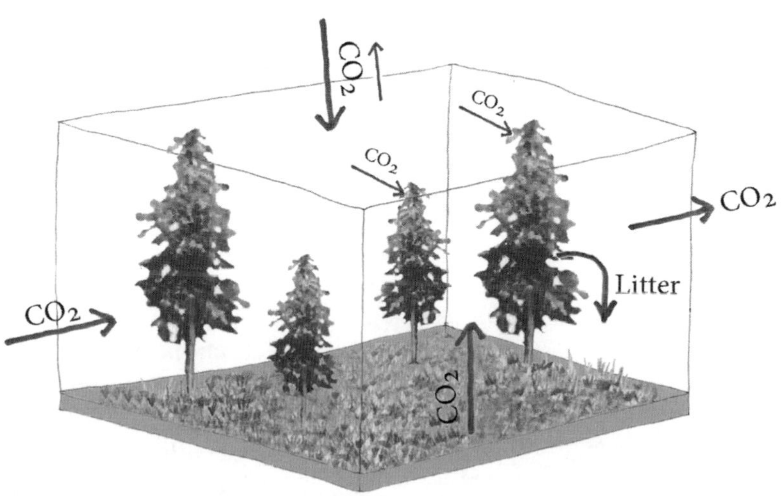

Fig. 9.10 The in and out fluxes of CO_2 in a hypothetical cube in forest

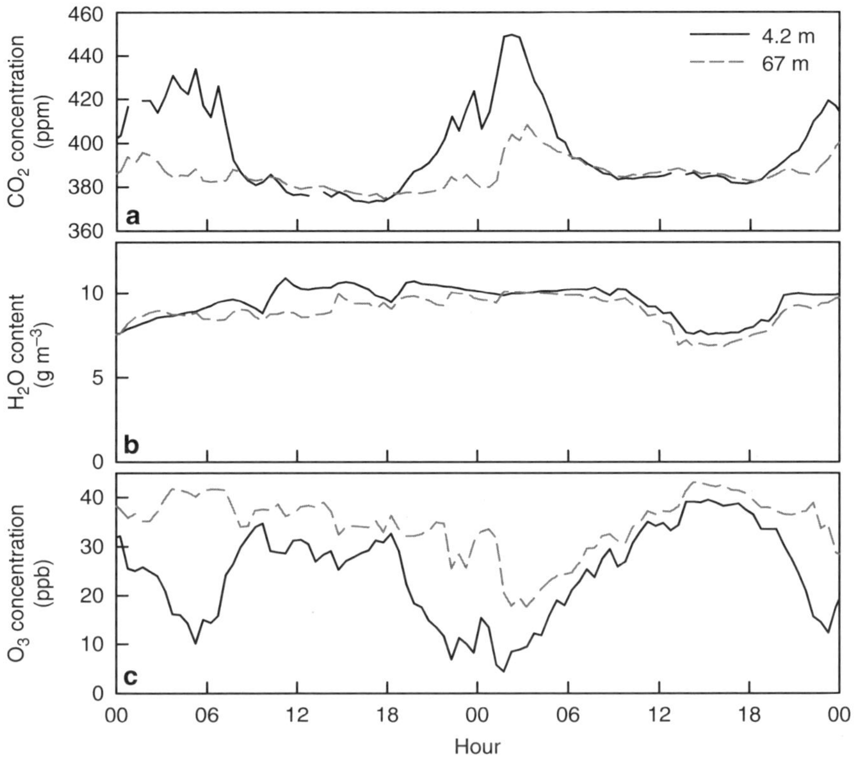

Fig. 9.11 Carbon dioxide (**a**), water vapour (**b**) and ozone (**c**) concentrations during 18–19 June 2011 at SMEAR II in the heights of 4.2 and 67 m

The vertical wind is very variable, it goes up and down, but above flat land area, the mean of the vertical wind is zero. There is CO_2 concentration variation connected with the vertical wind direction (Fig. 9.12); evidently the vertical wind carries different amounts of CO_2 upwards and downwards. Figure 9.12 shows that the vertical wind and CO_2 concentration seem to be negatively correlated. This indicates that the CO_2 concentration is increasing upwards. Thus, the forest is acting as a sink creating the downward flux. The exchange rates of CO_2 through a unit area are the concentrations in the air parcel multiplied by the wind speed. The CO_2 concentration is always positive but vertical wind is either positive or negative; thus, the flux as product of positive or negative number obtains also either positive or negative values. The sign convention used in meteorology is that the upward direction of the flow is positive and the downward one is negative. The instantaneous values of the flux are rather meaningless for practical information, and the mean flux is calculated for a prolonged period, typically 0.5 h.

In 1990s, the development of digital instruments enabled several vertical wind and CO_2 concentrations measurements within a second. This is fast enough to

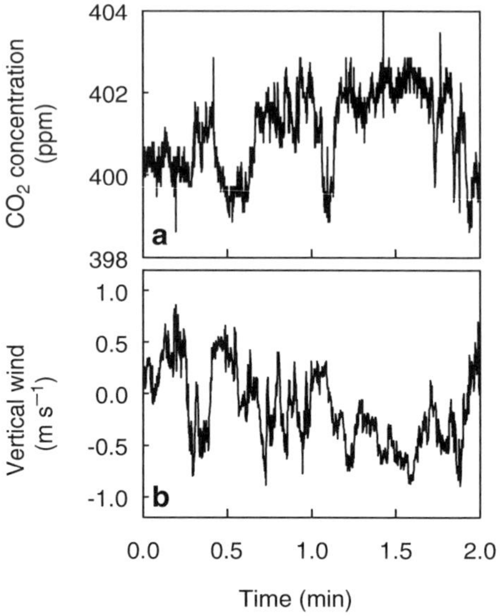

Fig. 9.12 The measured CO_2 concentration (**a**) and vertical wind velocity (**b**) during two minutes at SMEAR II starting at 14:00 during the day, 9 June 2001

monitor the air parcels going up and down, i.e. rapid turbulent velocity and concentration fluctuations. This method for measuring the vertical turbulent flux of mass is called Eddy Covariance. We started the Eddy Covariance measurements of fluxes of CO_2 at SMEAR II in1996. The measuring system consists of a 3-D anemometer (Fig. 9.13) and CO_2 analyser that operate at about 10 Hz measuring frequency.

We concluded previously, utilising conservation of CO_2, that the CO_2 flux above a hypothetical cube in a forest equals the CO_2 consumption or production in biological processes within the cube if the carbon dioxide concentration is temporally and horizontally constant. Most of the nights are calm and carbon dioxide accumulates into the trunk and canopy space and above the forest (Fig. 9.14). After a calm night in the morning, the heating by solar radiation starts the vertical wind and the turbulent mixing resulting in rather stable mean CO_2 concentration. Consequently, the measured carbon dioxide flux is useless during the night when the CO_2 storage in the air of the forest changes strongly, but during sufficient turbulence we obtain informative measurements.

We divide the period during which we want to determine the ecosystem flux into several subintervals of equal length. The instruments measure the concentrations of CO_2 and three components of wind during each subinterval. Let g_E denote the ecosystem flux during the period Δ (typically 0.5 h), t_i the moment of the ith measurements during the period in consideration, $w(t_i)$ the vertical wind velocity, w_{Mean} the mean vertical velocity over the period, $C_{CO2}(t_i)$ the CO_2 concentration, C_{Mean} the mean CO_2 concentration and n the total number of the instantaneous vertical wind and concentration measurements over the period. We obtain the ecosystem flux by summing the fluxes during each subinterval together:

Fig. 9.13 A sonic anemometer in summer and winter conditions. The ice in winter conditions destroys the measurement

$$g_E = \sum_{i=1}^{n} (w(t_i) - w_{mean})(C_{CO_2}(t_i) - C_{Mean})\frac{\Delta}{n} \tag{9.4}$$

Our Eddy Covariance measuring system of CO_2 flux is still in operation providing twice per hour carbon dioxide flux between the stand and the atmosphere day and night, summer and winter. If the measurement frequency is 10 Hz, then n

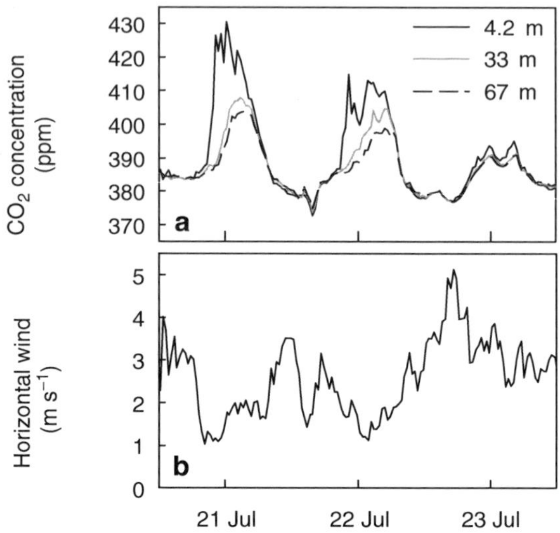

Fig. 9.14 CO_2 concentration at three different heights (**a**) and horizontal wind (**b**) at SMEAR II, southern Finland during 20–23 July 2010

in Eq. 9.1 is 18,000 for the period of 0.5 h. However, the measurements, when the turbulent mixing is weak and CO_2 accumulates into the air, are rejected.

The measuring principle of Eddy Covariance, described above for CO_2 flux, can be applied also for other gases such as water vapour, ozone, methane and volatile organic compounds and for heat. We have measured all these gas and heat fluxes at SMEAR II station. In addition we have applied so-called storage correction to reduce the data loss during calm nights.

9.7 Concluding Remarks

The development of instrumentation, characterised now by digital device, has been rapid during the past few decades and we can expect that this development will continue. There are plenty of novel possibilities to obtain feedback from the forest that should be utilised in forest ecological research. Also, the quality of the data has received little attention reducing the power of data in the testing of theories and in the model development. We should determine the precision and accuracy of the measurements and in this way improve the noise to signal ratio. The systematic measuring errors that decrease the accuracy are problematic in forest ecological measurements, especially when we measure water vapour.

Particular feature behind the development of previously described understanding of ecosystem functions and development of field measurements in our group has been the close collaboration between biologists, ecologists and physicists across the disciplines. Our long time collaborator, the late Academician Juhan Ross from Estonia, once described the intricate relationship between biologists and physicist in the development of science using a hunting analogue. If the hunter (biologist) is poor and cannot command the clever dog (physicist), the hunting fails. So the successful hunt requires adequate skills for both partners. Unique in our case is that we have taken the step to change roles.

We planned our measuring station SMEAR II in southern Finland to monitor all relevant material and energy fluxes, processes generating the fluxes and environmental factors affecting the processes in a forest ecosystem. Thus, the station measures the same quantities as our theory is dealing with, and the measuring system and the theory are coherent with each other. This fact makes the measurements at SMEAR II valuable for our physical and physiological theory of forest ecology. We planned and constructed SMEAR I 5 years before SMEAR II. However, the less systematically gathered data by SMEAR I are also valuable to support our theory.

The forests are responding and they will respond to the climate change and we need urgently data on the changes in the forests. A hierarchical network of measuring stations, as proposed by Hari et al. 2008, is able to provide the needed information.

References

Finnigan JJ, Clement R, Malhi Y, Leuning R, Cleugh HA (2003) A re-evaluation of long-term flux measurement techniques, Part I: averaging and coordinate rotation. Bound Layer Meteorol 107:1–48

Hari P, Nikinmaa E, Kulmala M (2008) Concluding remarks. In: Hari P, Kulmala L (eds) Boreal forest and climate change, vol 34, Advances in global change research. Springer, Dordrecht/London

Chapter 10
The Physical and Physiological Theory of Forest Ecology and Its Evaluation

Pertti Hari

Contents

Abstract We combine the results obtained in the previous chapters to create our physical and physiological theory of forest ecology. This is possible since the cover theory provided common ideas, concepts and methodology for the specific theories. Our theory gained strong corroboration in the tests dealing with different phenomena from metabolism to ecosystem development.

P. Hari (✉)
Department of Forest Sciences, University of Helsinki, P.O. Box 27, 00014 University of Helsinki, Helsinki, Finland
e-mail: pertti.hari@helsinki.fi

P. Hari et al. (eds.), *Physical and Physiological Forest Ecology*,
DOI 10.1007/978-94-007-5603-8_10,

Keywords Cover theory • Levels in hierarchy • Combining specific theories • Field test • Corroboration

10.1 Theory Compilation

In the preceding chapters, we have formulated our vision, concerning the energy and mass flows, the role of physics and physiology, conservation of mass and energy, regulation of organisms and evolution. The exactly defined and common concepts, with basic ideas at different levels, were the backbones in the gaining an understanding of the regularities in forest ecosystems. When we combine the sub-theories at different levels in the hierarchy, we reach theoretical understanding of the mass and energy fluxes between forest ecosystems and their environment, and in this way, we obtain the physical and physiological theory of forest ecology. The comprehensive measurements at SMEAR II and I provide material for testing the specific theories at different levels in the ecosystem.

10.2 The Approach

The forests are complicated systems including very different living organisms from trees to microbes in the soil. The spatial scales vary from microbes to large trees across about six orders of magnitude, and temporal scales run from the very fast photosynthetic light reactions to the century-scale rotation of forests, about 18 orders of magnitude. We would like to be able to consider simultaneously this huge range of phenomena. This is only possible within a common framework to treat different phenomena with a coherent set of concepts and common methodology. We try to solve the simultaneous treatment of versatile phenomena with a cover theory that outlines common and general treatment of all aspects in forest ecosystems.

10.3 Cover Theory

We started the formulation of the cover theory with definition of basic concepts that are used throughout the whole book. The names of some concepts are problematic, especially process and regulation. These words are commonly used in scientific texts, but their exact meaning is often somewhat unclear. We needed 12 basic concepts to cover the most relevant aspects in forest ecosystems from details of metabolism to the behaviour of ecosystems. The concept of functional substances, that is, the enzymes, membrane pumps and pigments, is important for the utilisation of physiological knowledge in the theory construction.

We introduced 14 basic ideas to clarify our vision, utilizing background knowledge. The regulation, that is, changes in the activity and concentrations of functional substances caused by the action of the biochemical regulation system, plays a very important role in the metabolism and also in the formation of plant structure. The hierarchy enables treatment of phenomena in different scales. The emergent properties when moving towards more aggregated levels are characteristic for living systems, and they also enable utilisation of detailed information in condensed form at more aggregated level.

We constructed specific theories within the cover theory to deal with important phenomena in forest ecosystems. Each specific theory has its own basic concepts and ideas that must be coherent with the cover theory. We developed theoretical models to clarify the basic ideas in concrete form that enables the use of mathematical and statistical methods. The analysis of material and energy flows results in dynamic differential or difference models. Physical knowledge plays an important role in the derivation of the dynamic models since they are based on conservation of mass or energy.

We tested the models with extensive and versatile field data. We evaluated four aspects in the tests: (1) predictions of characteristic features, (2) adequacy of the model structure, (3) explaining power of the theory and of the model, and (4) comparison with other studies found in the literature. If the model passes successfully all the four aspects, then the model and the theory gain corroboration. In the case of failure, we should reject or improve the specific theory and the model based on it.

The final step in the formation of the physical and physiological theory of forest ecology is to combine the essential features of the specific theories to the comprehensive treatment of the interactions of forest ecosystem with its environment.

10.4 Specific Theories

10.4.1 Processes

Living organisms have several metabolic tasks, such as photosynthesis, respiration and growth. Each task has its specific anatomic structure, such as stomata and chloroplasts. The structure is rather stable, and it does not change appreciably after its formation in growth. The functional substances, that is, enzymes, membrane pumps and pigments, are non-stable compounds, and organisms have to maintain proper concentrations and activities.

Organisms have a biochemical regulation system that synthesises, activates, decomposes and deactivates functional substances. Proper action of the biochemical regulation system is crucial for the success of organisms, and this is why effective regulation has developed in evolution. The biochemical regulation system activates and deactivates metabolism, and in this way it, for example, controls the annual cycle of metabolism.

The biochemical regulation system controls the concentrations and activities of functional substances in the long chains of metabolic processes in a balanced way; thus, there is a reasonable balance between sequential steps of the process. In this way, an emergent property, called efficiency of the functional substances, is formed as result of the action of the biochemical regulation system. It determines the level of the metabolic steps involved into the process.

The specific steps in the metabolic process have their own dependence on environmental factors and on the available raw material for the reaction. The dependence of process rate on environmental factors and on the availability of raw material is built into the chain of reaction steps. For example, the light reactions of photosynthesis response to light intensity and carbon reactions response to the availability of energy in ATP and NADPH and of carbon dioxide.

The concentrations and activities of the functional substances change due to the action of the biochemical regulation system. This gives rise to another emergent property, the state of the functional substances that characterises concentrations and activities of the functional substances of the process under consideration. The biochemical regulation system changes slowly the state of the functional substances because it requires synthesis, decomposition, activation and deactivation of functional substances.

The efficiency of the functional substances depends on the concentration and activities of the functional substances. Thus, the efficiency of the functional substances depends on the state of the functional substances.

We applied the above approach to several processes taking place in trees. We were able to predict photosynthesis and transpiration during the annual cycle and bud burst in the spring and test the predictions with extensive field data. The analysis of other processes was more descriptive.

10.4.2 Fluxes

Two types of fluxes take place in forest ecosystems: (1) within individuals and between ecosystem components and (2) ecosystem itself and the atmosphere. Both these fluxes are generated either by metabolic or physical processes. Water fluxes are the biggest ones although carbon flows are more important for the development of the ecosystem. In addition, rather small fluxes of nutrients, especially nitrogen, are needed for the proper functioning of the forest ecosystem.

Living cells in vegetation need raw material from other cells for their metabolism. The locations of production and consumption can be rather far away from each other. This fact generates the need of transport within individuals, and there are specialised structures for the connections between locations of production and consumption. Sugars, formed in photosynthesis, and water, taken up by roots, dominate the transport within individuals in vegetation.

Osmosis and tension difference enable water uptake by fine roots. The evaporation from the surface of the mesophyll cells generates tension, and the tension difference between mesophyll surface and roots pulls water from the roots up to leaves in xylem. The water transport from roots to leaves is physical phenomenon and in principle easy to understand, although there are still some weakly understood aspects, for example, cavitation. The transport of sugars from leaves to everywhere in a tree is a demanding task and poorly understood. The transport mechanism of sugars is pressure difference generated with osmosis.

Photosynthesis and respiration generate CO_2 concentration differences in forest soil and in the canopy and transpiration water vapour differences in the canopy. These concentration differences generate diffusion in the soil and transport with convection in the canopy.

We formulated specific theories for water and CO_2 fluxes within vegetation and in ecosystem based on the theoretical ideas above.

10.5 Structural Regularities

The biochemical regulation system controls the growth of needles, fine roots and water pipes in the woody structures. Balanced structure of a tree, that is, such structure that the production of different materials and their transport match with each other, contributes to the success of living organisms, and in evolution, balanced structures have developed. The balanced structure of trees is an emergent outcome of the action of the biochemical regulation developed in the evolution.

We studied tree structure assuming that the xylem in roots, stem and branches forms a balanced water transport system. We used sap wood area as measure of water transport capacity, and it enabled formulation of hypothesis that can be tested. In addition, we studied the idea that water transport system is in balance with the need of water transport in branches and in roots. This is a good starting point for the studies of structural regularities in trees. When detailed description of water transport is introduced into the analysis, our result may look as rough simplification.

The biochemical regulation system of Scots pine trees allocates nitrogen to the functional substances in the growing tissues resulting stable nitrogen concentrations in the tree, and concentrations do not reflect the fertility of the site.

10.6 Ecosystem Dynamics

The flows of carbon and nitrogen determine the accumulation and release of material from forest ecosystems. Carbon enters the ecosystem in photosynthesis by trees and ground vegetation. The resulting sugars are used for energy needs of living cells, for construction of new structures and for root exudates. The living structures die, the leaves after one or several growing seasons, but wood lasts longer, even centuries. The dead organic matter enters the carbon pool in the soil where microbes decompose its macromolecules with extracellular enzymes. Microbes utilise the resulting small carbon molecules, like sugars and amino acids, in their metabolism and release the carbon into soil air as carbon dioxide. Thus, carbon flows through the ecosystem, and the lifetime of an atom in the ecosystem varies greatly, in the case of root exudates only days, and in wood it may be centuries.

Most of the nitrogen in a forest ecosystem is as nitrogen molecules in the air that is unavailable for metabolism of vegetation and of microbes. Reactive nitrogen, ions and nitrogen-rich amino acids, is available for vegetation and microbes. Thus, the separation of reactive and molecular nitrogen is important. The natural reactive nitrogen fluxes between forest ecosystem and its surroundings are small, usually less than 0.1 g m^{-2} $year^{-1}$. The reactive nitrogen enters forest ecosystems in the deposition and in the nitrogen fixation. Nitrogen flows out in leaching as ions or amino acids and as molecular nitrogen from denitrification. The uptake and release of nitrogen oxides by leaves is very small.

Most of the reactive nitrogen in forest ecosystem is in proteins. Proteins are the functional substances that enable metabolism of living cells. The amount of proteins in forest soil, either as molecules or bound into the microbes, is large, about 500 g m^{-2}. In contrast, the amount of proteins in needles per land area is small, about 50 g m^{-2}. These numbers refer to the ecosystem around SMEAR II in Hyytiälä, southern Finland. The annual amount of nitrogen needed for synthesis of protein in the functional substances is about 2 g m^{-2} $year^{-2}$. The annual need of nitrogen is about one magnitude higher than the deposition: evidently, the nitrogen is circulating effectively in the forest ecosystems.

The annual cycle dominates the formation of tree structure in northern forests, for example, Scots pine lays down buds in late summer and there are primordia for all needles to grow during the next summer in the buds. The growth is mainly based on sugars photosynthesised during the period of growth since the sugar pools are insufficient for the synthesis of all the macromolecules needed for new cells.

There are strong regularities in the structures of living organisms because the new structures have to be able to operate effectively. Water transport determines the properties of xylem in roots, stems and branches, and it is the key to understand the regularities in tree structure. In addition, the nitrogen concentrations are stable in the needles of Scots pine. The observed regularities in tree structure indicate that the biochemical regulation system of Scots pine is able to construct these emergent properties.

We introduced the action of the biochemical regulation system with carbon and nitrogen balance equations. Carbon balance equation states that all annual photosynthetic production is used for maintenance and growth. Need of water transport determines the area of new xylem in branches, stem and roots, and the area of new xylem is proportional to the increase in needle mass. The nitrogen balance equation is based on specific nitrogen concentrations in needles, xylem and fine roots. It states that the nitrogen uptake and reuse from dying needles equals the consumption to the new cells in needles, xylem and fine roots.

The carbon and nitrogen balance equations include two unknown numbers, that is, the growth of needles and fine roots. The structural regularities determine the growth of xylem from needle growth. We have two equations that include two unknown factors; thus, we can solve them. Evidently, the biochemical regulation system is able to solve the allocation problem resulting in an effective metabolism in trees and the nice regularities in the structure of trees. The solution of the mass balance equations is our description of the action of the biochemical regulation system.

The litter input to the pool of organic material in soil is often less than 500 g m^{-2} $year^{-1}$, and the pool of organic material in forest soil is about 20 kg m^{-2} indicating slow decomposition. The lifetime of macromolecules in forest soil varies, but it is long, decades. The decomposition of proteins is very important for the dynamics of forest ecosystem since it provides reactive nitrogen for vegetation. Nitrogen flow from proteins in soil determines the growth of new cells because the functional substances in the new cells are synthesised from the nitrogen taken up from soil.

10.7 Test of the Physical and Physiological Theory of Forest Ecology

We have versatile data available measured at SMEAR II and SMEAR I and some additional measurements from Estonia and Eddy Covariance sites. We can use these data sets to test our theory and obtain feedback from nature to our thinking including very different aspects from cover theory to measurements. The cover theory provided a general backbone for the research at different levels. The material and energy fluxes provided a common interest on different levels of the forest ecosystem from leaf element to the whole ecosystem. The important role of dynamic modelling is also a consequence of the fluxes since analysis of material or energy in a container based on fluxes results in dynamic differential or difference equation. The cover theory was useful in the research, since it provided structure for the research and smooth knowledge flow from detailed level to more aggregated level.

10.7.1 Processes

We construct specific theories for important processes, photosynthesis, respiration and growth. The specific theories of processes provided exact predictions of the responses of processes to environmental factors characterised by strong diurnal and annual cycle. We tested these theories with field data measured at SMEAR I and SMEAR II.

Prediction of Characteristic Features. The specific theories were able to predict the characteristic features in the behaviour of the processes both at diurnal and annual levels. The data used in the test are massive in the case of photosynthesis and transpiration, about 700,000 measurements. The number of bud burst dates and daily height increment measurements are rather low, about 100.

Adequacy of the Model Structures. The functional substances, that is, enzymes, membrane pumps and pigments, enable the processes in living structures. The functional substances are non-stable proteins; thus, cells must have system to maintain proper concentrations and activities of the functional substances, that is, they are under the control of the biochemical regulation system. These are physiological facts, and our theories of processes are based on these facts, and our approach concerning processes is sound from the physiological point of view.

The residuals were small, and they showed only weak systematic features.

Explaining Power. The concentrations and activities of functional substances change slowly in the timescale of days or weeks. Thus, during a day or couple of days, the functional substances do not change and they are stable, and also the relationships between environment and the process rate are stable. Thus, variation in the environmental factors explains the variation in the process rates.

During longer periods, that is, several days or weeks, the action of the biochemical regulation system becomes visible and the relationships between process rates change due to the changes in the functional substances. This phenomenon is clear during the annual cycle of the processes.

The specific theories explained well the variations in the metabolic rates; the proportions of the explained variation in the prediction data were about 90%. The respiration is an exception because the measuring noise dominates the measured values.

Comparison with Other Theories. The theoretical argumentation dominates our development of dynamic models of process rates, and physiology plays an important role in the argumentation. Physiological knowledge is omitted to great extent in the present forest ecological research. The models by Farquhar et al. (1980) and Noe and Giersch (2004) are exceptions since they use physiological knowledge in their argumentation.

Statistical analysis of field measurements of process rates is characteristic for most of the forest ecological research. Consequently, the measurements of responses are related to the environmental factors, and argumentation referring to nonmeasurable properties of the research object is strongly avoided. Descriptions of these nonmeasurable properties in the applied models have often been called as 'fudge factor'. The same approach was used in human physiology, called behaviourism, in the first half of the twentieth century. Later, behaviourism has been rejected since it omits the most relevant features of human behaviour. However, this narrow-minded thinking has dominated forest ecological research for decades, but now a more quantitative approach is emerging.

10.7.2 Fluxes

The locations of production and consumption can be rather far away from each other. This fact generates the need of transport within individuals, and there are specialised structures for the connections between locations of production and consumption. Sugars, formed in photosynthesis, and water, taken up by roots, dominate the transport within individuals in vegetation. Plants have special structures for water and sugar transport, xylem for water and phloem for sugars. Evaporation from the water on mesophyll cells generates tension that pulls water from soil. The high sugar concentration, generated by the action of membrane pumps, gives rise to osmosis and overpressure in the phloem cells next to photosynthesizing cells. The overpressure moves sugars in the phloem.

Photosynthesis and respiration generate CO_2 concentration differences in forest soil and in the canopy. These concentration differences generate diffusion in the soil and transport with convection in the canopy.

We formulated specific theories for water and CO_2 fluxes within vegetation and in ecosystem based on the above theoretical ideas.

Prediction of Characteristic Features. Our predictions were only partially successful. The predictions of large patterns were rather satisfactory, but there were clear failures in the details. For example, the tree diameter responded too quickly to the decrease in transpiration in the evenings, and evidently root exudates caused large efflux of CO_2 from soil in late summer and autumn. In addition, the systematic discrepancy between predicted and measured ecosystem water vapour flux is evidently caused by unknown systematic measuring errors.

Adequacy of the Model Structures. The structures of our models have a rather sound physiological and physical basis, but there are evidently phenomena missing in the models that generate the observed discrepancy. This conclusion is supported by the systematic behaviour of the residuals.

Explaining Power. The physical and physiological basis of our analysis is sound; however, the missing details in the models remain as disturbing aspects. The proportions of the explained variance in the predictions remained rather low due to the shortcoming in the model structures, but the photosynthesis of the ecosystem is an exception.

Comparison with Other Theories. The analog models proposed by H. T. Odum have dominated the analysis of fluxes. Valuable results have been obtained, but here we meet the problem that explanation is missing and the models remain only as description of the phenomenon. The description of the diffusion of CO_2 into leaves through stomatal pores is an example of successful analog model, since physical analysis of Brownian movement of molecules results in the model that is analogous to the Ohm's resistance model. Also, we use resistance or conductance model to describe diffusion in our optimal stomatal control model.

Electrical analog models have been used to describe water flow in xylem. When the capacitors describe the water storage in wood, then the connection to the amount of water is lost and the interpretation becomes problematic. Another example is the concept of turbulent diffusion that deals with the vertical transport of gases. It is very problematic to understand the meaning of the concept since it deals with two scales and two phenomena, that is, diffusion in the scale less than 1 mm and turbulence that occurs in the scale of metres.

The proper understanding of fluxes requires that the argumentation is based on physical and physiological concepts and theories.

10.8 Regularities in Structure

Regularities in tree structure have received little attention in forest research, and also our material is rather limited. The trees have several processes running simultaneously, and transport within trees connects the processes with each other and provides the necessary raw materials for the processes. In an efficient system, the transport capacities and the needs of processes have to match with each other.

Prediction of Characteristic Features. We found the expected linear relationships between sapwood areas of roots, stems and branches and the mass of fine roots and sapwood area of the root as well as between the sapwood area of branch and its needle mass.

Adequacy of the Model Structures. More rigorous analysis of the water transport capacity of xylem is needed, for example, we should include the reduced water conductivity in the juvenile wood into the analysis. The limited material did not reveal any systematic behaviour in the residuals.

Explaining Power. The need to understand the regularities in tree structure is evident, and the efficient water transport system is one of the keys to explain the tree structure. The linear regressions explained high proportions of variances in the sapwood areas along water paths in xylem. Evidently, the low precision and accuracy of fine root mass measurements reduced the proportion of the explained variance by linear regression between sapwood area and fine root mass.

Comparison with Other Theories. The regularities in tree structure have received very little attention in the forest research. This is evidently caused by the behaviouristic thinking dominating the field that denies the role of internal properties of trees in the research. The action of the biochemical regulation system generates the regularities in the structure of trees, and they are considered as mystic features.

10.9 Dynamics of Carbon and Nitrogen Fluxes and Pools in Forest Ecosystem

We utilised widely the results obtained on processes, fluxes and regularities in tree structure to the theory of dynamics of carbon and nitrogen pools and fluxes in forest ecosystem. We combined the results to the forest ecosystem model, called MicroForest, to obtain exact predictions that enable tests of the theory.

Prediction of Characteristic Features. The testing of the theory of carbon and nitrogen dynamics requires data covering decades of the properties of an ecosystem. We have no data covering the whole ecosystem over long period of time, and this fact limits the analysis of prediction power of our theory. Evidently, the only option is to make retrospect measurements of tree growth in selected ecosystems. Our theory was able to predict the formation of size classes in the measured stands near SMEAR II and in Estonia from the initial states at the age about 5 years. Also, the prediction of the diameter and height growth patterns was successful.

Adequacy of the Model Structures. The theory of ecosystem carbon and nitrogen dynamics is based on the results obtained in the previous chapters, especially on photosynthesis, and regularities in the structure. The emergent properties play a key role in the formulation of the annual changes in the fluxes and pools. Quite strong idealisations were needed that may introduce inaccuracies which should be improved.

We could study only the residuals of diameter and height growth due to the lack of time series describing ecosystem behaviour in the timescale of decades. Only, the residuals in respect to the initial state showed systematic behaviour indicating problems in the measurements of the smallest seedlings when their diameter was less than 5 mm.

Explaining Power. Our theory explains the ecosystem dynamics with metabolism of trees, of ground vegetation and of microbes and with the regularities in tree structure. This gives theoretical and causal explanation for several phenomena in the ecosystem, for example, tree growth.

The proportion of explained variance of predicted diameters at the end of measurements was high, nearly 90%; the height prediction was less successful, only about 50%.

Comparison with Other Theories. Our theory of carbon and nitrogen dynamics includes several novel aspects, such as treatment of the chemical character of the macromolecules, simultaneous consideration of carbon and nitrogen and regularities in tree structure. Thus, the comparison with other models is problematic; we can compare only some components of our model with those in literature.

The tree component of our model is a continuation of the carbon balance models introduced by de Wit et al. (1970). We have utilised the rather long tradition at our department of dynamic carbon-based models of Scots pine stands (Hari et al. 1982, 1985, 2008; Mäkelä and Hari 1986; Nikinmaa 1992; Perttunen et al. 1996; Mäkelä

1997, 2002; Nissinen and Hari 1998; Mäkelä and Mäkinen 2003). Our expansion to the simultaneous nitrogen balance is a novel aspect, and it opens the possibility to connect the vegetation and soil in a well-justified way.

Traditionally, decomposition models have used aggregated soil organic matter types, such as rapidly, slowly and extremely slowly decaying organic matter, for which they have produced aggregated decomposition rates (e.g. Aber et al. 1991; Rastetter et al. 1991; Running and Gover 1991; Johnson 1999). The problem with this approach of rapidly, slowly and extremely slow soil organic matter types is that there are no means to measure the state variables describing the soil properties. Therefore, also the testing of the model is extremely problematic. The introduction of the chemical character of the macromolecules in the soil (Liski et al. 1998) enables direct measurements of the different carbon pools in the soil. We expanded this chemical approach to the proteins and in this way the nitrogen dynamics in the soil.

10.10 Evaluation of the Physical and Physiological Theory of Forest Ecology

We described the research as a never-ending loop of developing ideas and their tests in Fig. 2.2 and the flow of knowledge from processes to ecosystem research in Fig. 2.5. When we combine these two figures, we obtain Fig. 10.1 that stresses the knowledge flow and the importance of obtaining feedback from the nature.

We have only one single comprehensive ecosystem measuring system available, and this generates problems to apply the results obtained at SMEAR II to other types of forests. Sporadic short measurements of processes and fluxes dominate the forest ecology. However, ecosystems are complex systems, and to be able to combine phenomena in the system, we need versatile data that cover the relevant aspect of the ecosystem for prolonged periods. Thus, we need more comprehensive measuring systems of forest ecosystems and their relevant processes.

The measurements of protein, cellulose, lipid and starch pools provide quantitative information for the carbon circulation in the forest ecosystem. These measurements open new possibilities to study the carbon dynamics in forest soil. Especially, the proteins and amino acids are important in the nitrogen circulation between vegetation and soil organic matter. The measurements of macromolecules in forest soil should be taken as normal praxis.

We based the test of the dynamics of carbon and nitrogen fluxes and pools in forest ecosystem on retrospective measurements of tree diameters and heights in 11 stands. Although the predictions by the theory were very successful, the number of stands is so low that it reduces the obtained corroboration. In addition, the test material covers very limited area of the carbon and nitrogen dynamics in forest ecosystems. The treatment of forest soil in our analysis is novel, and its testing is important. The available measurements of forest soils are rather scarce because the measuring tradition is focusing on ions in the soil. The chemical character of soil organic matter should be measured, especially dynamics of proteins, amino acids

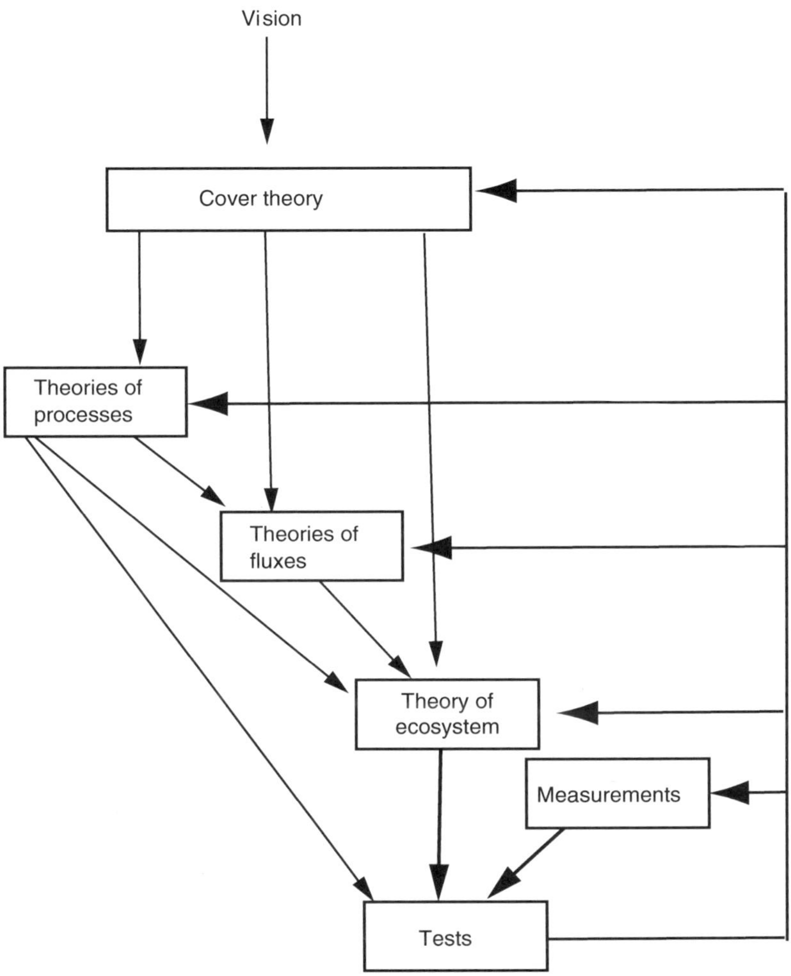

Fig. 10.1 Flow of knowledge in the formation and in the testing of the physical and physiological theory of forest ecology. *Boxes* describe theory, *arrow lines* flow of knowledge, *thick lines* testing and *thin lines* formation

and ammonium. The structure of the model MicroForest enables a great number of different tests with field data. Special measurements should be arranged to obtain additional tests.

We tested the emergent regularities in tree structure generated by the action of the biochemical regulation system. We assumed that the water transport capacities along the water path from fine roots to needles match with each other and with the need of water transport. We have rather old data from measurements in one single stand. We should urgently expand these data in several other stands. In addition, we should introduce the effect of the fine structure of wood into the measurements.

The tests of the fluxes within trees and on the ecosystem level revealed evident shortcomings in the applied model structures. Additional water storage should be introduced into the model describing the water transport in xylem. The lack of the annual cycle in root exudates evidently causes to high CO_2 efflux from soil in winter and autumn.

Extensive measurements of high quality enabled severe testing of the theories of photosynthesis and transpiration. The predictions passed successfully the tests, and they resulted in a clear corroboration of the theories. The shortcomings in the theories of photosynthesis are beyond our measuring precision and accuracy. The measuring system should be improved to be able to detect discrepancies between the theory prediction and measurements. The daily height increment data are old, and additional measurements are needed.

The cover theory provided a general backbone for the research at different levels. The material and energy fluxes provided a common interest on different levels of the forest ecosystem from leaf element to trees and to the whole ecosystem. The important role of dynamic modelling is also a consequence of the fluxes since material or energy balance in a container results in dynamic differential or difference equation. The cover theory was useful in the research, since it provided structure for the research and smooth knowledge flow from detailed level to more aggregated level.

We tested our physical and physiological theory of forest ecology with several phenomena in the hierarchy of forest ecosystem. The predictions based on specific theories of the phenomena passed successfully the severe tests, although we detected also some shortcomings in the applied models. Our theory gained strong corroboration in the tests.

10.11 Final Conclusion

The main results of our physical and physiological theory are:

- The material and energy flows are the key to understand the interaction between forest ecosystem and its environment and also phenomena within forest ecosystems.
- Forest ecosystems involve very different phenomena from metabolism of a single cell to the growth of the ecosystem during 100 years. We can study this versatility only within a common framework, called cover theory, that outlines the concepts and introduces the main ideas and methodology. We specify the common cover theory to meet the properties of each studied phenomenon in the forest ecosystem.
- We study the forest ecosystem within a hierarchy from the cell level, via trees to ecosystems, during decades. The emergent properties, when moving from detailed level to more aggregated considerations, play an important role in the understanding of ecosystem development.

- Metabolism of vegetation and microbes generates concentration, pressure and temperature differences that give rise to material flows between ecosystems and their surroundings and also within ecosystems
- Enzymes, membrane pumps and pigments, collectively called functional substances, enable metabolic processes within vegetation and microbes. Functional substances are non-stable molecules, and living organisms have to synthesise, activate, decompose and deactivate them. We call the system controlling the concentrations of active functional substances the biochemical regulation system.
- The action of the biochemical regulation system generates emergent regularities in the functional substances and in tree structure.
- Carbon and nitrogen fluxes together with regularities in tree structure determine the development of forest ecosystems, especially tree growth.
- We tested the specific theories with extensive and high-quality measurements done at SMEAR II and SMEAR I measuring stations and with additional measurements from other stands near SMEAR II stand and in Estonia.
- The predictions based on specific theories were successful, although in some cases, the residuals indicated shortcomings in some model structures. Our physical and physiological theory of forest ecology gained strong corroboration in the severe tests.
- Our physical and physiological theory of forest ecology has a wide range of applications from forestry to the interactions between climate change and forests.
- Systematic development and expansion of the data used in the tests will increase our knowledge on and our understanding of the interactions between forest ecosystems and their environment and of carbon sequestration in forests.

References

Aber J, Mellilo J, Nadelhofer J, Pastor J, Boone D (1991) Factors controlling nitrogen cycling and nitrogen saturation in northern temperate forest ecosystem. Ecol Appl 1:303–315

De Wit CT, Brouwer R, Penning de Vries FWT (1970) The simulation of photosynthetic systems. In: Setlik I (ed) Prediction and measurement of photosynthetic production. Proceedings of the IBP/PP technical meeting, Trebon 1969. Centre for Agricultural Publishing & Documentation, Pudoc, Wageningen

Farquhar GD, von Caemmerer S, Berry JA (1980) A biochemical model of photosynthetic CO_2 assimilation in leaves of C3 species. Planta 149:78–90

Hari P, Kellomäki S, Mäkelä A, Ilonen P, Kanninen M, Korpilaahti E, Nygren M (1982) Dynamics of early development of tree stand. Acta For Fenn 177:1–42

Hari P, Kaipiainen L, Korpilahti E, Mäkelä A, Nilsson T, Oker-Blom P, Ross J, Salminen R (1985) Structure, radiation and photosynthetic production in coniferous stands, vol 54, Department of silvicultural research notes. University of Helsinki, Helsinki, pp 1–233

Hari P, Salkinoja-Salonen M, Liski J, Simojoki A, Kolari P, Pumpanen E, Kähkönen M, Aakala T, Havimo M, Kivekäs R, Nikinmaa E (2008) Growth and development of forest ecosystems; the MicroForest model. In: Hari P, Kulmala L (eds) Boreal forest and climate change, vol 34, Advances in global change research. Springer, Dordrecht

Johnson D (1999) Simulated nitrogen cycling response to elevated CO_2 in *Pinus taeda* and mixed deciduous forests. Tree Physiol 19:321–327

Liski J, Ilvesniemi H, Mäkelä A, Starr M (1998) Model analysis of the effects of soil age, fires and harvesting on the carbon storage of boreal forest soils. Eur J For Sci 49:407–416

Mäkelä A (1997) A carbon balance model of growth and self-pruning in trees based on structural relationships. For Sci 43:7–24

Mäkelä A (2002) Derivation of stem taper from the pipe theory in a carbon balance framework. Tree Physiol 22:891–905

Mäkelä A, Hari P (1986) Stand growth model based on carbon uptake and allocation in individual trees. Ecol Model 33:205–229

Mäkelä A, Mäkinen H (2003) Generating 3D sawlogs with a process-based growth model. For Ecol Manage 184:337–354

Nikinmaa E (1992) Analyses of the growth of Scots pine; matching structure with function. Acta For Fenn 235:1–68

Nissinen A, Hari P (1998) Effects of nitrogen deposition on tree growth and soil nutrients in boreal Scots pine stands. Environ Pollut 102:61–68

Noe SM, Giersch C (2004) A simple dynamic model of photosynthesis in oak leaves: coupling leaf conductance and photosynthetic carbon fixation by a variable intracellular CO_2 pool. Funct Plant Biol 31:1195–1204

Perttunen J, Sievänen R, Nikinmaa E, Salminen H, Saarenmaa H, Väkevä J (1996) LIGNUM: a tree model based on simple structural units. Ann Bot 77:87–98

Rastetter E, Ryan M, Shaver G, Mellilo J, Nadelhoffer K, Hobbie J, Aber J (1991) A general biochemical model describing the responses of the C and N cycles in terrestrial ecosystems to changes in CO_2, climate and N deposition. Tree Physiol 9:101–126

Running S, Gover S (1991) FOREST-BGC, a general model of forest ecosystem processes for regional applications. II dynamic carbon allocation and nitrogen budgets. Tree Physiol 9: 147–160

Appendices

Appendix 1: Annual Cycle of Radiation

Ü. Rannik (✉)
Department of Physics, University of Helsinki, P.O. Box 64, 00014 University of Helsinki, Helsinki, Finland
e-mail: ullar.rannik@heuristica.ee

P. Hari
Department of Forest Sciences, University of Helsinki, P.O. Box 27, 00014 University of Helsinki, Helsinki, Finland
pertti.hari@helsinki.fi

The solar flux density reaching the top of the atmosphere is characterised by the *solar constant* (I_0). The solar constant is equal[1] to 1.37 kW m^{-2}. This is the flux of solar radiation through a tangential plane that represents the average distance between the Earth and the Sun. The distance varies during the year because the Earth's orbit is eccentric with an eccentricity value around $e = 0.0167$. The difference between the maximum and minimum distances during the year is equal to $2\ e\ r_0$ which is about 5×10^6 km. The mean distance between the Earth and the Sun (r_0) is 1.5×10^8 km. As a result, the distance (r) of the Earth from the Sun depends on the time of the year, and solar radiation (I) input at the top of the atmosphere varies according to

$$I = I_0 \frac{{r_0}^2}{r^2}, \tag{A1.1}$$

where r_0 is the average value of the distance.

The solar irradiance through plane parallel to the Earth's surface depends in addition on the angle between the Sun and Earth's surface—called the *solar height*.

[1]Total solar output measurement varies by approximately 0.1% or about 1.3 W m^{-2} peak-to-trough of the 11-year sunspot cycle.

P. Hari et al. (eds.), *Physical and Physiological Forest Ecology*,
DOI 10.1007/978-94-007-5603-8,

The solar height $h_{\oplus}$ is different for each geographic location on Earth and depends on the declination angle of the Sun (δ), that is, the angle between the rays of the Sun and the plane of the Earth's equator, latitude of the location (φ) and time of day as the following:

$$h_{\oplus} = \arcsin\left(\sin\phi \sin\delta + \cos\phi \cos\delta \cos\psi\right), \tag{A1.2}$$

where ψ is the hour angle.

The declination angle can be evaluated approximately as

$$\delta = \frac{23.45^{\circ} \sin\left(2\pi\left(284 + \text{DOY}\right)\right)}{365.24} \tag{A1.3}$$

where DOY denotes day of the year (day number since the beginning of the year). The declination angle varies between −23.45° during the northern winter solstice and 23.45° in the northern summer solstice with small deviations attributable to the wobble of the polar axis. The exact value of δ for any time can be found from astronomical tables.

The hour angle is determined as

$$\psi = \frac{2\pi\left(t - 12 + \frac{\Delta t}{60}\right)}{24}, \tag{A1.4}$$

where t is the local time in hours and the equation of time, Δt (in minutes),

$$\begin{aligned}\Delta t = 0.017 + 0.4281\cos\left(\frac{2\pi\,\text{DOY}}{365.242}\right) - 7.351\sin\left(\frac{2\pi\,\text{DOY}}{365.242}\right)\\ -3.349\cos\left(\frac{4\pi\,\text{DOY}}{365.242}\right) - 9.371\sin\left(\frac{4\pi\,\text{DOY}}{365.242}\right).\end{aligned} \tag{A1.5}$$

Thus, the amount of solar flux at the top of the atmosphere through a unit area perpendicular to the Earth's surface, by including the effect of solar height, is given by

$$I_S = I_0 \frac{{r_0}^2}{r^2} \sin(h_{\oplus}). \tag{A1.6}$$

The parameters related to the Earth's positioning and orientation with respect to the Sun-eccentricity and angle-vary themselves over different but very long time scales. These variations are caused mainly by gravitational interference of the planets of the solar system. Time scales of variation are from thousands to millions of years.[2]

[2]Variations in eccentricity, axial tilt and precession of the Earth's orbit have been related to influence on climate system and ice ages. These variations occur with cycles from 21,000 to 41,000 years, resulting in 100,000-year ice age cycles.

Appendix 2: List of Parameter Values and Units in MicroForest

Parameter	Value	Unit	Equation	Source
P_o	13.6	$g(CO_2)$ g(dry weight)$^{-1}$	7:2,3,5	SMEAR II
$a_{s\,2}$	0.00000027	g(dry weight)$^{-1}$	7:4,5	Simulations with MicroForest
$a_{n\,r}$	2	$g(CO_2)$ g(dry weight)$^{-1}$	7:14	SMEAR II
$a_{b\,r}$	0.02	$g(CO_2)$ g(dry weight)$^{-1}$	7:14	SMEAR II
$a_{s\,r}$	0.02	$g(CO_2)$ g(dry weight)$^{-1}$	7:14	SMEAR II
$a_{t\,r}$	0.02	$g(CO_2)$ g(dry weight)$^{-1}$	7:14	SMEAR II
$a_{r\,r}$	17	$g(CO_2)$ g(dry weight)$^{-1}$	7:14	Simulations with MicroForest
a_b	0.03	cm^2 g(dry weight)$^{-1}$	7:7	SMEAR II
a_s	0.033	cm^2 g(dry weight)$^{-1}$	7:8	SMEAR II
a_t	0.015	cm^2 g(dry weight)$^{-1}$	7:9	SMEAR II
$a_{b\,1}$	1.5	–	7:10	Simulations with MicroForest
d_b	0.4	g(dry weight) cm^{-3}	7:12	SMEAR II
d_s	0.4	g(dry weight) cm^{-3}	7:12	SMEAR II
d_t	0.2	g(dry weight) cm^{-3}	7:12	SMEAR II
$a_{t\,1}$	0.85	–	7:13	Simulations with MicroForest
$a_{t\,2}$	500	cm	7:13	Simulations with MicroForest
$a_{n\,gr}$	1.35	g(sugar) g(dry weight)$^{-1}$	7:14,23	Penning de Vries et al. (1974)
$a_{w\,gr}$	1.3	g(sugar) g(dry weight)$^{-1}$	7:14	Penning de Vries et al. (1974)
$a_{r\,gr}$	1.35	g(sugar) g(dry weight)$^{-1}$	7:14,23	Penning de Vries et al. (1974)
n_n	0.0134	g(N) g(dry weight)$^{-1}$	7:15,27	SMEAR II
n_w	0.00024	g(N) g(dry weight)$^{-1}$	7:15,27	SMEAR II
n_r	0.0134	g(N) g(dry weight)$^{-1}$	7:15	SMEAR II
c	0.575	g(N) g(dry weight)$^{-1}$	7:15	SMEAR II
u	0.002	m^2 g(dry weight)$^{-1}$	7:15	Simulations with MicroForest
$a_{h\,1}$	50	–	7:20	Mäkelä and Sievänen (1992)
$a_{h\,2}$	2,400	–	7:20	Simulations with MicroForest
ρ_n	0.00025	g(dry weight) cm^{-3}	7:21	SMEAR II
p_g	30	g(sugar) g(dry weight)$^{-1}$	7:22	SMEAR II
g_1	80	g(dry weight) m^{-2}	7:22	SMEAR II
$b_{1\,4}$	0.0012	m^2 year^{-1} g^{-1}	7:25	Stable behaviour
$b_{1\,5}$	0.0002	m^2 year^{-1} g^{-1}	7:25	Stable behaviour
$b_{2\,1}$	1	year^{-1}	7:26	Stable behaviour
$b_{2\,2}$	1	year^{-1}	7:26	Stable behaviour

(continued)

(continued)

Parameter	Value	Unit	Equation	Source
$b_{2\,3}$	1	year^{-1}	7:26	Stable behaviour
$b_{2\,4}$	1	year^{-1}	7:26	Stable behaviour
$b_{2\,5}$	1	yr^{-1}	7:26	Stable behaviour
$b_{4\,1}$	0.004	year^{-1}	7:27	Stable behaviour
$b_{4\,2}$	0.0003	year^{-1}	7:27	Stable behaviour
$b_{4\,3}$	0.0003	year^{-1}	7:27	Stable behaviour
$b_{4\,4}$	0.00006	year^{-1}	7:27	Stable behaviour
$b_{4\,5}$	0.0003	year^{-1}	7:27	Stable behaviour
$b_{5\,1}$	0.004	year^{-1}	7:28	Stable behaviour
$b_{5\,2}$	0.007	year^{-1}	7:28	Stable behaviour
$b_{5\,3}$	0.003	year^{-1}	7:28	Stable behaviour
$b_{5\,4}$	0.004	year^{-1}	7:28	Stable behaviour
$b_{5\,5}$	0.002	year^{-1}	7:28	Stable behaviour
c_1	1	year^{-1}	7:29	Stable behaviour
c_2	0.55	year^{-1}	7:30	Stable behaviour
c_3	6.7	year^{-1}	7:31	Stable behaviour
c_4	10	year^{-1}	7:32	Stable behaviour
c_5	0.00043	year^{-1}	7:33	Stable behaviour

Appendix 3: List of Variables in MicroForest Model

Symbol	Variable	Unit
A	Cross-sectional area of stem segment	cm^{-2}
d	Diameter	cm
h	Tree height	cm
C_C	Crown length	cm
C_L	Amount of large carbon compounds in soil	g m^{-2}
C_S	Amount of small carbon compounds in soil	g m^{-2}
C_{SM}	Amount of small carbon compounds in microbes	g m^{-2}
d_e	Destruction of enzyme	g m^{-2} a^{-1}
d_m	Death of microbes	(dry weight) m^{-2} a^{-1}
A_E	Area of stem base enlargement	cm^2
h	Height growth	cm
r	Radial growth	cm
E	Enzymes breaking large carbon molecules	g m^{-2}

(continued)

(continued)

Symbol	Variable	Unit
G_{Ab}	Growth of sapwood area in branches for a whorl	cm^2
G_{As}	Growth of sapwood area in stem for a whorl	cm^2
G_{Ar}	Growth of sapwood area in transport	Roots for a whorl cm^2
G_{Ab}	Growth of sapwood area in a whorl	cm^2
g_m	Growth of microbes	g(dry weight) m^{-2} a^{-1}
G_n	Growth of needles	g(dry weight)
h	Tree height	cm
i	Index for size class	
I	Light intensity	μmol m^{-2} s^{-1}
i_P	Photosynthetic interaction of a whorl	
I_{pi}	Mean interaction	
j	Index for a whorl	
k	Index for age	
M_b	Mass of water transport structures in branches	g(dry matter) for a whorl
M_B	Mass of growth of stem base enlargement	g(dry mass)
M_n	Needle mass of a whorl	g(dry matter)
M_{nG}	Leaf mass of ground vegetation	g(dry weight) m^{-2}
M_M	Mass of microbes	g(dry weight)
M_s	Mass of water transport structures in stem	g(dry matter) for a whorl
M_t	Mass of water transport structures in transport	g(dry matter)
M_r	Fine root mass	g(dry weight)
M_{rG}	Fine root mass of ground vegetation	g(dry weight)
M_S	Shading leaf mass	g(dry matter) ha^{-1}
M_{sn}	Shading leaf mass in the stand	g(dry weight) ha^{-1}
M_T	Mass of growth of water transport system	g(dry weight)
N	Number of trees	ha^{-1}
n	Index for microbe type	
N_a	Amount of tree-available nitrogen in soil	g(N) m^{-2}
p	Photosynthetic rate	$g(CO_2)$ dry $weight^{-1}$ s^{-1}
P	Annual photosynthetic production	$g(CO_2)$ m^{-2}
p_e	Production of enzyme	g m^{-2} a^{-1}
P_G	Photosynthetic production of ground vegetation	$g(CO_2)$ m^{-2}
P_w	Photosynthetic production of a whorl	$g(CO_2)$
R	Maintenance respiration	$g(CO_2)$ g(dry $matter)^{-1}$
r_a	Release of ammonium	g m^{-2} a^{-1}
r_m	Microbial respiration	$g(CO_2)$ m^{-2} a^{-1}
S	Stage of development	°C
s_p	Splitting of large carbon molecules	g m^{-2} a^{-1}

(continued)

(continued)

Symbol	Variable	Unit
T_A	Mass of carbohydrates allocated for the top	$g(CO_2)$
T	Temperature	°C
u_p	Microbial uptake of small molecules	$g\ m^{-2}\ a^{-1}$
v	Water vapour deficit	$g(H_2O)\ m^{-3}$
V	Stand volume	$m^3\ ha^{-1}$
V_w	Volume of stem segment	cm^3
V_s	Volume of stem	m^3

Appendix 4: Transport

T. Vesala (✉) • Ü. Rannik
Department of Physics, University of Helsinki, P.O. Box 48, 00014 University of Helsinki, Helsinki, Finland
e-mail: timo.vesala@helsinki.fi

P. Hari
Department of Forest Sciences, University of Helsinki, P.O. Box 27, 00014 University of Helsinki, Helsinki, Finland
pertti.hari@helsinki.fi

The system of the tree-atmosphere continuum is spatially large, which sets challenges for trees to be capable to transport various chemical compounds within them and to exchange compounds with the atmosphere. Transport phenomena and resulting mass and energy flows are governed only by few fundamental physical laws giving rise to different transport mechanisms. The mutual effectiveness of different transport mechanisms depends on the spatial scale. In a way, trees utilise effectively this frame of physical laws in their metabolism and interactions with environment.

By transport, we especially mean the movement of a certain quantity of mass or energy between the storage locations of the quantity. Storages may be apparent as sources and sinks of the quantity and unequal amounts of any entity in the storage locations lead to the concept of a non-zero gradient of the quantity. In practical terms, ecosystems store materials as mass (such as carbon, nutrients, water) and heat, and any differences in the amounts of mass and heat tend to disappear by physical mechanisms called transport phenomena. The third fundamental quantity besides *mass* and *heat* is *momentum* which is defined as the product of the mass and velocity of a homogenous fluid (fluid means either gas or liquid) parcel. All of these can be transported by two physically different mechanisms, which are *molecular transport* and *bulk transport*, also called *convection*. Transport of heat may happen also by means of a third heat transport mechanism which is *electromagnetic radiation*. Next we consider three transport mechanisms starting with the molecular one.

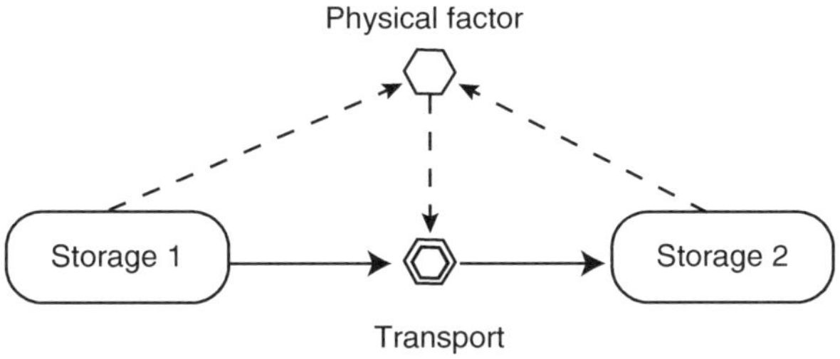

Fig. A4.1 Transport occurs between two storages creating sinks or sources. Transport is governed by pressure, temperature, density or/and concentration differences (The symbols are introduced in Fig. 2.3)

Molecular Transport

Molecular transport refers to spontaneous and irreversible transport of momentum, energy (heat) and mass of various chemical species along the gradients (profiles) of velocity, temperature and concentration and from the higher velocity, temperature and concentration to the lower one. Molecular transport stems from random molecular motion. Besides velocity, temperature and concentration, pressure and density and their spatio-temporal behaviour are the elemental quantities governing physics of transport phenomena (Fig. A4.1).

Conduction and diffusion are examples of molecular transport related to the movement of heat and mass, respectively. They are easily recognised and they are familiar concepts, whereas the idea of momentum transport may not be so easily identified. However, it is the movement of fluids, which involves the transport of momentum, and the capability of gas or liquid to transport momentum that give rise to the *viscosity* of a fluid, analogously with heat conductivity and diffusivity. If we consider a collision of two billiard balls, moving at different speeds, the faster one will have more momentum, part of which it will give off to the slower moving ball. Fluid is not a body, like a billiard ball, but it contains areas of higher and slower flow velocities and, similar to the transfer of momentum from a faster body to a slower one, momentum is transported from areas of higher flow velocities to those of slower movement.

Momentum associated with moving molecules is also the molecular cause for the macroscopic emergent concept of *pressure*. Let us consider bombardment of gas molecules onto a surface. When a single molecule hits the surface and rebounds, an impulse is exchanged between the molecule and the surface and this produces a *force* on the surface during the collision. In normal conditions near the Earth's surface, the bombardment rate is very high and the net result of many hits is the macroscopic force, which is defined as pressure when divided by the surface area. Pressure is a fundamental quantity for flows of fluids: pressure difference in the fluid produces flow, and on the other hand, surfaces nearby the flowing fluid "absorb" momentum from the flow and give rise to a pressure drop in the fluid. For air, at standard conditions the number of molecules is 2.5×10^{19} cm^{-3} and their average velocity is 460 m/s. The rate of molecular collisions with any surface can be further estimated to be 2.9×10^{23} cm^{-2} s^{-1} (Hinds 1982), at standard conditions.

Conductivity, diffusivity and viscosity are macroscopic concepts characterizing a fluid, but they naturally stem from molecular structure of the substance and the associated intermolecular forces. The fundamental driver of conduction, diffusion and molecular momentum transport is the random thermal motion of atoms/molecules modified by interactions due to molecular collisions. If we simplify without losing much from the inherent physics, the collisions resemble elastic collisions between moving billiard balls. Irrespective of whether we consider billiard balls or molecules, if the objects are moving randomly, the random motion tends to remove any inhomogeneities. Lack of homogeneity is caused by (macroscopic) sinks and sources. Note that a single molecule under thermal motion does not "know" in what direction it should move (it is determined by the last collision with its partner) in a mixture of two substances, but a net movement of a large group of molecules is along the gradient from the higher to lower concentration and is predictable.

The measure of the transport is the *flux density* giving the amount of energy, mass (or moles) or momentum crossing a unit area per unit time. The flux density is naturally a *vector,* a quantity possessing both magnitude and direction (like velocity). In contrast to the vector, a quantity possessing only magnitude is called a *scalar* (such as speed). At the *equilibrium*, all flux densities are zero and no transport occurs. In the case of small disturbances from the equilibrium,[3] the transport is measured by the gradients in temperature, concentration and velocity components. The flux density **I** due to the molecular transport is proportional to the gradient

$$\mathbf{I} \propto \nabla S \equiv \frac{\partial S}{\partial x}\mathbf{i} + \frac{\partial S}{\partial y}\mathbf{j} + \frac{\partial S}{\partial z}\mathbf{k}, \tag{A4.1}$$

where S refers to the temperature (conduction), concentration (diffusion) or velocity component (viscous flow). The mathematical concept *gradient* means the spatial rate of change of a quantity (scalar), which is a vector pointing in the direction of the steepest change, the components of which are the *partial derivatives* of the quantity. The partial derivatives in Eq. A4.1 are represented by the x-y-z coordinates and the corresponding *unit vectors* along x, y and z axis are **i**, **j** and **k**, respectively. If the flux density is integrated over a certain area, the resulting surface-averaged flux density is called *flux* and its dimension is mass, energy or momentum per unit time. The concepts of flux density and flux are used loosely in the literature. Molecular transport refers to diffusive transfer described above. Note that "diffusion" commonly means diffusion of mass and diffusive transport of heat is called conduction. As was explained above, momentum can be transported also by diffusion.

The physical meaning of the diffusion law can be illustrated by a one-dimensional case starting from the given concentration difference $\Delta C = C(x + \Delta x) - C(x)$ over the distance Δx (see also, e.g. Nobel 2005 for similar reasoning). Let us set a plane perpendicular to the x axis between the points x and $x + \Delta x$. Since the movement by diffusion is a random phenomenon, it is

[3]For applications in this book the assumption of small disturbance is always valid enough.

Fig. A4.2 Diffusive (molecular) transfer giving rise to the net transport from higher concentration $C(x + \mathrm{x})$ towards the lower one $C(x)$

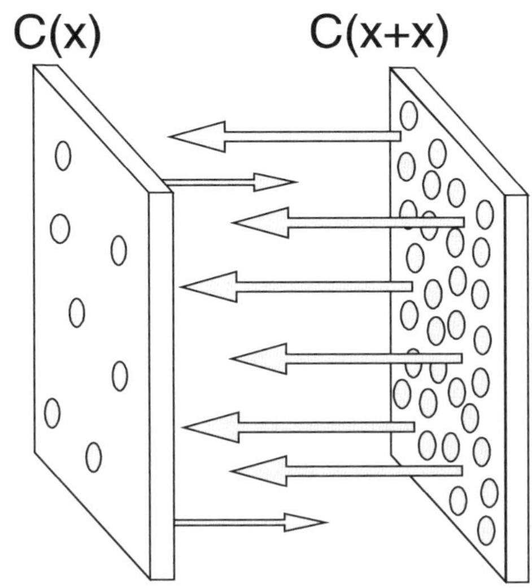

reasonable to assume that the flux of molecules, across the plane from two sides of it, is proportional to the concentration. Thus, the flux from the direction of the point $x + \Delta x$ is proportional to $C(x + \Delta x)$ and the flux from the opposite direction to $C(x)$ (Fig. A4.2). Let us assume further that x is positive and $C(x + \Delta x) > C(x)$. There is now a greater probability for molecules to move from the concentrated to the dilute region, and the flux g_A is

$$g_A \propto C(x + \Delta x) - C(x). \tag{A4.2}$$

However, the magnitude of the flux is naturally also affected by the distance over which the concentration difference occurs. Namely, if the given concentration difference exists over a long distance or the same ΔC exists over a shorter distance, the flux must be accordingly smaller in the former case. So we get

$$g_{\mathrm{A}} \propto \frac{C(x + \Delta x) - C(x)}{\Delta x}. \tag{A4.3}$$

At the limit $\Delta x \to 0$, the ratio

$$\frac{C(x + \Delta x) - C(x)}{\Delta x} = -\frac{C(x) - C(x + \Delta x)}{\Delta x} \to -\frac{\mathrm{d}C}{\mathrm{d}x}, \tag{A4.4}$$

which is the x derivative of C (one-dimensional gradient), and we obtain that $g_{\mathrm{A}} \propto -\frac{\mathrm{d}C}{\mathrm{d}x}$.

The result is easy to interpret: the derivative tells the steepness of the concentration profile (the steeper the profile, the larger is the flux), and the minus sign says that the transport is negative along the x axis, that is, from larger x values to lower ones. By convention, a net movement in the direction of increasing x is positive. The formula derived is still just the proportionality, and the final form requires the coefficient, which is known to be the diffusivity D_{AB}.

The most rigorous and general transport equations are rather complex, and we refer to the book by Bird et al. (2002). For applications considered in this book, the approximated formulas of the mass flux density **g**, heat flux density **q** and the momentum flux density τ_{yx} can be simplified. The mass flux density is given by

$$\mathbf{g}_A = -\rho\, D_{AB} \nabla \omega_A, \tag{A4.5}$$

(Fick's first law)

which states that the mass transport rate per unit area is proportional to the gradient of the mass fraction ω_A of substance A. The mass fraction characterises the concentration of A. When the density ρ of the mixture of two compounds A and B is multiplied by the mass fraction of A, the mass concentration C_A is obtained.

The three-dimensional extension of the one-dimensional case is obtained when the x axis is replaced with the gradient of the concentration, which tells the direction of steepest concentration drop, to which the diffusive transport is directed (in analogy, a ball would roll to a direction of the steepest slope characterised by ∇z where z is the elevation of the surface).

Similarly to diffusion the heat is transported by conduction according to

$$\mathbf{q} = -k\ \nabla T. \tag{A4.6}$$

(Fourier's law for conduction)

This is fully analogical to Fick's law of diffusion: heat is transferred from the higher temperature to a lower one, and the magnitude is governed by the heat conductivity.

Osmosis is important in many biological systems. For example, the transport and cycling of water in a tree is partly driven by osmosis, which is the mechanism of the water and sugar movement downwards in the phloem in Münch's cycle (Sect. 5.1). Osmosis can be considered as a special case of ordinary diffusion, but its treatment requires the usage of activity rather than concentration (see, e.g. Nobel 2005). Molecules of solute species interact with each other as well as with other solute and solvent molecules, and this influences the behaviour of all compounds in the solution. The use of concentration for describing the thermodynamic properties of the solution is generally only an approximation and generally the concept of *activity* is needed. The activity can be regarded as a concentration corrected for the interactions between solution constituents. For a so-called ideal solute, the activities are equal to concentrations.

The classical example of osmosis is the transport of water through a membrane impermeable to the dissolved solutes but permeable to water (Fig. A4.3). Solutes in an aqueous solution decrease the activity of water, which leads to the concept of *osmotic pressure*. In the first approximation, for dilute solutions, the

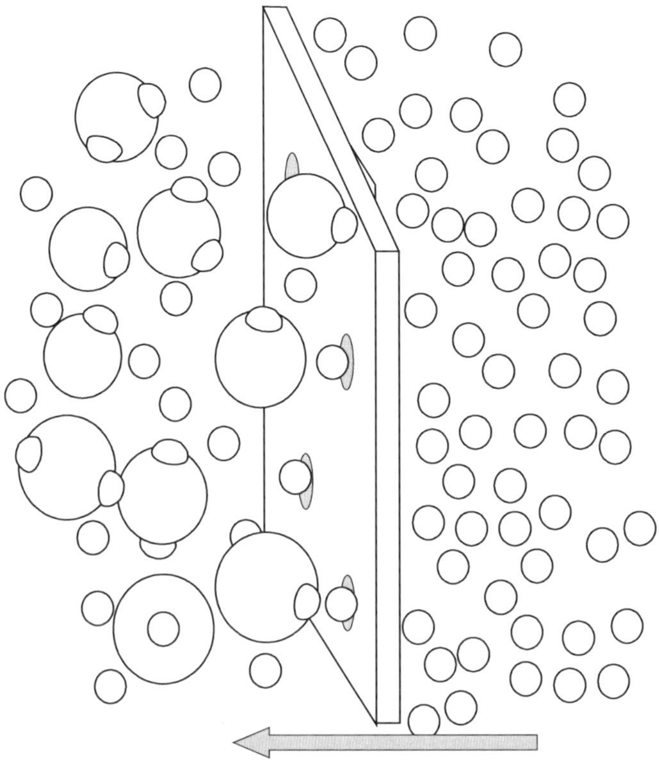

Fig. A4.3 Osmotic flux of water. Small molecules can penetrate the holes, but large molecules are unable to penetrate the holes. Small molecules also interact with larger ones. The *ball size* in the figure refers to the size of the molecules

pressure is directly related to concentration of solutes and for the osmotic flux density g_{H_2O} of water through the membrane (e.g. Nobel 2005), we have the proportionality

$$g_{H_2O} \propto -\sigma_j R\ T\ \Delta C_j, \tag{A4.7}$$

where R is the gas constant and ΔC_j is the concentration difference of the dissolved species j across the membrane. T is the temperature in Kelvin. σ_j is the reflection coefficient of the species j, which is the phenomenological coefficient[4] describing the relative ease with which the solutes cross the membrane. It is unity when the membrane is completely impermeable to solutes and is zero when the membrane is nonselective.

[4]Linked to Onsager reciprocity relations of the general thermodynamic framework of transport phenomena (see Onsager 1931a, b).

When the membrane is not very permeable to water, rather high osmotic pressure difference can accumulate across the membrane. If the permeability increases (e.g. water conducting pores become more opened), large water flow is induced. The flow naturally dilutes the more concentrated solution and decreases the driving force and the flow, if no mechanism produces more solutes.

Convection

The spontaneous transport along the gradients by molecular movements and collisions is not the only transport. It is intuitively clear that any flow of the material has a capability to carry, and thus transport, its own mass and momentum and heat associated to it. Hence, besides the molecular transport of energy (conduction), mass (diffusion) and momentum (viscosity), the quantities can be transported by the bulk flow of the fluid, which is called convective transport. Note that in fields of physics and engineering sciences of fluid mechanics and of transport phenomena, convection refers generally to the bulk flow of the fluid, but in meteorology, it refers only to the vertical flow of air resulting from density variations due to the temperature field.[5]

Generally, the flow appears in two "forms": *laminar* or *turbulent*. Laminar flow is orderly and turbulent flow is chaotic and dispersive with the quantities (velocity components, pressure, temperature, concentrations) violently fluctuating. Atmospheric flows are practically always turbulent, while, for example, the sap flow in a tree is laminar because the flow velocities are low. With respect to the causes for flows, they are classified into two types: *forced* and *free* (or natural) convections. Free convection refers to the flow determined by the buoyant forces arising from the fact that heated air rises because it is less dense.[6] Forced convection refers to all other situations where some external force generating pressure difference generates the flow. In some cases, both effects are important and this situation is called mixed convection. Both forced and free convections can be either laminar or turbulent.

Again, we measure the amount of transport by flux densities. The convective mass flux density due to the flow velocity **v** is given by

$$\mathbf{g}_{\mathrm{A,\,conv}} = \rho\, \mathbf{v}\, \omega_A. \tag{A4.8}$$

(convective mass flux)

The interpretation of this formula is simple: ρ gives the amount of mass per unit volume, and if the mass is transported by the velocity **v**, $\rho\,\mathbf{v}$ gives the amount moved per unit time per unit area perpendicular to the velocity vector. Namely, the volume

[5]Which is only a special case of fluid physicists' various convective mechanisms.

[6]Note that the concept of convection in meteorology is free convection in terms of fluid physicists.

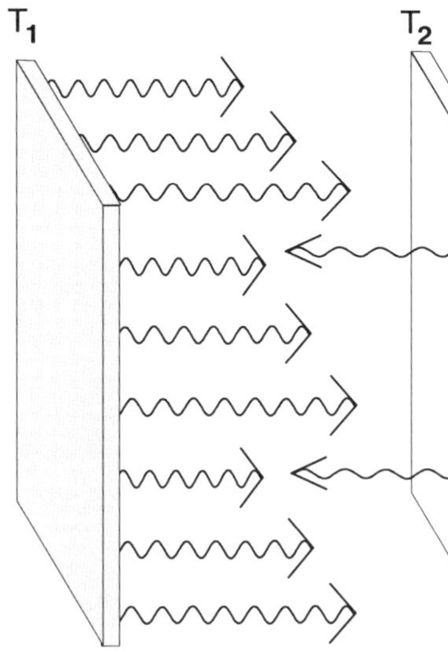

Fig. A4.4 Radiative heat flux. $T_1 > T_2$. The intensity of the radiation is proportional to the fourth power of the temperature, that is, $\propto T^4$

rate of flow across the surface element dS perpendicular to, for example, the x axis is v_xdS and the volume rate multiplied by the density gives the mass flow rate. Multiplication by the mass fraction ω_A gives the amount of the compound A carried by the bulk flow.

Radiative Transport

Finally, we consider the electromagnetic radiation (Fig. A4.4) which is the third mode of energy transport, for which there is no analogy in mass and momentum transport. Radiation does not even require the existence of the material but it transports energy at the speed of the light in a vacuum. In practice, we can ignore radiation as a transport mechanism for a specific point in a transparent medium, and it should be taken into account only for opaque surfaces. It also differs drastically from molecular modes of transport, which are proportional to gradients, while radiative transfer between two bodies is proportional to the difference of the fourth powers of their absolute temperatures. The books by Thomas and Stamnes (1999) and Ross (1981) cover all elemental as well as advanced aspects of radiative transfer in the atmosphere and plant stands.

Summary of Transport Phenomena

We summarise the main classification of the transport phenomena (Fig. A4.5). The fundamental quantities of mass, heat and momentum are all carried by convection

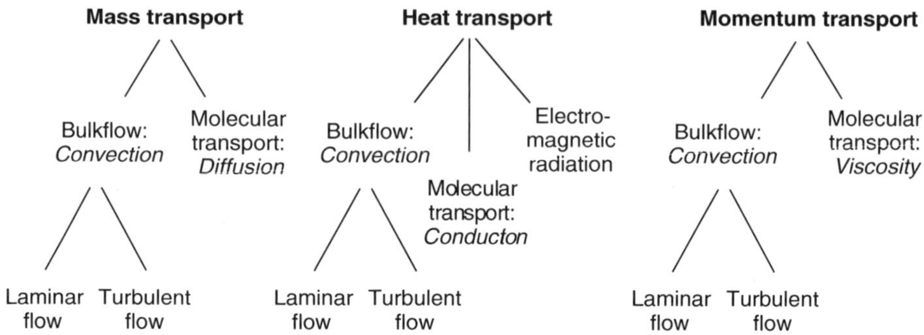

Fig. A4.5 General classification of transport phenomena of three basic quantities

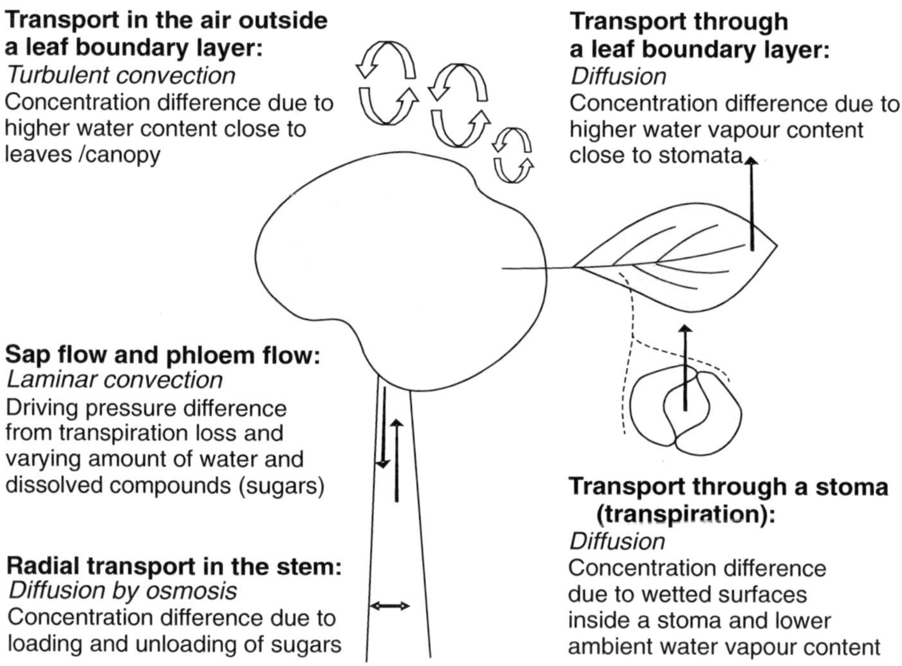

Fig. A4.6 Main water transport phenomena in the tree-atmosphere continuum

(bulk flows) and molecular transport (diffusion, conduction and viscous flow, respectively). Heat is carried also by radiation. Flows have two distinct modes: laminar and turbulent ones.

Transport phenomena are fundamental drivers of many processes in the plant-atmosphere continuum. As an example, we decompose the tree water cycling and water exchange with the atmosphere in the terms of transport phenomena

(Fig. A4.6). For the gas phase, (air) there are three major components: diffusion inside a stoma, diffusion across the leaf boundary layer and turbulent transport in the canopy air. Inside a tree, water is transported in vertical and radial directions in the stem. Detailed description of transport in twigs and leaves is omitted here as they resemble those in the stem.

Glossary

Absolute humidity Water vapour concentration in the air.

Absorptivity A property of a material, the ratio of the absorbed radiation to the incident one; in equilibrium, absorptivity and emissivity are the same.

Acclimation Changes in the biochemical regulations system, in functional substances and in fine structure matching vegetation with irregular changes in the environment.

Accuracy Magnitude of systematic error in measurements.

Adaptation Results of evolution to a specific environment.

Aerosol The combination of solid or liquid aerosol particles and the gaseous medium they are suspended in.

Air seeding Gas can be sucked from embolised conduits to water-conducting ones through pores in the membranes separating the conduits when water tension rises through increasing transpiration demand.

Albedo The fraction of solar radiation reflected by a surface of an object, often expressed as a percentage.

Allocation to something The share of available resources used to the growth of the specified component.

Anemometer Device for measuring wind speed.

Annual cycle of environmental factors Exact annual cycle in all environmental factors caused by the Earth's orbiting around the Sun, seasonality.

Annual cycle of metabolism Changes in the biochemical regulation system, in functional substances and in fine structure matching vegetation with the annual cycle of environment.

Apoplastic Transport of water and solutes outside the plasma membrane, along the cell walls and intercellular space.

Aquaporin Integral membrane protein enabling selective penetration of water molecules across the membrane.

ATP Adenosine-5′-triphosphate transports chemical energy within cells for metabolism.

Attenuation of light Reduction of light due to absorption.

P. Hari et al. (eds.), *Physical and Physiological Forest Ecology*,
DOI 10.1007/978-94-007-5603-8,

Base saturation The proportion of cation-exchange sites occupied by the so-called basic cations (Na^+, K^+, Mg^{2+}, Ca^{2+}) rather than acidic cations (H^+, Al^{3+}).

Biochemical regulation system Formed by substances that synthetise, decompose, and activate the functional substances.

Biome A biotic community of plants and animals.

Black body A perfectly efficient emitter and absorber of radiation.

Boreal zone Terrestrial division consisting of the northern coniferous parts.

Bulk flow Ordinary flow of fluid, but 'bulk' stresses the contrast to molecular motion.

BVOC Biogenic volatile organic compounds; a diverse group of volatile compounds evaporating from plant tissues, e.g. isoprenoids, acetone and methanol.

Cambium Lateral meristem tissue which gives rise to secondary thickness growth; vascular cambium forms phloem and xylem tissues, and cork cambium forms cork tissues.

Carbon balance Carbon fixed in photosynthesis during time unit is used for the maintenance and growth.

Carbon reaction (dark reaction, Calvin cycle) Joins a carbon atom to five-carbon sugar in chloroplast stroma utilizing energy released from ATP and NADPH.

Carbon sequestration Amount of carbon fixed in vegetation and soil organic matter during a year.

Carboxylation Reaction where RuBisCo molecule binds a CO_2 molecule.

Casparian band A suberised section of cell wall, located on the radial and transverse walls of the endodermis of a plant, impermeable to water and solutes.

Cell Basic structural and functional unit in all living organisms.

Coagulation (aerosol dynamic process) The process where two particles collide and coalesce, forming a new larger particle.

Cohesion A physical property of a substance. Cohesion is caused by the intermolecular attraction between similar molecules within a body or substance that acts to unite them. Water, for example, is strongly cohesive.

Cohesion theory The hypothesis that water is pulled upwards along a tension gradient during transpiration by the electric forces between water molecules. According to this theory, the evaporation of water at the leaf surfaces pulls a continuous water column against a gravitational gradient through a continuous pathway of xylem conduits reaching all the way down to the roots.

Condensation The phase transition where a vapour is changing to liquid; it is accompanied by the release of (latent) heat.

Conduction Molecular transport mechanism of energy (heat) arising from random motion of molecules.

Convection Bulk flow of the fluid. Flowing gas or liquid carries mass, heat and momentum associated to it. In meteorology, convection refers to the vertical bulk flow of air.

Cortex Outer layer of the stem or root of a plant, important in storage and diffusion of solutes.

Cover theory Common vision, background and methodology for a set of specific theories dealing with a large entity.

Decomposition of soil organic matter A process where organic macromolecules are disassembled by soil microorganisms leads to release of nutrients and CO_2.

Denitrification Respiratory bacterial reduction of nitrate $\left(NO_3^-\right)$ or nitrite $\left(NO_2^-\right)$ to nitrogen oxides or molecular nitrogen.

Density of mass Amount of material in a volume element divided with the volume of the element.

Deposition (aerosol dynamic process) The process of aerosol particles being transported and deposited onto existing surfaces and thus away from the airborne phase.

Dermal tissue The outermost cell layer facing the atmosphere or soil.

Diffusion Transport of atoms and molecules by random thermal movement.

Dormancy Period in an organism's annual cycle when growth, development and (in animals) possible physical activity are temporarily suspended.

Dynamic model Mathematical model having time in a central role.

Ecosystem A spatial unit composed of interacting biotic (animals, plants, microorganisms) and abiotic (nonliving physical and chemical) factors.

Eddy Flow pattern in the turbulent flow with some coherent structure.

Eddy covariance Meteorological technique to measure exchange of mass, energy and momentum between the surface and the atmosphere, based on the direct determination of the turbulent transport.

Embolism Introduction of air bubbles into the water transporting xylem tissue, as individual water-conducting vessels are filled by water vapour or air.

Emergent property New phenomena or properties at higher level that are generated by interactions at lower level.

Emission scenario A plausible representation of the future development of emissions of substances that are potentially radiatively active (e.g., greenhouse gases, aerosols), based on a coherent and internally consistent set of assumptions about driving forces (such as demographic and socioeconomic development, technological change) and their key relationships. In many climate model simulations, some of the so-called Special Report on Emissions Scenarios (SRES) (Nakićenović and Swart 2000) scenarios have been used.

Emissivity A property of a material, the ratio of the emitted radiation by the material to that of a black body (perfectly efficient) radiator.

Environmental factor Such feature in the environment that effects on some process.

Enzyme Large protein molecules acting as biological catalysts in all metabolic processes that occur in living tissue.

Evaporation The phase transition where liquid is changing to vapour; it is accompanied by the consumption of (latent) heat.

Evapotranspiration The sum of evaporation and transpiration.

Evolution New metabolic processes, structures and biochemical regulations and improvements in the existing ones emerge, and these novel features are tested against the existing ones in the process of natural selection. If the novel feature contributes to the success of the species, then it has the tendency to become more common. This process results in slow evolution of the species as it becomes adapted to the environment of its location.

Extracellular enzyme Enzyme emitted by microbes outside the cell to split macromolecules.

FAD^+ Flavin adenine dinucleotide.

Flux density Amount of flow of material or energy through a plane element during time element divided with the product of the area and the length of the time element.

Forced convection Convection (bulk flow) arising from any other reasons excluding density (temperature) deviations in the fluid, especially flow arising from pressure difference.

Free convection Natural convection; convection (bulk flow) arising from density deviations in the fluid, especially flow arising from temperature differences.

Freezing tolerance Ability to survive freezing temperatures.

Functional balance The uptake of carbon and nitrogen are balanced with each other.

Functional substance Enable metabolic processes in living organisms. Enzymes, membrane pumps and pigments are examples of functional substances.

GCM General circulation models or general climatic models.

Global radiation The sum of direct solar radiation and diffuse sky radiation received by a unit horizontal surface.

GPP Gross primary production of an ecosystem, i.e. photosynthesis of the ecosystem.

Greenhouse effect Greenhouse gases absorb thermal infrared radiation, emitted by the Earth's surface, by the atmosphere itself due to the same gases and by clouds. As a result of this, only a small part of the thermal radiation emitted by the Earth's surface and the lowest atmospheric layers escape directly to space. Although greenhouse gases by themselves emit thermal radiation to all directions, the upward-emitted radiation that finally escapes to space originates from air layers that are much colder than the surface, and the intensity of this radiation is therefore smaller. Consequently, the Earth's surface can maintain an average temperature of about +14 °C, which is 33 °C warmer than the temperature (−19 °C) that would be observed without the greenhouse effect. As the concentrations of greenhouse gases in the atmosphere are increasing, the greenhouse effect is becoming stronger and this is expected to lead to an increase in the Earth's surface temperature.

Greenhouse gas Greenhouse gases are gases that absorb and emit radiation at wavelengths within the spectrum of thermal infrared radiation emitted by the Earth's surface, by the atmosphere itself and by the clouds. Water vapour (H_2O), carbon dioxide (CO_2), nitrous oxide (N_2O), methane (CH_4) and ozone (O_3) are the primary greenhouse gases in the Earth's atmosphere. In addition, there are a number of entirely human-made greenhouse gases in the atmosphere, such as halocarbons including CFC-11 and CFC-12.

Ground tissue Present throughout the plant body and formed of parenchyma, collenchyma (supporting) or sclerenchyma (protective and storage) cells.

Growth respiration Processes supplying energy for formation of new tissue; CO_2 is released.

Hartig net Network of fungal hyphae between epidermal and cortical cells in the root, where most nutrient exchange between mycorrhiza and plant probably occurs.

Hierarchy Ordered structure.

Humidity see 'Relative humidity' and 'Absolute humidity'.

Interception of light Absorption of light in vegetation.

Isoprenoid A large and diverse class of naturally occurring organic chemicals. Isoprenoids are derived from five-carbon isoprene units assembled and modified in thousands of ways. See 'Terpenes'.

LAI Leaf area index, ratio of total upper leaf surface of vegetation divided by the surface area of the land on which the vegetation grows.

Laminar flow A form of fluid flow in which fluid flows in an ordered manner with regular paths; laminar flow occurs when the flow velocities are low and always close to surfaces; otherwise, the flow is turbulent.

Latent heat Heat absorbed or released when a substance changes its physical state (phase), here especially related to condensation/evaporation and freezing/melting.

Latent heat flux The flux of heat from the Earth's surface to the atmosphere by evapotranspiration.

Light reactions Processes in chloroplast thylakoids converting absorbed light quanta into chemical energy (ATP, NADPH) and liberating O_2 from H_2O.

Longwave radiation Radiation emitted from the Earth, typical wavelengths from 4 to 30 μm.

LUE Light use efficiency.

Macrofauna Soil or benthic organisms which are at least one millimetre in length.

Macromolecule Large molecules, commonly formed of more than 1,000 atoms, e.g. nucleotides, lipids, proteins and carbohydrates.

Maintenance respiration Formation of ATP (respiration) to supply energy for cellular household tasks such as synthesis of macromolecules or active transport; CO_2 is released.

Mantle Layer of fungal hyphae on the surface of mycorrhizal root tip.

Mass density Mass of vegetation elements in a volume element divided by the volume of the element.

Mass-specific process rate Amount of products produced in a volume element during short time element divided with product of the volume of the element with the length of the time element.

Mean (arithmetic mean) The sum of a property over all members in a population divided by the number of items in the population.

Measuring noise Random error in measurements.

Membrane pump Intrinsic proteins which facilitate the movement of polar molecules through membranes either against their electrochemical gradient (active pumps) or in the direction of the electrochemical gradient (channels and carriers). Active pumps need energy provided by ATP for their operation.

Meristematic tissue Undifferentiated cells in places where new tissue is formed.

Mesophyll Leaf parenchymatous tissue situated between dermal and vascular tissues, site of photosynthesis.

Methanogenesis Acetate or CO_2 are converted into CH_4 and CO_2, while hydrogen is consumed.

Microfauna Microscopic or very small animals (usually including protozoans and very small animals such as rotifers).

Mineralisation A process in which organic molecules in soil are broken down into inorganic forms, often ions such as $NH_3{}^+$, K^+ and Ca^{++} are released.

Mitochondrion Cell organelle housing the essential respiratory machinery which generates ATP by the citric acid cycle and electron transfer chain.

Molecular transport Transport of mass, heat and momentum arising from random motion of molecules in the presence of gradients. Leads to concepts of diffusion, conduction and viscosity.

Momentum The product of the mass and velocity of an object.

Momentum absorption Momentum absorbed from the flowing fluid on the surface.

Momentum flux Transport rate of momentum in a fluid.

Münch hypothesis Water is drawn in from the xylem to the phloem at the top of the tree and is pushed back into the xylem at the bottom.

Mycorrhiza A symbiotic association between fungal hyphae and plant roots, which greatly improves the plant water and nutrient uptake.

NAD^+, NADH Nicotinamide adenine dinucleotide, reduced nicotinamide adenine dinucleotide.

$NADP^+$, NADPH Nicotinamide adenine dinucleotide phosphate, reduced nicotinamide adenine dinucleotide phosphate.

Near-infrared radiation (NIR) Radiation in the wavelength band 750–3,000 nm that contributes to about half of the incoming solar radiation.

NEE Net ecosystem exchange, photosynthesis minus respiration in an ecosystem.

Net radiation The sum of the downward and upward solar and thermal radiations.

Nitrification Oxidation of ammonium $\left(NH_4^+\right)$, ammonia (NH_3) or organic N compounds to nitrite $\left(NO_2^-\right)$ and nitrate $\left(NO_3^-\right)$.

Nitrogen balance All nitrogen taken up by roots and released from the senescencing structures during a year are used for growth during that year.

Nitrogen fixation A biological process in which nitrogen from the atmosphere is converted to ammonium form.

NPP Net primary production, formation of new organic matter in an ecosystem.

Nucleation event Regional formation and growth of new atmospheric aerosol particles created and grown by nucleation and condensation of atmospheric vapours.

Organelle A cellular subunit enclosed in a membrane, e.g. plastid, mitochondrion or nucleus.

Osmosis Diffusion through a membrane, which is partly or fully impermeable to dissolved solutes but permeable to water.

PAR Photosynthetically active radiation.

Parameter Constant in a model.

Parenchyma Thin-walled cells of the ground tissue that make up the bulk of most nonwoody structures.

Particle formation (aerosol dynamic process) The formation of new secondary aerosol particles from vapour phase.

Penumbra Occasion when light coming from the solar disc is partially covered by vegetation.

Proportion of explained variance (PEV) Commonly used as a measure of the goodness of fit of the model. It gives the proportion of the variance of the measured response that can be explained by the model.

Phase transition A change in a feature that characterises a physical system; here, especially, changes from solid to liquid and liquid to vapour and the reverse changes.

Phenology Study of timing of phenomena associated with the annual cycle of vegetation.

Phloem Living vascular tissue where translocation of sugars takes place.

Photorespiration Analogous to carbon reactions (Calvin cycle); catalysed by RuBisCo but using O_2 as a substrate instead of CO_2; should not be mixed with 'true' respiration although it eventually leads into CO_2 release as respiration.

Photosynthesis Process where organic carbon molecules (sugars) are formed from CO_2 and H_2O in chloroplasts of green plants using solar radiation as energy source.

Photosynthetic pigment A group of macromolecules which participate in absorbing selective wavelengths of light quanta and transferring the excitation into processes where it is chemically bound, e.g. chlorophylls, carotenoids and xanthophylls.

Photosynthetic reaction centre A complex of several proteins, pigments and other macromolecules; the site where molecular excitations originating from sunlight are transformed into a series of electron transfer reactions.

Photosynthetically active radiation (PAR) Number of light quanta in the visible range (400–700 nm) of solar radiation.

Phytoelement Vegetation element.

Plasmodesmata Microscopic channels which traverse the cell walls and membranes of plant cells, enabling flow of water and solutes between cells.

Plastid Cellular organelles functioning in photosynthesis (chloroplasts), other biosynthetic processes (leucoplasts) and energy storage (amyloplasts).

Podsol Typical soils that develop in temperate to cold moist climates of coniferous forests.

Podsolisation Formation of podsol soil.

Precision Random variation in the model or measurement.

Pressure Force per unit area from gas or liquid to solid material, liquid or gas.

PRI Photochemical Reflectance Index, a reflectance index that captures changes in leaf carotenoid contents associated to the acclimation of photosynthesis.

Process Converts material and/or energy into a new form or moves material through a membrane.

Protein Large macromolecules built from amino acid units and folded in a complex three-dimensional structure.

Radiation (electromagnetic radiation) The third mode of heat transport beside conduction and bulk flow (convection). Transports heat by the speed of the light in a vacuum.

Radiation energy Energy carried in the form of electromagnetic radiation by photons.

Radiative forcing Radiative forcing is the change in the net, downward minus upward, irradiance (expressed in W m^{-2}) at the tropopause due to a change in an external driver of climate change, such as a change in the concentration of carbon dioxide or the output of the Sun. Generally, radiative forcing is defined as the global and annual average change in net radiation resulting from changes in external factors relative to the year 1750.

Residual The difference between measured and modelled values.

Respiration (dark, aerobic) Formation of ATP utilizing stored chemical energy. Simultaneous release of CO_2 takes place.

Relative humidity (RH) The ratio (%) of actual to saturated water vapour content in an air parcel.

Rhizomorph Linear aggregations of parallel-oriented hyphae growing from the mycorrhiza to surrounding soil, specialised in transporting of water and nutrients.

Rhizosphere The zone that surrounds the roots of plants.

Root exudate Molecules or ions emitted by roots.

RuBisCo Ribulose-1,5-bisphosphate carboxylase/oxygenase (RuBisCO) is an enzyme that is used in the Calvin cycle to catalyse the first major step of carbon fixation.

Runoff The amount of water per unit of area and time leaving an ecosystem as liquid.

Sapwood Youngest and outermost xylem tissue in stems, consists of living cells involved in conducting water.

Senescence The process of aging of an organelle, cell, organ or whole plant individual.

Sensible heat Heat flux associated to the movement of warmer/cooler air.

Shortwave radiation A term used to describe the radiant energy in the visible, near-ultraviolet and near-infrared wavelengths. Shortwave radiation may be as broadly defined as between 0.1 and 5.0 micrometres.

SMEAR Station for measuring ecosystem-atmosphere relations.

Soil CO_2 efflux The flow of carbon dioxide generated by respiring soil organisms such as plant roots, microbes and fauna.

Soil water storage Amount of water in top layer up to 1 m of soil per 1 m^2 (unit litres m^{-2} or mm^{-2}).

Solar constant Amount of the Sun's incoming electromagnetic radiation per unit horizontal area, at the outer surface of Earth's atmosphere. The solar constant includes all types of solar radiation, not just the visible light. It is roughly 1,366 W m^{-2} though this fluctuates during a year.

SOM Soil organic matter denotes all living and dead biologically derived organic material at various stages of decomposition in the soil or on the soil surface, excluding the above-ground portion of living plants.

Space element So small spatial volume that environmental factors do not vary within it.

Spectrum Distribution of electromagnetic radiation as a function of wavelength or frequency.

Stable System or property that does not change.

State of functional substances Regularities in the functional substances generated by the action of the biochemical regulation system.

Stomatal cavity Air space in mesophyll tissue below a stoma.

Stomatal conductance Parameter in models of photosynthesis, transpiration and other gaseous fluxes between leaf and atmosphere introducing diffusion of gas molecules.

Succession Orderly changes in the composition or structure of an ecological community. Succession that occurs after the initial succession has been disrupted is called secondary succession.

Symplastic The continuum of cell interiors joined together by plasmodesmata, in which water and low-molecular-weight solutes can freely diffuse.

Taiga Earlier referred to coniferous forest by Siberians, nowadays a more general meaning, referring to different forests of the northern hemisphere, boreal forest.

Tensiometer Device used to determine matrix water tension in the soil.

TER Total ecosystem respiration.

Terpene A large group of volatile and non-volatile isoprenoids; the major components of resin (turpentine) in coniferous trees. See 'Isoprenoids'.

Terrestrial radiation Longwave radiation emitted by the Earth's surface and its atmosphere.

Thermal radiation Electromagnetic radiation emitted from the surface of an object which is due to the object's temperature.

Thylakoid Membrane structures in chloroplasts, housing the pigment system used in capturing solar radiation; grana thylakoids are stacked, while stroma thylakoids are unstacked.

Time element So short time interval that environmental factors do not vary during it.

Tissue A group of closely connected cells having similar functions within an organ.

Trace gas Gas or gases which make up less than 1% by volume of the Earth's atmosphere.

Transpiration Flow of water vapour through stomata into the atmosphere.

Transport Moves material or energy in space.

Turbulent flow A form of gas or liquid flow in which gas or liquid flows in a disordered manner in irregular paths; turbulent flow occurs when the flow velocities are high; otherwise, the flow is laminar.

Turgor pressure Hydrostatic pressure of water in the cells.

Unstable System or property that changes.

UV Ultraviolet, electromagnetic radiation with a wavelength between X-rays and visible light (UVA = 315–400 nm, UVB = 280–315 nm).

Variance Average of the squared distance of random variable from the population mean.

Vascular tissue Issues where substances (e.g. water, sugars, mineral nutrients) are transported longitudinally over long distance; xylem, phloem.

Visible light Electromagnetic radiation with a wavelength between 380 and 750 nm.

VOC Volatile organic compounds.

Water tension Force per unit surface area from solid material to liquid (negative pressure).

WUE Water use efficiency.

Xeromorphic plant Plants structurally adapted to dry environments can withstand prolonged drought and have wilting point lower than –1.5 MPa.

Xylem Vascular tissue where transport of water and solutes occur.

References

Bird RB, Stewart WE, Lightfoot EN (2002) Transport phenomena, 2nd edn. Wiley, New York

Hinds WC (1982) Aerosol technology: properties, behavior, and measurement of airborne particles, 1st edn. Wiley, New York

Mäkelä A, Sievänen R (1992) Height growth strategies in open-grown trees. J Theor Biol 159:443–467

Nobel PS (2005) Physicochemical and environmental plant physiology, 3rd edn. Academic/Elsevier, San Diego

Onsager L (1931a) Reciprocal relations in irreversible processes I. Phys Rev 37:405–426

Onsager L (1931b) Reciprocal relations in irreversible processes II. Phys Rev 38:2265–2279

Penning de Vries FWT, Brunsting AHM, van Laar HH (1974) Products, requirements and efficiency of biosynthesis: a quantitative approach. J Theor Biol 45:339–377

Ross J (1981) The radiation regime and architecture of plant stands. Kluwer Academic, The Hague

Thomas GE, Stamnes K (1999) Radiative transfer in the atmosphere and ocean. Cambridge University Press, Cambridge

Index

P. Hari et al. (eds.), *Physical and Physiological Forest Ecology*,
DOI 10.1007/978-94-007-5603-8,

Printed by Printforce, the Netherlands